Inside Rhinoceros® 5

RON K. C. CHENG

Australia • Brazil • Mexico • Singapore • United Kingdom • United States

Inside Rhinoceros® 5

Ron K. C. Cheng

Vice President, Careers & Computing:
Dave Garza

Director of Learning Solutions: Sandy Clark

Acquisitions Editor: Kathryn Hall

Director, Development-Career and Computing:
Marah Bellegarde

Managing Editor: Larry Main

Senior Product Manager: John Fisher

Editorial Assistant: Kaitlin Murphy-Schlict

Brand Manager: Kristin McNary

Market Development Manager: Erin Brennan

Senior Production Director: Wendy Troeger

Production Manager: Mark Bernard

Content Project Management and Art Direction:
PreMediaGlobal

Cover Image(s): © Michael Tanamachi

© 2014, 2008 Cengage Learning

ALL RIGHTS RESERVED. No part of this work covered by the copyright herein may be reproduced, transmitted, stored, or used in any form or by any means graphic, electronic, or mechanical, including but not limited to photocopying, recording, scanning, digitizing, taping, Web distribution, information networks, or information storage and retrieval systems, except as permitted under Section 107 or 108 of the 1976 United States Copyright Act, without the prior written permission of the publisher.

For product information and technology assistance, contact us at
Cengage Learning Customer & Sales Support, 1-800-354-9706.
For permission to use material from this text or product,
submit all requests online at **www.cengage.com/permissions**.
Further permissions questions can be e-mailed to
permissionrequest@cengage.com.

Library of Congress Control Number: 2013931541

ISBN: 978-1-111-12491-5

Cengage Learning
200 First Stamford Place, 4th Floor
Stamford, CT 06902
USA

Cengage Learning is a leading provider of customized learning solutions with office locations around the globe, including Singapore, the United Kingdom, Australia, Mexico, Brazil, and Japan. Locate your local office at **www.cengage.com/global**.

Cengage Learning products are represented in Canada by Nelson Education, Ltd.

To learn more about Cengage Learning, visit **www.cengage.com**

Purchase any of our products at your local college store or at our preferred online store **www.cengagebrain.com**

Notice to the Reader
Publisher does not warrant or guarantee any of the products described herein or perform any independent analysis in connection with any of the product information contained herein. Publisher does not assume, and expressly disclaims, any obligation to obtain and include information other than that provided to it by the manufacturer. The reader is expressly warned to consider and adopt all safety precautions that might be indicated by the activities described herein and to avoid all potential hazards. By following the instructions contained herein, the reader willingly assumes all risks in connection with such instructions. The publisher makes no representations or warranties of any kind, including but not limited to, the warranties of fitness for particular purpose or merchantability, nor are any such representations implied with respect to the material set forth herein, and the publisher takes no responsibility with respect to such material. The publisher shall not be liable for any special, consequential, or exemplary damages resulting, in whole or part, from the readers' use of, or reliance upon, this material.

Printed in the United States of America
1 2 3 4 5 6 7 17 16 15 14 13

CONTENTS

Introduction vi

MODELING PROJECTS 1

NURBS Curves, NURBS Surfaces, Polygon Meshes, and Solids 3 • Design Development 3 • Surface Modeling Projects 4

MODELING PROJECT 1: BUBBLE CAR CHASSIS 5

Introduction 5 • Objectives 6 • Overview 6 • Constructing the Mechanical Components 7 • Surface Crease 8 • Surface Creases 9 • Summary 26

MODELING PROJECT 2: BUBBLE CAR BODY 27

Introduction 27 • Objectives 27 • Overview 28 • Bodywork Construction 29 • Curves for Making the Surfaces 30 • Continuity Issue 31 • Mirroring and Surface Continuity 34 • Summary 51

MODELING PROJECT 3: SMALL OBJECTS 52

Introduction 52 • Objectives 52 • Overview 52 • Models 53 • Summary 67

MODELING PROJECT 4: TOY CAR ASSEMBLY 68

Introduction 68 • Objectives 68 • Overview 68 • Components and Assembly 69 • Summary 78

CHAPTER 1 RHINOCEROS—WHAT IS IT? 79

Introduction 79 • Objectives 79 • Overview 79 • Computer Modeling and Rhinoceros 79 • Rhino's User Interface 88 • Utilities and Help 113 • Filing 114 • Consolidation 116 • Review Questions 116

CHAPTER 2 FAMILIARIZING RHINOCEROS 117

Introduction 117 • Objectives 117 • Overview 117 • Multiple Construction Planes Concept 117 • Using Coordinates Input 119 • Using Drawing Aids 121 • Manipulating Geometric Objects 132 • Holding Down the Alt Key While Dragging 138 • History Manager 150 • Layer Manipulation 153 • Object Properties 161 • Construction Plane Manipulation 164 • Consolidation 172 • Review Questions 172

CHAPTER 3 RHINOCEROS NURBS SURFACE MODELING 173

Introduction 173 • Objectives 173 • Overview 173 • Surface Modeling Concepts 173 • Surface Modeling Approaches 174 • Concepts of Surfaces and Curves Continuity 177 • Rhino's Major Advantage 179 • Rhino's NURBS Surface 179 • Common Free-Form Surfaces 179 • Other Kinds of Rhino Free-Form Surfaces 204 • Planar Surfaces 210 • Derived Surfaces 213 • Lightweight Extrusion Objects 218 • Consolidation 219 • Review Questions 219

CHAPTER 4 FREE-FORM NURBS CURVES AND POINT OBJECTS 220

Introduction 220 • Objectives 220 • Overview 220 • Curves for Surface Modeling 220 • Rhino Curves 222 • Spline Segment, Polynomial Degree, and Control Point 230 • Point Editing 231 • Points and Point Clouds 240 • Consolidation 245 • Review Questions 245

CHAPTER 5 CURVES OF REGULAR PATTERN 246

Introduction 246 • Objectives 246 • Overview 246 • Line 246 • Polyline 252 • Rectangle 254 • Polygon 256 • Circle 258 • Arc 261 • Ellipse 263 • Parabola 264 • Hyperbola 265 • Conic 266 • Helix 266 • Spiral 268 • Degree of Polynomial and Point Editing 270 • Consolidation 270 • Review Questions 270

CHAPTER 6 CURVE MANIPULATION 271

Introduction 271 • Objectives 271 • Overview 271 • Manipulating a Curve's Length 271 • Treating Two or More Separate Curves 277 • Curve Refinement Methods 286 • Curves and Points from Existing Objects 293 • Consolidation 312 • Review Questions 313

CHAPTER 7 NURBS SURFACE MANIPULATION 314

Introduction 314 • Objectives 314 • Overview 314 • Surface Boundary Manipulation 314 • Treating Two or More Separate Surfaces 328 • Surface Profile Manipulation 340 • Surface Edge Manipulation 350 • Consolidation 355 • Review Questions 355

CHAPTER 8 RHINOCEROS POLYSURFACES AND SOLIDS 356

Introduction 356 • Objectives 356 • Overview 356 • Rhino's Solid Modeling Method 356 • Solids of Regular Geometric Shapes 357 • Constructing Free-Form Solid Objects 386 • Combining Rhino Solids 390 • Detailing a Solid 393 • Editing Solids 404 • Lightweight Extrusion Objects 417 • Consolidation 417 • Review Questions 417

CHAPTER 9 POLYGON MESHES 418

Introduction 418 • Objectives 418 • Overview 418 • Constructing Polygon Mesh Primitives 418 • Constructing Polygon Meshes from Existing Objects 434 • Mapping Polygon Meshes 439 • Combining and Separating 441 • Manipulating Mesh Faces 445 • Consolidation 462 • Review Questions 462

CHAPTER 10 ADVANCED MODELING METHODS—TRANSFORMATION 463

Introduction 463 • Objectives 463 • Overview 463 • Translation 463 • Deformation 476 • Translate and Deform 490 • Consolidation 496 • Review Questions 496

CHAPTER 11 RHINOCEROS DATA ANALYSIS 497

Introduction 497 • Objectives 497 • Overview 497 • General Tools 497 • Dimensional Analysis Tools 503 • Surface Analysis Tools 506 • Mass Properties Tools 510 • Diagnostics Tools 513 • Consolidation 515 • Review Questions 515

CHAPTER 12 GROUP, BLOCK, AND WORK SESSION 516

Introduction 516 • Objectives 516 • Overview 516 • Group 516 • Block 518 • Worksession Manager and Design Collaboration 535 • Drag and Drop 540 • Textual Information 540 • Consolidation 540 • Review Questions 541

CHAPTER 13 2D DRAWING OUTPUT AND DATA EXCHANGE 542

Introduction 542 • Objectives 542 • Overview 542 • Engineering Drawing 542 • Data Exchange 564 • Consolidation 568 • Review Questions 568

CHAPTER 14 RENDERING 569

Introduction 569 • Objectives 569 • Overview 569 • Digital Rendering and Animation 570 • Camera Setting 582 • Digital Lighting 586 • Digital Material 604 • Environment Objects 625 • Render Properties 630 • Consolidation 632 • Review Questions 633

Index   634

INTRODUCTION

INTRODUCING RHINOCEROS

Rhinoceros, also known as Rhino, is a 3D surface modeling application. It has already gained popularity in the industry, including industrial design, marine design, jewelry design, CAD/CAM, rapid prototyping, reverse engineering, graphic design, and multimedia. It is based on the popular NURBS (nonuniform rational B-spline) mathematics, which enables the construction of free-form organic surfaces that are compatible with most other computer models used in the industry.

To cope with rapid prototyping and some animation and visualization applications that use faceted approximation instead of NURBS surfaces to represent 3D free-form objects, Rhino also provides a set of polygon mesh tools.

Making NURBS surfaces usually requires a set of curve frameworks. Therefore, Rhino has a set of curve tools. To enable modification and improvisation of surface models, Rhino incorporates a comprehensive set of editing, transformation, and analysis tools. Apart from that, it has tools for outputting 2D engineering drawings and highly photorealistic rendered images and animations.

AUDIENCE AND PREREQUISITES

This book is intended for students and practitioners in the following industry: industrial design, marine design, jewelry design, CAD/CAM, rapid prototyping, reverse engineering, graphic design, and multimedia. There is no prerequisite needed for several reasons, because the book is written in a nonlinear manner, enabling novices to learn the application in a short period of time, and allowing experienced users to discover more advanced use of Rhino in design.

PHILOSOPHY AND APPROACH

With learning-by-doing in mind, this book stresses application, using Rhinoceros as a tool to produce the digital model of a design in mind for downstream processes. In addition, this book has a balanced emphasis on modeling projects and computer modeling concepts, bridging the theoretical and software-oriented approaches to modeling in the computer, because theory is relatively useless without hands-on experience and vice versa.

To enable both novices and more advanced users to learn about Rhino in a nonlinear way, this book has two major components: a set of hands-on modeling projects and the main chapters. The set of modeling projects is placed at the beginning of the book, allowing all readers to have hands-on experience on using Rhino to construct some real-world objects. In order to provide a clear picture about what can be produced by using Rhino, the main chapters are written logically, with concepts explained together with simple tutorials focusing on individual elements.

INTRODUCTION

This book is written to Rhinoceros Release 5, addressing the new tools provided as well as tools that are already provided in previous releases.

At the beginning of this book, there is a set of four easy-to-follow, hands-on modeling projects. In the first project, guides are provided to construct the chassis and mechanical components of the famous bubble car. The second project completes the bubble car by making the free-form bodywork. The third project relates to a set of small objects to highlight the use of Rhino in various ways. Finally, in the fourth project, a way to assemble a set of components is suggested.

There are 14 main chapters. Chapter 1 provides an overview about digital modeling, explains the concepts of surface modeling, and introduces Rhinoceros as a digital modeling tool, as well as an examination of the Rhino user interface, including its key functional components. Chapter 2 familiarizes you with the concepts of construction plane, basic geometry construction and modification methods, layer management, and surface display methods. Chapter 3 provides you with a clear understanding on the kind of surfaces that you can construct by using Rhino and the framework of points and curves required for building such surfaces. Chapters 4 through 6 delineate the ways to construct and manipulate points and curves for making surfaces. After discussing curves, in Chapter 7 you learn more detailed surface manipulation methods. Chapter 8 covers the manner in which Rhino represents solids in the computer and various ways of solid construction. In Chapter 9, you will learn the ways to construct and manipulate polygon meshes, which are significantly used for rapid prototyping.

Chapter 10 explains various methods to transform curves, surfaces, and polygon meshes. Chapter 11 deals with analyzing curves, surfaces, and polygon meshes. In Chapter 12, you learn about grouping drawing objects, construction of data blocks, and use of work sessions. Chapter 13 addresses producing 2D engineering drawings from 3D models, as well as import and export of files. Chapter 14 explains how rendered images and animations can be produced from 3D models.

SUPPLEMENTS TO THIS BOOK

Student Companion Site

A Student Companion Website is available containing Rhinoceros files for the project exercises.

Accessing a Student Companion Website from CengageBrain:

1. Go to http://www.cengagebrain.com
2. Enter author, title, or ISBN in the search window.
3. When you arrive at the Product page, click on the Access Now tab.
4. Click on the resources listed under Book Resources in the left navigation pane to access the project files.

Instructor Site

An Instruction Companion Website containing supplementary material is available. This site contains an Instructor Guide containing answers to the end-of-chapter review questions. Contact Cengage Learning or your local sales representative to obtain an instructor account.

Accessing an Instructor Companion Website from SSO Front Door

1. Go to http://login.cengage.com and login using the Instructor email address and password.
2. Enter author, title, or ISBN in the Add a title to your bookshelf search box and click on Search button.
3. Click Add to My Bookshelf to add Instructor Resources.
4. At the Product page click on the Instructor Companion site link.

New Users

If you are new to Cengage.com and do not have a password, contact your sales representative.

ACKNOWLEDGMENTS

I sincerely thank the Cengage team members for their efforts in association with this project, in particular, Senior Product Manager John Fisher, Acquisitions Editor Kathryn Hall, and Project Manager Soumya Nair.

ABOUT THE AUTHOR

Highly influenced and inspired by Jean Piaget's constructivism theory, Seymour Papert's constructionism theory, and Jerome Seymour Bruner's cognitive learning theory, Cheng, with master degree in education as well as master degree in engineering, designed and pioneered the integrative project learning approach in the Hong Kong Polytechnic University over a decade ago, training its undergraduate as well as postgraduate students. After serving the university for over thirty-five years at the age of sixty, leading engineering design and integrative project learning in the Industrial Center, Cheng now began his second phase of life, continuing to assert and advocate the learning approach that he pioneered in various education institutions, and working on a number of industrial automation projects as well as engineering design projects, leading to commercialization of scientific research.

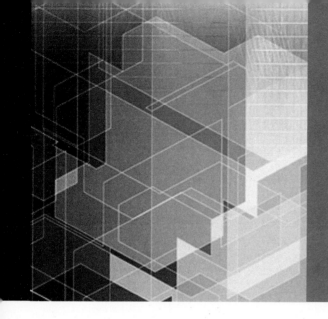

Modeling Projects

We will begin with a set of modeling projects, which are intended to let you appreciate how to use Rhinoceros as a tool in computer modeling. These modeling projects are followed by the main chapters, delineating various computer modeling concepts and Rhinoceros commands in a logical way. It is hoped that you will find these modeling projects interesting and not too difficult to follow. However, if you are a novice in surface modeling or using Rhinoceros and find these modeling projects too difficult, it is suggested that you should first go to the main chapters and then come back to the modeling projects later. Figures P–1 through P–8 show the rendered views of these projects.

FIGURE P–1 *Bubble car chassis modeling project*
© 2014 Cengage Learning®. All Rights Reserved.

FIGURE P–2 *Bubble car body modeling project*
© 2014 Cengage Learning®. All Rights Reserved.

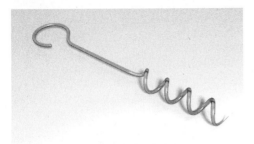

FIGURE P-3 *Small objects modeling project 1*
© 2014 Cengage Learning®. All Rights Reserved.

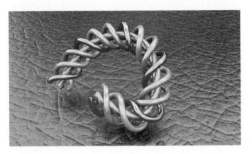

FIGURE P-4 *Small objects modeling project 2*
© 2014 Cengage Learning®. All Rights Reserved.

FIGURE P-5 *Small objects modeling project 3*
© 2014 Cengage Learning®. All Rights Reserved.

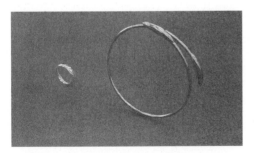

FIGURE P-6 *Small objects modeling project 4*
© 2014 Cengage Learning®. All Rights Reserved.

FIGURE P–7 *Small objects modeling project 5*
© 2014 Cengage Learning®. All Rights Reserved.

FIGURE P–8 *Toy car modeling project*
© 2014 Cengage Learning®. All Rights Reserved.

NURBS CURVES, NURBS SURFACES, POLYGON MESHES, AND SOLIDS

As will be explained in detail in the main chapters of this book, Rhinoceros is a 3D modeling tool and we can use it to construct three major kinds of geometric objects: NURBS curves, NURBS surfaces, and polygon meshes. Among them, we use curves as framework on which surfaces are constructed. As for NURBS surfaces and polygon meshes, they are two distinct ways of composing a 3D surface model, with NURBS surfaces used for making accurate models and polygon meshes used for approximated representation of 3D objects.

In Rhinoceros terms, by joining two or more contiguous NURBS surfaces together, a polysurface is formed. If a single NURBS surface, such as a sphere or an ellipsoid, or a polysurface forms a closed loop enclosing a volume without any gap or opening, a solid is implied.

DESIGN DEVELOPMENT

Prior to working on the modeling projects, let us have a quick review on the two-stage iterative design development process.

Top-Down Thinking Process

With an idea or concept in mind, you start deconstructing the 3D object into discrete surface elements by identifying and matching various portions of the object with various types of surfaces. Unless the individual surfaces are primitive surfaces, like box, sphere, or ellipsoid, you need to think about the shapes and locations of curves and points required to build the surface, as well as the surface construction commands to be applied on the curves and points. This is the top-down thinking process.

Bottom-Up Construction Process

After this thinking process, you create the curves and points and apply appropriate surface construction commands. By putting the surfaces together properly in 3D space, you obtain the 3D object. This is the bottom-up construction process. If the surfaces you construct do not conform to your concept, you think about the curves and points again.

Detailing

With practice you gain the experience that makes this process more efficient. Logically, you construct points and curves, and from these you construct surfaces and solids. In detailing your design, you add surface features. To construct surface features that conform in shape to existing surfaces, you create points and curves from existing objects.

SURFACE MODELING PROJECTS

The prime goal of surface modeling is to construct a set of surfaces and put them together to represent an object depicting a design. Constructing surfaces inevitably requires curves and points. In comparison to surface construction, curve construction is a tedious job.

Because the location and shape of the curves have a direct impact on the shape of surfaces constructed from them, the particulars of most curves in these modeling projects are given to you. While working on these modeling projects, you should try to relate the 3D curves to the 3D surfaces. It is hoped that you reverse the process, seeing the curves when a surface is given.

MODELING PROJECT 1

Bubble Car Chassis

INTRODUCTION

This modeling project, together with the next one, concerns making a 1- to 10-scale model of the famous bubble car, which was first designed and produced by an Italian refrigerator manufacturer about six decades ago. The design was later on franchised to a number of car manufacturers, who built the car in huge quantities, with a variety of configurations. The computer model that you are going to construct, as shown in Figures P1–1 and P1–2, is based on an earlier model.

FIGURE P1–1 *Right-front view of the Rhinoceros bubble car model rendered by using Flamingo nXt*

FIGURE P1–2 *Rear-right view of the Rhinoceros bubble car model rendered by using Flamingo nXt*

OBJECTIVES

After working on this modeling project, you should be able to

- Demonstrate a general understanding on using Rhinoceros's simple surface and solid modeling tools

OVERVIEW

This bubble car is chosen as a Rhinoceros computer modeling project because the car has simple engineering structure and a stylish body. Besides, we may find a lot of information about it by searching over the Internet, using the keyword "Isetta." For the sake of simplicity, construction of this model will be carried out in two unique modeling project sessions. In the first session, we will work on the engineering components. Making the engineering components can help you understand the process of constructing geometric shapes of regular pattern. In the second session, we will concentrate on body panel parts. Building the body panels lets you experience how free-form surfaces can be constructed. The components that we will construct here are shown in Figure P1–3. The main body panels, consisting of an egg-like shell (Figure P1–4) and a refrigerator-like front door (Figure P1–5), will be dealt with in the next session.

FIGURE P1–3 *Bubble car mechanical components to be constructed in this project*

FIGURE P1–4 *Rendered bubble car body panels to be constructed in the next project*

FIGURE P1–5 *Bubble car's refrigerator-like front door panel to be constructed in the next project*

CONSTRUCTING THE MECHANICAL COMPONENTS

Figure P1–6 shows the bottom views of the mechanical components we will make in this project. They are wheels, tires, rear axle, chassis, rear suspension, engine, exhaust system, intake system, gas tank, front suspension, drive controls, floor panels, and seat.

FIGURE P1–6 *Bottom views of the mechanical components of the bubble car model*

Template Files

To build the mechanical components, we need NURBS curves as wireframe skeletons, upon which we will build the NURBS surfaces, polysurfaces, and solids. Because curve building is a tedious task, we have constructed all the required curves and placed them in various layers of a file located on the Student Companion Website. Consult the preface of this text for information on how to access the Student Companion Website and download all of the Rhinoceros files. As the computer model is a 1- to 10-scale model of the real car, we have simplified most of the component parts appropriately. Now open the template file.

1. Select File > Open and select the file Bubble Car 1.3dm from the Modeling Project 1 folder that you downloaded from the Student Companion Website.

    In the folder, you will find other template files depicting various stages of model construction. Therefore, at any stage if you want to have a break, you do not need to save your file. Next time when you continue with your tutorial, you need only to select one of the templates in the folder.

Bubble Car's Tires and Wheels

A default Rhinoceros file has four viewports; each has its own construction plane. As shown in Figure P1–6, the Top viewport is maximized and there are several curves, residing on three separate layers. Curves X, Y, and Z serve to indicate the rear track, wheelbase, and front track, respectively. They are used for reference purpose, positioning the tires and wheels. Curves A and D depict the cross-section of the tire and wheel. We will revolve them to construct a set of revolved surfaces.

SURFACE CREASE

Because the curves you will use to construct revolved surfaces have kink points and you have to fillet the surface at the kinks, prior to surface construction, you need to perform the following steps to split the resulting surface at creases in order to produce a polysurface instead of a single surface.

2. Type CreaseSplitting at the command area.

    A way to execute a command is to type its command name at the command area.

3. Select Yes at the command area.

    Subsequently constructed surfaces from curves with kink points will be constructed as polysurfaces. Otherwise, they will be constructed as a single surface with creases.

Now you can use the curves to construct revolved surfaces.

4. Select Surface > Revolve.
5. Select curve A, indicated in Figure P1-7, and press the ENTER key.
6. Set object snap mode to End and then select endpoints B and C.
7. Select the FullCircle option on the command area.

    A polysurface depicting the wheel of the bubble car is constructed on the current layer.

8. Set current layer to Tire.

    You will construct the next object on this layer.

9. Select Surface > Revolve.
10. Select curve D, indicated in Figure P1-7, and press the ENTER key.
11. Select endpoints B and C and select the FullCircle option on the command area.

    Another polysurface representing the tire is constructed.

12. Turn off layers Curves Wheel and Curves Tire.

    The curves are not required anymore.

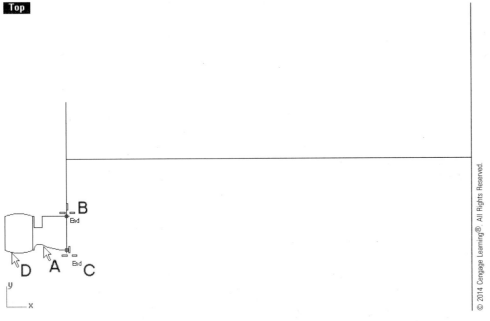

FIGURE P1-7 *Curves for making the tire and wheel and curves indicating the wheelbase and track*

Because curves A and D are sets of curve elements joined together, the outcome is sets of surfaces joined together or a single surface, depending on the CreaseSplitting setting. As each of these outcome surfaces encloses a volume without any opening, it is regarded as a solid in Rhinoceros terms.

The edges of the polysurface representing the tire are to be rounded by filleting. Basically, there are two different ways to construct a filleted edge: by using the fillet command from the Surface menu or the fillet command in the Solid menu. Because the surfaces are already joined as a polysurface during their construction, we will use the solid filleting command to round off two edges, as follows.

13. Select Solid > Fillet Edge > Fillet Edge.
14. Select the Current Radius option on the command area and set it to 2 mm.
15. Select edges A and B, indicated in Figure P1–8, and press the ENTER key twice.

    Fillet surfaces are constructed at the edges.

SURFACE CREASES

If the revolved surface is a single surface because prior setting of CreaseSplitting is not Yes, the resulting surfaces are single surfaces, not a polysurface. Naturally, you cannot fillet along a single surface; therefore, you have to select Surface > Surface Edit Tools > Divide Surface; select A, indicated in Figure P1–8, and repeat steps 13 through 15 again.

We will now copy the tire and wheel from the rear axis to the front axis of the car, using the front and rear track lines as references.

16. Select Transform > Copy.
17. Select the tire B and wheel C (Figure P1–8) and press the ENTER key.
18. Select endpoints D and E and press the ENTER key.

    The tire and wheel are copied.

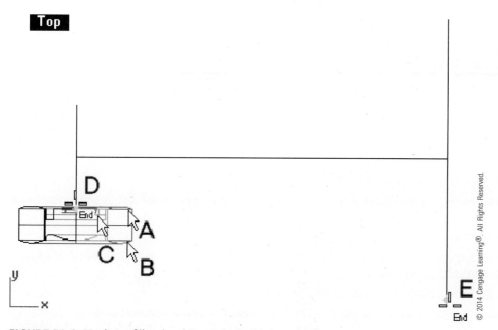

FIGURE P1–8 *Tire being filleted and tire and wheel being copied*

As the car is symmetric left and right, we will make a mirror copy of the tires and wheels, as follows.

19. Select Transform > Mirror.
20. Select A, B, C, and D, indicated in Figure P1–9, and press the ENTER key.

21. Click on endpoints E and F, indicated in Figure P1-9.

    The tires and wheels are complete. For your reference, Figure P1-9 also shows the perspective rendered view of the tires and wheels.

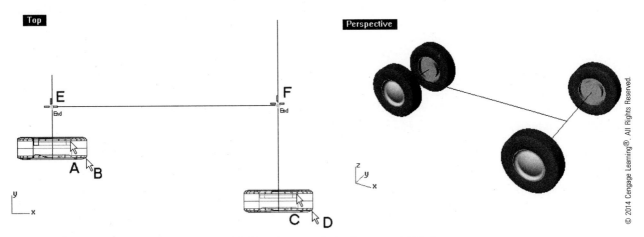

FIGURE P1-9 *Tires and wheels being mirrored (left) and tires and wheels mirrored (right)*

Rear Axle

In our model, the rear axle, as shown in Figure P1-10, will be simplified and constructed by first making two closed polysurfaces and then combining them together. To construct these polysurfaces, we need to extrude one curve and revolve another curve.

FIGURE P1-10 *From left to right: rear axle and two polysurfaces*
© 2014 Cengage Learning®. All Rights Reserved.

Now we will construct an extrude solid and a revolved surface.

22. Turn on layer Curves Rear Axle and set current layer to Rear Axle.

    You will use the curves residing on the Curves Rear Axle layer to construct surface/solid objects on the Rear Axle layer.

23. Select Solid > Extrude Planar Curve > Straight.

    Using this command will extrude the curve as well as make two planar surfaces at both ends of the extruded surface, thus producing a closed polysurface—a solid.

24. Select curve A, indicated in Figure P1-11, and press the ENTER key.
25. If the option BothSides=Yes, click on it to change it to No. Otherwise, proceed to the next step.
26. Click on endpoint B.

    A closed polysurface is constructed.

27. Select Surface > Revolve.
28. Select curve C, indicated in Figure P1-11, and press the ENTER key.
29. Select endpoints D and E and select the FullCircle option on the command area.

    A polysurface is constructed.

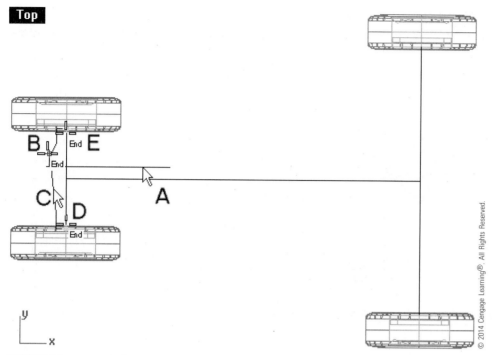

FIGURE P1–11 *A curve being extruded and another curve being revolved*

Now we have two closed polysurfaces (Rhinoceros solids), and we will use the Boolean Union operation to combine them as one, as follows.

30. Turn off layer Curves Rear Axle.
31. Select Solid > Boolean Two Objects.
32. Select polysurfaces A and B, indicated in Figure P1–12.
33. Click on the screen to iterate through various Boolean operations until you find the two selected Rhinoceros solids are united as one, shown in Figure P1–12.

    Clicking on the screen cycles through various operations.

34. Press the ENTER key.

    The rear axle is complete.

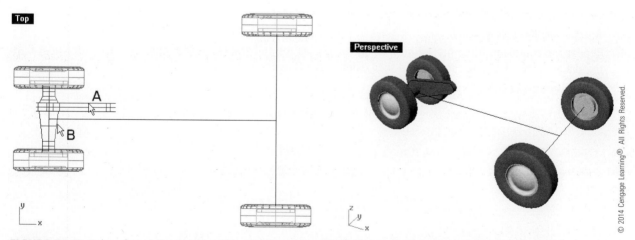

FIGURE P1–12 *Polysurfaces being combined by using Boolean Union (left) and perspective view of the model (right)*

Chassis

Here, the chassis of the bubble car includes not only the tubular framework, but also the mud guards. For the sake of simplicity, the hollowness of the frame and the mounting bolts and nuts are disregarded. For your reference, Figure P1–13 shows the chassis and mud guards as well as the chassis, mud guards, and components that are already constructed.

FIGURE P1–13 *Chassis of the bubble car (left) and chassis and other components already constructed (right)*

So far, we have been only extruding and revolving curves. To make the chassis, we will perform sweeping operations as well. Again, curves for making these objects are already constructed and residing on a layer. Now we will perform sweeping operations, as follows.

35. Turn on layer Curves Chassis; set current layer to Chassis; and turn off layers Curves Wheelbase, Curves Rear Axle, Tire, Wheel, and Rear Axle.

    Now only the curves for constructing the chassis are displayed, and you will use these curves to construct the chassis on the Chassis layer.

36. Maximize the Perspective viewport, by double-clicking the Top viewport's label and then double-clicking the Perspective viewport's label.
37. Select Surface > Sweep 1 Rail.
38. Select curve A near B (Figure P1–14) as the rail, select circle B as the section, and press the ENTER key twice.

    Because the location where we click on the rail curve has an effect on the direction of the resulting swept surface, here when selecting the rail curve A, we have to click near circle B.

39. Click on the OK button of the Sweep 1 Rail Options dialog box.

    A swept surface is constructed.

40. Repeat the command three more times, using curve C as the rail and circle D as the section, using curve E as the rail and circle F as the section, and then using curve G as the rail and circle H as the section.

    Three swept surfaces are constructed.

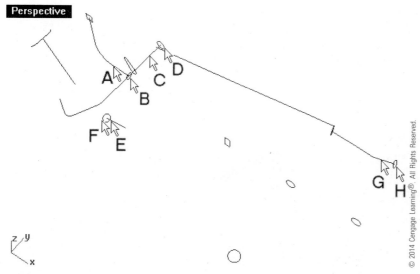

FIGURE P1–14 *Swept surfaces being constructed*

If you take a closer look, you will find that both ends of the swept surfaces are opened. In order to use the Boolean Union command to combine these surfaces with other polysurfaces that we will construct later on, we will cap their planar holes, causing the objects to become closed polysurfaces. In Rhinoceros terms, they will become solids. We will continue with the following steps.

41. Select Solid > Cap Planar Holes.
42. Select surfaces A, B, C, and D, indicated in Figure P1–15, and press the ENTER key.

    The swept surfaces are capped and become closed polysurfaces.

Let us construct a few more polysurfaces, as follows.

43. Select Solid > Extrude Planar Curve > Along Curve.
44. Select curve E (Figure P1–15) and press the ENTER key.

    You may have to zoom in more closely in order to select curve E.

45. Select curve F as the path curve.

    An extruded solid is constructed.

46. Repeat the command, using curve G as the curve to extrude and curve H as the path curve.

    Because curve G is an open curve with two sharp corners, the resulting object is an open polysurface consisting of three surfaces joined together. This polysurface will represent the mud guard of the bubble car.

47. Select Solid > Extrude Planar Curve > Straight.
48. Select curve J and press the ENTER key.
49. If the option BothSides=No, click on it to change it to Yes. Otherwise, proceed to the next step.
50. Type 30 at the command area.

    The curve is extruded 30 mm in both directions. In other words, the total distance of extrusion is 60 mm.

51. Repeat the command to extrude curve K a distance of 1 mm in both directions.

    Similar to the previous step, the total extrusion distance is two times the specified distance, 2 mm.

52. Repeat the command to extrude curve L a distance of 33 mm in both directions.

    Total distance is 66 mm.

53. Repeat the command to extrude curve M a distance of 46 mm in both directions.

    Total distance is 92 mm.

54. Repeat the command to extrude curve N a distance of 3 mm in both directions.

    Total distance is 6 mm.

55. Repeat the command to extrude curve P a distance of 42 mm in both directions.

    Total distance is 84 mm.

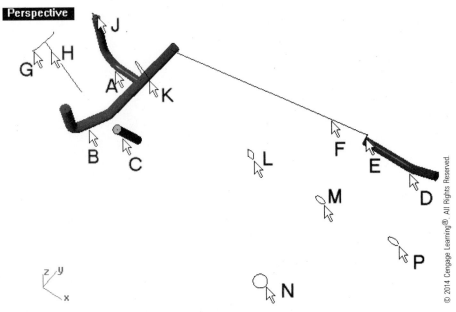

FIGURE P1–15 *Swept surfaces being capped and curves being extruded*

To complete the components of the chassis, we will construct three mirrored objects, as follows.

56. Turn off layer Curves Chassis and turn on layer Curves Wheelbase.
57. Select Transform > Mirror.
58. Select polysurfaces A, B, C, and D, indicated in Figure P1–16, and press the ENTER key.
59. Select endpoints E and F.
    The polysurfaces are mirrored.

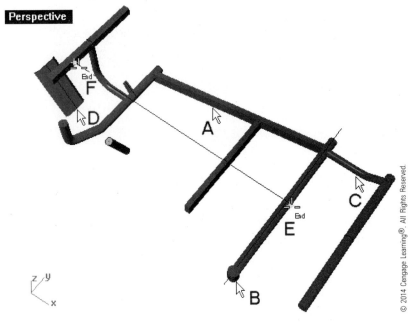

FIGURE P1–16 *Polysurfaces being mirrored*

To complete the chassis, we will combine the chassis elements by using the Boolean Union operation, as follows.

60. Select Solid > Union.
61. Select all the surface objects, except A and B, indicated in Figure P1–17, and press the ENTER key.
    The selected polysurfaces are combined, and the chassis of the bubble car is complete.

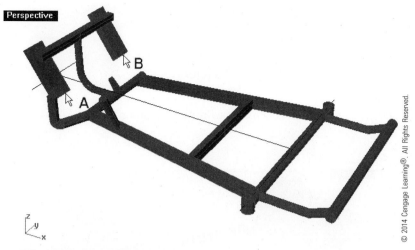

FIGURE P1–17 *Chassis elements being combined*

Rear Suspension

The rear suspension of the car consists of two sets of leaf springs and shock absorbers. To depict these objects, we will use the curves already built to construct a revolved polysurface and an extruded polysurface, as follows.

62. Turn on layer Curve Rear Suspension, set current layer to Rear Suspension, and turn off layers Chassis and Curves Wheelbase.

    Now the curves for constructing the rear suspensions elements are displayed, and the constructed surface will be residing on the Rear Suspension layer.

63. Select Solid > Extrude Planar Curve > Straight.
64. Select curve A (Figure P1–18) and press the ENTER key.
65. If the option BothSides=Yes, click on it to change it to No. Otherwise, proceed to the next step.
66. Type 5 at the command area.

    The curve is extruded 5 mm in one direction. The direction of extrusion is determined by the normal direction.

67. Select Surface > Revolve.
68. Select curve B (Figure P1–18) and press the ENTER key.
69. Select endpoints C and D and select the FullCircle option on the command area.

    A polysurface is constructed

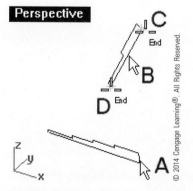

FIGURE P1–18 *Leaf spring and shock absorber being constructed*

Now we will mirror the polysurfaces representing the leaf spring and shock absorber, as follows.

70. Turn off layer Curve Rear Suspension and turn on Curves Wheelbase.
71. Select Transform > Mirror.
72. Select polysurfaces A and B, indicated in Figure P1-19, and press the ENTER key.
73. Select endpoints C and D.

    The polysurfaces are mirrored. For your reference, all objects constructed so far are also shown in Figure P1-19 *(right).*

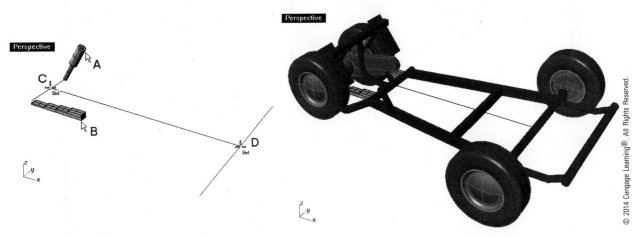

FIGURE P1-19 *Leaf spring and shock absorber being mirrored (left) and all components constructed so far (right)*

Engine Unit

The engine unit of this bubble car model will be simplified and represented by two polysurfaces produced by making a set of extruded and revolved surfaces/polysurfaces combined together and then filleted. Figure P1-20 shows the engine unit added to the model.

FIGURE P1-20 *Engine unit added to the model*
© 2014 Cengage Learning®. All Rights Reserved.

Now we will continue with the following steps to revolve two curves.

74. Turn on layer Curves Engine, set current layer to Engine, and turn off layers Curves Wheelbase and Rear Suspension.
75. With reference to Figure P1-21, revolve curves A and D (Figure P1-21) separately about axes BC and EF to construct two polysurfaces.

Because revolving curve A produces two separate objects, we have to join them for performing Boolean operations in later stage.

76. Select Edit > Join.
77. Select endpoints G and H and press the ENTER key.

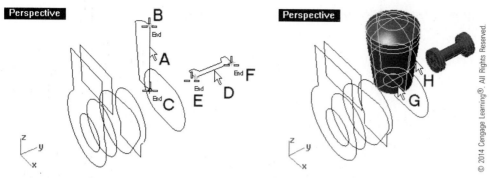

FIGURE P1–21 *Curves being revolved (left) and revolved polysurfaces (right)*

Now we will continue with the following steps to construct four closed polysurfaces.

78. Select Solid > Extrude Planar Curve > Straight to extrude curves A, B, C, and D (Figure P1–22) separately to construct four closed polysurfaces. Extrusion distance for A is 16 mm, B is 15 mm, C is 18 mm, and D is 10 mm.
79. Combine the polysurfaces E, F, G, H, J, and K to form a single polysurface by using Boolean Union.

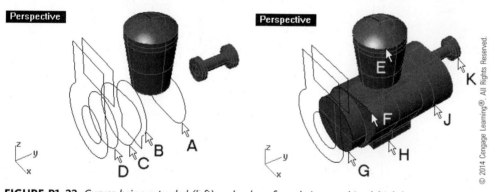

FIGURE P1–22 *Curves being extruded (left) and polysurfaces being combined (right)*

The polysurface representing the engine's main body is complete. We will continue with the following steps to construct two polysurfaces and combine them into one.

80. Extrude curve A (Figure P1–23) a distance of 33 mm to build a closed polysurface.
81. Repeat the command to extrude curves B and C together for a distance of 8 mm, respectively, to another closed polysurface.

    As can be seen, extruding two curves (one inside another) produces a hollow object.

82. Use Boolean Union to combine polysurfaces D and E to form a single polysurface.
83. Select Solid > Solid Edit Tools > Faces > Merge All Faces and click on the combined polysurface D to remove unwanted overlapped faces.

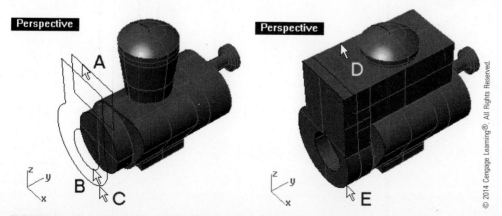

FIGURE P1–23 *Curves being extruded (left) and two polysurfaces being combined (right)*

To complete the engine, we will fillet an edge and change the color property of a polysurface, as follows.

84. Select Solid > Fillet Edge > Fillet Edge.
85. Select the Next Radius option on the command area.
86. Change the radius to 5 mm.
87. Select edge A (Figure P1–24) and press the ENTER key twice.
   The edge is filleted.
88. Select Edit > Object Properties.
89. Select the filleted polysurface B, indicated in Figure P1–24, and press the ENTER key.
90. In the Properties dialog box, change the display color to Magenta.
   The engine is complete.

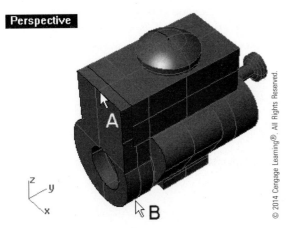

FIGURE P1–24 *Edge being filleted*

Exhaust System

The exhaust system of the bubble car is represented by two capped swept surfaces and an extruded polysurface. Figure P1–25 shows the exhaust system incorporated in the model.

FIGURE P1–25 *Exhaust system included in the model*

Now we will continue with the following steps.

91. Turn on layer Curves Exhaust, set current layer to Exhaust, and turn off layers Curves Engine and Engine.

    Now only the curves for making the exhaust system elements are displayed, and the constructed surface will be residing on the Exhaust layer.

92. Construct a Sweep 1 Rail surface, using curve A (Figure P1–26) as rail and circle B as cross-section.
93. Construct another Sweep 1 Rail surface, using curve C as rail and circle D as cross-section.
94. Extrude planar ellipse E along curve F to construct a solid.
95. Cap the planar holes of swept surfaces G and H to form two solids.
96. Combine polysurfaces G, H, and J to form a single polysurface by using Boolean Union.

    The exhaust system is complete.

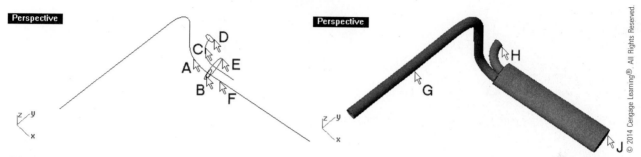

FIGURE P1–26 *Curves being used to construct polysurfaces (left) and swept surfaces being capped and combined with extruded polysurface (right)*

Intake System

The polysurface that we will construct to represent the intake system consists of four solid elements, which will be constructed by revolving, sweeping, and extruding. The intake system, together with other objects already constructed, is shown in Figure P1–27.

FIGURE P1–27 *Intake system included in the model*

Now we will continue with the following steps.

97. Turn on layer Curves Intake, set current layer to Intake, and turn off layers Curves Exhaust and Exhaust.

    Now only the curves for making the intake system are displayed and you will use these curves to construct the surfaces on the Intake layer.

98. Revolve curve A (Figure P1–28) about endpoints B and C to construct a revolved polysurface.
99. Construct a Sweep 1 Rail surface with curve D as rail and circle E as cross-section.
100. Extrude planar curve F a distance of 36 mm to construct a solid.

101. Extrude planar curve G a distance of 10 mm to construct another solid.
102. Cap the planar holes of the swept surface H to convert it to a solid.
103. Combine solids H, J, K, and L to form a single solid by using Boolean Union.

    The intake system is complete.

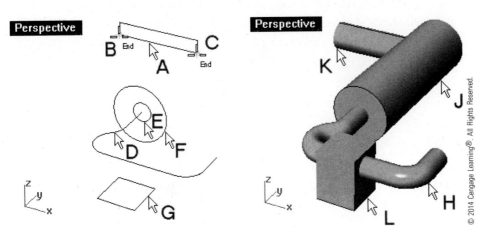

FIGURE P1–28 *Curves for making four solids (left) and four solids being combined (right)*

Fuel Tank

Design the polysurface representing the fuel tank is quite simple. We need to only construct a revolved solid and an extruded solid and then combine them to form a single solid. Figure P1–29 shows the fuel tank added to the model.

FIGURE P1–29 *Fuel tank added to the model*

Now we will continue with the following steps.

104. Turn on layer Curves Fuel, set current layer to Fuel, and turn off layers Curves Intake and Intake.

    Now you will use the curves displayed to construct the fuel tank on the Fuel layer.

105. Revolve curve A (Figure P1–30) about endpoints B and C to construct a solid.
106. Extrude curve D a distance of 19 mm in both directions.

    The total extrusion distance is 38 mm.

107. Combine solids E and F by using Boolean Union.

    The fuel tank is complete.

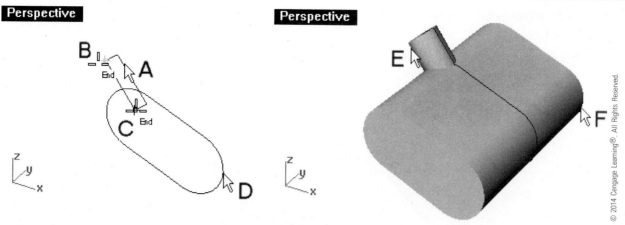

FIGURE P1–30 *Curves for making the fuel tank (left) and solids being combined (right)*

Front Suspension

The objects representing the front suspension are a set of extruded polysurfaces. We will use the curves already constructed as framework. After making the extruded solids, we will combine them by using Boolean Union and round off some corners by filleting. Figure P1–31 shows the front suspension added to the model.

FIGURE P1–31 *Bottom view showing the front suspension added to the model*

Now we will continue with the following steps.

108. Turn on layer Curves Front Suspension, set current layer to Front Suspension, and turn off layers Curves Fuel and Fuel.

    Now you will use the curves displayed on the screen to construct the front suspension on the Front Suspension layer.

109. Construct six solid objects by extruding planar curves A, B, C, D, E, and F (Figure P1–32) in one direction. Extrusion distance for A is 3 mm, B is 5 mm, C is 15 mm, D is 6 mm, E is 2 mm, and F is 2 mm.

110. Turn off layer Curves Front Suspension.

111. Round off edges G, H, J, and K with a radius of 2 mm.

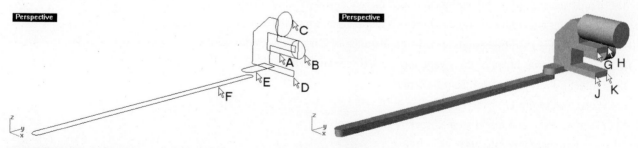

FIGURE P1–32 *Curves being extruded (left) and edges being filleted (right)*
© 2014 Cengage Learning®. All Rights Reserved.

112. Combine solids A, B, C, D, and E (Figure P1–33) by using Boolean Union.
113. Turn on layer Curves Wheelbase.
114. Mirror the combined solid about endpoints F and G.

    The front suspension is complete.

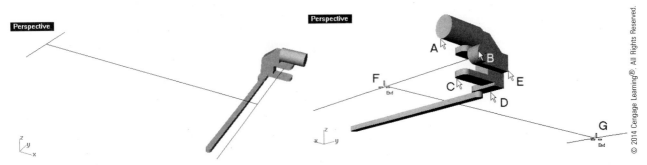

FIGURE P1–33 *Two views showing solids being combined and mirrored*

Drive and Steering

Figure P1–34 shows the drive and steering unit added to the model.

FIGURE P1–34 *Drive and steering added to the model*
© 2014 Cengage Learning®. All Rights Reserved.

Now we will continue with the following steps.

115. Turn on layer Curves Drive, set current layer to Drive, and turn off layers Front Suspension and Curves Wheelbase.

    Now only the curves for making the drive and steering systems are displayed and you will construct surfaces on the Drive layer.

116. Construct a Sweep 1 Rail surface, using curve A (Figure P1–35) as rail and curve B as cross-section.
117. Construct another Sweep 1 Rail surface, using curve C as rail and curve D as cross-section.
118. Construct a solid by extruding curve E (the smaller circle) along path curve F.
119. Construct a solid by extruding curve G a distance of 4 mm in one direction.
120. Construct a solid by extruding curve H a distance of 3 mm in both directions (total 6 mm).
121. Construct a solid by extruding curve J a distance of 2 mm in both directions.
122. Cap the planar holes of the swept surface K to construct a solid.
123. Select Solid > Sphere > Center, Radius.
124. Select endpoint L.
125. Type 4 at the command area to specify the diameter.

    A solid sphere is constructed.

126. Select Transform > Copy.

127. Select solid M and press the ENTER key.
128. Click on any location in the viewport to specify the point to copy from.
129. Type r14<90 at the command area.

    The select object is copied 14 mm in 90 degree direction from the reference point.
130. Type r6<270. Another copy of the selected object is constructed.
131. Press the ENTER key to terminate the command.

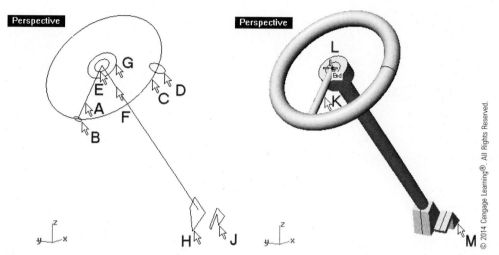

FIGURE P1–35 *Curves being used for surface building (left) and swept surface being capped, sphere being made, and solid being copied (right)*

Now we will construct an array, as follows.

132. Right-click on the Perspective viewport's label and select Wireframe to set the display to wireframe.

    The display is changed to wireframe.
133. Select Solid > Cap Planar Holes.
134. Select A (Figure P1–36) and press the ENTER key.
135. Select Transform > Array > Along Curve.
136. Select A (Figure P1–36) and press the ENTER key.
137. Select curve B.
138. In the Array Along Curve Options dialog box, set the number of items to 3 and click on the OK button. The polysurface is arrayed.
139. Combine all the solids, except C, D, and E, by using Boolean Union.

    The drive and steering unit is complete.

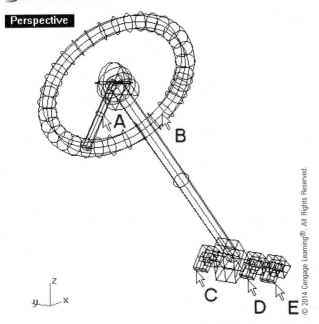

FIGURE P1–36 *A solid being arrayed and solids being combined*

Floor Panel and Seat

To finish this modeling project session, we will add floor panels and a bench seat. Because the outer boundary of the floor panels needs to fit with the external body panels, we will construct a floor panel a bit wider than required and trim it in the next modeling project. Figure P1–37 shows the floor panel (trimmed by the external body panels) and the bench seat.

FIGURE P1–37 *Bench seat and trimmed floor panels added to the model*

Now we will continue with the following steps to construct the floor panels.

140. Turn on layer Curve Floor, set current layer to Floor Panel, and turn off layers Curves Drive and Drive.
     Now you use the curves displayed on the Curve Floor layer to construct surfaces on the Floor Panel layer.
141. Construct a solid by extruding curve A (Figure P1–38) a distance of 100 mm in one direction.
142. Construct another solid by extruding curve B a distance of 60 mm in one direction.
143. Construct a fillet edge of 11 mm radius at edge C.
144. Select Edit > Explode.
145. Select solids D and G.
     They are decomposed into individual surfaces.
146. Delete faces D, E, F, and G.

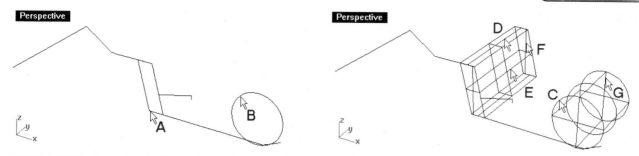

FIGURE P1–38 *Curves for the floor panels being extruded (left) and polysurfaces being exploded and surfaces being deleted (right)*
© 2014 Cengage Learning®. All Rights Reserved.

147. Set the display to Rendered.
148. Turn on layer Curve Wheelbase.
149. Mirror surfaces A, B, and C about endpoints D and E.
150. Extrude curve F (Figure P1–39) along path G.
151. Extrude curves H, J, and K separately a distance of 100 mm in both directions (total distance 200 mm).
152. Turn off layers Curves Wheelbase and Curves Floor.

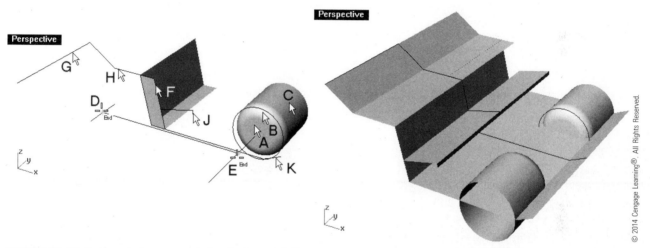

FIGURE P1–39 *Surfaces being constructed and mirrored (left) and surfaces constructed and mirrored (right)*

153. Turn off layer Curves Wheelbase and Curves Floor.
154. With reference to Figure P1–40, trim away the unwanted parts of the surfaces.
155. Join the trimmed surfaces together to form a polysurface.

The floor panels are complete.

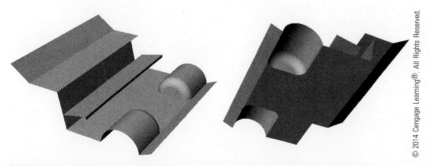

FIGURE P1–40 *Two views showing floor panel surfaces trimmed and joined*

To complete the model, we will work on the seat of the model, as follows.

156. Turn on layer Curves Seat and set current layer to Seat.

    You will construct the seat on the Seat layer.

157. Construct two solids by extruding curves A and B (Figure P1–41) a distance of 49 mm in both directions.

158. Turn off layer Curves Seat and turn on layers Wheel, Tire, Rear Axle, Chassis, Rear Suspension, Engine, Exhaust, Intake, Fuel, Front Suspension, and Drive.

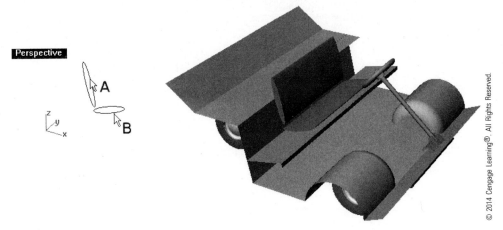

FIGURE P1–41 *Curves for the seat being extruded (left) and bench seat constructed (right)*

Apart from the exterior body panels, the model car is complete. We will continue with making the body panels and use the body panels to trim away unwanted parts of the floor panels in the next modeling project, shown in Figures P1-4 and P1-5.

SUMMARY

In this modeling project, we have experienced using the curves as framework to construct surfaces, polysurfaces, and solids of regular geometric shapes. To let you appreciate how to use Rhinoceros as a tool to construct 3D surfaces and polysurfaces quickly, we have done the tedious work of curve construction and saved these curves for 3D surfaces/polysurfaces building in a template file. If you were to construct similar objects from scratch, you have to think about the objects' cross-section profile and location and then construct them by using the curves tools.

Practically, Rhinoceros polysurfaces and solids are derived from NURBS surfaces by joining a set of contiguous surfaces together. If a polysurface or a single surface, such as a sphere or an ellipsoid, encloses a volume without any opening, a solid is formed. Opposite to joining, we can explode a polysurface to become individual surfaces.

MODELING PROJECT 2

Bubble Car Body

INTRODUCTION

This is the second part of the bubble car modeling project. Here, we will work on the bubble car's bodywork, which mainly concerns building free-form surfaces and making necessary modifications to the boundaries of these surfaces. Making free-form surface inevitably requires a set of curves to depict various cross-sections of the surface. Unlike what we have done in modeling the internal structure of the car, in which we need only to extrude a curve along a straight line, revolve a curve about an axis, or sweep a single curve along a path, we have to use various surface-making techniques. Obviously, a thorough prior understanding on what kinds of surface profiles can be used, how these surfaces can be made, and what kinds of curves are required for making such surfaces is essential. Figure P2–1 shows two preliminary sketches of the bubble car body, depicting the car's general profile and silhouette.

FIGURE P2–1 *Sketches showing the front and rear views of the bubble car*

OBJECTIVES

After working on this modeling project, you should be able to

- Demonstrate a general understanding on various Rhinoceros free-form surface modeling tools and the techniques in using these tools

OVERVIEW

The car's bodywork can be divided into two major components: main body and skirt. We will first make the main body, which consists of an egg-shaped body shell, a front opening door, and a number of window sections. To make these components, we need to construct a number of contiguous surfaces. To ensure proper surface profile continuity between these surfaces, we will treat them holistically when designing the curves. To help construct these curves, we have, based on the photos from various sources, prepared some sketches. Some of these sketches are shown in Figure P2-2. With these and other sketches, a wireframe of curves is produced and a set of contiguous surfaces can then be made, as shown in Figure P2-3.

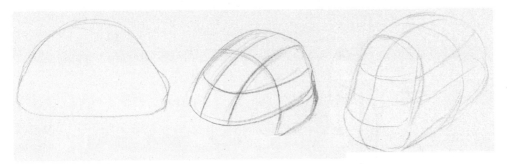

FIGURE P2–2 *Sketches made while thinking about various cross-sections of the main body*
© 2014 Cengage Learning®. All Rights Reserved.

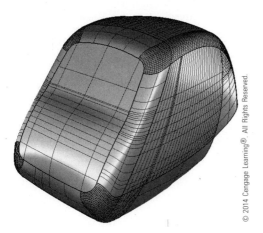

FIGURE P2–3 *Surfaces representing the bubble car's main body panels*

Upon completion of the main body, we will proceed to make the skirt, by producing five intersecting surfaces, trimming away the unwanted portions, and rounding off the edges. Figure P2-4 shows the preliminary sketches, and Figure P2-5 shows the surfaces and trimmed surfaces. To finish the bodywork, we will cut the window openings, trim away the unwanted portions of the surfaces, and add lights and other accessories, as shown in Figure P2-6.

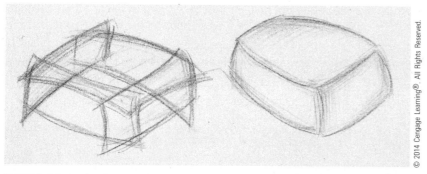

FIGURE P2–4 *Sketches showing four intersected surfaces (left) and surfaces trimmed (right)*

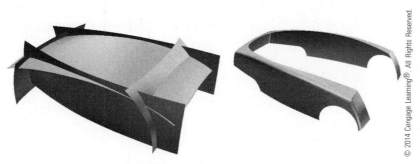

FIGURE P2–5 *Surfaces for making the skirt of the bubble car (left) and skirt surfaces trimmed (right)*

FIGURE P2–6 *Main body, door, and skirt (left) and car body with accessories (right)*

BODYWORK CONSTRUCTION

Using the curves provided, you will first make the bubble car's main body. Then we will proceed to make the door panel, skirt, and window sections. Finally, we will complete the model by adding lights and other small fittings.

Template File

Although this is the second part of the bubble car modeling project, we can skip the previous modeling project and work directly on this because we have prepared a template, in which all the work needed to be done in Modeling Project 1 is saved. In addition, the template file has all the curves required to build the bodywork. Now open the template file.

1. Select File > Open and select the file Bubble Car 2.3dm from the Modeling Project 2 folder on the Student Companion Website.

Like the previous modeling project, you will find other template files in the folder, depicting various stages of model building, enabling you to take a break while working on this project and restart your project without having to save your work.

As can be seen in Figure P2–7 (left), in addition to the work that we have done in the previous modeling project, we have turned on the Curve Body Panel layer. Here is where we will proceed on.

We will now turn off those layers within which previous work resides and set the current working layer, as follows.

2. Select Edit > Layers > One Layer On.
3. In the Layer to leave on dialog box, click on the Curves Body Panel and then the OK button.

   All the layers are turned off except the Curves body panel layer.

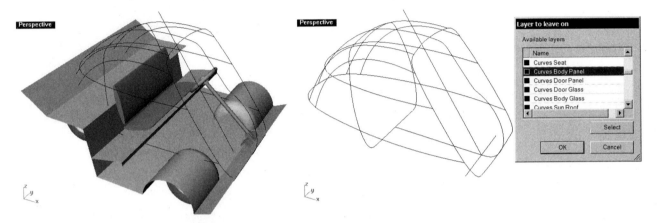

FIGURE P2–7 *From left to right: template file, curves for making the body panel, and the Layer to leave on dialog box*
Source: Robert McNeel and Associates Rhinoceros® 5

CURVES FOR MAKING THE SURFACES

The curves that are shown in Figure P2–7 (right) for making the main body panel of the bubble car are the results of quite a number of trials and errors on various possibilities and combinations. They are smooth continuous curves with G2 continuity, and some of them are split at their intersections with other curves, in order to build surfaces with smooth G2 continuity. You may refer to main chapters for a more detailed explanation on what is meant by G2 continuity.

Main Body

We will now construct the surfaces for making the main body panel of the bubble car, as follows.

4. Set the current layer to Body Panel.

   You will work on this layer.

5. With reference to Figure P2–8, rotate the display.
6. Select Surface > Curve Network.
7. Select curves A, B, C, D, E, F, and G (Figure P2–8) and press the ENTER key.

   In accordance with the selected curves' proximity, they are automatically sorted into two sets of curves in U and V directions.

8. In the Surface From Curve Network dialog box, click on the Position buttons for all four surface edges and click on the OK button.

   A curve network surface is constructed. By checking the Position boxes in the dialog box, this curve network surface's edges will coincide exactly with the input curves A, C, D, and G.

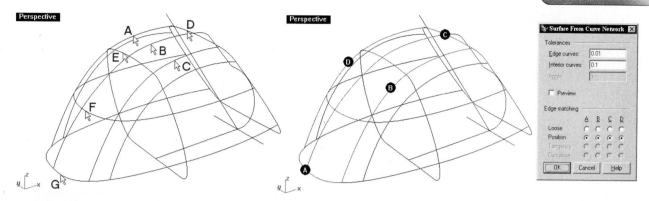

FIGURE P2–8 *From left to right: curves being selected, curve network surface being constructed, and the Surface From Curve Network dialog box*
Source: Robert McNeel and Associates Rhinoceros® 5

Second Surface of the Main Body

Now we will construct the second surface. Because surface continuity is very important, we will use an existing edge as one of the input curves. Let us perform the following steps.

9. Select Surface > Curve Network.
10. Select surface edge A and curves B, C, D, E, and F (Figure P2-9), and press the ENTER key.

CONTINUITY ISSUE

The effects of selecting curve A and surface edge A are different: Using surface edge A can provide a smooth continuity (G2) between the existing surface and the surface to be constructed.

11. In the Surface From Curve Network dialog box, click on the Curvature check box for the edge connecting with the previous surface and click on the OK button.

    Another curve network surface is constructed, and its edge contiguous to the previous surface has a G2 continuity.

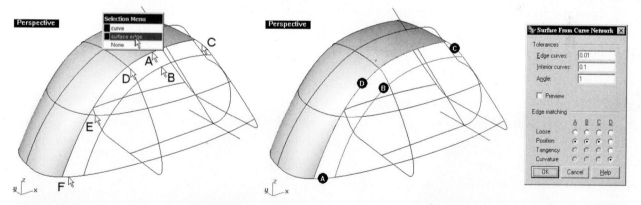

FIGURE P2–9 *Surface edge and curves being selected (left) and curve network surface being constructed (right)*
Source: Robert McNeel and Associates Rhinoceros® 5

Edge Splitting

Because the next and subsequent surfaces that we will construct will use a segment of the second surface's edge as input curve, we will first split its edge into three segments, as follows.

12. Select Surface > Edge Tools > Split Edge.
    Osnap should be set to END.
13. Select surface edge C (Figure P2-10), endpoints P and Q, and press the ENTER key.
    The surface edge C is split into three segments, at points P and Q.

Using Split Surface Edges

Now we will construct a sweep 2 rails surface with a segment of the split edge of the second surface as one of the input curves, as follows.

14. Select Surface > Sweep 2 Rails.
15. Select curves A and B, surface edge segment C, and curves D and E (Figure P2–10), and press the ENTER key.
16. In the Sweep 2 Rail Options dialog box, accept the default of Do not simplify and click on the OK button.

    Here curves A and B are rails, and edge segment C and the two other curves are cross-sections. A sweep 2 rails surface is constructed.

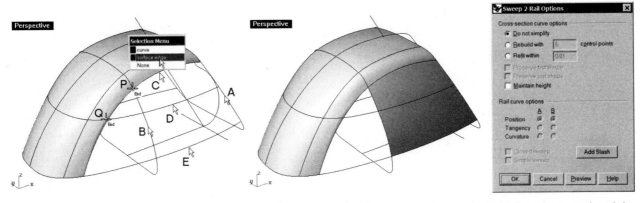

FIGURE P2–10 *From left to right: surface edge split and curves and edges being selected, sweep 2 rails surface constructed, and the Sweep 2 Rail Options dialog box*
Source: Robert McNeel and Associates Rhinoceros® 5

Again, for the purpose of using a segment of the edge of the surface that you just constructed as input curves, we will now split one of the edges of the surface and then construct another sweep 2 rails surface, as follows.

17. Split surface edge A at endpoint X (Figure P2–11).

    Now the surface edge has two segments.

18. Construct a sweep 2 rails surface, using curves C and D as rails, and surface edge segments A and B (not curves) as cross-sections, as shown in Figure P2–11. Remember to click on the Curvature buttons on the Sweep 2 Rails Options dialog box to ensure a G2 continuity between the surfaces to be constructed and the existing surface edges.

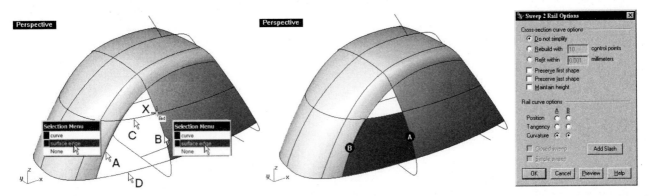

FIGURE P2–11 *From left to right: surface edge split and curves and surface edges selected, and sweep 2 rails surface being constructed, and the Sweep 2 Rail Options dialog box*
Source: Robert McNeel and Associates Rhinoceros® 5

To fill up the triangular opening among the surfaces, we will construct a patch surface, as follows.

19. Turn off layer Curves Body Panel.

    This way, we can select surface edges more easily.

20. Select Surface > Patch.
21. Select surface edges A, B, and C (Figure P2–12) and press the ENTER key.
22. Click on the OK button on the Patch Surface Options dialog box.

    A patch surface is constructed.

23. Set current layer to Curves Body Panel.

    We will construct some curves on this layer for further surface construction.

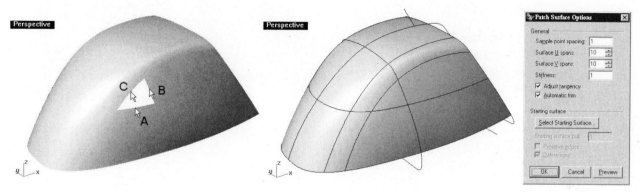

FIGURE P2–12 *From left to right: surface edges selected, and patch surface constructed, and the Patch Surface Options dialog box*
Source: Robert McNeel and Associates Rhinoceros® 5

Curve Construction from Existing Objects

Now we will construct two curves for surface construction, by projecting a curve onto a surface and blending two curves, as follows.

24. Maximize the Front viewport.

    We will work on the Front viewport because the subsequent curve projection is viewport dependent.

25. Select Curve > Curve from Objects > Project.
26. Select curve A (Figure P2–13) and press the ENTER key.
27. Select surface B and press the ENTER key.

    A curve is projected on the surface.

28. Maximize the Perspective viewport and change its display to wireframe.
29. Turn off layer Body Panel.
30. Select Curve > Blend Curves > Quick Curve Blend.
31. Select curves C and D (Figure P2–13).

    A blend curve is constructed between the two selected curves.

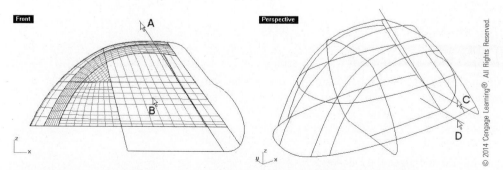

FIGURE P2–13 *Curve being projected (left) and blend curve being constructed (right)*

More Surfaces

Now we will construct three swept surfaces, as follows.

32. Set current layer to Body Panel and set the display to Rendered.
33. With reference to Figure P2–14, split surface edge A at endpoint P.
34. Construct a sweep 1 rail surface, using surface edge segment A as rail and curves B and C as cross-sections.

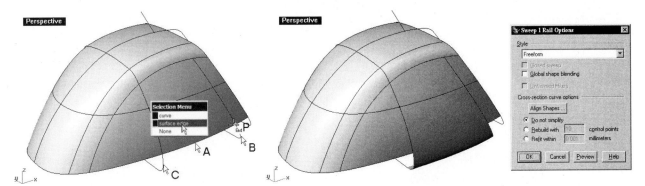

FIGURE P2–14 *From left to right: surface edge split and surface edge and curves being selected, sweep surface being constructed, and the Sweep 1 Rail Options dialog box*
Source: Robert McNeel and Associates Rhinoceros® 5

35. Rotate the display with reference to Figure P2–15.
36. Construct a sweep 2 rails surface, using curves A and B as rails and curve C as cross-section.

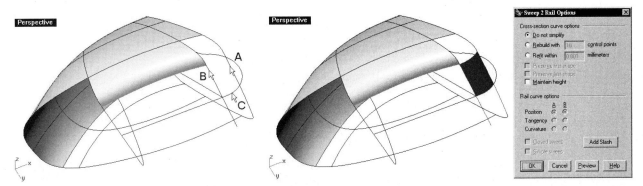

FIGURE P2–15 *From left to right: curves being selected, and sweep surface being constructed, and the Sweep 2 Rail Options dialog box*
Source: Robert McNeel and Associates Rhinoceros® 5

37. Construct a sweep 2 rails surface, using surface edges A and B as rail and surface edge C and curve D (Figure P2–16) as cross-sections. Remember to click on the Curvature buttons on the Sweep 2 Rails Options dialog box to obtain a G2 continuity.

MIRRORING AND SURFACE CONTINUITY

We are not going to mirror surface E because the contiguous edge between the mirrored surface and the original surface will have only a G0 continuity, which is not we wanted. Its only purpose is to provide a surface edge for constructing a surface in step 37. Now, we will delete it.

38. Select surface E (Figure P2–16) and press the DEL key.

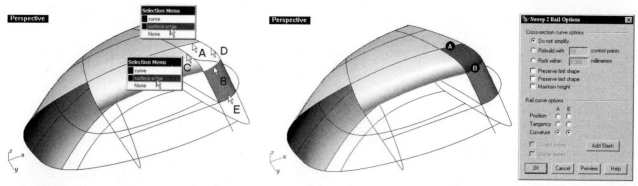

FIGURE P2–16 *From left to right: curves being selected, and sweep surface being constructed, and the Sweep 2 Rail Options dialog box*
Source: Robert McNeel and Associates Rhinoceros® 5

Now we will mirror a set of surfaces, as follows.

39. With reference to Figure P2–17, rotate the display.
40. Select Transform > Mirror.
41. Select surfaces A, B, C, D, E, and F (Figure P2–17) and press the ENTER key.
42. Type w0,0 at the command area.

   This specifies the origin point with reference to the world construction plane.

43. Type w1,0 at the command area.

   This point, together with the previous point, specifies a plane about which selected surfaces are mirrored. The surfaces are mirrored.

44. Rotate the display to see the result of mirroring.

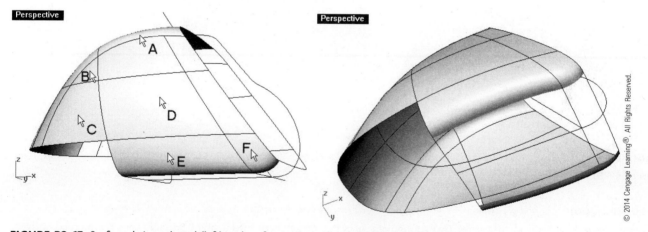

FIGURE P2–17 *Surfaces being selected (left) and surfaces mirrored and display rotated (right)*

Now we will construct the final piece of surface for the car's main body panel.

45. Construct a sweep 2 rails surface, using curves A and B (Figure P2–18) as rails and surface edge C, curve D, and surface edge E as cross-sections.

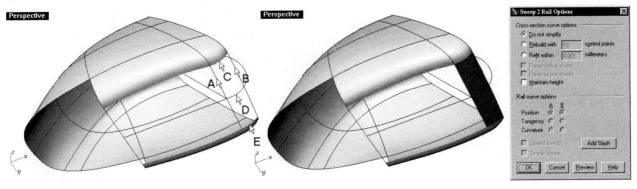

FIGURE P2–18 *From left to right: curves and surface edges being selected, sweep surface being constructed, and the Sweep 2 Rail Options dialog box*
Source: Robert McNeel and Associates Rhinoceros® 5

To complete the body panel, we will join the surfaces together by joining edges of all the contiguous surfaces.

46. Select Surface > Edge Tools > Join 2 Naked Edges.
47. Select edge A (Figure P2-19) twice.
48. In the Edge Joining dialog box, click on the Yes button.

    The surfaces are joined. Note that the tolerance value shown here may be different from yours.

49. Repeat the command to join all the contiguous edges.

    It is essential to go through all the contiguous edges because we are going to split the body into windows and metal panels at the later stage of this tutorial.

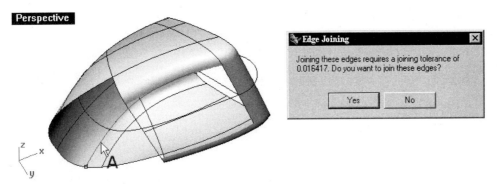

FIGURE P2–19 *Edges of two contiguous surfaces being joined (left) and the Edge Joining dialog box (right)*
Source: Robert McNeel and Associates Rhinoceros® 5

Frontal Door

To make the door, we will first make two contiguous surfaces, merge them into a single surface, trim the merged surface, split the edge of the trimmed surface, and construct a set of blend surfaces. Now let us continue with the following steps to construct a sweep 1 rail surface and a curve network surface.

50. Turn on layer Curves Door Panel, set current layer to Door Panel, and turn off layers Body Panel and Curves Body Panel.

    Now only curves for making the door panel are displayed, and you will construct the door panel on the Door Panel layer.

51. Select Surface > Sweep 1 Rail.
52. With reference to Figure P2-20, select curves A, B, and C and press the ENTER key.

    The first selected curve will be the rail, and subsequently selected curves will be sections.

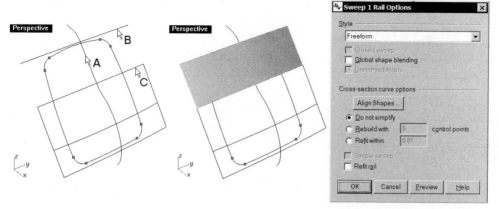

FIGURE P2–20 *From left to right: curves selected, sweep 1 rail surface being constructed (right), and the Sweep 1 Rail Options dialog box*
Source: Robert McNeel and Associates Rhinoceros® 5

53. With reference to Figure P2–21, use surface edge A and curves B, C, D, E, and F to construct a curve network surface.

    A curvature continuity (G2) is needed for the edge contiguous with the previous surface.

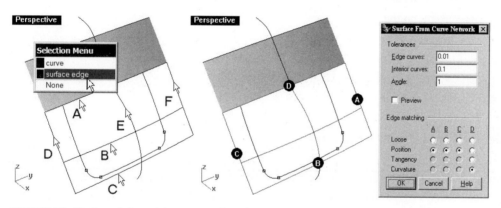

FIGURE P2–21 *From left to right: curves selected, curve network surface being constructed, and the Surface From Curve Network dialog box*
Source: Robert McNeel and Associates Rhinoceros® 5

Surface Merging

We will now merge the contiguous surfaces into one single surface.

54. Select Surface > Edit Tools > Merge.
55. Select surfaces A and B (Figure P2–22) and press the ENTER key.

    The surfaces are merged into one.

Setting Construction Plane and Trimming

Now we will set the construction plane orientation for subsequent operations.

56. Select View > Set CPlane > 3 Points.
57. Select endpoints C and D and press the ENTER key.

    The construction plane is set. To help you realize the effect of change of construction plane orientation, grid lines are shown in Figure P2–22 (right). However, for the sake of clarity in illustration, grid lines are not shown in other figures.

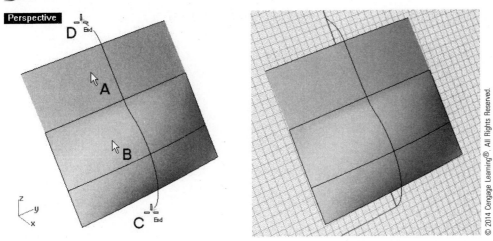

FIGURE P2–22 *Surfaces being merged and CPlane being set (left) and surface merged and CPlane set (right)*

Now we will use a curve to trim the merged surface. Prior to trimming, we will project the endpoints of the trimming curve to the merged surface. The projected points will be used for subsequent splitting of the trimmed surface's edge.

58. Set the display to wireframe.
59. Set the current layer to Curves Door Panel.

    We will construct some curve objects on this layer.

60. Select Curve > Curve from Objects > Project.
61. Select points A, B, C, D, E, F, G, and H (Figure P2–23) and press the ENTER key.
62. Select surface J.

    The selected points are projected onto the surface.

63. Select Edit > Trim.
64. Select curve K (Figure P2–23) and press the ENTER key.
65. Select location L and press the ENTER key.

    The surface is trimmed. Note that projection and trimming directions are construction plane dependent. They work in a direction perpendicular to the active construction plane. That is why we have to set the construction plane prior to performing this task.

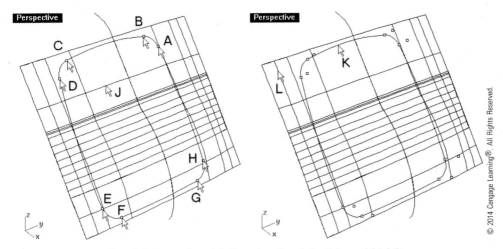

FIGURE P2–23 *Endpoints being projected (left) and surface being trimmed (right)*

Now we will use the projected points to split the trimmed surface's edge into eight segments.

66. Set Osnap to Point.
67. Split surface edge A at points B, C, D, E, F, G, H, and J (Figure P2–24).

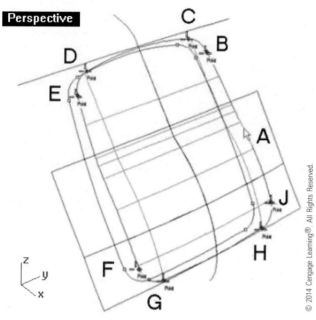

FIGURE P2–24 *Surface edges being split*

Surface Edge Splitting and Surface Construction

We will now use the split edge segments, together with the surface edges of the main body panel, to construct blend surfaces and curve network surface, as follows.

68. Set the current layer to Door Panel, turn off layer Curves Door Panel, and turn on layer Body Panel.
69. With reference to Figure P2–25, rotate the display and set the display to Shaded.
70. Select Surface > Blend Surface.
71. Select surface edges A and B.

    It is necessary that we select both edges near the right or left side because the location where we select on the edge has a direct impact on the final outcome.

72. In the Adjust Surface Blend dialog box, click on the OK button.

    A blend surface is constructed.

73. Repeat the command to construct two more blend surfaces, between surface edges C and D and surface edges E and F.

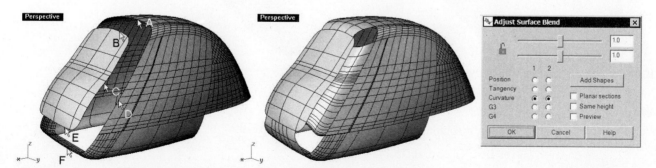

FIGURE P2–25 *From left to right: surface edges being selected, blend surfaces constructed, and the Adjust Surface Blend dialog box*
Source: Robert McNeel and Associates Rhinoceros® 5

74. With reference to Figure P2–26, construct two curve network surfaces. The first surface uses surface edges A, B, C, and D as input curves, and the second surface uses surface edges E, F, G, and H. Edge matching must be curvature.

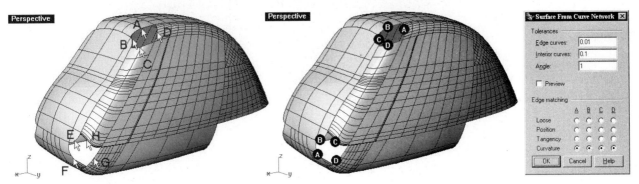

FIGURE P2–26 *From left to right: surface edges being selected, curve network surfaces constructed, and the Surface From Curve Network dialog box*
Source: Robert McNeel and Associates Rhinoceros® 5

Now we will mirror three surfaces and join the surfaces into a single polysurface.

75. Select Transform > Mirror.
76. Select surfaces A, B, and C (Figure P2–27) and press the ENTER key.
77. Type w0,0 at the command area.
78. Type w1,0 at the command area.
    The surfaces are mirrored.
79. Turn off layer Body Panel.
80. Select Edit > Join.
81. Select all the door panel surfaces and press the ENTER key.
    The door panel is complete.

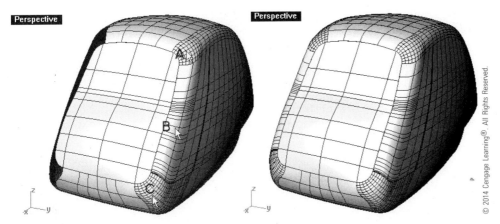

FIGURE P2–27 *Surfaces selected (left) and surfaces mirrored (right)*

Windows

We will make the window sections by splitting the body panel and the door panel.

82. Turn on layer Curves Door Glass and Door Glass.
    Curves for window construction are displayed, and you will work on the Door Glass layer.
83. Set the display to wireframe.

84. Select Edit > Split.
85. Select polysurface A, indicated in Figure P2–28, and press the ENTER key.
86. Select curve B and press the ENTER key.

    The polysurface representing the door panel is split into two, one representing the door panel itself and the other one representing the windscreen.

As the split surface is still residing on the Door Panel layer, now you will change its layer.

87. Select Edit > Layers > Change Object Layer.
88. Select polysurface C and press the ENTER key.
89. In the Layer for objects dialog box, select Door Glass and click on the OK button.
90. Shade the display.
91. Select View > Set CPlane > World Top.

    The door window is complete, and the construction plane is reset to its default, which is the same as the Top viewport.

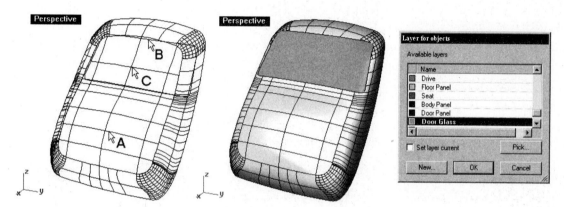

FIGURE P2–28 *From left to right: polysurface being split, polysurface split and one polysurface's layer being changed, and the Layer for objects dialog box*
Source: Robert McNeel and Associates Rhinoceros® 5

Car Body Window and Sun Roof

We will continue to split the polysurface representing the body panel into window and sun roof elements, as follows.

92. Set the current layer to Body Panel; turn on layers Curves Body Glass and Curves Sun Roof, Body Glass, and Sun Roof; and turn off layers Door Panel, Door Glass, and Curves Door Glass.
93. Maximize the Front viewport and set the display to wireframe.

    To reiterate, splitting by using line elements is view dependent.

94. Select Edit > Split.
95. Select polysurface A (Figure P2–29) and press the ENTER key.
96. Select curves B, C, D, and E and press the ENTER key.

    The polysurface representing the body panel is split into seven polysurfaces.

97. With reference to Figure P2–29, change polysurface F to layer Sun Roof, and change all polysurfaces except A and F to layer Body Glass.
98. Maximize the Perspective viewport and rotate it around to appreciate the change.

    The window and sun roof sections are complete.

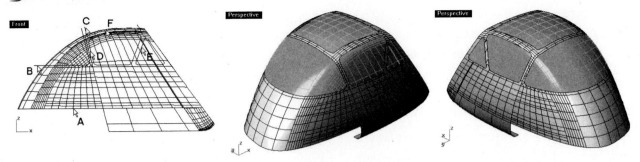

FIGURE P2–29 *From left to right: polysurface being split, right rear view of the body, and left rear view of the body*
© 2014 Cengage Learning®. All Rights Reserved.

Skirts

We will make the skirt by using five intersecting surfaces, trimming the surfaces to remove unwanted portions, and rounding off the edges. Now we will continue with the following steps to construct five surfaces.

99. Set current layer to Skirt; turn on layer Curves Skirt; and turn off layers Body Panel, Body Glass, Sun Roof, Curves Sun Roof, and Curves Body Glass.

    Now only the curves for making the skirts are displayed, and you will work on the Skirt layer.

100. Construct a sweep 1 rail surface, using curve A as rail and curves B, C, and D (Figure P2–30) as cross-sections.
101. Construct another sweep 1 rail surface, using curve E as rail and curves F and G as cross-sections.
102. Select Surface > Loft.
103. Select curves H, J, and K and press the ENTER key. When selecting the curves, select either their upper ends or lower ends. Otherwise, you may have to click on the Align Curve button on the Loft Options dialog box and make necessary adjustment.
104. Click on the OK button.
105. Extrude curve L a distance of 80 mm in both directions (total 160 mm).
106. Mirror sweep surface M about w0,0 and w1,0.

    The surfaces for making the main body of the skirt are complete.

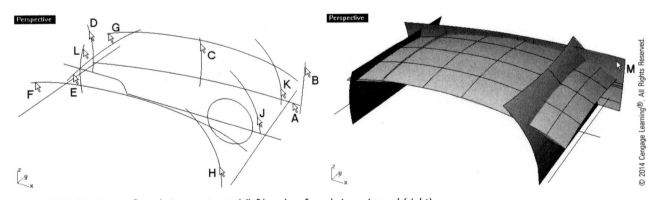

FIGURE P2–30 *Five surfaces being constructed (left) and surfaces being mirrored (right)*

Now we will trim the surfaces and then join them together, as follows.

107. Maximize the Top viewport and shade the display.
108. Use curves A and B to (Figure P2–31) trim away the outer portions of the surfaces at C, D, E, F, G, and H.

    Because a longer surface can trim a shorter surface but not vice versa, it is necessary to use two lines to first trim away some portions of the surfaces to make their lengths identical. Now they can trim each other.

109. Maximize the Perspective viewport.
110. Using surfaces J, K, and L as cutting objects, trim away portions J, K, L, and M (Figure P2–31).

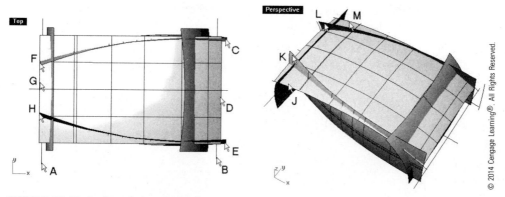

FIGURE P2–31 *Surfaces being trimmed*

111. Maximize the Front Viewport.
112. With reference to Figure P2–32, use curves A and B to trim away the lower portions of all the surfaces.
    Remember that trimming by using line elements is view dependent.

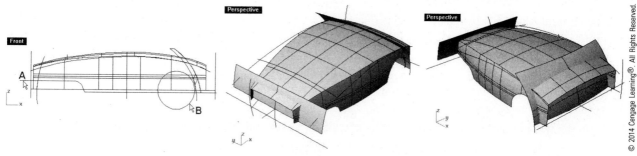

FIGURE P2–32 *From left to right: cutting objects, rear view of trimmed surfaces, and front view of trimmed surfaces*

113. Turn off layer Curves Skirt.
114. With reference to Figure P2–33, trim the surfaces accordingly.
115. Join all the surfaces into a single polysurface.

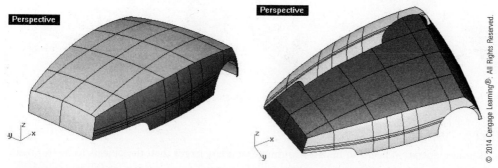

FIGURE P2–33 *Surfaces trimmed, joined, and curves turned off*

116. Select Solid > Fillet Edge > Blend Edge.
117. Select the NextRadius option on the command area and change it to 6 mm.
118. Select edges A and B (Figure P2–34) and press the ENTER key twice.
    Blend surfaces are constructed along the edges.

119. Repeat the command to construct blends of radius 3 mm on the remaining edges in accordance with Figure P2–34.

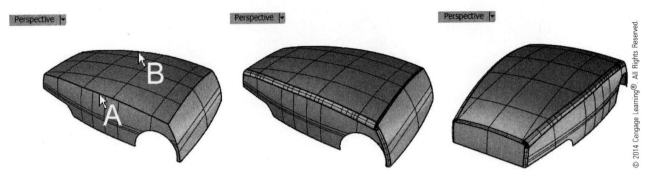

FIGURE P2–34 *From left to right: two edges being filleted, frontal filleted, and rear portion filleted*

Trimming Body, Skirt, and Floor Panels

We will now finish the main bodywork by trimming away unwanted portions of the main body, the skirt, and the floor panel that we constructed in the previous modeling project.

120. Set current layer to Body Panel.
121. With reference to Figure P2–35, use the two polysurfaces as cutting objects to trim away the unwanted portions.

    If you encounter any problem in trimming the unwanted portions, you may try exploding the skirt's polysurface into individual surfaces and try again. After trimming, you have to join them back into a single polysurface. Alternatively, you may open one of the template files on the Modeling Project 2 folder on the Student Companion Website.

122. Join the two polysurfaces into a single polysurface by first selecting the main body and then the skirt.

    Note that the selection sequence determines in which layer the joined polysurface will reside. Here, because we select the body panel first, the joined polysurface will reside in layer Body Panel.

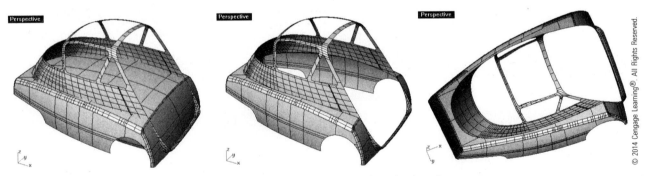

FIGURE P2–35 *From left to right: two polysurfaces and two views of the trimmed polysurfaces*

Trimming the Floor Panel

Because the cutting object in a trimming operation needs to have a trimming boundary larger than the objects to be trimmed, in order to trim away some portions of the skirt, we will construct a planar surface at the door opening, as follows.

123. Select Solid > Cap Planar Holes.
124. Select polysurface A (Figure P2–36) and press the ENTER key.

    A planar surface is constructed and joined to the polysurface. Note that this command caps only planar openings; therefore the other nonplanar openings are not capped.

If you encounter any problem in constructing a planar surface by capping, you may try using the PLANARSRF command (by selecting Surface > Planar Curves) and selecting the edges where the body panel meets the door panel.

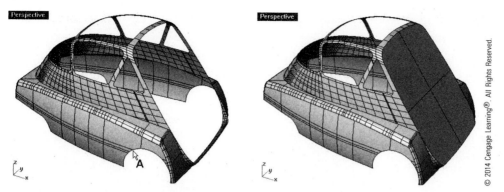

FIGURE P2–36 *Polysurface selected (left) and polysurface capped (right)*

Now we will use the capped body and skirt polysurface to trim away the unwanted portions of the floor panels. After trimming, we will remove the capped planar surface.

125. Turn on layer Floor Panel.
126. With reference to Figure P2-37, use polysurface A as cutting object and trim away portions B, C, and D of the floor panels.
127. Select Solid > Extract Surface.
128. Select element E and press the ENTER key.

    The selected surface element is extracted from the polysurface.

129. Select surface E and press the ENTER key.

    The extracted surface is deleted, and the car body panels are complete.

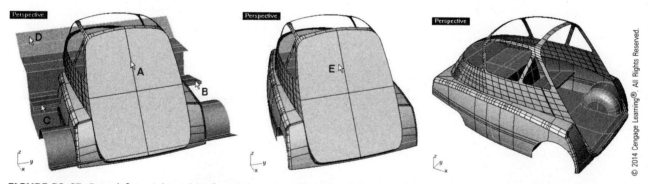

FIGURE P2–37 *From left to right: polysurfaces being trimmed, surface being extracted, and extracted surface deleted*

Lights, Gas Cap, and Engine Vents

We will now finish the car model by adding lights, gas cap, and engine vents. We will first construct a revolved polysurface, an extruded solid, and a sphere, as follows.

130. Set current layer to Engine Misc, turn on layer Curves Engine Misc, and turn off all other layers.
131. Revolve curve A (Figure P2-38) full circle about endpoints B and C.
132. Select Surface > Planar Curves.
133. Select curves D and E and press the ENTER key.

    A planar surface is constructed.

134. Select Surface > Extrude Curve > Tapered.
135. Select curves D and E and press the ENTER key.
136. Select the DraftAngle option on the command area and set it to -10 degree.

137. Type 15 at the command area.

    Two tapered surfaces are constructed.

138. Join the planar surface and tapered surfaces constructed from curves D and E to form a polysurface.

139. Select Solid < Cap Planar Holes and then select the joined polysurface to form a closed polysurface.

140. Construct a sphere with center at point F (use Point Osnap) and a radius of 10 mm.

Now we have three objects: a sphere, a revolved polysurface, and a closed polysurface. The sphere will be used for making the gas cap, the revolved polysurface for making the engine's intake vent, and the closed polysurface for making the engine's exhaust vent.

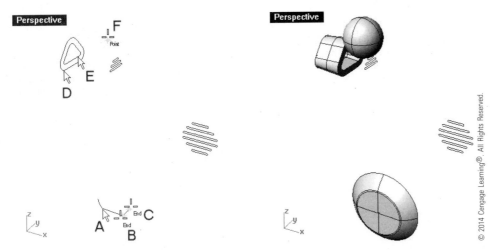

FIGURE P2–38 *Curves and point selected (left) and revolved surface, extruded solid, and sphere constructed (right)*

Detailing the Intake Vent, Exhaust Vent, and Gas Cap

We will now cut some holes in the revolved polysurface in the Front viewport, as follows.

141. Maximize the Front viewport and set its display to Wireframe.

142. With reference to Figure P2–39, use the curve A to trim polysurface B.

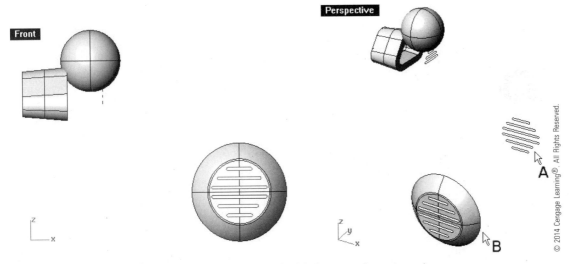

FIGURE P2–39 *Front viewport (left) and right viewport (right) showing a face trimmed*

We will rotate the revolved polysurface and fillet the edges of the extruded polysurface, as follows.

143. Maximize the Perspective viewport and rotate the view in accordance with Figure P2–40.
144. Select Transform > Rotate.
145. Select polysurface A (Figure P2–40) and press the ENTER key.
146. Select endpoint B as the base point.
    This is the center of the circular face.
147. Type −10 at the command area.
    The polysurface is rotated.
148. Use the solid filleting command to fillet edges C and D with a radius of 1 mm.

> **NOTE**
> To reiterate, many commands are construction plane dependent. Rotate command is also one of them. Therefore, we must not forget to reset the construction plane of the Perspective viewport to World Top. Otherwise, the effect of rotation is very different.

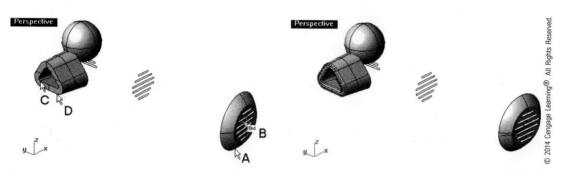

FIGURE P2–40 *Polysurface being rotated and edges being filleted (left) and polysurface rotated and edges filleted (right)*

Car's Lights

We will now construct the front and rear lights of the car, as follows.

149. Turn on layer Curves Lights and set current layer to Lights (Figure P2–41).

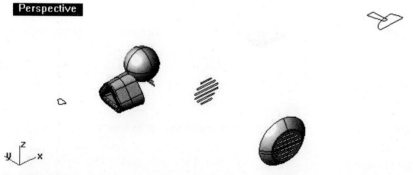

FIGURE P2–41 *Curves for constructing lights turned on and display rotated*
© 2014 Cengage Learning®. All Rights Reserved.

150. With reference to Figure P2–42, zoom in as may be necessary and construct two revolved polysurfaces by revolving curve A about endpoints B and C and curve D about endpoints E and F.
151. Combine polysurfaces G and H by Boolean Union operation.
152. Delete surface J.

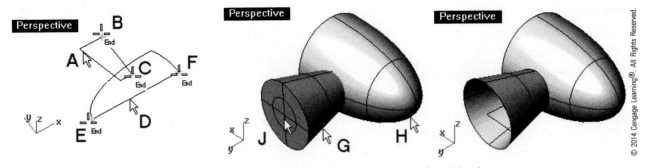

FIGURE P2–42 *From left to right: curves being rotated, polysurface being united, and surface deleted*

153. With reference to Figure P2–43, rotate curve A about endpoints B and C.

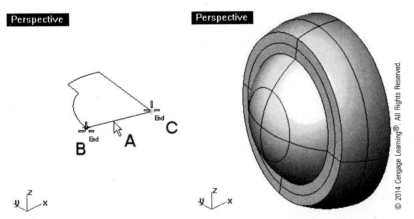

FIGURE P2–43 *Curves for the rear light (left) and rear light constructed (right)*

Combining the Intake Vent, Exhaust Vent, Gas Cap, and Lights with the Body

The objects for making the lights, gas cap, and engine vents are ready. We will perform some trimming operations to remove some unwanted portions, as follows.

154. Turn on layer Body Panel and Skirt.

    Depending on the sequence of prior to joining the body and skirt, the joined body and skirt might be residing on layer Body Panel or Skirt.

155. Select Solid > Boolean Two Objects.
156. Select A and then B and press the ENTER key.
157. Click to repeat until you get the result similar to Figure P2–44.

    Because these two polysurfaces reside in different layers, we have to select A and then B to have the combined polysurface placed in layer Body Panel.

158. Construct a solid fillet of 1 mm radius at edge C.

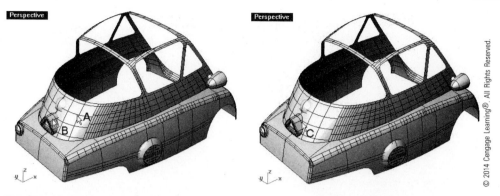

FIGURE P2–44 *Rear vent being joined to the main body (left) and edge being filleted (right)*

159. With reference to Figure P2–45, mirror light objects A and B about w0,0 and w1,0.

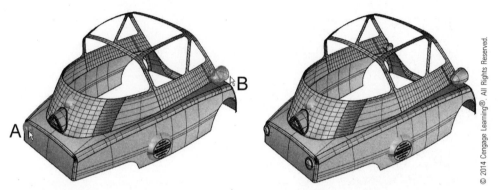

FIGURE P2–45 *Light objects selected (left) and light objects mirrored (right)*

160. Use Boolean Two Objects commands six times to join object A to objects B, C, D, E, F, and G, each time clicking to iterate until you get the desired result, as shown in Figure P2–46.

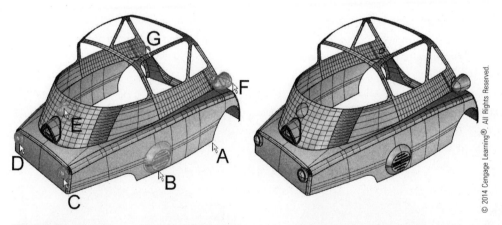

FIGURE P2–46 *Objects to be joined (left) and objects joined together as a single polysurface (right)*

Finally, we will construct a fillet edge and cut several holes, as follows.

161. With reference to Figure P2–47, use solid filleting tool to construct a fillet at edge A.

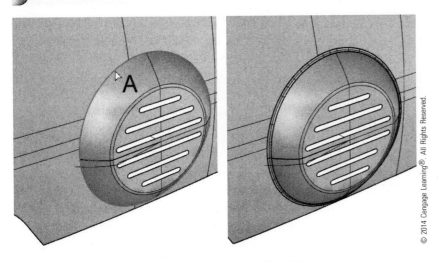

FIGURE P2–47 *Edge being selected (left) and edge filleted (right)*

162. Maximize the Right viewport.
163. With reference to Figure P2–48, use curves A, B, and C to cut three holes on the main body.

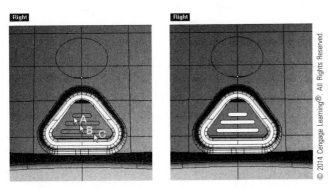

FIGURE P2–48 *Curves selected (left) and polysurface trimmed (right)*

164. Turn off layers Curves Engine Misc and Curves Lights and turn on the layers with surface/solid objects.

    The model is complete. Figure P2–49 shows the shaded images of the rear right, front right, and front right of the car with its door opened by rotating the door panel and the door glass. Rendered images of the bubble car are shown in Figures P2–50 and P2–51.

165. Do not save your file.

    You can find a completed model of the bubble car on the Modeling Project 2 folder on the Student Companion Website.

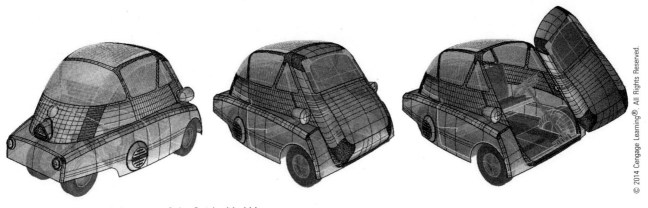

FIGURE P2–49 *Shaded images of the finished bubble car*

FIGURE P2–50 *Render images of the finished bubble car*

FIGURE P2–51 *Bottom view*

SUMMARY

To construct a model of a 3D free-form object in the computer, we use a set of free-form surfaces. Before making the surfaces, we first study and analyze the 3D object to determine what types of surfaces are needed. We consider various types of primitive surfaces, basic free-form surfaces, and derived surfaces. Among these surfaces, free-form surfaces are those most commonly used. All free-form surfaces have one thing in common: They all need to be constructed from smooth curves and/or point objects.

It is natural to start thinking about the surfaces but not the points and curves while we are designing and making a surface model. However, because the computer constructs basic free-form surfaces from defined curves and/or point objects, the first task we need to tackle in making free-form surfaces is to think about what types of curves and/or points are needed and how they can be constructed. After making the curves and/or points, we then let the computer generate the required surfaces.

MODELING PROJECT 3

Small Objects

INTRODUCTION

Unlike the previous case studies, in which all the curves for making the models are provided, we will practice how to work on curves prior to constructing surfaces.

OBJECTIVES

After working on this modeling project, you should be able to

- Demonstrate a general understanding on using Rhinoceros's curve tools

OVERVIEW

Because curve construction is a tedious job, we will work on several simple objects. The first object is a simple bottle opener, the second is an electric cable, the third is an infant spoon, and the fourth case is a set of three jewelry objects. (See Figures P3–1 and P3–2.)

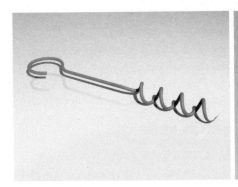 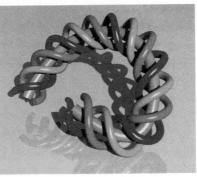 

FIGURE P3–1 *From left to right: bottle opener, cable, and spoon*

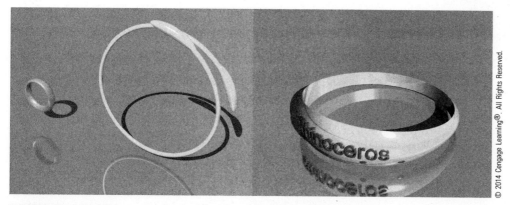

FIGURE P3–2 *Jewelry*

MODELS

We will construct the models one by one, starting from the bottle opener, then the cable, spoon, and finally the jewelry set.

Bottle Opener

Practically, the bottle opener is made from a wire. To model it, we will construct a curve to depict its spine and a circle around the curve to depict the wire's cross-section. With these curves, we construct two sweep surfaces, join them together, and cap a planar hole to form a Rhinoceros solid. Now we will open the template file.

1. Select File > Open and select the file BottleOpener.3dm from the Modeling Project 3 folder on the companion CD.

    In this project, we will make use of a set of points provided in a template file to construct a number of curves and use the curves to construct the surface model.

Curves Construction

You will now construct the bottle opener's curves.

2. Set Osnap mode to Point and End.
3. Select Curve > Arc > Center, Arc, Angle and click on points A, B, and C (Figure P3-3) to construct an arc.
4. Select Curve > Line > Single Line and select points D and E (Figure P3-3) to construct a line.
5. Select Curve > Helix and select points F and G (Figure P3-3), set the number of turns to 4, and select point H to construct a helix.

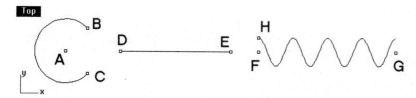

FIGURE P3–3 *From left to right: arc, line, and helix constructed from the given points*
© 2014 Cengage Learning®. All Rights Reserved.

6. Maximize the Perspective viewport.
7. Turn off the Points layer.
8. Select Curve > Blend Curves > Quick Curve Blend and select A and B, indicated in Figure P3-4, to construct a blend curve.
9. Construct two blend curves, bridging the gaps between A and B and between C and D, shown in Figure P3-4.
10. Repeat the previous step to construct another blend curve between C and D (Figure P3-4).

11. Select Edit > Join and select all the curves one by one to form a single curve.
12. Select Edit > Change Degree, select the joined curve, and press the ENTER key.
13. If Deformable=No, click on it to change it to Yes. Otherwise, proceed to the next step.
14. Type 3 at the command area to change the degree of polynomial.

    Degree of polynomial will be explained in the main chapters.

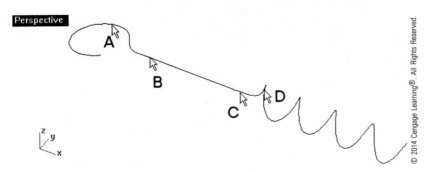

FIGURE P3–4 *Blend curves constructed*

15. Select Curve > Circle > Center, Radius and select the AroundCurve option.
16. Referring to Figure P3–5, select curve A and click on location B (exact location is unimportant here).
17. Type 2 at the command area.

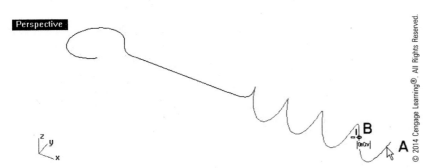

FIGURE P3–5 *Curves joined, simplified, and circle around it constructed*

Surface and Solid

Now you will use the curves that you constructed to build two swept surfaces. Then you will join them together and cap its end to form a Rhinoceros solid.

18. Set the current layer to Surface.
19. Select Surface > Sweep 1 Rail.
20. Select curves A and B and press the ENTER key twice (Figure P3–6).
21. Check the Refit within button and click on the OK button on the Sweep 1 Rail Options dialog box.

    Because only one section curve is used and the curve is not at the end of the rail, the location where you click on the rail has a direct impact on which part of the rail will be used. Here you should click on A.

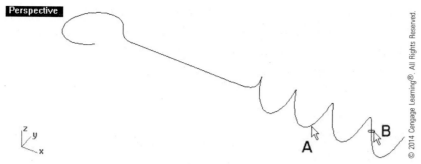

FIGURE P3–6 Sweep 1 rail surface being constructed

22. Repeat the command.
23. Select curves A and B (Figure P3–7) and press the ENTER key.
24. Select the Point option and click on endpoint C.
25. Press the ENTER key.
26. Click on the OK button on the Sweep 1 Rail Options dialog box.

A sweep 1 rails surface with one of its ends converging to a point is constructed.

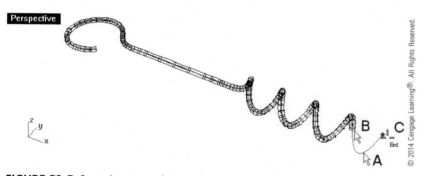

FIGURE P3–7 Sweeping to a point

Now you will join the surfaces and cap its end.

27. Turn off layer Curves.
28. Select Edit > Join.
29. Select the two sweep surfaces and press the ENTER key.
30. Select Solid > Cap Planar Holes and select the joined surface.
    Select surface edge C as the profile to construct the surface.
    The model is complete. (See Figure P3–8.)

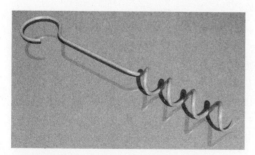

FIGURE P3–8 Completed bottle opener
© 2014 Cengage Learning®. All Rights Reserved.

Electric Cable

The electric cable is a set of four wires: a central core and three additional wires wound on the core. In essence, these surfaces are sweep surfaces with both ends capped. To make the surfaces, we need a free-form curve to define the path for making the central core. Around the curve, we will construct three helixes to define the path for the wires winding on the core. To properly phase the helix on the core, we will construct a circle around the path of the core and divide the circle into three equal parts by placing three points on it. After making the helix curves, we will make several circles around the curves for use as cross-sections of the sweep surfaces. Now let us perform the following steps.

1. Select File > Open and select the file Cable.3dm from the Modeling Project 3 folder on the companion CD.

   There are two bitmap images placed in the Top and Front viewports, respectively. These images serve as references for making a 3D curve, defining the path of the central core.

Curves Construction

Now you will construct a 3D curve by holding down the CONTROL key while picking points on the screen.

2. Select Curve > Free Form Curve > Interpolate Points.
3. Hold down the CONTROL key.
4. Referencing Figure P3-9, click on location A in both the Top and Front viewports. (Background bitmaps are placed in the Top and Front viewports.)

   A 3D point is selected.

5. While still holding down the CONTROL key, continue to click on locations B, C, D, and E in both the Top and Front viewports.
6. Press the ENTER key.

   A 3D curve is constructed.

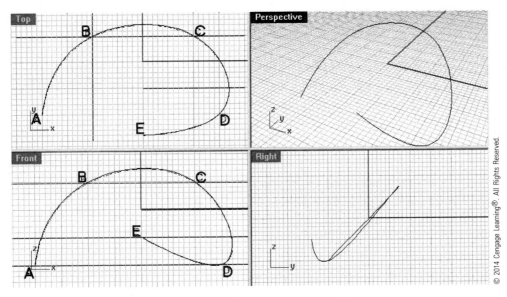

FIGURE P3-9 *3D curve based on two background bitmaps constructed*

Now you will construct a circle around the curve at one of its ends.

7. Maximize the Perspective viewport and turn off grid lines and grid axes by clearing appropriate boxes of the Options dialog box.
8. Set Osnap mode to End.
9. Select Curve > Circle > Around Curve.

10. Select curve A (Figure P3-10) and then click on endpoint A.
11. Type 4 at the command area.

    A circle with a radius of 4 units is constructed around the curve at its end.

Now you will construct three-point objects on the circle that you constructed.

12. Select Curve > Point Object > Divide Curve By > Number of Segments.
13. Select the circle and press the ENTER key twice.
14. Type 3 at the command area.

    Three points are constructed and equally spaced on the circle.

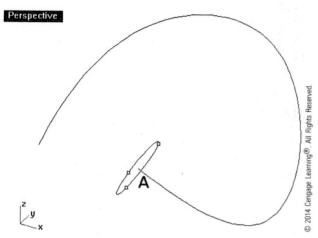

FIGURE P3-10 *Circle around and at the endpoint of the curve and points along the circle constructed*

Now you will construct three helices around the 3D curve with their starting points located at the point objects around the circle at the 3D curve's end.

15. Select Curve > Helix.
16. Click on the Around Curve option.
17. Select curve A, indicated in Figure P3-11.
18. Type 5 at the command area.
19. Click on point B.

    A five-turn helix with starting point at B is constructed around curve A.

20. Repeat steps 15 through 19 twice to construct two more helices at points C and D.

Now you will construct four around-curve circles: three at the end of the helices and one at the 3D curve.

21. Select Curve > Circle > Around Curve.
22. Select curve E (Figure P3-11) and then click on endpoint E.
23. Type 1 at the command area.

    A circle with a radius of 1 unit is constructed around the curve E at its end.

24. Repeat steps 21 through 23 twice to construct two more circles of 1 unit radius around curves F and G.
25. Repeat steps 21 through 23 once to construct one more circle of 1.5 unit radius around curve H.

    A total of four circles are constructed.

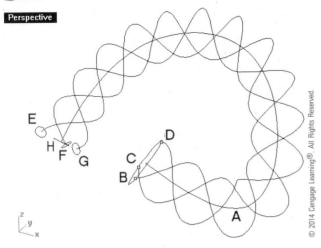

FIGURE P3–11 *Helices and circles around curve constructed*

Now you will complete the model by constructing four sweep 1 rail surfaces and then cap their ends.

26. Change the current layer to Surface.
27. With reference to Figure P3–12, construct four sweep 1 rail surfaces.
28. Cap their endpoints to change them to closed polysurfaces.

The electric cable is complete.

FIGURE P3–12 *Completed cable*

Spoon

This model can be divided into two components: spade and handle. Although some of the curves that define the shape and silhouette of a spoon are given, we have to perform some curve manipulation tasks: We need to remap a curve from one construction plane to another; construct a 3D curve from two planar curves; split two curves; and construct cross-section profiles. Now let us perform the following steps.

1. Select File > Open and select the file Spoon.3dm from the Modeling Project 3 folder on the companion CD.

Spoon's Spade

As you can see, there are three curves and they are constructed on the Front viewport. This is done intentionally so that they can accurately intersect each other. Snap mode was turned on when the curves were produced. Now you will remap a curve from the Front viewport to the Top viewport.

2. Click on the label of the Front viewport to make it the current viewport.
3. Select Transform > Orient > Remap on CPlane.

4. Select curve A, indicated in Figure P3-13, and press the ENTER key.
5. Click on B.

   Curve A is remapped to the Top viewport.

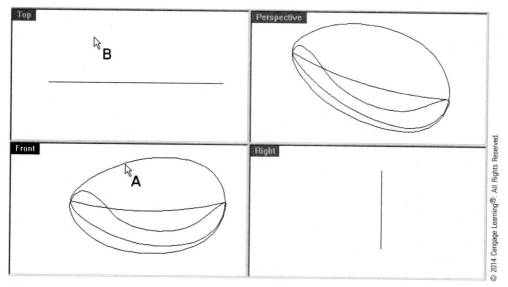

FIGURE P3-13 *A curve being remapped from the Front viewport to the Top viewport*

Now you will construct a 3D curve from two 2D curves, one residing on the Top viewport and the other residing on the Front viewport.

6. Select Curve > Curve From 2 Views.
7. Select curves A and B, shown in Figure P3-14.

   A 3D view is constructed.

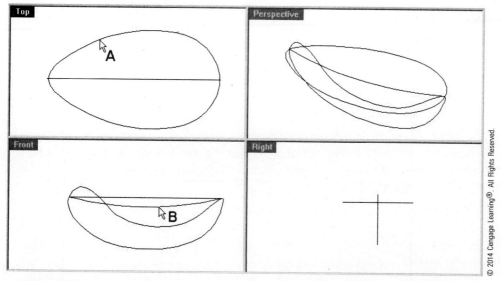

FIGURE P3-14 *A 3D curve being constructed from two curves residing on two construction planes*

Now you will split two curves into four. The reason for first making a single curve and then splitting is to ensure better continuity between the curves.

8. Turn off the CurveSpadeCurveOriginal layer.

   Now the planar curves for making the 3D curve are hidden.

9. Select Edit > Split.
10. Click on the Point option.
11. Select curve A (Figure P3–15).
12. Select points B and C and press the ENTER key.

    Curve A is split at B and C into two curves.

13. Repeat steps 9 through 12 to split curve D at B and C.

    Now you have four curves.

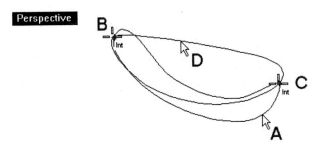

FIGURE P3–15 *Curves being split at their intersecting points*
© 2014 Cengage Learning®. All Rights Reserved.

Now you will construct cross-section curves from the four curves that you split from two.

14. Select Curve > Cross-Section Profiles.
15. Select curves A, B, C, and D (Figure P3–16) in order and press the ENTER key.
16. Click on L, M, J, K, G, H, E, and F (exact location is unimportant) and press the ENTER key.

    Four cross-section profiles are constructed.

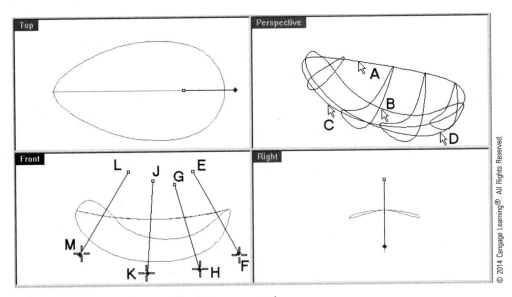

FIGURE P3–16 *Cross-section profiles being constructed*

Now you will use the curves to construct a curve network surface.

17. Set Surface as the current layer and maximize the Perspective viewport.
18. Select Surface > Curve Network.
19. Referring to Figure P3–17, click on A and B to select all the curves and press the ENTER key.
20. Click on the OK button on the Surface From Curve Network dialog box.

    The surface representing the spade of the spoon is constructed.

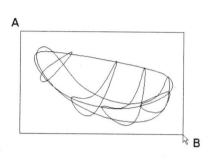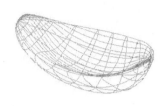

FIGURE P3–17 *Curve network surface constructed*
Source: Robert McNeel and Associates Rhinoceros® 5

Spoon Handle

Now you will work on the spoon's handle.

21. Turn off layers CurveSpadeTopProfile and CurveSpadeFrontProfile and turn on the CurveHandle layer.
    You should still be working on the Surface layer.
22. Select Surface > Curve Network.
23. Referring to Figure P3–18, click on A and B to select all the curves and press the ENTER key.
24. Click on the OK button on the Surface From Curve Network dialog box.
    A surface representing the handle of the spoon is constructed.

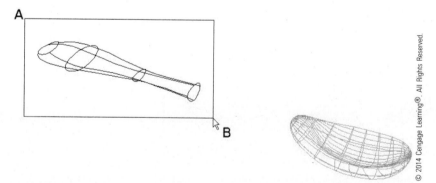

FIGURE P3–18 *Another curve network surface being constructed*

Now you will transform the surface.

25. Turn off the CurveHandle layer and turn on the CurveHandleSpline layer.
26. Select Transform > Flow along Curve.
27. Select surface A (Figure P3–19) and press the ENTER key.

28. Set Copy=No and Stretch=Yes.
29. Select curve B and then C.

    The surface representing the handle of the spoon is transformed and stretched. The reason for this transformation is that it is easier to construct a surface along a straight axis and then flow it along a curve.

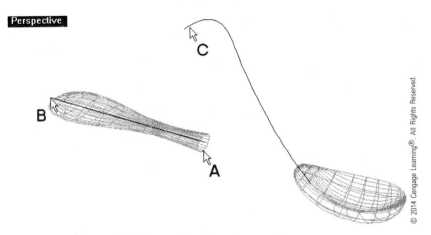

FIGURE P3–19 *A surface being made to flow along another curve*

Now you will construct a patch surface to fill the opening at the end of the spoon's handle, do some trimming on the handle and the spade, and construct a blend surface to connect the handle to the spade smoothly.

30. Maximize the Front viewport.
31. Turn off the CurveHandleSpline layer and turn on the TrimCurves layer.
32. Select Surface > Patch.
33. Select edge A (Figure P3–20) and press the ENTER key.
34. Click on the OK button on the Patch Surface Options dialog box.

    A patch surface is constructed, filling the opening.

35. Select Edit > Trim.
36. Select curves B and C (Figure P3–20) and press the ENTER key.
37. Click on the handle surface and the spade surface at a location between curves B and C and press the ENTER key.

    The surfaces are trimmed.

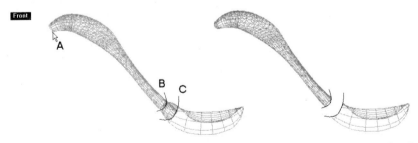

FIGURE P3–20 *Surface flown along a curve (left) and patch surface constructed and surfaces trimmed (right)*
© 2014 Cengage Learning®. All Rights Reserved.

38. Maximize the Perspective viewport and turn off layer TrimCurves.
39. Select Surface > Blend Surface.
40. Select edges A and B, shown in Figure P3–21.
41. On the Adjust Surface Blend dialog box, select G3 for both edges and click on the OK button.

42. Select Edit > Join.
43. Select all the four surfaces.

    The model is complete.

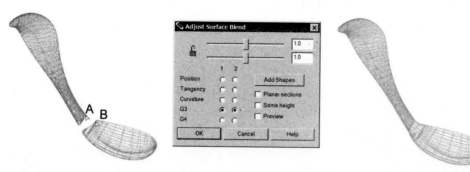

FIGURE P3–21 *From left to right: display shaded, the Adjust Surface Blend dialog box, and blended surface constructed*
Source: Robert McNeel and Associates Rhinoceros® 5

Jewelry

We will now construct two jewelry objects by deforming objects of regular shape. Let us perform the following steps.

1. Select File > Open and select the file Jewel.3dm from the Modeling Project 3 folder on the companion CD.

Ring

Now you will work on a ring by first constructing an extrude surface.

2. Set Osnap mode to point.
3. Select Surface > Extrude Curve > Straight.
4. Select curve A, shown in Figure P3–22, and press the ENTER key.
5. Set Both Sides=No.
6. Click on point B.
7. Select Edit > Explode.
8. Select the extruded object and press the ENTER key.

    The extrude object is changed to an extrude surface.

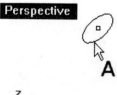

FIGURE P3–22 *A curve being extruded*
© 2014 Cengage Learning®. All Rights Reserved.

Now you will transform the extrude surface by twisting.

9. Select Transform > Twist.
10. Select surface A (Figure P3–23) and press the ENTER key.
11. Click on points B and C to define the axis of twist.
12. Type 360.

    The surface is twisted.

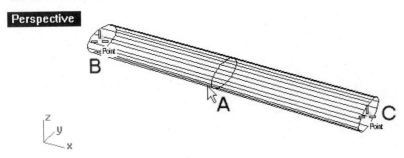

FIGURE P3-23 *Surface being twisted*
© 2014 Cengage Learning®. All Rights Reserved.

Now you will flow the twisted extrude surface along a circle to complete the ring.

13. Turn on the CurveFlow1 layer and turn off CurveRing layer.
14. Select Transform > Flow along Curve.
15. Select surface A (Figure P3-24) and press the ENTER key.
16. Set Rigid=No and Stretch=Yes.
17. Select curve B as the base curve and then curve C as the target curve.

    The ring is complete.

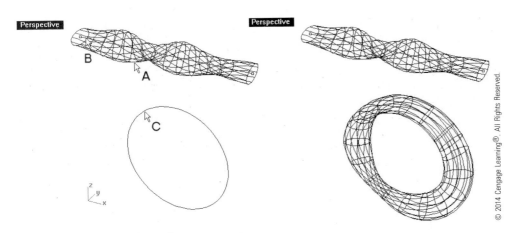

FIGURE P3-24 *Surface being flown*

Hand Lace

Now you will work on the hand lace.

18. Turn off layers CurveFlow1 and CurveRing and turn on layers CurveFlow2 and Surface2.
19. With Stretch=Yes, flow polysurface A (Figure P3-25) from base curve B to target curve C.

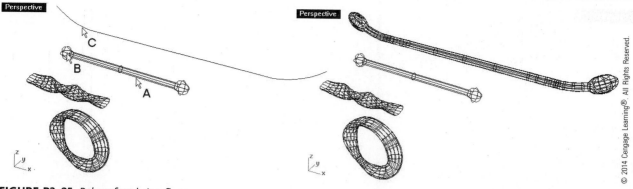

FIGURE P3–25 *Polysurface being flown*

20. Turn off the CurveFlow2 layer and turn on the CurveFlow3 layer.
21. With reference to Figure P3–26, flow polysurface A with Stretch=Yes.

    When selecting the base and target curves, the location where you select has an effect on the final outcome. Here, you should click on B and C.

    The models are complete.

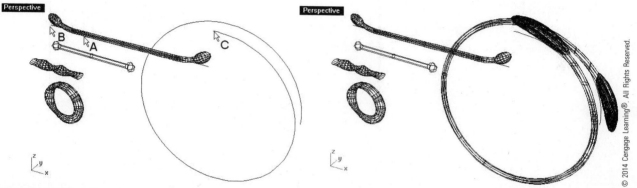

FIGURE P3–26 *Polysurface being flown*

Expanding the work performed for the jewelry objects, this hand lace is also made by the deformation of regular objects. Now let us perform the following steps.

1. Select File > Open and select the file HandLace.3dm from the Modeling Project 3 folder on the companion CD.
2. Select Transform > Flow along Curve.
3. Select polysurface A (Figure P3–27) and press the ENTER key.
4. Set Right=No and Stretch=Yes.
5. Select base curve B and then target curve C.

    The polysurface is flown. Note the locations where you select the backbone curves.

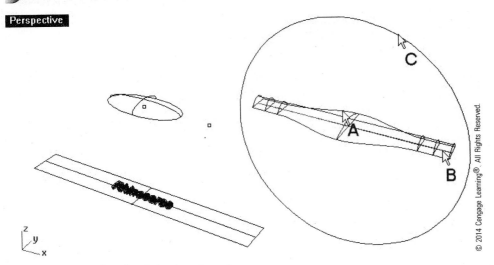

FIGURE P3–27 *Polysurface being transformed*

6. Select Transform > Flow along Surface.
7. With reference to Figure P3-28, click on A and B to select the text objects and press the ENTER key.
8. Set Right=No.
9. Select C as the base surface and D as the target surface.

   Note that the location where you click on the base surface and the target surface has a direct impact on the outcome of flowing.

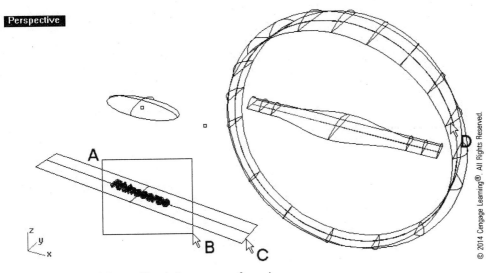

FIGURE P3–28 *Solid text object being space conformed*

10. Type Splop at the command area.
11. Select surface A, indicated in Figure P3-29, and press the ENTER key.
12. Click on point B as the reference sphere's center.
13. Click on point C to define the reference sphere's radius.
14. Select several pick points (E, F, and G) on the upper face of the polysurface (exact locations are unimportant) and drop points on the surface.
15. Turn off all layers except layers Sporph and OriginalSurface.

    Several copies of surfaces are transformed. The model is complete.

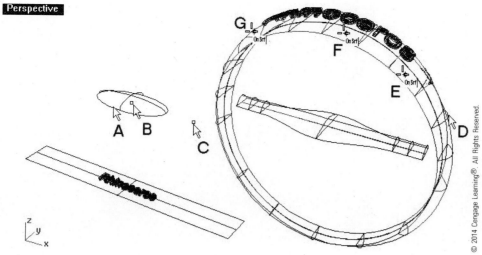

FIGURE P3–29 *Ellipsoid surface being space dropped*

SUMMARY

Curve construction is a tedious task, but surface modeling is semiautomatic. We only need to use an appropriate surface modeling command, and the system will construct the surface for us. Through working on several small objects in this modeling project, we practiced how to use some of the Rhinoceros curve tools. The main chapters will help you to discover more about curve and surface construction and manipulation tools.

MODELING PROJECT 4

Toy Car Assembly

INTRODUCTION

In this modeling project, we will assemble a simple toy car model with three components: body, wheels, and axles. Unlike the bubble car modeling project, in which all the components are constructed in different layers of the same file, the parts of this toy car reside in different files.

OBJECTIVES

After working on this modeling project, you should be able to

- Demonstrate a general understanding on using Rhinoceros's block management tools

OVERVIEW

We will construct the body in terms of Rhinoceros solid and then assemble the body and the other two supplied components together in another file, by inserting the components as block definitions. Figure P4–1 shows rendered images of the finished assembly.

FIGURE P4–1 *Rendered images of the assembly*

COMPONENTS AND ASSEMBLY

The assembly of the toy car consists of a car body, two axles, and four wheels. There are many ways to put the components together. One way is to make use of subassembly, in which two wheels and one axle are first assembled as a subassembly and then inserted into the main assembly.

You will first make the solid body from the curves provided and then assemble it with two preconstructed components.

Toy Car Body

We will make the body in two stages. In stage 1, we will build the exterior surfaces, shown in Figure P4–2, using the curves provided in a template file. In stage 2, we will build a solid by adding a few more surfaces to form a closed polysurface. In addition, we will cut two holes.

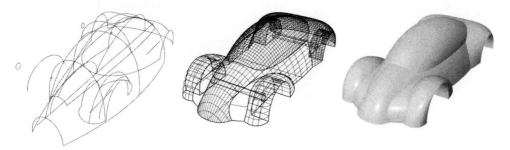

FIGURE P4–2 *From left to right: curves, solid constructed, and rendered model of the solid model*
© 2014 Cengage Learning®. All Rights Reserved.

Now let us perform the following steps.

1. Select File > Open and select the file CarBody.3dm from the Modeling Project 4 folder on the companion CD.

   In this file, there are five sets of curves residing on various layers. You will turn on these layers one by one to construct various surface elements of the model car.

2. Select Surface > Curve Network.

3. Referencing Figure P4–3, click on A and drag to B to select all the curves and press the ENTER key.

   Basically, this command will try to divide the selected curves into two sets of elements to produce a network from which a surface is generated. However, if there are any ambiguously positioned curves, you will be required to do the selection manually.

4. Click on the OK button on the Surface From Curve Network dialog box.

   A curve network surface is constructed.

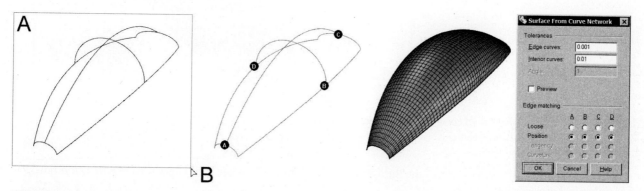

FIGURE P4–3 *From left to right: curve being selected, curve selected, surface constructed, and the Surface From Curve Network dialog box*
Source: Robert McNeel and Associates Rhinoceros® 5

Now you will hide the surface that you constructed and continue with constructing the second surface.

5. Select Edit > Visibility > Hide.
6. Select the surface that you constructed and press the ENTER key.
   The curve network surface is hidden.
7. Turn on the Rear layer and turn off the Upper layer.
   Remember to keep Surface as the current layer.
8. Repeat steps 2 through 4 to use the curves indicated in Figure P4–4 to construct another curve network surface.

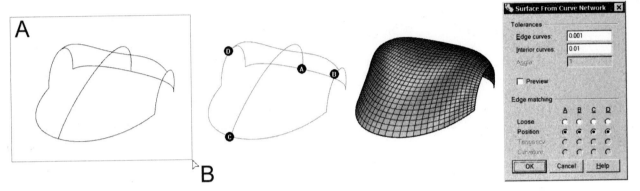

FIGURE P4–4 *Construction of the second curve network surface*
Source: Robert McNeel and Associates Rhinoceros® 5

Again, you will hide the previous surface and then construct the third surface.

9. Hide the surface that you constructed.
10. Turn on the Main layer and turn off the Rear layer.
    Again, keep Surface as the current layer.
11. Referencing Figure P4–5, Repeat steps 2 through 4 to construct the third curve network surface.

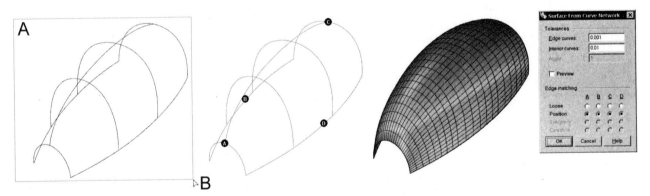

FIGURE P4–5 *Construction of the third curve network surface*
Source: Robert McNeel and Associates Rhinoceros® 5

In order to achieve the desired surface shape, the third surface was intentionally made a bit larger than required, and now you will trim away its lower portion.

12. Turn off the Main layer, turn on the Main1 layer, and maximize the Right viewport.
    Again, keep Surface as the current layer.
13. Select Edit > Trim.

14. Select A (Figure P4–6) and press the ENTER key.
15. Select B (Figure P4–6) twice and press the ENTER key.

    You have to click on B twice because there are two sides of the surface to be trimmed.

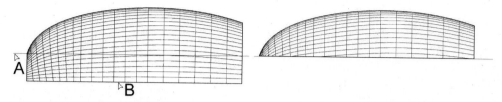

FIGURE P4–6 *Right viewport maximized and surface being trimmed (left) and surface trimmed (right)*
© 2014 Cengage Learning®. All Rights Reserved.

Now you will hide the third surface and construct the forth surface. This time, you will construct a loft surface.

16. Hide the surface constructed, turn on the Front layer, turn off the Main1 layer, and maximize the Perspective viewport.

    The current layer is still Surface.

17. Select Surface > Loft.
18. Select A, B, and C (Figure P4–7) and press the ENTER key.
19. In the Loft Options dialog box, click on the Rebuild with check box and click on the OK button.

    A loft surface is constructed.

FIGURE P4–7 *The forth surface, a loft surface, being built*
Source: Robert McNeel and Associates Rhinoceros® 5

Now you will hide the loft surface and construct the fifth surface, which is a sweep 2 rails surface.

20. Hide the surface constructed, turn on the FrontWheel layer, and turn off the Front layer. Remember to keep Surface as the current layer.
21. Select Surface > Sweep 2 Rails.
22. Referencing Figure P4–8, select A, B, C, D, and E and press the ENTER key.

    Here, the first two selected curves A and B will be used as rails, and any subsequently selected curves will be used as section profiles. While constructing the sweep surface, click on the Maintain height button.

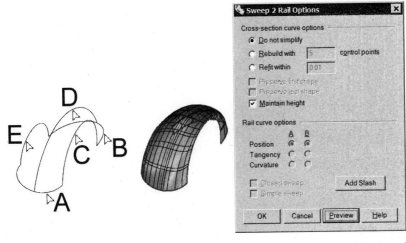

FIGURE P4–8 *Curves and surface constructed (left) and the Sweep 2 Rail Options dialog box (right)*
Source: Robert McNeel and Associates Rhinoceros® 5

In order to obtain a much smoother surface, you will rebuild the sweep 2 rails surface that you constructed. Then you will make a mirror copy of this surface.

23. Turn off the FrontWheel layer.
24. Select Edit > Rebuild.
25. Select A (Figure P4–9) and press the ENTER key.
26. In the Rebuild Surface dialog box, set point count in U direction to 10, point count in V direction to 20, and degree of polynomial in U and V directions to 3, and click on the OK button.
    Rebuilding will produce a smoother surface in the expense of some deviation from the original surface.
27. Select Transform > Mirror.
28. Select A (Figure P4–9) and press the ENTER key.
29. Type w0,0 and press the ENTER key to define the start of mirror plane.
30. Type w0,1 and press the ENTER key to define the end of mirror plane.
31. Select Edit > Visibility > Show.
    The main surfaces are complete.

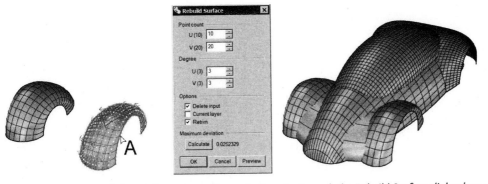

FIGURE P4–9 *From left to right: sweep surface rebuilt and mirrored, the Rebuild Surface dialog box, and hidden surfaces unhidden*
Source: Robert McNeel and Associates Rhinoceros® 5

Now you will trim away the unwanted portions of the surfaces and join them together.

32. Select Edit > Trim.
33. Select all the surfaces and press the ENTER key.
    All the surfaces are cutting objects.

34. Referencing Figure P4–10, rotate the viewport and click on the unwanted portions to trim the surfaces.
35. Select Edit > Join.
36. Select all the surfaces contiguously to join them together to form a polysurface.

    A polysurface is constructed. If you encounter any difficulties in performing the trimming action, you may refer to the file CarBody 15.3dm from the companion CD.

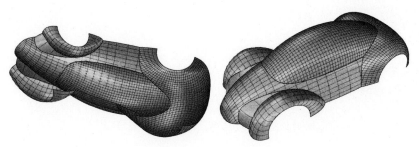

FIGURE P4–10 *Viewport rotated, unwanted portions trimmed away, and surfaces joined*
© 2014 Cengage Learning®. All Rights Reserved.

Now you will construct two extrude surfaces to form the part of the wheel arches of one side of the model car.

37. Select Surface > Extrude Curve > Straight.
38. Set Bothside=No.
39. Type -15.

    Extrude surfaces are constructed (Figure P4–11).

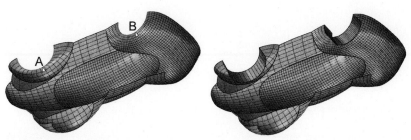

FIGURE P4–11 *Edges being extruded (left) and extrude surfaces constructed*
© 2014 Cengage Learning®. All Rights Reserved.

Now you will complete one side of the car's wheel arches by constructing two line segments and two planar surfaces.

40. Set the current layer to 0 and set Osnap mode to End.
41. Select Curve > Line > Single Line.
42. Select endpoints A and B (Figure P4–12).
43. Repeat steps 41 and 42 to construct another line segment from endpoint C to endpoint D.
44. Set the current layer to Surface.
45. Select Surface > Planar Curves.
46. Referencing Figure P4–12, select lines E and F and edges G and H and press the ENTER key.

    Two planar surfaces are constructed.

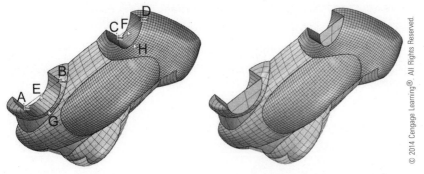

FIGURE P4–12 *Line segments constructed and planar surface being constructed*

Now you will construct the wheel arches of the car's other side by mirroring.

47. Turn off layer 0.
48. Select Transform > Mirror.
49. As shown in Figure P4–13, select surfaces A, B, C, and D and press the ENTER key.
50. Type w0,0.
51. Type w0,1.
    Surfaces are mirrored.
52. Select Edit > Join.
53. Select the polysurface and all the surfaces and press the ENTER key.

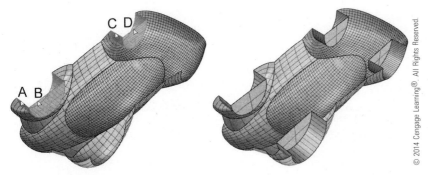

FIGURE P4–13 *Surfaces being mirrored (left) and surfaces mirrored and joined to the polysurface*

Now you will construct a Rhinoceros solid by constructing a planar surface.

54. Select Solid > Cap Planar Holes.
55. Select A (Figure P4–14) and press the ENTER key.
    A closed polysurface, a Rhinoceros solid, is constructed.

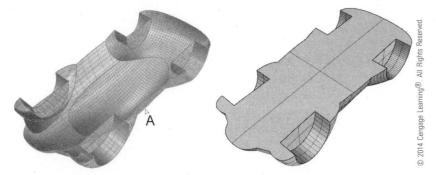

FIGURE P4–14 *Planar hole joined (left) and polysurface capped (right)*

To finish the model, you will construct two holes.

56. Turn on the HoleCurve layer and stay with using Layer Surface as the current layer.
57. Select Solid > Edit Solid Tools > Holes > Make Hole.
58. Select curves A and B (Figure P4–15) and press the ENTER key.
59. Select solid C.
60. Press the ENTER key.

    Two holes are constructed to cut through the solid.

61. Turn off the HoleCurve layer.
62. Do not save your file.

    The model is complete.

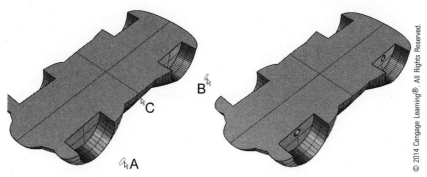

FIGURE P4–15 *Holes being placed (left) and holes placed (right)*

Subassembly

The next thing to do is to construct an assembly of two wheels and an axle, the models of which are already constructed for you. Let us perform the following steps.

1. Start a new file, using the Small Objects, Millimeters.3dm template.
2. Maximize the Perspective viewport.
3. Select Edit > Blocks > Insert Block Instance.
4. In the Insert dialog box, click on button A (Figure P4–16) and then select the file CarAxle.3dm from the Modeling Project 4 folder on the companion CD.
5. In the Insert as area, click on Block Instance. In the Insertion point area, clear the Prompt check box. In the Scale area, check the Uniform check box and click on the OK button.
6. In the Insert File Options dialog box that follows, check the Link and Reference check boxes and click on the OK button.

    The file CarAxle is inserted as a block instance.

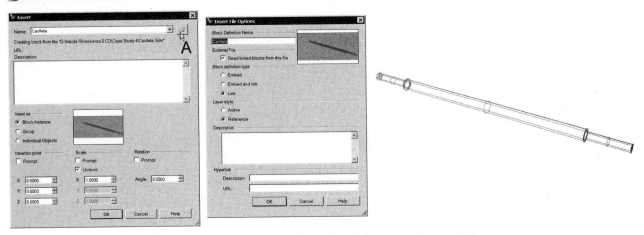

FIGURE P4–16 *From left to right: Insert dialog box, Insert File Options dialog box, and inserted file*
Source: Robert McNeel and Associates Rhinoceros® 5

Now you will insert another file and then make a copy of the inserted file.

7. Repeat steps 3 through 6 to insert the file CarWheel.3dm from the Modeling Project 4 folder on the companion CD.
8. Select Transform > Copy.
9. Select A (Figure P4–17) and press the ENTER key.
10. Click on B and C and press the ENTER key.
    The file CarWheel.3dm is inserted, and the instance is copied.

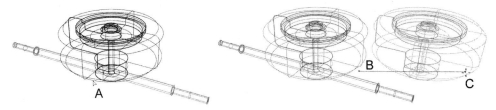

FIGURE P4–17 *Car wheel inserted and being selected (left) and car wheel being copied (right)*
© 2014 Cengage Learning®. All Rights Reserved.

Now you will orient the wheels three dimensionally by selecting three reference points on the wheel and selecting three corresponding points on the axle.

11. On the Osnap dialog box, clear the Project button (if it is checked) and check the cen and quad buttons.
12. Select Transform > Orient > 3 Points.
13. Select A (Figure P4–18) and press the ENTER key.
14. Select cen B, quad C, and quad D as references.
    Note that the three selected points establish a plane, with the first point as the origin and the other two points as two directions on the plane.
15. Select cen E, quad F, and quad G as targets.
    Similarly there are three target points, with the first point as the origin for the object to orient to and the other two to establish a plane to match the reference plane.

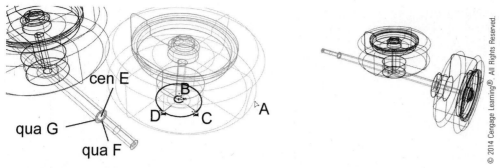

FIGURE P4–18 *A wheel being oriented (left) and wheel oriented (right)*

16. Repeat steps 12 through 15 to orient the other wheel in accordance with Figure P4–19.
17. Do not save your file.

    The subassembly is complete. Understanding that you need this file for constructing the overall assembly of the car, a file CarAxleSubAssembly.3dm is prepared and can be found in the Modeling Project 4 folder on the companion CD.

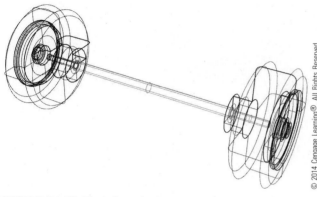

FIGURE P4–19 *The other wheel oriented*

Assembly

Now you will complete the car model by starting a new file, inserting the car body and wheel and axle assembly, and orienting the inserted blocks. Perform the following steps to insert the toy car's components into a new file.

1. Start a new file, using the Small Objects, Millimeters.3dm template.
2. Maximize the Perspective viewport.
3. Select Edit > Blocks > Insert Block Instance.
4. In the Insert dialog box, click on button A (Figure P4–16) and then select the file CarBodyFinal.3dm from the Modeling Project 4 folder on the companion CD.
5. In the Insert as area, click on Block Instance. In the Insertion point area, clear the Prompt check box. In the Scale area, check the Uniform check box and click on the OK button.
6. In the Insert File Options dialog box that follows, check the Link and Reference check boxes and click on the OK button.

    The car body is inserted as a block.

7. Repeat steps 3 through 6 to insert the file CarAxleSubAssembly.3dm from the Modeling Project 4 folder on the companion CD. This time, check the Prompt box in the Insertion point area of the Insert dialog box.
8. After clicking the OK button on the Insert File Options dialog box, click anywhere on the screen to specify a location.
9. Select Transform > Copy.
10. Select the CarAxleSubAssembly and press the ENTER key.
11. Click anywhere on the screen and pick another location on the screen. (Exact location is unimportant.)

    The car axle subassembly is inserted as another block, and the instance is copied, as shown in Figure P4–20.

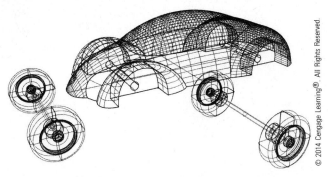

FIGURE P4–20 *Car body and car axle subassembly inserted and car axle subassembly copied*

Now you will complete the toy car by orienting the car axle subassembly.

12. Select Transform > Orient > 3 Points.
13. Select A (Figure P4–21) and press the ENTER key.
14. Select cen B, quad C, and quad D as references.
15. Select cen E, quad F, and quad G as targets.
    A car axle subassembly is oriented.

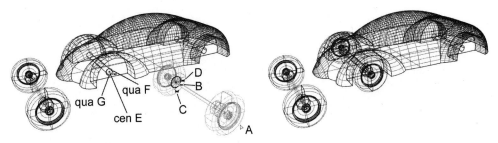

FIGURE P4–21 *A car axle subassembly being oriented*
© 2014 Cengage Learning®. All Rights Reserved.

Now you will complete the model.

16. Repeat steps 11 through 14; referencing Figure P4–22, put the components together.
    The assembly is complete.
17. Do not save your file.

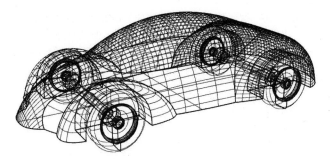

FIGURE P4–22 *Finished toy car assembly*
© 2014 Cengage Learning®. All Rights Reserved.

SUMMARY

There are many ways to represent a solid in the computer. Rhinoceros uses a single closed surface or a closed polysurface to represent a solid. In this modeling project, we constructed a solid using this approach. Although the toy car modeling project has few components, we practiced a way of managing the components by treating them as block insertions.

CHAPTER 1

Rhinoceros—What Is It?

INTRODUCTION

This chapter explains computer modeling concepts and introduces Rhinoceros 5's functions, user interface, help system, and file saving methods.

OBJECTIVES

After studying this chapter, you should be able to

- Evaluate the use of Rhinoceros in design and computer modeling
- Comment on the key functions of Rhinoceros
- Use Rhinoceros's user interface and help system and save files

OVERVIEW

Rhinoceros (also known as Rhino) is a 3D digital modeling application that enables you to construct computer models in terms of NURBS (nonuniform rational B-spline) surfaces, polysurfaces (sets of contiguous NURBS surfaces joined together), solids ("water-tight" surfaces and polysurfaces), and polygon meshes (an approximation of an object by using a set of small contiguous planar faces). It also enables you to produce photorealistic rendered images from surfaces, polysurfaces, solids, and polygon meshes. To facilitate downstream computerized operations and reuse of existing computer models constructed using some other computer application, you can export Rhino models to various file formats and import various file formats into Rhino. To enable human interpretation, you can construct 2D engineering drawings.

COMPUTER MODELING AND RHINOCEROS

To facilitate the process of designing a product or system, you use various media to capture your design ideas. Initially, you may use sketches to represent the subject of the design. To elaborate on the design, you may use models, which can be physical or digital. (Figure 1–1 shows the clay model of a car.) Making a physical model can often be time consuming or simply not feasible if the size is too big. To construct a computer model, you use computer technology and computer-aided design applications. Naturally, you need to know related computer modeling concepts and acquire technical know-how.

FIGURE 1–1 *Clay model of a car*

Essence of Computer Modeling

In essence, the purpose of constructing a computer model of an object is to represent it in the computer in digital form in order to facilitate design, analysis, and downstream processes—in particular computerized operations. Using computer-aided design and rendering applications, you represent the object's geometry, texture, and color. To make better use of the computer and computer-aided design applications, you need to know the various ways models are represented in the computer, the types of modeling tools available, and the techniques for using these tools.

Wireframe, Surface, and Solid Models

Basically, you can represent a 3D object in the computer in three ways: as a wireframe model, as a surface model, or as a solid model.

Comparing Wireframe, Surface, and Solid Models

Contrary to a surface model that has a set of surfaces to depict the boundary face of an object and a solid model that is a comprehensive set of data, a wireframe model is simply a set of unassociated curves that represent an object by defining the object's edges. The object in Figure 1–2 is shown as a wireframe model.

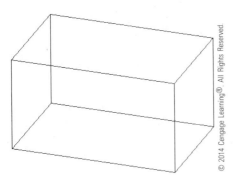

FIGURE 1–2 *Wireframe model representing a rectangular block*

In particular, this model has eight separate, unrelated curves. Between the curves, there is no information. Because a wireframe model is simply a set of unrelated curves, no surface and volume information is stored. As a result, a wireframe model has limited application in computerized downstream operations, such as analysis, CNC machining, and rapid prototyping. Because of the sophisticated data structure, a solid model, although superior in terms of data integrity, is generally less flexible in free-form surface manipulation. Therefore, if your design has more emphasis on exterior form and shape, in particular, free-form shape, and you want to realize your design in the computer quickly, you should consider using a surface modeling tool, such as Rhinoceros. Even if you want to have the advantage of integrated comprehensive data of a solid model, you can still start with using a surface modeling tool initially, to gain a quick insight into the final form and shape or to appreciate alternative form and shape. Once you have finalized the exterior shape and silhouette, you can proceed to constructing a solid model through exportation to other solid modeling tools. To summarize, surface model is best suited for representing free-form objects, and solid model is more suitable for making engineering parts. In the following sections, we will focus our attention on surface modeling.

What Is Rhinoceros?

In essence, Rhino is a very flexible and user-friendly 3D surface modeling tool, enabling you to construct NURBS surfaces as well as polygonal meshes for making 3D models of free-form objects. To facilitate NURBS surface construction, it provides a comprehensive set of tools for making and manipulating NURBS curves and point objects.

Constructing the model of an object usually concerns the making of two or more surfaces with different surface patterns. For easy handling of surface objects, you may join two or more contiguous surfaces sharing common edges to form a polysurface. Among the many ways to represent a solid in the computer, Rhino's solid is a surface or a polysurface enclosing a volume without any gaps, openings, or intersections among the individual surfaces. To obtain special form and shape effects from surfaces that are already constructed, you can use Rhino's transformation tools.

To help improvise your design, Rhino provides a set of analysis tools. To facilitate the management of models of products or systems with a number of components, Rhino enables you to construct block definition and block instances. To cope with other upstream and downstream computerized operation, Rhino enables you to import and export various file formats. To facilitate human interpretation of design, Rhino enables you to output 2D engineering drawing. In terms of rendering and animation, you can use Rhino's basic rendering tool as well as other plug-in tools, such as Flamingo and Bongo.

Wireframe

In the history of computer modeling, the 3D wireframe model is the earliest type of 3D model. It is the most primitive type of 3D object. In essence, a wireframe model is a set of unassociated curves assembled in 3D space. The curves serve only to give the pattern of a 3D object. There is no relationship between the curves. Therefore, the model does not have any surface information or volume information. It has only data that describe the edges of the 3D object. Because of the limited information provided by the model, the use of wireframe models is very confined. Figure 1–3 shows a wireframe model of a complex object and the same model in a cutaway view.

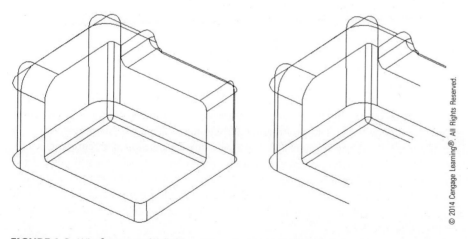

FIGURE 1–3 *Wireframe model (left) and its cutaway view (right)*

Rhino's Wireframe Tools

Although wireframe models by themselves have limited utility in design and manufacture, curves and points are required in many surface and solid construction operations. Therefore, you need to learn how to construct curves and points for the purpose of making surfaces and solids. Using Rhino, you can easily construct points and various types of 3D curves. Using the points and curves as framework, you construct various kinds of free-form surfaces. You will learn about points and curves in Chapters 4 through 6. Figure 1–4 shows a free-form object and the curves for making it.

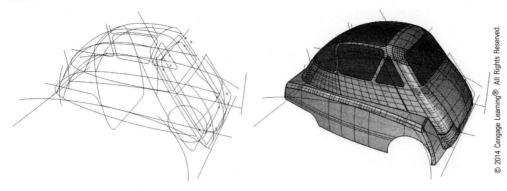

FIGURE 1–4 *NURBS curves (right) for making free-form NURBS surface (left)*

Surface Model

In the computer, a surface is a mathematical construct represented as a thin sheet without thickness. A surface model is a set of contiguous surfaces assembled in 3D space to represent a 3D object. When compared to a 3D wireframe model, a surface model has, in addition to edge data, information on the contour and silhouette of the 3D object. Surface models are typically used in computerized manufacturing systems and in the generation of photorealistic rendering or animation. Figure 1–5 shows a surface model of a car body, a cross-section of the model, and a rendered image of the sectioned model.

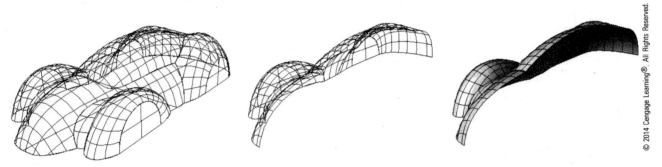

FIGURE 1–5 *From left to right: Surface model, cross-section of the model, and a rendered image of the sectioned model*

Surface Representation Methods

In the computer, there are two basic ways of representing a surface: using a set of planar polygon meshes to approximate a surface and using complex mathematics to obtain an exact representation of the surface.

Polygon Meshes

A polygon mesh is a method of approximately representing a surface. It reduces a smooth, free-form surface to a set of planar polygonal faces and curved edges, and reduces silhouettes to sets of straight line segments. Accuracy of representation is inversely proportional to the size of the polygon faces and line segments. A mesh with smaller polygon size represents a surface better, but the memory required to store the mesh is larger. Figure 1–6 shows the polygon mesh of a scale model car body. A severe drawback of the polygon mesh method is that, despite using a very small polygon, it can never represent a surface accurately because the surface is always faceted. As a result, this method can be used only for visualization of the real object, and cannot be used in most downstream computerized manufacturing systems. Application and construction of polygon meshes will be discussed in more detail in Chapter 9.

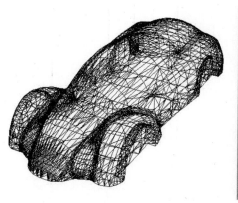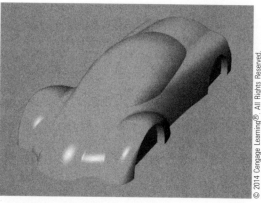

FIGURE 1–6 *Polygon mesh (left) and rendered image of the polygon mesh (right)*

NURBS (Nonuniform Rational B-Spline) Surfaces

To accurately represent free-form smooth surfaces in 3D design applications and computerized manufacturing systems, a higher-order spline surface is used. This is a surface that uses NURBS mathematics to define a set of control vertices and a set of parameters (knots). The distribution of control vertices, together with the values of the parameters, controls the shape of the surface. The use of NURBS mathematics allows the implementation of multipatch surfaces with cubic surface mathematics, and maintains full continuity control even with trimmed surfaces. Figure 1–7 shows a NURBS surface and its control vertices. Because a NURBS surface accurately represents smooth, free-form surfaces in the computer, it is the most appropriate tool for aesthetic and engineering design.

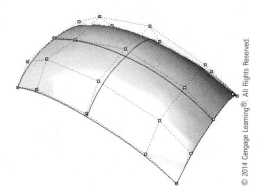

FIGURE 1–7 *NURBS surface and its control vertices*

Rhino's NURBS Surfaces and Polygon Meshes Tools

Naturally, the prime objective of using Rhino as a tool in digital modeling is to construct free-form objects. Using Rhino, you can construct both NURBS surfaces and polygon meshes. You will learn about NURBS surfaces in Chapters 3 and 7, and about polygonal meshes in Chapter 9. Figure 1–8 shows a NURBS surface and Figure 1–9 shows a polygonal mesh of a car.

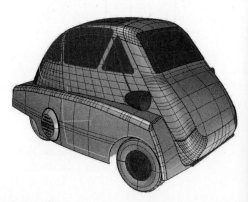

FIGURE 1–8 *NURBS surface model of a car (left) and its rendered image (right)*

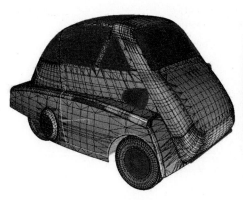 

FIGURE 1–9 *Meshed model of the same car (left) and its rendered image (right)*

Rendering and Rendering Mesh

You may notice in Figures 1–2 and 1–3 that the rendered images for both (NURBS surfaces and polygon meshes) are more or less the same. This is the case because Rhino first generates a set of meshes from the NURBS surface prior rendering. The mesh that is produced is called render meshes. To fine-tune the mesh, you may change the settings in the Mesh tab of the Rhinos Options dialog box accessible from the Tools pull-down menu.

Advantages of NURBS Surfaces

A NURBS surface model contains all surface data of a 3D object, and you can retrieve the coordinates of any point on the surface of the object. Hence, you can use a NURBS surface model in most downstream computerized manufacturing operations, as well as in visualization of the object. For example, you generate cross-sections for CNC machining. (See Figure 1–10.) In visualization, you generate hidden-line projection views, shaded images, and photorealistic rendered images.

FIGURE 1–10 *CNC milling of a free-form surface*

Surface Model's Limitation

A surface has no thickness. Hence, a surface model is simply a thin shell with no volume, and therefore represents no information other than surface data. This produces difficulties such as detecting collision between two surface models in situations in which one surface model lies completely inside another. Figure 1–11 shows the components of a scale model car constructed in surfaces and polysurfaces. Collision and interference of components are difficult to detect.

Apart from collision detection, volume and mass evaluation is limited only to objects of thin shell, by multiplying the surface area with the thickness of the shell.

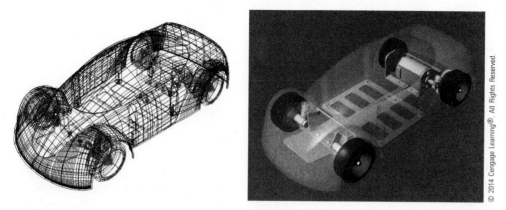

FIGURE 1–11 *A set of components represented by surfaces in the computer (left) and rendered image of the surfaces (right)*

Solid Model

In regard to information, a 3D solid model is superior to the other two models because a solid model in a computer is a complete representation of the object. It integrates mathematical data that includes surface and edge data as well as data on the volume of the object the model describes. In addition to visualization and manufacturing, solid modeling data is used in design calculation. Figure 1–12 shows a solid model of a toy car and a cross-section of the model.

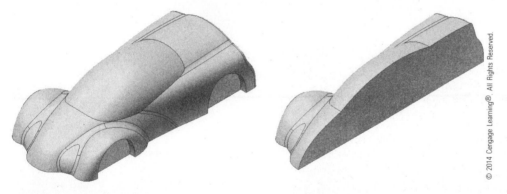

FIGURE 1–12 *Rendered solid model (left) and half of the same model removed (right)*

Rhino's Polysurface and Solid Tools

One major advantage of using surface modeling tool rather than solid modeling tool in free-form modeling is its flexibility, in terms of constructing individual surfaces to represent unique facets of an object, because individual surfaces are independent of each other in the database of the file, and there is no relationship between them. To help handle a set of contiguous surfaces collectively, you can join them together at their connecting edges to form a polysurface. Figure 1–13 shows sets of contiguous surfaces joined together to form various parts of the model of a watch.

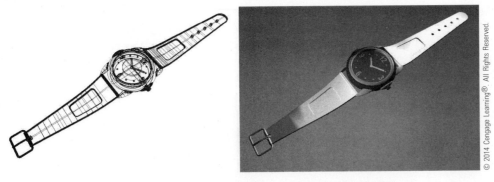

FIGURE 1–13 *Polysurfaces depicting a watch (left) and its rendered image (right)*

In Rhino, if a polysurface or any single surface (for example, a sphere or an ellipsoid having its edges collapsed to a point) encloses a watertight volume (without any gaps, openings, or intersections among the individual surfaces), a solid is formed. You can construct Rhino solids in two basic ways: directly, using the Rhino solid modeling tools, or by converting a set of contiguous NURBS surfaces into a solid volume by joining them. You will learn more about Rhino's solid modeling in Chapter 8. Figure 1–14 shows a Rhino solid, which consists of a set of NURBS surfaces joined together without any gaps, openings, or intersection among the surfaces.

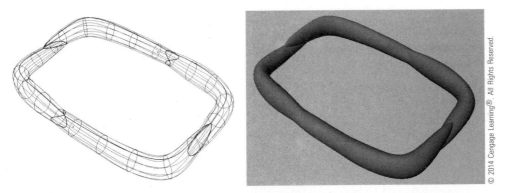

FIGURE 1–14 *A closed polysurface (a solid in Rhinoceros) (left) and its rendered image (right)*

Rhino's Transformation and Analysis Tools

To change the shape of your design by manipulating surfaces that are already constructed, you use the transformation tools. Figure 1–15 shows how a complex shape can be obtained by transforming objects of simple shapes. To analyze objects in order to improvise the design, you use the analysis tools. You will learn about transformation in Chapter 10 and analysis in Chapter 11. Figure 1–16 shows the use of zebra lines to help visualize the smoothness of a polysurface.

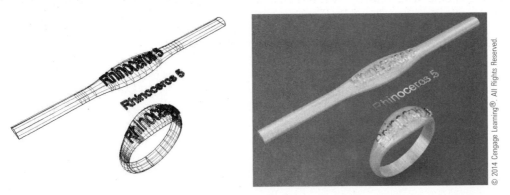

FIGURE 1–15 *Polysurfaces in the form of text objects first transformed to the body of a ring and then both transformed to form a ring (wireframe display at the left and rendered image at the right)*

FIGURE 1–16 *Zebra stripes displaying on a model to help visualize its smoothness*

Assembly Tools

To facilitate evaluation of how various component parts of a product or system can or should be put together, you assemble individual components in an assembly model. Using Rhinos, an assembly can be simulated by manipulating and inserting individual files representing various components into a single file, using groups and blocks that you will learn about in Chapter 12. Figure 1–17 shows the assembly of a set of components.

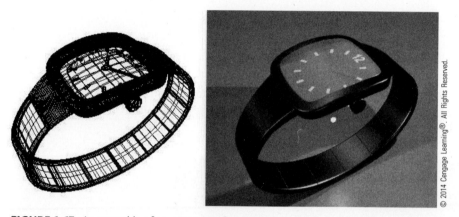

FIGURE 1–17 *An assembly of components of a watch, including case, bracelet, dial, and hands*

Data Exchange and Engineering Drawing Tools

To reuse digital data constructed by using other applications and to facilitate downstream computerized processes, you exchange data in various formats. To represent a 3D object in a 2D drawing sheet, you use an orthographic engineering drawing. If you already have a 3D digital model, you use the computer to generate orthographic views of the model. Using Rhino, you can open files saved in other data formats, insert them into a Rhino file, export Rhino file to other data formats, and generate a 2D drawing from the digital model of the 3D object and add appropriate dimensions and annotations to the drawing. You will learn about data exchange and generation of 2D drawings from digital models in Chapter 13. Figure 1–18 shows a 2D engineering drawing generated from the model shown in Figure 1–16.

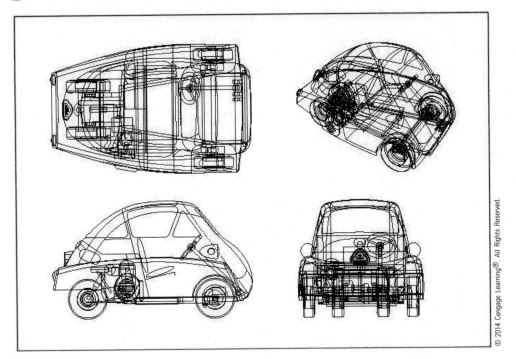

FIGURE 1–18 *A 2D orthographic engineering drawing generated from a 3D surface model*

Rendering and Animation Tools

Rendering is a way to produce photorealistic images from 3D surfaces, polysurfaces, and polygon meshes. To add reality to the images, you have to include material and lighting information. Figure 1–19 shows the two rendered images of the model shown in Figure 1–16.

FIGURE 1–19 *Rhino rendering (left) and Flamingo nXt rendering (right)*

To produce a rendered image in Rhino, you can use the basic renderer or the Flamingo nXt renderer. To further realize the 3D model in terms of photorealistic animation, you can use the Bongo animator. Note that you need to install Flamingo before you can use the Flamingo nXt renderer and install Bongo before you can produce photorealistic animations. You will learn about rendering and animation in Chapter 14.

RHINO'S USER INTERFACE

If you are a novice, then have a bit of hands-on examination of Rhinoceros 5's user interface. Perform the following step.

1. Start Rhino by selecting the Rhinoceros 5 icon from your desktop.
   The Rhino Startup dialog box (Figure 1–20), together with the user interface, will display (Figure 1–21).

FIGURE 1–20 *Startup dialog box*
Source: Robert McNeel and Associates Rhinoceros® 5

The Startup dialog box has three panes. The first pane serves to provide you with a choice of template file upon startup. The second pane displays a list of recently opened files and the third pane enables you to browse through the computer to open an existing file. After a while, if no action is taken on this dialog box, it will disappear automatically.

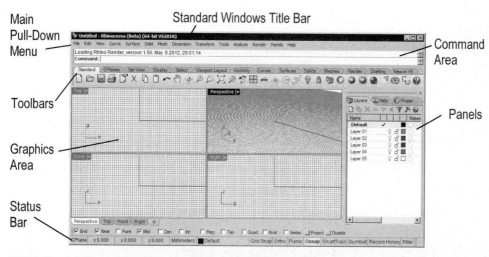

FIGURE 1–21 *Rhino application window*
Source: Robert McNeel and Associates Rhinoceros® 5

In the Application window, you will find five major areas: standard Windows title bar, main pull-down menu, command area, graphics area, and status bar. In addition, there are a number of toolbars and a Rhinoceros dialog box. These components are described in the sections that follow.

Standard Windows Title Bar

At the top of the Application window, there is the standard Windows title bar. This title bar functions no differently and contains nothing different from the basic Windows title bar.

Rhinoceros Panel

By default, the Rhinoceros Panel is situated at the right-hand side of the graphics area, displaying three panes: Properties, Layers, and Help. To discover what other panes you can display in this dialog box, right-click on one of the pane's label or select Panels from the pull-down menu to display a list of panes and click on the pane you want to display.

By double-clicking the title area of the Rhinoceros dialog box, it will be floated, as shown in Figure 1–22. If you close all the panes, this dialog box will disappear. To display this dialog box, you can activate one of the panes, for example, typing Properties at the command area to open the Properties pane.

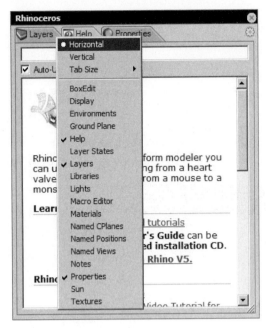

FIGURE 1–22 *Rhinoceros panel*
Source: Robert McNeel and Associates Rhinoceros® 5

Main Pull-down Menu

Below the standard Windows title bar is the main pull-down menu, which contains fourteen options: File, Edit, View, Curve, Surface, Solid, Mesh, Dimension, Transform, Tools, Analyze, Render, Panels, and Help. The functions of these options are outlined in Table 1–1.

TABLE 1–1 *Pull-down Menu Options and Their Functions*

Pull-down Menu Options	Function
File	For working on files and templates
Edit	For editing points, curves, surfaces, and solids
View	For manipulating display settings and establishing construction planes
Curve	For constructing and manipulating points and curves
Surface	For constructing and manipulating NURBS surfaces
Solid	For constructing and manipulating solids and polysurfaces
Mesh	For constructing and manipulating polygon meshes
Dimension	For constructing 2D drawings and adding annotations
Transform	For transforming objects you have constructed
Tools	Provides various types of useful tools
Analyze	For analyzing objects you have constructed
Render	For shading and rendering
Panels	For displaying various panes in the Rhinoceros panel
Help	Provides useful help information

One way to perform commands and operations is to select options from pull-down menu. To appreciate how to use the pull-down menu, continue with the following steps to construct a cone and a sphere.

2. Select Solid > Cone. (Select Cone from the Solid pull-down menu.)
3. Click on locations A and B as shown in Figure 1–23 (exact location is unimportant here because it is only a simple practice on how to use the pull-down menu) in the Top viewport (to indicate the base center and to specify the radius of the cone) and then location C in the Front viewport (to indicate the vertex of the cone).

   A cone is constructed.

Note that the orientation of objects that you construct by picking points in different viewports will not be the same because there is a construction plane in each of the viewports and picking a point in a viewport with a pointing device means that you are picking a point on that particular construction plane. Rhinoceros's multiple construction planes concept will be discussed in Chapter 2.

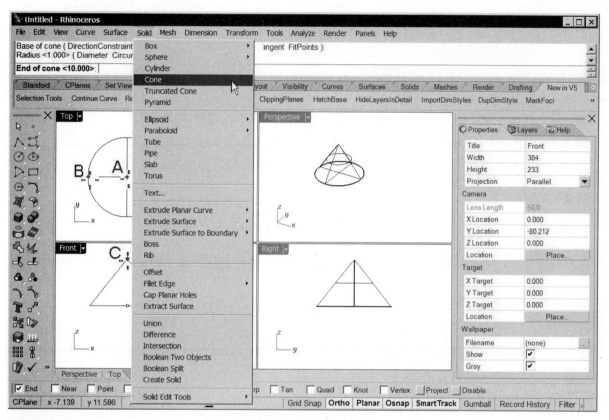

FIGURE 1–23 *Cone command activated from the Solids pull-down menu*
Source: Robert McNeel and Associates Rhinoceros® 5

4. Select Solids > Sphere > Center, Radius.

   The Center, Radius command from the Sphere cascading menu of the Solids pull-down menu is used to construct a sphere by specifying its center and radius.

5. Click on two locations A and B indicated in Figure 1–24 (again, exact location is unimportant here) to specify the center and a point on the sphere.

   A sphere is constructed.

> A cone is a polysurface consisting of a slant surface and a flat surface joined together. A sphere is a single surface with edges collapsed to a single point. Because cone and sphere are regarded as solids in Rhino, these commands are grouped under the Solid menu.

NOTE

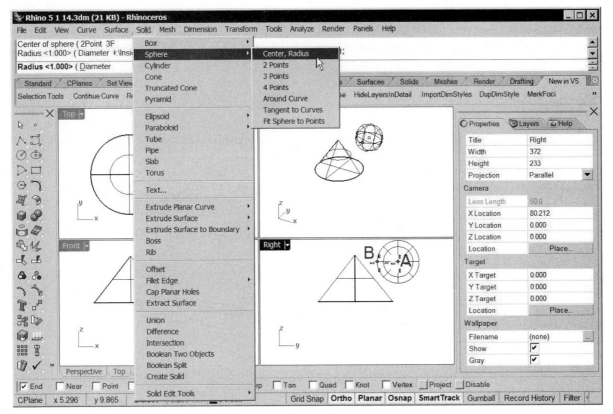

FIGURE 1–24 *Sphere cascading menu for constructing a center, radius sphere*
Source: Robert McNeel and Associates Rhinoceros® 5

 NOTE If you are using the Evaluation version, the number of times you can perform a save is limited. To be able to see the result of your work, do not save your files unless you are working on the case studies. Opening a new file, starting a new file, or exiting Rhino automatically closes your current file.

Command Area

Below the main pull-down menu is the command area (Figure 1–25), which provides a place for textual interaction. By default, the command area is docked below the pull-down menu. However, you may select and drag it to anywhere on the screen.

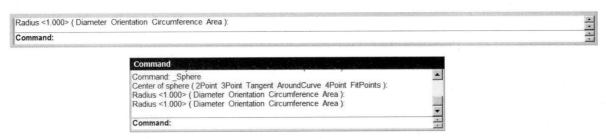

FIGURE 1–25 *Command area docked below the pull-down menu (above) and floating command area (below)*
Source: Robert McNeel and Associates Rhinoceros® 5

Case Sensitivity

Command names are NOT case sensitive, and therefore you can use any combination of small and capital letters to specify a command name. In the command area, you run a command by typing the command name or alias of the command and then pressing the ENTER key or the space bar. After a command is run, further prompts or instructions will appear in this area or in any associated pop-up dialog boxes.

Command List

If you are not too sure about the spelling of a command name, you may simply type the first letter and then move the cursor to the command area to display the list of command beginning with the letter that you just typed.

6. Type A at the command line and move the cursor to the command area.

   A list of command beginning with the letter A is displayed. You may click on a command to select it. (See Figure 1–26.)

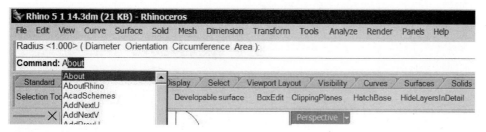

FIGURE 1–26 *List of commands displayed*
Source: Robert McNeel and Associates Rhinoceros® 5

Auto-Complete Feature

An auto-complete function is incorporated at the command line. That is, if you type a command in the command line, you do not have to type the full name. For example, after you type AR, the system will automatically complete the typing to display the word "Arc." Continue with the following steps to learn more about the command area.

7. Type R at the command area.

   Together with the letter A that you typed previously, you have two letters AR typed. Full name of the command 'Arc' is displayed. (This is the auto-complete feature.)

8. Press the ENTER key.

   The command is executed. (See Figure 1–27.)

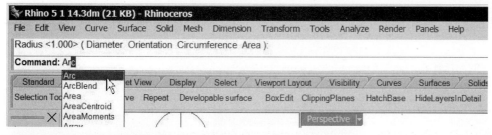

FIGURE 1–27 *Auto-complete feature*
Source: Robert McNeel and Associates Rhinoceros® 5

Clickable Options in the Command Area

Options in the command area are clickable, which means that you can use the mouse to click on an option as well as typing the keyword of the options. Continue with the following steps.

9. Use the mouse to click on the Deformable option (A in Figure 1–28) in the command area.

   The Deformable option is selected. (Note: A deformable arc is a degree 3 curve in the shape of an arc. You will learn more about the degree of polynomial in Chapter 4.)

10. Click on locations B, C, and D (Figure 1–28).

    An arc is constructed.

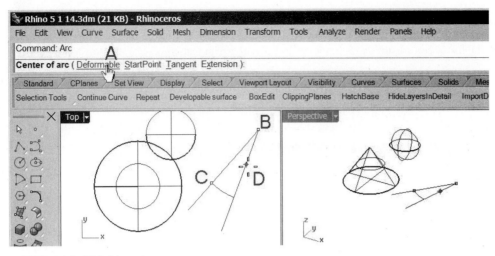

FIGURE 1–28 *Clickable options*
Source: Robert McNeel and Associates Rhinoceros® 5

Displaying Recently Used Commands

By right-clicking on the command area, a pop-up menu showing all recently used commands is displayed, providing a quick access to commands that are used. Continue with the following steps.

11. Right-click (A in Figure 1–29) on the command area.

    A pop-up menu appears, displaying all the recently used commands. Depending on what commands you have used previously, your screen may look a bit different. (See Figure 1–29.)

12. Click on Cone (B in Figure 1–29) from the pop-up menu.

    The recently used command, cone, is executed again.

13. Press the ESC key to terminate the command.

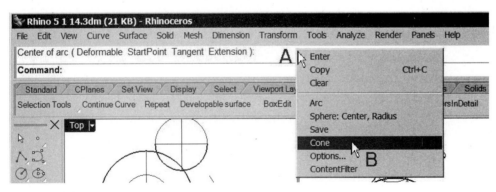

FIGURE 1–29 *Pop-up menu*
Source: Robert McNeel and Associates Rhinoceros® 5

Displaying Most Used Commands

Another way to display a set of previously used commands is to use the PopupPopular command, as follows:

14. At the command area, type PopupPopular.

    A list of fifteen most used commands sorted by the number of times used is displayed.

Displaying Command History

Apart from displaying the command name, you may display the command line history in a dialog box. History can be saved in a text file for reference or for construction of a script file (see scripting in "Command Scripting" section). Continue with the following steps.

15. Select Tools > Commands > Command History.

    The Command History dialog box is displayed. (See Figure 1–30.)

In the Command History dialog box, there are three buttons. Clicking the Copy All button will copy the command line history to the Windows clipboard for subsequent pasting to another file. Clicking on the Save As button will save the command line to a text file.

16. Click on the Close button.

    The Command History dialog box is closed.

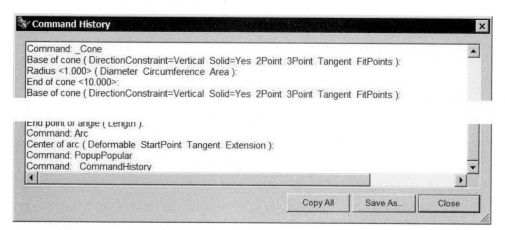

FIGURE 1–30 *Command History dialog box*
Source: Robert McNeel and Associates Rhinoceros® 5

Complete List of Commands

Another way of knowing Rhino command names is to list them or save them to a text file, as follows:

17. Type Command List at the command area.

    The Print Window dialog box is displayed, as shown in Figure 1–31.

18. Click on the Close button. Do not save the list if you are using the Evaluation version.

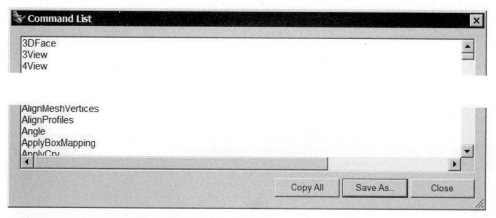

FIGURE 1–31 *Print Window dialog box*
Source: Robert McNeel and Associates Rhinoceros® 5

Command Scripting

All Rhino commands are scriptable. It means that if you write down the command sequence in a text file with an extension of .txt, you can then run the script file to have all the command sequence executed automatically. Simply speaking, a script file consists of a number of lines that you will type at the command area to execute an action. In the script file, you may use a [Space] or [Return] to execute a command. If a command calls for a dialog box, you can prefix the command name with the character [-]. At the beginning of the script file, you may use the exclamation mark [!] and a space to terminate any previous commands. Continue with the following steps.

19. Select Tools > Commands > Read from File.

20. In the Open Text File dialog box, select the file "Rhino Script" from the Chapter 1 folder that you downloaded from the Student Companion Website and click on the Open button.

   A circle is constructed by the script file. (See Figure 1–32.)

To view the script file, you may use any text editor.

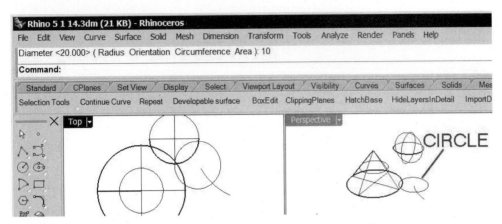

FIGURE 1–32 *Opening the script file and a circle constructed by the script file*
Source: Robert McNeel and Associates Rhinoceros® 5

Command Aliases

To speed up calling a command, you may use command aliases that are already defined by default or define your own command aliases. To view the command aliases available, you may open the Options dialog box. Continue with the following steps.

21. Select Tools > Options to display the Options dialog box.
22. In the Options dialog box shown in Figure 1–33, select Rhino Options > Aliases from the left pane.

NOTE In Aliases tab, there are a list of aliases and four buttons, enabling you to import and export aliases to a text file, define a new alias, and delete an existing alias.

23. Click on the Cancel button to close the Options dialog box without changing anything.

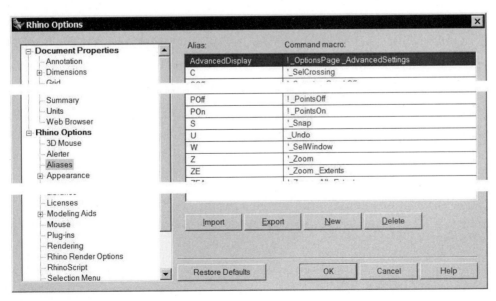

FIGURE 1–33 *Aliases tab of the Options dialog box*
Source: Robert McNeel and Associates Rhinoceros® 5

Graphics Area

Down below the command area is the graphics area, where you construct your model. This area can be divided into a number of viewports, which can be docked or floating. The default file that you are working on has a default four-viewport configuration (i.e., Top, Front, Right, and Perspective viewports). You may regard each viewport as a view window of an imaginary camera. Through each imaginary camera, you can see the three-dimensional working space from different directions. You can double-click the viewport's label (at the upper-left corner of the viewport) to maximize it. To return to the previous viewport configuration, you double-click it again. Continue with the following steps.

24. Double-click the label of the Perspective viewport.

    The viewport is maximized. (See Figure 1–34.)

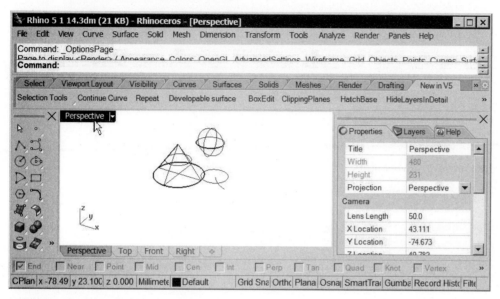

FIGURE 1–34 *Perspective viewport maximized*
Source: Robert McNeel and Associates Rhinoceros® 5

25. Double-click the label again.

    The previous viewport configuration is restored.

> To reiterate, if you are using the Evaluation version, do not save your file. **NOTE**

Viewport Configuration and Using Template

Although the number of viewports and their orientation are determined by the viewport setting of the template file you use to start a new file, you may configure it, and the number of viewports is unlimited. To use a template, you start a new file by selecting New from the File pull-down menu. This accesses the Open Template File dialog box (shown in Figure 1–35), in which you select a template. Perform the following steps:

1. Select File > New.
2. Do not save your file. In the dialog box that pops up, click on the No button.
3. In the Open Template File dialog box (Figure 1–35), select the file "Small Objects - Millimeters" and click on the Open button.

    A new file is started.

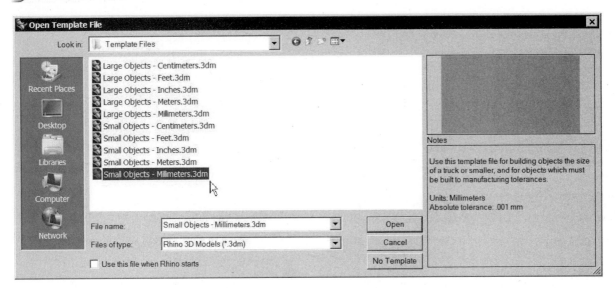

FIGURE 1–35 *Open Template File dialog box*
Source: Robert McNeel and Associates Rhinoceros® 5

4. Select View > Viewport Layout > 3 Viewports.

   The display is set to a three-viewport display. (See Figure 1–36.)

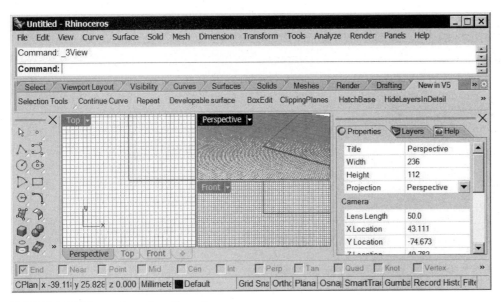

FIGURE 1–36 *Three-viewport configuration*
Source: Robert McNeel and Associates Rhinoceros® 5

Continue with the following steps to manipulate viewports.

5. Click on the Perspective viewport to set it as the current viewport.
6. Select View > Viewport Layout > Split Vertical.

   The selected viewport is split vertically into two viewports. (See Figure 1–37.)

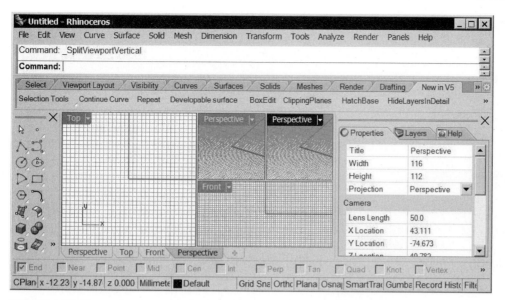

FIGURE 1–37 *Perspective viewport split vertically*
Source: Robert McNeel and Associates Rhinoceros® 5

7. Select View > Viewport Layout > New Viewport.
   A new viewport is constructed. (See Figure 1–38.)

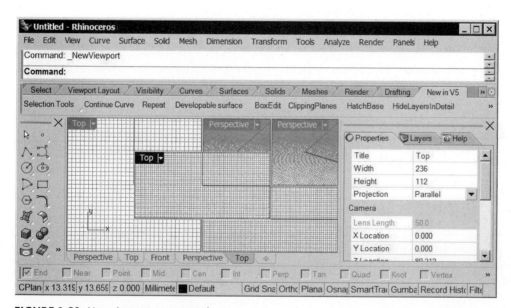

FIGURE 1–38 *New viewport constructed*
Source: Robert McNeel and Associates Rhinoceros® 5

8. Double-click the Top viewport's label to maximize it.

Viewport's Gradient Background

A viewport's background can be set to gradient color, as follows:

9. Type-GradientView at the command area.
10. If the View option is All, select it from the command area to change it to Active.
11. Select the State option to change it to On; the active viewport has a gradient color. (See Figure 1–39.)
12. Type GradientView at the command area.

The viewport's background is reset.

This command has two versions: Adding the hyphen sign [-] displays command line options.

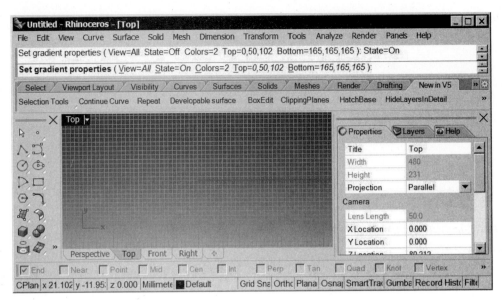

FIGURE 1–39 *Viewport's background changed*
Source: Robert McNeel and Associates Rhinoceros® 5

Viewport Properties

You can manipulate a viewport's properties by continuing with the following steps:

13. Select View > Viewport Properties or right-click the viewport's label and select Viewport Properties.

    The Viewport Properties dialog box is displayed. (See Figure 1–40.)

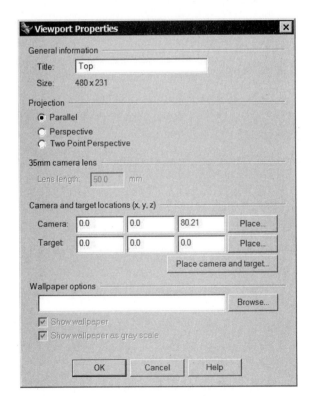

FIGURE 1–40 *Viewport Properties dialog box*
Source: Robert McNeel and Associates Rhinoceros® 5

As shown in Figure 1–40, the Viewport Properties dialog box has five areas, as follows:

- The General Information area tells the size of the viewport in terms of pixel size and enables you to set the viewport's name.
- The Projection area sets the viewport's display method: Parallel, Perspective, or Two Point Perspective projection.
- If the Projection is set to Perspective, you can set the camera's lens length in the third area.
- By setting the camera and target locations in the fourth area, the direction of viewing is determined.
- To enhance the display, the fifth area allows you to include and display a wallpaper in the viewport.

14. Try out various buttons to experience how to set the viewport's properties.
15. When you are done, click on the OK button to close the dialog box.

If you simply want to learn about the properties of a viewport, continue with the following steps:

16. Do not select any object.
17. You may press the ESC key once or twice to stop any command running and clearing any selection of objects.

Another way to alter a viewport's properties is to use the Properties pane of the Rhinoceros dialog box, as follows:

18. Select Edit > Object Properties.

    The Properties pane of the Rhinoceros dialog box is displayed, showing the properties of the active viewport. (See Figure 1–41.)

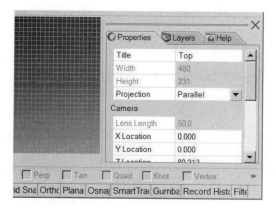

FIGURE 1–41 *Viewport properties*
Source: Robert McNeel and Associates Rhinoceros® 5

Floating and Docked Viewports

The viewports that you constructed are called docked viewports because they are restricted within the graphics window. Apart from these viewports, you can construct floating viewports that can float outside the main Rhino window. The advantage of having floating viewports is that, if you are using two display units together with extended display mode, you can display Rhino's application in one display unit and a floating viewport in another display unit.

Floating viewports and docked viewports are interchangeable; you can dock a floating viewport or make a docked viewport floating. Continue with the following steps.

19. Click on the New Floating Viewport button on the Viewport Layout toolbar.
20. Press the ENTER key to accept the default Perspective viewport.

    A new floating viewport showing the perspective view is constructed. (See Figure 1–42.)

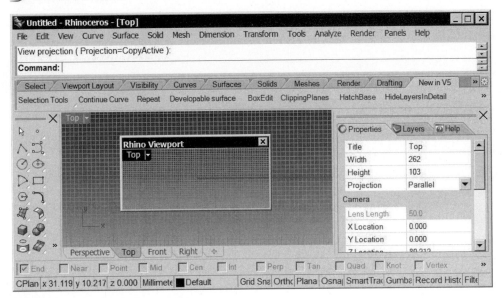

FIGURE 1–42 *New floating viewport constructed*
Source: Robert McNeel and Associates Rhinoceros® 5

21. Click on the Toggle Floating Viewport State button on the Viewport Layout toolbar.

    The new floating viewport, being the active viewport, is docked.

22. Double-click on the Top viewport's label.

23. Click on the Toggle Floating Viewport State button on the Viewport Layout toolbar.

    The Top viewport becomes a floating viewport.

Viewport Tab

You may display a set of viewport tabs alongside the viewports, at the right, left, top, or bottom. By right-clicking on one of the tabs, a menu is displayed, as shown in Figure 1–43. Viewport tab can be manipulated by selecting the Viewport tab Controls button on the Viewport Layout toolbar or typing Viewport Tabs at the command area. To simply toggle turning on or off the viewport tab, right-click on the Toggle viewport tabs button on the Viewport Layout toolbar.

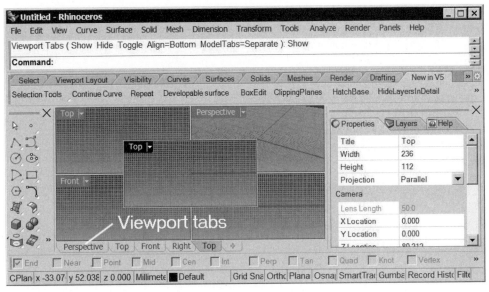

FIGURE 1–43 *Viewport tab and its right-click menu displayed*
Source: Robert McNeel and Associates Rhinoceros® 5

Viewport Synchronization

A way to help visualize objects' size in various viewports, you synchronize them such that their display scales are the same for all the viewports, as follows:

1. Select File > Open and select the file Synchron.3dm from the Chapter 1 folder on the Student Companion Website.
2. Right-click on the Zoom Extents All Viewports button on the Standard toolbar.

    All the viewports are zoomed to their extent. However, their zoom scales are different. (See Figure 1–44.)

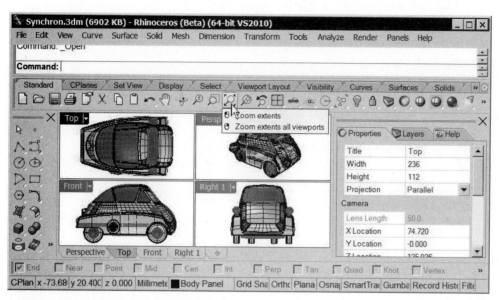

FIGURE 1–44 *Viewports zoomed to their extent*
Source: Robert McNeel and Associates Rhinoceros® 5

3. Click on the Perspective viewport to make it the current viewport.
4. Select View > Set Camera > Synchronize.

    The viewports are synchronized. All the viewports' zoom scales are synchronized to the zoom scale of the Perspective viewport, as shown in Figure 1–45.

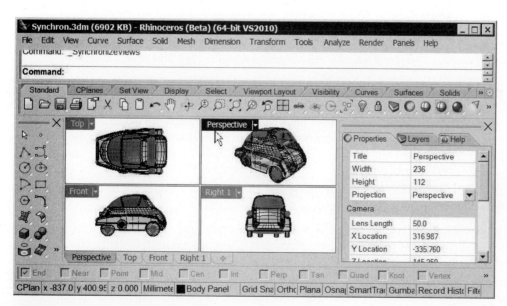

FIGURE 1–45 *Viewports synchronized to the Perspective viewport*
Source: Robert McNeel and Associates Rhinoceros® 5

Status Bar

At the bottom of the Application window is the status bar (Figure 1–46). The status bar contains eight panes, showing the location of the cursor marker, the layer manipulator, and five buttons controlling the state of the drawing aids (Snap, Ortho, Planar, and Osnap) and history recording. By right-clicking these buttons, you can bring out a context menu. The functions of these panes are outlined in Table 1–2. (Details of them will be explained later in this chapter.)

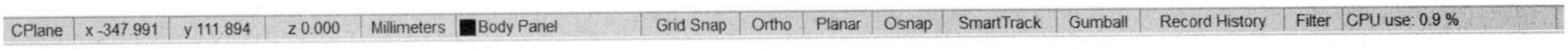

| CPlane | x -347.991 | y 111.894 | z 0.000 | Millimeters | Body Panel | Grid Snap | Ortho | Planar | Osnap | SmartTrack | Gumball | Record History | Filter | CPU use: 0.9 % |

FIGURE 1–46 *Status bar*
Source: Robert McNeel and Associates Rhinoceros® 5

TABLE 1–2 *Status Bar Panes and Their Functions*

Pane	Function
World/CPlane toggle switch	Toggles the coordinate display to show either World coordinates or construction plane coordinates
Coordinate Display	Shows the coordinates of the mouse pointer
Distance	Optionally displays the distance from the mouse pointer to the last picked point
Layer Manipulator	Shows the current layer and, when clicked on, displays the layer manager
Grid Snap	Toggles on/off constraining the cursor movement to the specified snap intervals
Ortho	Toggles on/off constraining the cursor movement to be orthogonal or some other preset direction
Planar	Toggles on/off constraining the cursor movement to be parallel to the current construction plane from the last point
Osnap	Toggles on/off the display of the Osnap dialog box (Osnap stands for "object snap")
SmartTrack	Toggles on/off smart tracking of cursor point
Gumball	Toggles on/off gumball transformation control
Record History	Toggles on/off History recording
Filter	Toggles on/off Filter dialog box
CPU and Save	Shows CPU usage, absolute tolerance, time from last save, memory usage, and available physical memory

Toolbars

Buttons on toolbars represent commands in a graphical way. To run a command, you select a button on a toolbar.

Docking and Floating Toolbars

Rhino toolbars can be docked at the top or left side of the application window, and can also be made floating anywhere on the screen. You can drag a floating toolbar to the top or left side of the window to dock it; you can drag a docked toolbar to change it to floating. Alternatively, you may double-click on a toolbar to change it from docking to floating or vise versa.

Displaying Toolbars

There are many commands and toolbars. By default, the Side Bar One is docked to the left side and a series of toolbars are docked at the top with their labels displayed as tabs along the top.

- To find out which toolbars are available, and to display them on the screen, you type Toolbar at the command area or select Tools > Toolbar Layout. In the Toolbars dialog box, shown in Figure 1–47 (left), check the box next to a toolbar to display that toolbar.
- Another way of displaying toolbars not shown on the screen is to right-click on the blank space of a docked toolbar and select Show Toolbar to display the complete list of toolbars, as shown in Figure 1–47 (middle).

Manipulating Toolbars

By selecting Toolbars > Size and Styles from the left pane of the Rhino Options dialog box, as shown in Figure 1–47 (right), you can manipulate toolbar's various aspects.

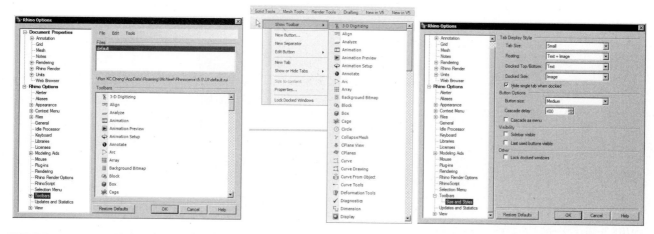

FIGURE 1–47 *From left to right: Rhino options' Toolbar tab, right-click menu, and Rhino options' Toolbar size and styles tab*
Source: Robert McNeel and Associates Rhinoceros® 5

1. Start a new file, using the Small Objects, Millimeters template file.
   Do not save the previously opened file.
2. Select Tools > Toolbar Layout.
3. In the Toolbars tab of the Rhino Options dialog box, click on the Line box.
   The Line toolbar is displayed.
4. Click on the X mark at the upper-right corner of the Toolbars dialog box to close it.
5. Click on the X mark at the upper-right corner of the Line toolbar to close it also.

Cascading Toolbars

Some buttons on the toolbar have a small triangular mark at their lower-right corner. Clicking on this mark will bring out a cascading toolbar. Selecting and dragging the header of the cascading toolbar will cause it to stay on the screen.

6. With reference to Figure 1–48, click on the small triangular mark at the lower-right corner of the 4 Viewports/ 4 Default Viewports button of the Standard Toolbar.
   The Layer toolbar is brought out.
7. Select the Viewport Layout toolbar's header and drag it away from the button.
   The Layer toolbar is displayed, as shown in Figure 1–48.
8. Click on the X mark at the upper-right corner of the Layer toolbar.
   The toolbar is closed.

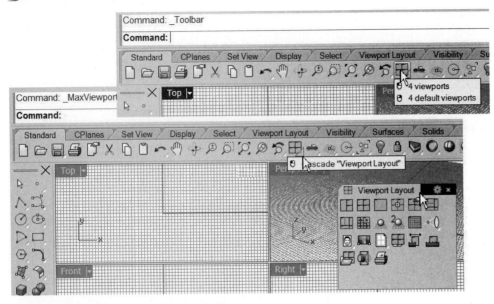

FIGURE 1–48 *Bringing out a cascading toolbar*
Source: Robert McNeel and Associates Rhinoceros® 5

Command Interaction

To summarize, there are three ways to run a Rhino command:

- Make selections from the main pull-down menu
- Type a command at the command area
- Select a button on the toolbar

Mouse Left-Click and Right-Click

Normally, your pointing device (mouse) has two buttons (left and right). You use the left button to select an item from the main pull-down menu, a button on a toolbar, or a check box in a dialog box, as well as to specify a location on the graphics area. Depending on where you place your cursor, right-clicking has different effects. Some toolbars have two commands sharing a single button. For example, the commands 4 Viewports and 4 Default Viewports, shown in Figure 1–48, share a single button on the Standard toolbar. Left-clicking activates the 4 Viewports command, and right-clicking starts the 4 Default Viewports command.

Several commands are accessed by placing the cursor over the graphics area and right-clicking in one of three ways.

1. Simply right-clicking repeats the last command.
2. Right-clicking and holding down the mouse button for a while (or holding down the SHIFT key) accesses the Pan command.
3. Right-clicking and holding down the CONTROL key accesses the Zoom command.

Middle Mouse Button

If your mouse has the third (middle) button, pressing it will access a menu or toolbar based on the settings in the Mouse tab of the Options dialog box (accessible by selecting Tools > Options).

If the third button can be slided forward and backward, doing so will zoom in and out on the display.

Viewport Display Mode

To better visualize the surfaces in the graphics area, you can right-click a viewport's label and select one of the following display modes: wireframe, shaded, rendered, rendered shadows, ghosted, X-ray, technical, artistic, and pen. Perform the following steps:

1. Select File > Open and select the file Display.3dm from the Chapter 1 folder on the Student Companion Website.

2. Click on the small triangle at the right side of the Perspective viewport's label or right-click on the label to display a list of commands.
3. Select one of the nine display modes one by one to discover various ways of displaying objects, as shown in Figure 1–49.

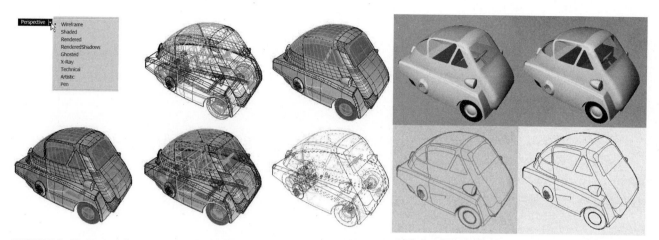

FIGURE 1–49 *From left to right and from top to bottom: selection of display mode from the viewport's menu, wireframe, shaded, rendered, rendered shadows, ghosted, X-ray, technical, artistic, and pen display modes*
Source: Robert McNeel and Associates Rhinoceros® 5

Display Control

The 3D space within each viewport is unlimited but the physical size of the viewport is limited. Therefore, you may need to zoom in, zoom out, pan, and rotate the viewport, as follows:

4. Click on the Zoom Dynamic button on the Standard toolbar.
5. Click on anywhere on the graphics area, hold down the left button, and drag up and down.
   The display is zoomed in and out.
6. Click on the Pan button on the Standard toolbar.
7. Click on anywhere on the graphics area, hold down the left button, and drag left and right.
   The display is paned to the left and to the right.
8. Click on the Rotate View button on the Standard toolbar.
9. Click on anywhere on the graphics area, hold down the left button, and drag around.
   The display is rotated.
10. Right-click on the Perspective viewport and select Set View > Two Point Perspective.
    The view is set to perspective, see Figure 1–50.

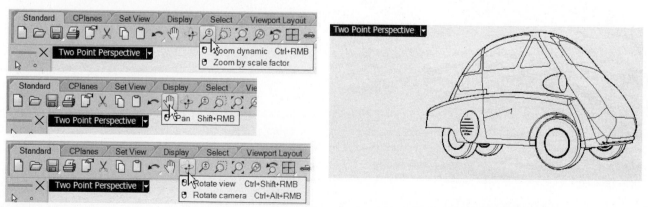

FIGURE 1–50 *Zooming in and out, panning around, and rotating commands and two point perspective display*
Source: Robert McNeel and Associates Rhinoceros® 5

Apart from using the pull-down menu or the appropriate toolbar button, you may perform the followings:

- Hold down the CONTROL and SHIFT keys, right-click, and drag to rotate the display
- Hold down the SHIFT key, right-click, and drag to pan the display
- Hold down the CONTROL key, right-click, and drag, or spin the mouse wheel to zoom in and out

Panning in the Perspective Viewport
Panning in the perspective is carried out by moving the location of the imaginary camera. Continue with the following steps.

11. Double-click on the Perspective viewport's label to maximize the viewport.
12. Select View > Set Camera > Rotate Camera.
13. Click on the viewport and drag the mouse.

    The perspective viewport is rotated. This is equivalent to panning in a parallel projection viewports.

Zooming in the Perspective Viewport
Zooming in the perspective viewport is done by adjusting the lens length of the camera, as follows:

14. Select View > Set Camera > Adjust Lens Length.
15. Click on the viewport and drag the mouse.

    As the lens length changes, the viewport appears to be zoomed in and out.

Dollying in the Perspective Viewport
Dollying is to change the lens length of the camera and at the same time move the camera nearer or farther away from the object, keeping the displayed size of the object in the viewport more or less constant. Continue with the following steps.

16. Select View > Set Camera > Adjust Lens Length and Dolly.
17. Click on the viewport and drag the mouse.

    Because both the lens change in length and the location of the camera is moving nearer or farther away from the objects, the perspective effect is changed but the object remains more or less the same size in the viewport. See Figure 1–51.

FIGURE 1-51 *Dollying in the perspective viewport*

Transparent Zoom and Pan
To zooming in and out and panning during drafting and design, the zoom and pan commands can work transparently inside other commands. The term transparent means that you can, in the middle of another command, run the zoom and pan command and, upon finishing the zoom and pan command, continue with the original command.

Isometric View

Apart from setting to perspective view, you can set the display to isometric views, as follows:

18. Click on one of the buttons on the Isometric toolbar or type Isometric at the command area and select NE, NW, SE, and SW one by one to find out various isometric display angles. (See Figure 1-52.)

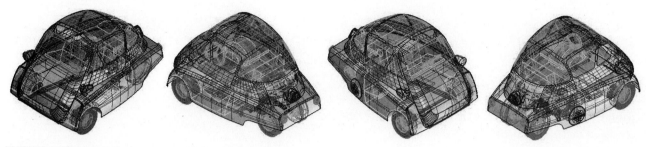

FIGURE 1-52 *Isometric views: NE, NW, SE, and SW*
© 2014 Cengage Learning®. All Rights Reserved.

Undo View Change and Set View Speed

The Undo command does not undo view display changes.

- To undo or redo viewport changes, select View > Undo View Change or select View > Redo View Change.
- To change the speed of view manipulation, such as pan, rotate, and zoom, use the SetViewSpeed command and set a value.

Full-Screen Display

Contrary to displaying all the major areas, you can set the display to full screen, showing only the graphics area and occupying the entire screen display. This can be useful in a presentation.

- To set full-screen display, type FullScreen at the command area. (A hyphenated version of the command adds command line options.)
- To return to normal display mode, press the ESC key.

Viewport Zoom Aspect Ratio

To help visualize the effect of scaling an object nonuniformly, you may set the viewport's horizontal and vertical zooming scale differently, as follows:

19. Maximize the Front viewport and set the display to Wireframe.
20. Select Tools > Options.
21. In the Rhino Options dialog box, select Rhino Options > Appearance
22. Suppose we would like to change the aspect ratio for the shaded display, click on the + sign prefixing the Shaded folder under the Advanced Settings folder and then click on Other Settings.
23. Set the horizontal scale to 1.5, as shown in Figure 1-53 and click on the OK button.
    The horizontal display scale of the Shaded viewport is changed. (See Figure 1-53.)
24. Open the Advanced settings tab of the Options dialog box and click on the Restore Defaults button.

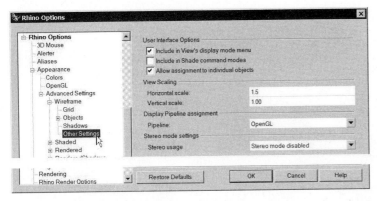 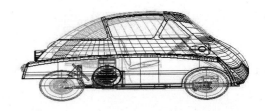

FIGURE 1–53 *Advanced settings (left) and Viewports' aspect ratio changed (right)*
Source: Robert McNeel and Associates Rhinoceros® 5

It is important to note that this is only an illusion of what would happen if the objects are stretched in one direction and in essence the objects are not changed at all.

Turntable Display

You may rotate the viewport continuously as if it is a turntable, as follows:

25. Maximize the Perspective viewport and set its display to X-ray.
26. Select View > Set Camera > Turntable.
    The Perspective viewport, being the active viewport, will rotate continuously.
27. You may set the rotation speed and the number of revolution in the Turntable dialog box. (See Figure 1–54.)
28. If you want to rotate only once, set the number of revolutions to 1 in the Turntable dialog box after the turntable is activated or right-click on the Turntable/Turntable: one cycle button.
29. To stop rotating, press the ESC key.

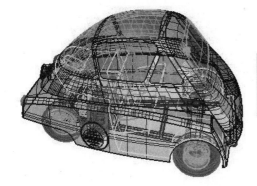

FIGURE 1–54 *Turntable dialog box*
Source: Robert McNeel and Associates Rhinoceros® 5

Clipping Planes

Clipping planes are imaginary planes set in a 3D space, with which you can clip away objects in front of it, similar to a section view. Now continue with the following steps:

30. Select View > Viewport Layout > 4 Viewports to set to 4 Viewport Display.
31. Right-click on the Zoom Extent/Zoom Extent 4 Viewports button on the Standard toolbar.
32. Select View > Clipping Plane and click on A and B indicated in Figure 1–55.
33. Repeat the command by pressing the ENTER key and then click on C and D indicated in Figure 1–55.

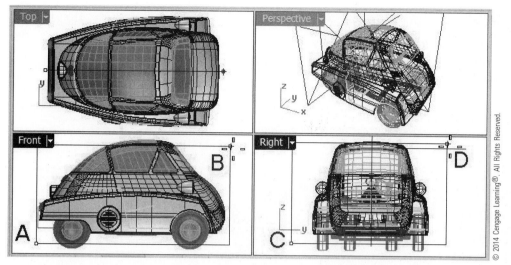

FIGURE 1–55 *Clipping planes in the front and right viewports being constructed*

34. Maximize the Perspective viewport by double-clicking the viewport's label.
35. Left-click on the Disable Clipping Plane/Enable Clipping Plane button on the Visibility toolbar and select clipping plane A in Figure 1–56.
36. Right-click on the Disable Clipping Plane/Enable Clipping Plane button on the Visibility toolbar and select clipping plane B in Figure 1–56.

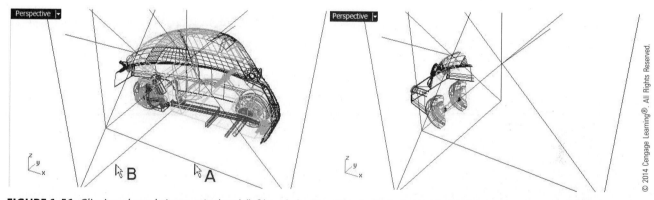

FIGURE 1–56 *Clipping planes being manipulated (left) and clipping planes' effect changed in the Perspective viewport (right)*

More Advanced Display Methods

As mentioned earlier, a NURBS surface is an accurate representation of a smooth surface and producing a rendered image of such surfaces requires the existence of a set of polygon meshes representing the surfaces. To appreciate the effect of the density of the meshes in relation to the display quality, perform the following steps.

1. Select File > Open and select the file Mode.3dm from the Chapter 2 folder on the Student Companion Website.
2. As shown in Figure 1–57, click on the small triangle at the lower-right corner of the Shaded Viewport/Wireframe Viewport button on the Standard toolbar to bring out the Display toolbar and then dock it to the top of the graphics area.
3. Click on the Render Mesh settings button on the Display toolbar.
4. On the Mesh tab of the Document Properties dialog box, click on the Custom button, set Density to 0.01, and click on the OK button.

   Mesh density is changed. Note the jagged edges shown in the right of Figure 1–58.

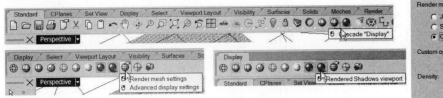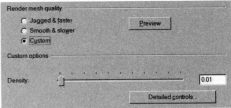

FIGURE 1–57 *Bringing out and docking the Display toolbar and setting mesh density*
Source: Robert McNeel and Associates Rhinoceros® 5

FIGURE 1–58 *Fine (left) and coarse (right) mesh density*
Source: Robert McNeel and Associates Rhinoceros® 5

View Capture

A Rhino active viewport can be captured as a bitmap file to the Windows clipboard for subsequent pasting to other applications. Continue with the following steps if you want to capture the viewport.

5. Select View > Capture > To Clipboard.

    The current viewport is captured, and the captured image is placed in the Windows clipboard. You can now paste it into another application that supports copy and paste.

Display Order

Typically, when two objects are exactly at the same location in the 3D space, such as two curves or a curve at the edge of a surface, the display of one of them will dominate. To find out how draw order can be changed, perform the following steps:

1. Select File > Open and select the file Display Order.3dm from the Chapter 1 folder on the Student Companion Website.
2. Click on Display Order Bring to Front button on the Draw Order toolbar and select A (Figure 1–59) to bring the rectangle to the front in the draw order.
3. Click on Display Order Send to Back button on the Draw Order toolbar and select A to bring it to the back in the draw order.
4. Click on Display Order Send Backward button on the Draw Order toolbar and select B to bring it backward in the draw order.
5. Click on Display Order Bring Forward button on the Draw Order toolbar and select C to bring it forward in the draw order.
6. Click on Clear Draw Order on the Draw Order toolbar and click on D and then E to select all the rectangles to return curve draw order to the default settings.

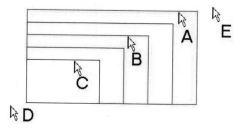

FIGURE 1–59 *Draw order being manipulated*
© 2014 Cengage Learning®. All Rights Reserved.

UTILITIES AND HELP

The help system contains several categories: Command Help dialog box, Help Topics, Frequently Asked Questions, Learn Rhino, Help on the Web, Command List, Feature Overview, and Technical Support. The sections that follow describe Help Topics and Frequently Asked Questions.

Web Browser

You can open a stay on top web browser window that you can read while working in Rhino.

1. Select Tools > Web Browser.
2. Select the Open option.
3. Type a web address: www.rhino3d.com

   A web browser window is displayed on top of the Rhino window.
4. To close the web browser window, click on the x mark at the top-right corner of the web browser window.

Hyperlink

To incorporate additional information to selected objects, you may attach a hyperlink. Naturally, unwanted hyperlinks can be removed. If a hyperlink is already attached to an object, you may open its hyperlink, view its hyperlink, and select it by specifying the hyperlink. You may perform these operations by selecting an item from the Tools > Hyperlink cascading menu.

Calculators

There are two calculators available in Rhino: the ordinary calculator and the reverse polish notation calculator. They can be accessed by selecting Tools > Calculator and selecting Tools > RPN Calculator. Both calculators provide a dialog box for inputting, and they output the results to the command line.

Rhino Options and Document Properties

By now you should have a general understanding of Rhino's user interface and various drawing aids. Before you proceed to the other chapters to learn how to construct various types of objects, spend some time learning the meaning of various settings in the Rhino Options and Document Properties dialog boxes, which share the same dialog box. To access this dialog box, select File > Properties (or Tools) > Options. Some of these options are discussed in the sections that follow.

Shortcut Keys

Using the Keyboard tab, you can assign shortcut keys to the commands you use most frequently.

Files Location

To set the template file and auto-save file location, you use the Files tab.

Appearance

In the Appearance tab, you set the color and appearance of the user interface.

Units

Before you start constructing a model, you should check the units of measurement so that the model you construct is compatible with any upstream and downstream operations. In the Units tab, you set the units of measurement and the tolerance of the model. In addition, you set the display tolerance.

Export and Import Options

Options that you saved in a Rhino file can be exported to an option file and then imported to another Rhino file, thus saving a lot of time to set options repeatedly.

- To export options set in a file to an option file, select Tools > Export Options. In the Save As dialog box, specify a file name and click on the OK button.
- To import options set in another file, select Tools > Import Options. In the Import Options dialog box, select an option file, click on the option items that you want to import, and click on the OK button.

Help Topics

Apart from using the Help pane in the Rhinoceros dialog box, you can display the Help dialog box in several ways: by pressing the F1 key, selecting Help Topics from either the Standard toolbar, the Help toolbar, or Help pull-down menu, you gain access to the comprehensive Help dialog box, shown in Figure 1–60. Here you will find all the information you need.

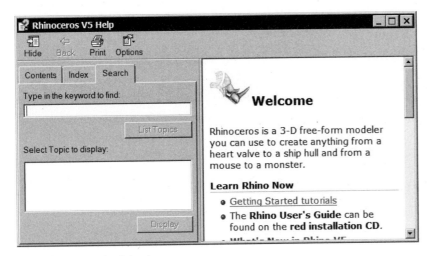

FIGURE 1–60 *Help dialog box*
Source: Robert McNeel and Associates Rhinoceros® 5

FILING

This section discusses save methods, file sending via email, way to add note, render mesh extraction, and view capture.

Save Methods

A Rhino file can be saved via five operations: Save, Save Small, Incremental Save, Save As, and Save As Template. Save file format will be discussed as follows:

Save

- Select File > Save.
  You save your file together with a preview and associated rendering meshes.

Save Small

- Select File > Save Small.
  You save your file without the preview and rendering meshes.

When you use the Render command to construct a rendered image, the computer automatically constructs a set of polygon meshes and uses the meshes for rendering purposes. (You will learn rendering in Chapter 14.) Normally, these polygon meshes are preserved when you save your file. The next time you render the object (if it is not modified), the computer will use the preserved meshes (saved in the file) for rendering. As a result, rendering time is reduced because the program does not have to construct the polygon meshes.

If render meshes are needed for any purpose, you may extract them from selected objects.

- To extract render meshes, type ExtractRenderMesh at the command area or right-click on the RenderMesh Settings/Extract Render Mesh button on the Render toolbar and select objects from which you want to extract render meshes.

The inclusion of polygon mesh data, however, makes file sizes larger. To minimize the storage requirement for polygon mesh data, you can use the Save Small command, which saves NURBS surfaces but not the polygon meshes required for construction of a rendered image. However, as a result, the next time you open the file rendering time will be longer because the computer will have to reconstruct the set of polygon meshes. Apart from that, Save Small also deletes the preview image.

Save with Preview Image but Not Rendered Meshes

- If you want to keep the preview image but do not want to save the rendered meshes, type ClearAllMeshes at the command area to clear the meshes and then save normally.

Incremental Save

- Select File > Incremental Save.
  You save your model in a set of versions so that you can experiment with changes and retrieve previous versions.

If the file has not been saved before, the IncrementalSave command will prompt for a file name. Running this command again will save the file incrementally with a file name suffixed by three numbers, starting from 001.

Save As

- Select File > Save As.
  You save your model with a different file name.

The first time you save a file, using the Save or Save As command will require you to specify a file name, and the file will be saved accordingly. If the file is already saved previously, using the Save command will simply save the file without any prompt. However, using the Save As command will prompt for a file name. After specifying a file name, Rhino will save the data from the current file to a new file, close the current file without saving, and let you work on the new file.

Save As Template

- Select File > Save As Template.
  The model is saved as a template.

If the objects constructed in a file are to be used repeatedly as a starting point for many other files, you may consider saving the file as a template in the template folder.

Send File via Email

You can send email with the current Rhino file as attachment by selecting File > Send. Naturally, you need to configure your email profile.

Add Notes to File

You can include textual information in a Notes window that is saved in the file by selecting File > Notes and add textual notes to the Note dialog box.

Revert

You can discard changes made to a file and revert to the previously saved document by using the Revert command.

CONSOLIDATION

Digital representation of an object in the computer can be in the form of wireframe, surface, or solid. Among them, the wireframe model is the oldest form of digital model and nowadays has very limited application in design and downstream computerized manufacturing systems. However, curves are extremely useful and important in constructing free-form surfaces and solids because you can use them as frameworks upon which surfaces and solids are built. Therefore, you need to have a good understanding of the characteristics of curves and the methods of constructing 3D curves.

A surface model is a set of 3D surfaces with different patterns assembled in 3D space to represent a 3D object. Each surface of a surface model is a mathematical expression depicting the profile and silhouette of that surface. Naturally, surface modeling requires prior thinking about how an object is deconstructed into a set of definable surfaces as well as thinking about how the individual surfaces can be constructed. When surfaces are put together, continuity between them is a concern.

There are two basic methods of representing a surface in the computer: using planar polygon meshes to approximate a surface and using complex spline surfaces to exactly represent a surface. A better way to represent a continuously smooth surface is to use spline mathematics. In most contemporary computer-aided design applications, NURBS surface mathematics is used.

Rhinoceros is a 3D digital modeling tool for constructing points, curves, NURBS surfaces, polysurfaces, solids, and polygon meshes. In addition, you output photorealistic renderings, 2D drawings, and file formats of various types for downstream computerized operations. On the other hand, you can reuse upstream digital models by opening various file formats.

The Rhino user interface contains five major areas: standard Windows title bar, main pull-down menu, command area, graphics area, and status bar. In addition, there are a number of toolbars and a Rhinoceros dialog box in which there are a number of command panes. You perform commands via the main pull-down menu, command area, toolbar, or Rhinoceros dialog box. The graphics area is where you construct digital models. To help visualizing 3D objects in different directions, there are several viewports. To contrast an object from its surrounding background, you shade or render the viewport in a number of ways, enhancing the visual representation of a 3D object by applying a color shade to the surface of the object. You can also zoom, pan, and rotate the display. In addition, there are a number of utility tools, such as web browser, hyperlink, and calculator.

REVIEW QUESTIONS

1. In how many ways can a 3D object be represented in a computer?
2. Is wireframe modeling tool still essential in computer modeling? Why?
3. What are the two ways of representing a surface in the computer? State their advantages and disadvantages.
4. What types of objects can you construct by using Rhino?
5. In how many ways can a command be executed in Rhino? What are they?
6. State the methods by which an object can be shaded in the Rhino viewport.

CHAPTER 2

Familiarizing Rhinoceros

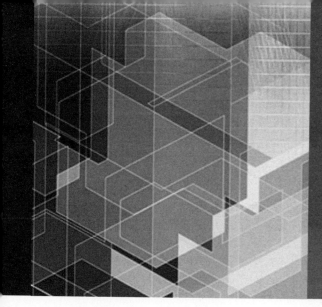

INTRODUCTION

This chapter familiarizes you with Rhino's basic operation methods.

OBJECTIVES

After studying this chapter, you should be able to

- Input precise coordinates and use various drawing aids to construct geometric objects
- Manipulate geometric objects that are already constructed
- Comment on the concepts of history management
- Manage layers and manipulate object's properties
- Use various kinds of construction planes in geometric object construction

OVERVIEW

Working in a computer-aided design system involves two categories of tasks: you construct objects and manipulate them. To construct objects accurately, you need to know what is meant by a construction plane, how precise coordinates can be input, and how various tools can be used in geometric object construction. To be able to manipulate objects that are already constructed, you need to have an understanding of various object translation tools: Rhino's history management concept, layer management methods, and object manipulation methods.

MULTIPLE CONSTRUCTION PLANES CONCEPT

To construct geometric objects, apart from using the options from the pull-down menu, toolbar, or command area, you must specify a location or a number of locations for the geometric object to be constructed. To specify a location using the pointing device, you can pick a point in one of the viewports. To accurately locate a point, you can use object snap utilities or key in a set of coordinates at the command area.

Default Construction Planes

In each viewport, there is an imaginary plane, called the construction plane. When you pick a point in a viewport on the screen with the pointing device, you are picking a point in this plane. In a default four-viewport display, there are three construction planes. With the exception of the Perspective viewport, which has the same construction plane as the Top viewport, each viewport has a construction plane parallel to itself and passing through the origin.

Among the viewports, only one of them is active and activation can be done simply by moving the cursor over the viewport. When you construct objects, their orientation depends on the active construction plane, and the active construction plane depends on which viewport you select. You can construct objects on more than one construction plane. For example, to construct two circles on two different construction planes, you would perform the following steps.

1. Start a new file. Use the Small Objects, Millimeters template.
2. From the main pull-down menu, select Curve > Circle > Center, Radius. Alternatively, select the Circle: Center, Radius option from the Circle toolbar, or type Circle at the command area.
3. Move the cursor over the Top viewport.

   Indicating that the viewport is active, the label of the viewport (top left-hand corner) should be highlighted and the labels in the other viewports should be grayed out, as shown in Figure 2–1.

4. Pick a point to specify the center location and pick to specify a point on the circumference. Exact location is unimportant.

   A circle is constructed on a construction plane parallel to the Top viewport and passing through the origin.

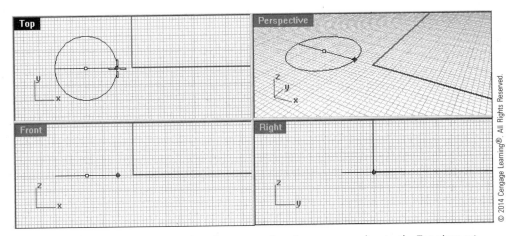

FIGURE 2–1 *Circle being constructed on the construction plane corresponding to the Top viewport*

5. Press the ENTER key to repeat the Circle command.
6. Move the cursor over the Front viewport.

   Now the label of the Front viewport is highlighted. You are working on a construction plane parallel to the Front viewport.

7. Pick a point to specify the center location and pick to specify a point on the circumference. Exact location is unimportant.

   A circle is constructed on a construction plane parallel to the Front viewport and passing through the origin, as shown in Figure 2–2.

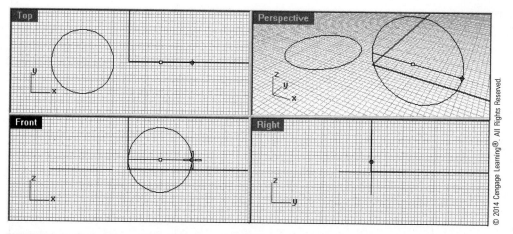

FIGURE 2–2 *Circle being constructed on a construction plane (the front construction plane) parallel to the Front viewport*

Manipulating Construction Planes

Apart from using the default construction planes, you can construct user-defined construction planes in each viewport as well as universal construction planes and mobile construction planes, which will be discussed later in this chapter.

USING COORDINATES INPUT

To specify a point precisely, you input its coordinates. There are two coordinate systems: the construction plane coordinate system (corresponding to the active viewport) and World coordinate system (independent of the active viewport). The construction plane coordinate system corresponds to the construction plane.

Coordinate Axes

In each viewport, a red line and a green line depict, respectively, the X and Y axes of the construction plane in the viewport. (The color of these lines and other color settings are configurable via the Color tab of the Rhino Options dialog box, accessed by selecting Tools > Options from the main pull-down menu.) The Z direction is perpendicular to the construction plane.

In addition to the red and green lines, there is a World Axes icon in the lower-left corner of the respective viewport. The World Axes icon is in the shape of a tripod. The lines on the tripod depict the absolute X, Y, and Z axes of the World coordinate system.

Cartesian and Polar Coordinate Systems

To specify a location using the command area, you indicate the nature of the Cartesian coordinate system or polar coordinate system you want to use via the nature of the coordinate input.

- For the Cartesian coordinate system options, you specify X and Y values or X, Y, and Z values (separated by a comma or commas, as shown in Table 2–1). If you specify only the X and Y values, the Z value is assumed to be zero.
- For the polar coordinate system options, you specify a distance value and an angular value, separated by a less-than (<) sign.
- For either coordinate system, if you want to specify a point relative to the last selected point, you prefix the coordinate with the letter r or the @ sign. If you want to use a World coordinate, you prefix the coordinate with the letter w. Coordinate systems that operate under Rhino, and what each system specifies, are summarized in Table 2–1.

TABLE 2–1 *Coordinate Systems and Their Functions*

Coordinate System	Example	Specifies
Construction plane Cartesian system	2,3	A point 2 units in the X direction and 3 units in the Y direction from the origin (Z is zero)
	2,3,4	A point 2 units in the X direction, 3 units in the Y direction, and 4 units in the Z direction from the origin
Construction plane polar system	2<45	A point 2 units at an angle of 45 degrees from the origin
Relative construction plane Cartesian system	r2,3 or @2,3	A point 2 units in the X direction and 3 units in the Y direction from the previous input point
Relative construction plane polar system	r4<60 or @4<60	A point 4 units at an angle of 60 degrees from the previous input point
World Cartesian system	w3,5	A point 3 units in the absolute X direction and 5 units in the absolute Y direction from the absolute origin, regardless of the location of the current construction plane
World relative Cartesian system	wr3,6	A point 3 units in the absolute X direction and 6 units in the absolute Y direction from the previous input point, regardless of the location of the current construction plane
World polar system	w4<30	A point 4 units at an angle of 30 degrees on the absolute XY plane from the absolute origin, regardless of the location of the current construction plane
World relative polar system	wr5<45	A point 5 units at an angle of 45 degrees on the absolute XY plane from the previous input point, regardless of the location of the current construction plane

To appreciate how precise coordinates can be input to construct Rhino objects, perform the following steps.

1. Start a new file. Use the Small Objects, Millimeters template.

> **NOTE**   Do not save the previous file.

2. Double-click on the Top viewport title to maximize the viewport.

    Double-clicking again on the maximized Top viewport title will set the graphics area to the previous four-viewport configuration.

3. Select Curve > Line > Single Line.
4. Type 10,10 at the command area to specify the first endpoint of the line segment, shown as A in Figure 2–3.

    This is a construction plane Cartesian coordinate, indicating that the endpoint is 10 units from the origin in the construction plane's X direction and 10 units from the origin in the construction plane's Y direction.

5. Type 20<170 at the command area to specify the second endpoint of the line segment, shown as B in Figure 2–3.

    This is a construction plane polar coordinate, indicating that the endpoint is 20 units away from the origin and at an angle of 170 degrees from the construction plane's origin.

6. Press the ENTER key to repeat the last command, select Curve > Line > Single Line, or click on the Line button on the Lines toolbar.
7. Type r10,0 at the command area to specify the first endpoint, shown as C in Figure 2–3.

    This is a relative construction plane Cartesian coordinate, indicating that it is 10 units from the last input point, shown as C in Figure 2–3, in the construction plane's X direction and 0 units from the last input point in the construction plane's Y direction.

8. Type r20<300 at the command area to specify the second endpoint, shown as D in Figure 2–3.

   This is a relative construction plane polar coordinate, indicating that it is 20 units from the last input point and at an angle of 300 degrees from the last input point.

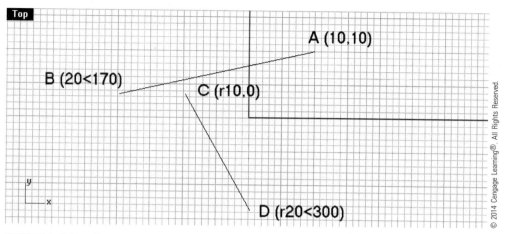

FIGURE 2–3 *Two line segments constructed*

Angle Constraint

If you want to construct a line at a certain angle, but have no definite idea on its length, you may first select a point to specify the start point of the line and then type <60 in the command area. The endpoint of the line will be constrained at an angle of 60 degrees. Continue with the following steps.

9. Select Curve > Line > Single Line.
10. Click on point A in the Top viewport.
    Exact location is unimportant.
11. Type <60 at the command area to specify an angle constraint.
12. While the cursor is being constrained at an angle of 60 degrees, click on B, shown in Figure 2–4.
    A line at an angle of 60 degrees is constructed.
13. Do not save your file.

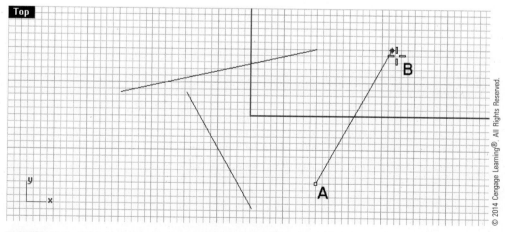

FIGURE 2–4 *A line constrained at an angle of 60 degrees constructed*

USING DRAWING AIDS

To specify locations for the construction of geometric objects, you use various drawing aids, such as grid meshes, snapping to grids, planar mode, elevator mode, ortho mode, and object snap.

Setting Grid Mesh

A grid mesh of known spacing on the screen gives you a sense of the actual size of the current viewport. Perform the following steps.

1. Start a new file. Use the "Small Objects, Millimeters" template.

NOTE Do not save the previous file.

2. Select View > Grid Options.
3. Change Grid line count to 100 millimeters. (See Figure 2–5.)
4. Click on the OK button.

Grid spacing is set.

In the Grid tab of the Document Properties dialog box, you control the display of grid lines and set their spacing. In addition, you control the display of the World Axes icon. To quickly turn on or off the grid mesh in a viewport, press the F7 key. An alternative way to set grid settings is to type Grid at the command area and make appropriate settings.

Snapping to Grid

The grid display shown in the viewport is for visual reference only. Without the use of further aids, it is virtually impossible to select these grid points precisely using the pointing device. To restrict the movement of the cursor so that it will stop only at the grid intervals, you use the Snap option, or select or deselect the Snap button on the status bar. (See Figure 2–5.) Continue with the following steps.

5. Click on the Snap button of the Status bar.

   Snap is turned on.

6. Move the cursor around in the graphics area.

   You will find that the cursor is restricted to grid locations.

7. Click on the Snap button of the Status bar.

   Snap is turned off.

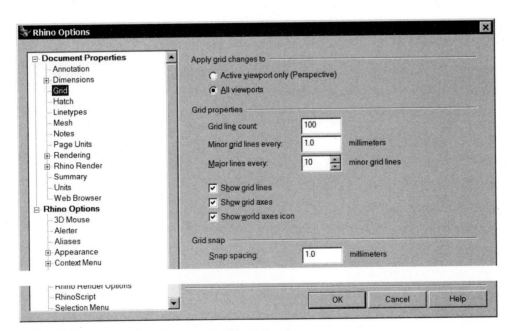

FIGURE 2–5 *Grid tab of the Document Properties dialog box*
Source: Robert McNeel and Associates Rhinoceros® 5

Using Planar Mode

As we have said, there are three preset construction planes in a default four-viewport Rhino file. By using the cursor to pick a point in the viewport, you are restricted to picking a location from one of these planes. To restrict the current construction plane to that of the last selected point, you can use planar mode (established via the Planar option), or you can select or deselect the Planar button on the status bar. Continue with the following steps.

8. Check the Planar button from the Status bar, if it is not already checked.
9. Select Curve > Polyline > Polyline.
10. Click on location A of the Front viewport, as indicated in Figure 2–6.
    Exact location is unimportant.
11. Move the cursor to the Top viewport and continue to click on locations B, C, and D. (See Figure 2–6.)
12. Press the ENTER key.
    A planar polyline is constructed.
13. Click on the Planar button from the Status bar to deselect planar mode.

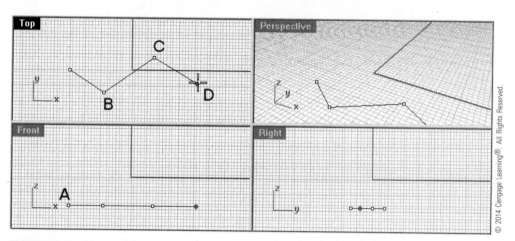

FIGURE 2–6 *Planar polyline constructed*

Using Elevator Mode

To help construct 3D points by first picking the X and Y coordinates of a point from a construction plane and then picking a point from an adjacent construction plane to depict the Z coordinate, you use the elevator mode, which is activated by holding down the CONTROL key while picking a point from the construction plane. Continue with the following steps.

14. Select Curve > Line > Single Line.
15. Hold down the CONTROL key and click on location A, shown in Figure 2–7.
    This specifies the X and Y coordinates relative to the Front viewport.
16. Move the cursor to location B in the Top viewport, and click on location B.
    This indicates the Z coordinate relative to the Front viewport's construction plane.
17. With the CONTROL key still held down, click on location C in the Front viewport and then location D in the Top viewport.
    This specifies the endpoint of the line.

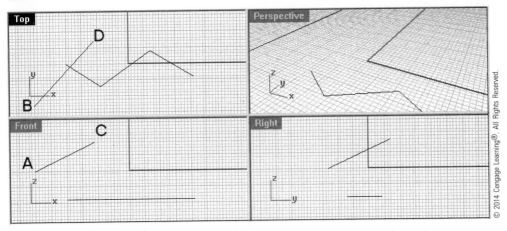

FIGURE 2–7 *Elevator mode being used to construct a line*

Using Ortho Mode

To restrict cursor movement in a specified angular direction, you select or deselect Ortho on the status bar. Continue with the following steps.

1. Start a new file. Use the Small Objects, Millimeters template.

> **NOTE** Do not save the previous file.

2. Double-click on the Top viewport title to maximize the viewport.
3. Click on the Ortho button from the Status bar to turn on ortho mode.
4. Select Curve > Line > Single Line.
5. Click on location A, shown in Figure 2–8, to specify the start point.
6. Move the cursor around; you will find that it is restricted to horizontal or vertical movements.
7. Click on location B, shown in Figure 2–8, to specify the endpoint of the line.

   A horizontal line is constructed.

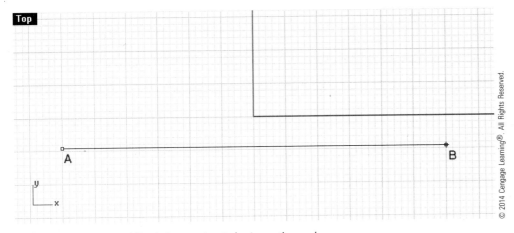

FIGURE 2–8 *Horizontal line being constructed using ortho mode*

Setting Ortho Angle

By default, the ortho angle is 90 degrees. In other words, the second selected point is restricted horizontally or vertically. However, you may set the ortho angle to any value other than 90 degrees by typing Orthoangle at the command area while constructing a line or curve, or by setting the ortho angle in the Modeling Aids tab of the Options dialog box. Continue with the following steps to set the ortho angle at the command area.

8. Select Curve > Line > Single Line.
9. Click on location A, indicated in Figure 2–9, to specify the start point.
10. Right-click on Ortho Toggle/Set Ortho Angle button on the Object snap toolbar.
11. Type 40 to specify an ortho angle of 40 degrees.
12. Click on location B, shown in Figure 2–9, to specify the endpoint.

    A line at an angle of 40 degrees is constructed.

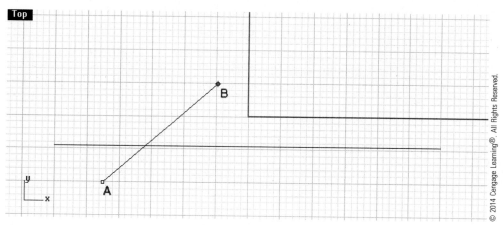

FIGURE 2–9 *Line at an ortho angle of 40 degrees being constructed*

To reset the ortho angle to 90 degrees in the Options dialog box, perform the following steps.

13. Select Tools > Options.
14. On the left pane of the Options dialog box, select Rhino Options > Modeling Aids.
15. Set Ortho Snap every 90 degrees.
16. Click on the OK button to close the dialog box.

    Ortho angle is set to 90 degrees.

17. Click on the Ortho button on the Status bar to turn off ortho mode.

Using Object Snap (Osnap) Tools

While constructing geometric objects, you can use object snap (Osnap) tools to help locate the cursor in relationship to selected features of existing geometric objects. Feature aspects you snap to are end, near, point, midpoint, center, intersection, perpendicular, tangent, quadrant, and knot.

There are two ways to set object snap. You can temporarily set object snap mode (i.e., for each snap to a feature) by specifying the snap mode before you select an object. In Rhino terms, this is one-shot object snap. Alternatively, you can set persistent object snap by selecting Osnap on the status bar to use the Snap option from the Osnap dialog box. Continue with the following steps to use one-shot object snap in geometry construction.

18. Select Curve > Circle > Center, Radius.
19. Select Tools > Object Snap > Intersection.
20. Move the cursor to location A, shown in Figure 2–10, until you see the Int symbol displaying at the cursor, indicating that the intersection between two line segments is found.
21. While Int is still displaying, click on the graphics area.

22. Type Mid at the command area to specify a midpoint object snap.
23. Move the cursor to location B, shown in Figure 2–10, until you see Mid displaying at the cursor.
24. While Mid is still displaying, click on the graphics area.

    A circle with its center at the intersection of two lines and passing through the midpoint of a line is constructed.

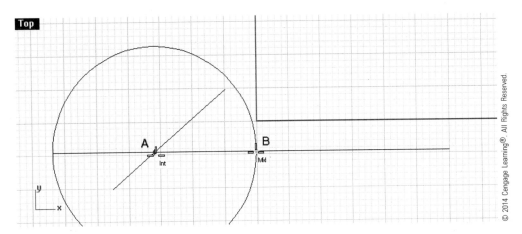

FIGURE 2–10 *Circle being constructed*

Osnap Dialog Box

If a certain object snap mode is required a number of times, it is more convenient to use persistent object snap via the Osnap dialog box. Continue with the following steps.

25. Check the Osnap button from the Status bar to display the Osnap dialog box, if it is not already displayed.

    As shown in Figure 2–11 (left), there are eleven persistent object snap modes: End, Near, Point, Mid (midpoint), Cen (center), Int (Intersection), Perp (perpendicular), Tan (tangent), Quad (quadrant), Knot, and Vertex. In addition, there are two buttons: Project and Disable.

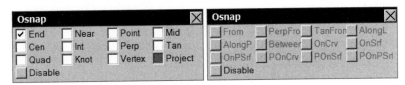

FIGURE 2–11 *Normal Osnap dialog box (left) and dialog box with CONTROL key held down (right)*
Source: Robert McNeel and Associates Rhinoceros® 5

Checking the Project button enables projection mode, checking the STrack button enables smart tracking, and checking the Disable button disables object snap mode specified in the dialog box. Right-clicking the Disable button clears all the persistent object snaps. If you hold down the CONTROL key, the Osnap dialog box displays its second page of icons, as shown in Figure 2–11 (right). Now continue with the following steps.

26. Check the End box of the Osnap dialog box to specify a persistent endpoint object snap.
27. Select Curve > Line > Single Line.
28. Click on endpoints A and B, shown in Figure 2–12, to specify the start and endpoints of the line.

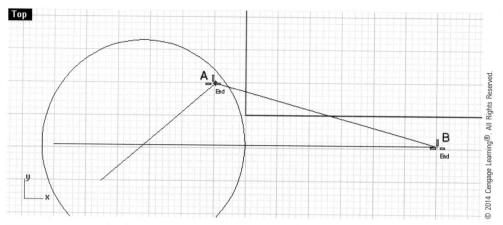

FIGURE 2–12 *Endpoints selected while constructing a line*

Enabling/Disabling Osnap by Using the ALT key

When Rhino is asking for a point, you can press the ALT key to temporarily enable Osnap check boxes if Osnap is disabled. On the other hand, you can press the ALT key to temporarily disable Osnap check boxes if Osnap is enabled.

Osnap: Along Line

Along Line is a snapping tool. It enables you to select two points to establish a tracking line, along which you can select any point. Perform the following steps.

1. Open the file DrawingAids1.3dm from the Chapter 2 folder that you downloaded from the Student Companion Website.

> **NOTE**
> Do not save the previous file.

2. Click on the Osnap button from the Status line.
3. In the Osnap dialog box, click on End and Point, and clear all other buttons.
4. Select Curve > Line > Single Line.
5. Select point A, indicated in Figure 2–13, to specify the start point of the line.
6. Select Tools > Osnap > Along Line.
7. Select endpoints B and C, indicated in Figure 2–13, to specify the start and end of the tracking line.
8. Click on point D. Endpoint E is constructed along the tracking line and closest to point D.

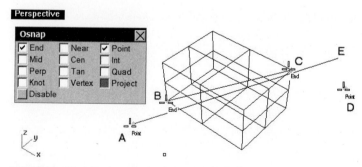

FIGURE 2–13 *A line's endpoint E constructed along tracking line BC*
Source: Robert McNeel and Associates Rhinoceros® 5

Osnap: Along Parallel

Along Parallel is a snapping tool. It establishes a tracking line passing through a point and parallel to a reference line. You can select a location along this parallel line. Continue with the following steps.

9. Select Curve > Line > Single Line.
10. Select point A, shown in Figure 2–14, to specify the start point of the line.
11. Select Tools > Osnap > Along Parallel.
12. Select endpoints B and C, shown in Figure 2–14, to specify the start and end of the reference line.
13. Select endpoint D to specify the start point of the parallel tracking line.
14. Click on point E. Endpoint F is constructed along the tracking line passing through point D, parallel to reference line BC, and closest to point E.

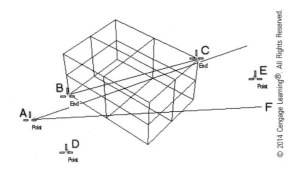

FIGURE 2–14 *A line's endpoint being constructed along a line parallel to a track line*

Osnap: Between Two Selected Points

Between is a snapping tool that enables you to specify a point midway between two selected points. Continue with the following steps.

15. Select Curve > Line > Single Line.
16. Select point A, shown in Figure 2–15.
17. Select Tools > Osnap > Between.
18. Select points A and B, shown in Figure 2–15.

    The endpoint of the line is constructed midway between two selected points.

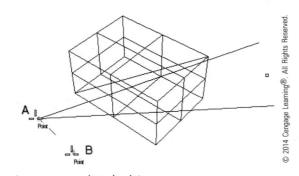

FIGURE 2–15 *Midway between two selected points*

Osnap: From, Perpendicular from, and Tangent from

These three snapping tools enable you to specify a point a distance from a selected point, perpendicular from a point, and tangent from a curve. Perform the following steps.

1. Open the file DrawingAids2.3dm from the Chapter 2 folder on the Student Companion Website.

> **NOTE**
> Do not save the previous file.

2. Check the End and Point buttons, and clear all other buttons in the Osnap dialog box.
3. Select Curve > Line > Single Line.
4. Select Tools > Object Snap > From.
5. Select endpoint A, shown in Figure 2–16.
6. Type 6 at the command area to specify the distance from the selected point.
7. Select point B, shown in Figure 2–16.

   The starting point of the line is specified. It is 6 units from the endpoint A and closest to point B.

8. Select Tools > Osnap > Perpendicular From.
9. Select curve C, shown in Figure 2–16.
10. Click on point D, indicated in Figure 2–16.

    A tracking line is defined. It is perpendicular to curve C and closest to point D.

11. Click on point E, indicated in Figure 2–16.

    The line's endpoint is defined. It resides on the defined tracking line and is closest to point E.

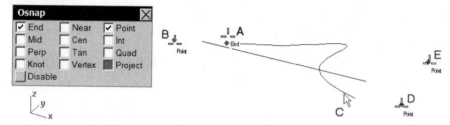

FIGURE 2–16 *A line being constructed by using From and Perpendicular From object snaps*
Source: Robert McNeel and Associates Rhinoceros® 5

12. Select Curve > Line > Single Line.
13. Select Tools > Osnap > Tangent From.
14. Select curve A, shown in Figure 2–17.
15. Select point B, shown in Figure 2–17.

    A tracking line tangent to curve A and closest to point B is defined.

16. Click on point B again.

    The start point of the line is defined. It is on the tracking line and closest to point B.

17. Select endpoint C, shown in Figure 2–17.

    A line is constructed.

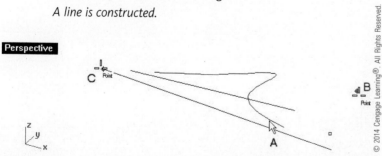

FIGURE 2–17 *A line being constructed by using the Tangent From object snap*

Osnap: On Curve and On Surface

These two snapping tools enable you to snap to a point on a curve and on a surface. Perform the following steps.

1. Open the file DrawingAids3.3dm from the Chapter 2 folder on the Student Companion Website.

NOTE Do not save the previous file.

2. Select Curve > Line > Single Line.
3. Select Tools > Osnap > On Curve.
4. Select curve A, indicated in Figure 2–18.
5. Click on endpoint B, shown in Figure 2–18.
   The line's start point is defined. It is along curve A and closest to endpoint B.
6. Select Tools > Osnap > On Surface.
7. Select surface C, shown in Figure 2–18.
8. Click on point D. (See Figure 2–18.)
   The line's second point is defined. It lies on surface C and is closest to point D.

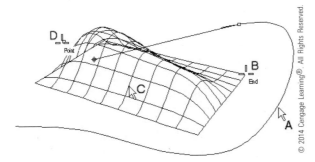

FIGURE 2–18 *A line constructed with its endpoints on a curve and a surface*

NOTE Lines on the surfaces are called isoparametric curves. You can use the intersection snapping tool to snap to the intersections of these curves.

Osnap: Project Mode

If the Project button on the Osnap dialog box is selected, the resulting snap point is projected onto the active construction plane. Perform the following steps.

1. Open the file DrawingAids4.3dm from the Chapter 2 folder on the Student Companion Website.

NOTE Do not save the previous file.

2. Click on the Osnap button from the Status bar.
3. Clear all the buttons except Mid and Project.
4. Select Curve > Line > Single Line.
5. Click on location A of the Top viewport, shown in Figure 2–19.

The midpoint of the curve that is projected onto the current construction plane is used as the starting point of the line.

6. Click on the Project button of the Osnap dialog box to clear it.
7. Click on location A of the Top viewport again.

The midpoint of the curve is used as the endpoint of the line. A line is constructed.

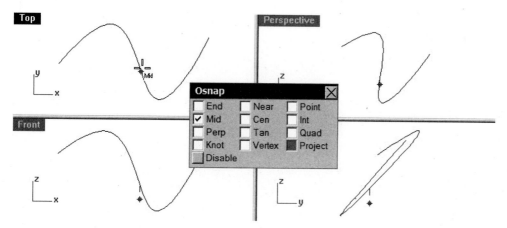

FIGURE 2–19 *A line being constructed*
Source: Robert McNeel and Associates Rhinoceros® 5

Right-Click Menu on the Toolbar's Snap or Grid Button

Right-clicking on the status bar brings out a context menu from which you may make various grid and snap settings.

Smart Tracking

Smart Tracking is a drawing aid, providing a set of temporary reference lines and points on the viewport to help indicate implicit relationships among various 3D points, geometry in space, and coordinate axes' direction.

> To quickly turn on/off Smart Tracking, you can check or clear the STrack button on the Osnap dialog box and select Tools > Object Snap > Smart Track.

NOTE

With Smart Tracking turned on, temporary infinite lines, called tracking lines, appear on the screen after a point or a geometric object is encountered. With two or more nonparallel tracking lines, you get temporary intersection points called smart points. You can snap to these smart points as if they are real point objects. Perform the following steps:

1. Open the file DrawingAids5.3dm from the Chapter 2 folder on the Student Companion Website .

> Do not save the previous file.

NOTE

2. Click on the Osnap button from the Status bar.
3. Clear all the buttons except End and STrack.
4. Select Curve > Line > Single Line.
5. Move the cursor to endpoint A, shown in Figure 2-20, but do not click on it.
6. Move the cursor to endpoint B, shown in Figure 2-20, but do not click on it.
7. Move the cursor vertically upward until two tracking lines and an intersection (C) between the two tracking lines appear.
8. While the tracking lines are still displayed, click on the intersection mark.

The start point of the line is specified.

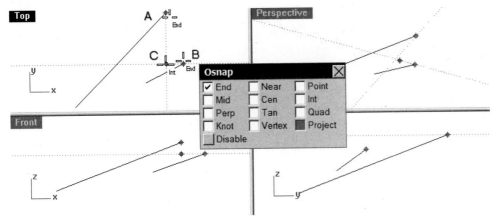

FIGURE 2–20 *Smart tracking tool being used to define the start point of a line*
Source: Robert McNeel and Associates Rhinoceros® 5

9. Move the cursor to endpoint D, shown in Figure 2–21, but do not click on it.
10. Move the cursor to endpoint E, shown in Figure 2–21, but do not click on it.
11. Move the cursor vertically upward until two tracking lines and an intersection (F) between the two tracking lines appear.
12. While the tracking lines are still displayed, click on the intersection mark.

    The endpoint of the line is specified.

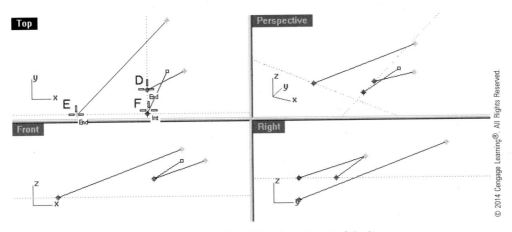

FIGURE 2–21 *Smart tracking tool being used to define the endpoint of the line*

MANIPULATING GEOMETRIC OBJECTS

To manipulate geometric objects that are already constructed, you have to select them and apply a command. The sequence of first selecting the objects and then applying a command and vice versa is unimportant.

Geometric Object Selection Methods

Obviously, the simplest way to select an object is to move the cursor over the object and click. If you want to select multiple objects, you can hold down the SHIFT key or the CONTROL key. However, the SHIFT key favors adding objects to the selection set, and the CONTROL key favors removing objects from the selection set.

Window and Crossing Selection

To select multiple objects, you can drag a rectangular zone. If you drag from right to left, you describe a crossing zone, with objects partly or fully within the zone being selected. If you drag from left to right, you describe a window zone. Only objects fully inside the zone are selected. If more than one object is located at where you drag, a pop-up selection menu will display. You can click one of the objects or None if none of the highlighted objects is correct. Now perform the following steps.

1. Open the file Selection 1.3dm from the Chapter 2 folder on the Student Companion Website.

> **NOTE**
> Do not save the previous file.

2. Hold down the SHIFT key and select A, B, and C. (See Figure 2–22.)
   Three circles are selected.
3. Hold down the CONTROL key and select B and C again.
   Two selected circles are deselected.
4. Drag from D to E. (See Figure 2–22.)
   A crossing zone is defined, and three circles are selected.
5. Drag from F to G. (See Figure 2–22.)
   A window zone is defined, and only one circle is selected.
6. Click on H. (See Figure 2–22.)
   Because there are two identical circles located at the same position, a pop-up selection menu is displayed for you to choose one of them.
7. Press the ESC key to clear all the selections.

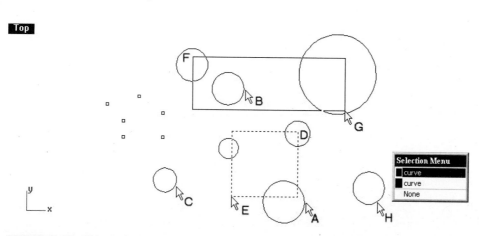

FIGURE 2–22 *Object selection*
Source: Robert McNeel and Associates Rhinoceros® 5

Lasso Selection

Continue with the following steps to select object by using the Lasso command:

8. Select Edit > Control Points > Select Control Points > Lasso.
9. Referencing Figure 2–23, drag an irregular sketch.
   A set of points is selected.

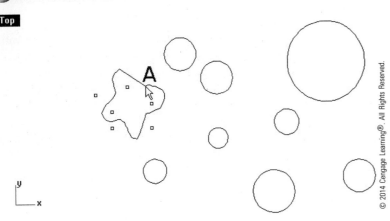

FIGURE 2-23 *Selecting a set of points*

Note that after executing the Lasso command and before you make the first click, you may change viewport and perform various zoom and pan operations.

Brushing Selection

You can drag the mouse in a way similar to a brush stroke to select objects. Continue with the following steps.

10. Select Edit > Select Objects > Area and Volume Select > Brush.
11. With reference to Figure 2-24, drag the mouse like a brush.
    Objects covered by the brush stroke are selected.

To select points and control points (you will learn about control points in Chapters 6 and 7) by using brush strokes, continue with the following steps.

12. Edit > Control Points > Select Control Points > Brush.
13. Also refer to Figure 2-24 to drag the mouse across the graphics area.
    Unlike the previous command, only point objects (control points are not turned on here) are selected.

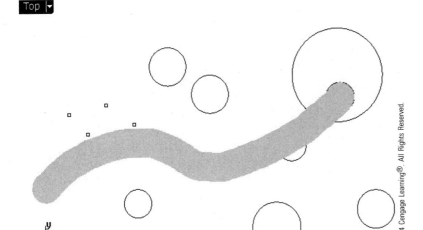

FIGURE 2-24 *Dragging the mouse like a brush across the graphics area*

Boundary and Pipe Selection

Various ways are used to select points. Perform the following steps to select objects by picking a closed curve:

1. Open the file Selection 1.3dm from the Chapter 2 folder on the Student Companion Website.

> **NOTE** Do not save the previous file.

2. Select Edit > Select Objects > Area and Volume Select > Boundary.
3. Referencing Figure 2–25, select curve A in the Top viewport.
   The red and white spheres are selected.
4. Press the ESC key to cancel the selection, repeat the previous command, and select curve A in the Perspective viewport.
   Nothing is selected, and obviously this command is view dependent.

Continue with the following steps to select objects by describing a pipe-shaped volume.

5. Press the ESC key to cancel the selection and select Edit > Select Objects > Area and Volume Select > Pipe.
6. Type S in the command area to choose the Selection Mode option.
   There are four options: window, crossing, invert window, and invert crossing, representing totally inside, crossed by, objects not totally inside, and objects not crossed by the imaginary pipe.
7. Select the Window option.
8. With reference to Figure 2–25, select curve B.
9. Type 6 in the command area to specify the radius of an imaginary pipe with axis along the selected curve.
   The white and green spheres are selected.

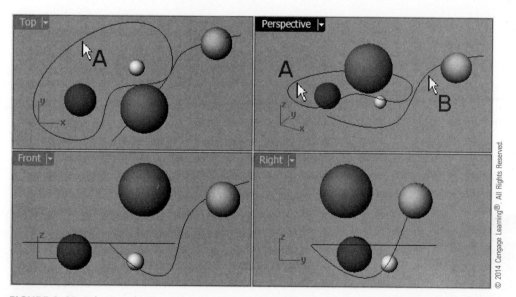

FIGURE 2–25 Selecting objects enclosed by a boundary curve and a pipe-shaped volume

Circle, Sphere, and Box Selection

You can select objects by describing a circular zone, spherical zone, and rectangular box zone. Again, there are four options: window, crossing, invert window, and invert crossing. Now continue with the following steps:

10. Press the ESC key to cancel the previous selection and select Edit > Select Objects > Area and Volume Select > Circular.
11. Referencing Figure 2–26, click on A and then B to describe a circular zone.
    The green sphere is selected. This command is view dependent.

12. Press the ESC key to cancel the previous selection and select Edit > Select Objects > Area and Volume Select > Sphere.
13. With reference to Figure 2–26, click on A and then B to describe a spherical zone.
    Nothing is selected because, if the window option is selected, nothing is totally inside the spherical zone.
14. Press the ESC key to cancel the previous selection and select Edit > Select Objects > Area and Volume Select > Box.
15. Referencing Figure 2–26, click on C, D, E, and F to describe a box zone.
    The green sphere is selected.

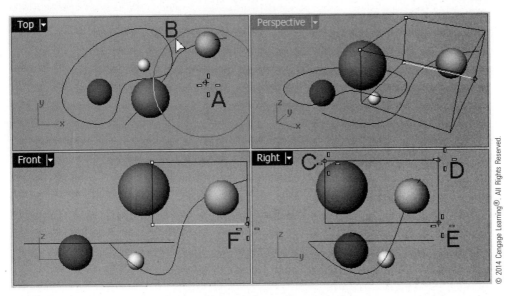

FIGURE 2–26 *Selecting objects by describing a circle, a sphere, and a box*

Selection Filter

You can use selection filter to select specific types of objects. Continue with the following steps:

16. Select Edit > Selection Filter.
17. In the Filter dialog box, shown in Figure 2–27, clear all the check boxes except Curves.
18. Select Edit > Select Objects > All Objects.
    Only the curves are selected.

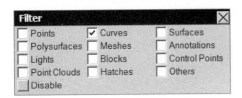

FIGURE 2–27 *Filter dialog box*
Source: Robert McNeel and Associates Rhinoceros® 5

Selection by Object Type

You can select objects by specifying their object type, such as clipping planes, dimension styles, extrusion objects, hatchings, lines, texture mapping widgets, no-manifold objects, picture frames, and objects' rendering color. These operations can be performed by selecting Edit > Select Objects and then selecting one of the available options.

Moving Objects

To relocate one or more geometric objects, you move them. Perform the following steps.

1. Select File > Open and select the file Movecopy.3dm from the Chapter 2 folder on the Student Companion Website.
2. Select Transform > Move.
3. Select surface A, shown in Figure 2–28, and press the ENTER key to terminate the selection process.
4. Click anywhere on the graphics area to specify the point to move from.
5. Type r20<0 at the command area to specify a point to move to.

   This is a relative construction plane coordinate, indicating that the second point is 20 units from the last input point and at an angle of 0 degree from it. As a result, the circle is moved a distance of 20 units in the 0 degree direction.

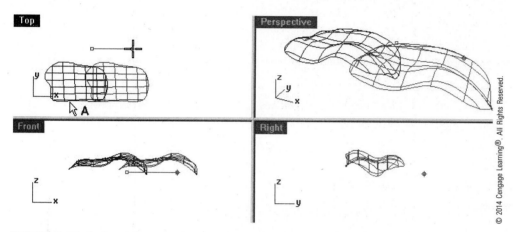

FIGURE 2–28 *Surface being moved*

Moving by Dragging and Drag Mode

In addition to using the Move command to move objects from one location to another location, you can select objects and drag them to a new location.

However, when you use the dragging method, you must have prior understanding of the dragging direction, which can be set via the Edit menu.

To set dragging mode, select Edit > Change Drag Mode. As can be seen at the command area, there are five options, providing five dragging modes, as follows:

- CPlane dragging mode: Dragging is parallel to the CPlane (construction plane) setting of the active viewport.
- World dragging mode: Dragging is parallel to the World coordinate system's XY plane.
- View dragging mode: Dragging is parallel to the active viewport.
- UVN dragging mode: Dragging is along the U, V, and N directions. (U and V are two orthogonal directions on the surface, which are parallel to the isocurves. N is the normal direction.) To drag along the normal direction, hold down the CONTROL key.
- Control Polygon mode: This mode concerns dragging of control point along the control polygons that connect to it.

> **NOTE**
> Next option sets the dragging mode to the next in the list. In other words, if the current dragging mode is CPlane, selecting the Next option will change the dragging mode to World. To remind you about any drag mode setting other than the default CPlane option, the cursor icons are slightly different, as shown in Figure 2–29.

FIGURE 2–29 *Cursor icons (left to right): CPlane, World, View, UVN, and Control Polygon*
Source: Robert McNeel and Associates Rhinoceros® 5

HOLDING DOWN THE ALT KEY WHILE DRAGGING

By holding down the ALT key while dragging an object, the object will be copied instead of moved. The ALT key also suspends the auto-close function of polyline, curve, and interpolated curve. In addition, it can force a window/crossing selection. Normally, window/crossing selection needs to have the first selection point picked on empty space. Furthermore, it suspends locked-object snap.

Copying Objects

To duplicate geometric objects, you copy them. Perform the following steps.

6. Select Transform > Copy.
7. Select surface A, shown in Figure 2–30, and press the ENTER key to terminate the selection process.
8. Click anywhere on the graphics area to specify the point to copy from.
9. Type r20<180 at the command area to specify the point to copy to.
10. Type r40<180 at the command area to specify the point to copy to.
11. Press the ENTER key to terminate the command.

    The selected circle is copied twice.

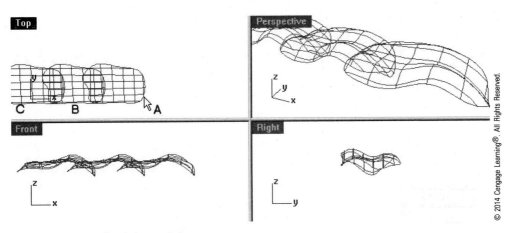

FIGURE 2–30 *Surface being copied*

Delete

To erase an unwanted geometric object, you delete it. Perform the following steps.

12. Hold down the SHIFT key and select objects B and C, indicated in Figure 2–26.
13. Press the DELETE key.

    The selected circles are deleted.

14. Do not save your file.

Rotating Objects

To change the orientation of a geometric object, you rotate it 2D or 3D. 2D rotation causes the selected objects to be rotated around a point on the construction plane. 3D rotation, on the other hand, causes the selected objects to be rotated around an axis defined in 3D space. Perform the following steps.

1. Select File > Open and select the file Rotate.3dm from the Chapter 2 folder on the Student Companion Website.
2. Select Transform > Rotate.
3. Select ellipsoid A, shown in Figure 2–31, and press the ENTER key.
   Exact location of A here is unimportant.
4. Type 10,10 at the command area to specify the center of rotation.
5. Type 45 to specify the angle of rotation.
   The ellipsoid is rotated at an angle of 45 degrees around a point located at 10,10.

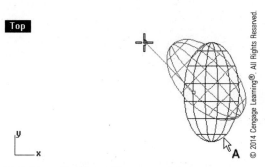

FIGURE 2–31 *Ellipsoid being rotated*

6. Double-click on the Top viewport title to return the display to a four-viewport configuration.
7. Select Transform > Rotate 3-D.
8. Select the ellipsoid and press the ENTER key.
9. Type w10,10 at the command area to specify the start-of-rotation axis.
10. Type w20,20,10 at the command area to specify the end-of-rotation axis.
11. Type 45 at the command area to specify the angle of rotation.
    The ellipsoid is rotated 45 degrees around an axis formed by the start and endpoints of the axis. (See Figure 2–32.)
12. Do not save your file.

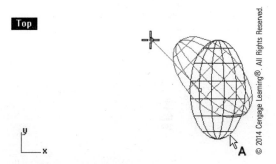

FIGURE 2–32 *Ellipsoid rotated in 3D*

> The direction of rotation follows a right-hand rule in which the thumb indicates the direction of the axis and the fingers indicate the direction of rotation. In other words, the direction of rotation will change to the opposite direction if the sequence of input of the axis endpoints is swapped.

Scaling Objects

To change the size and the overall proportion of a selected geometric object, you can scale it in five ways, using the Scale, Scale1D, Scale2D, ScaleByPlane, and ScaleNU commands.

Scale Uniformly in 3D

Perform the following steps.

1. Select File > Open and select the file Scale.3dm from the Chapter 2 folder on the Student Companion Website.
2. Select Transform > Scale > Scale 3-D.
3. Select box A, shown in Figure 2–33, and press the ENTER key.
4. If copy = no, click on it at the command area to set copy option to yes.
5. Type 0,0 at the command area to specify the origin point.
6. Type 0.5 at the command area to specify the scale factor.

   The box is scaled uniformly in 3D, and the original box is retained.

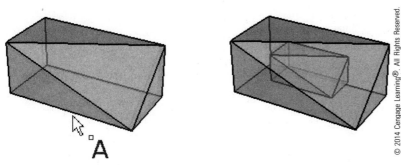

FIGURE 2–33 *Box being selected (left) and box copied and scaled uniformly in 3D (right)*

Scale in 2D

Scaling 2D changes the size of the geometric object in two directions only. Continue with the following steps.

7. Select Transform > Scale > Scale 2-D.
8. Select box A, shown in Figure 2–34, and press the ENTER key.
9. If copy = no, click on it at the command area to set copy option to yes.
10. Type 0,0 at the command area to specify the origin point.
11. Type 0.75 at the command area to specify the scale factor.

    The box is scaled in 2D, with the third axis unchanged.

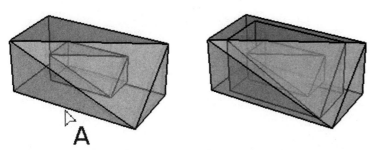

FIGURE 2–34 *Box being selected (left) and selected box copied and scaled in two directions (right)*
© 2014 Cengage Learning®. All Rights Reserved.

Scale in 1D

Scaling 1D stretches a geometric object in a specified direction. Continue with the following steps.

12. Select Transform > Scale > Scale 1-D.
13. Select box A, shown in Figure 2–35, and press the ENTER key.

14. If copy = no, click on it at the command area to set copy option to yes.
15. Type 0,0 at the command area to specify the origin point.
16. Type r20<0 to specify the first reference point.
17. Type r40<0 to specify the second reference point.

    The two reference points together define a direction and scale factor.
18. Press the ENTER key to terminate the command.

    The box is stretched in 1D.

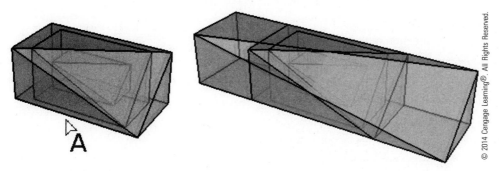

FIGURE 2–35 *Box being selected (left) and selected box copied and scaled in one direction (right)*

Scale Nonuniformly

To scale an object nonuniformly, continue with the following steps:

19. Select Transform > Scale > Non-Uniform Scale.
20. Select box A, shown in Figure 2–36, and press the ENTER key.
21. If copy = yes, click on it at the command area to set copy option to No.
22. If World coordinates = no, click on it at the command area to set World coordinates option to yes.
23. Type 0,0 at the command area to specify the origin point.
24. Type 0.3 at the command area to specify the X-axis scale.
25. Type 1.2 at the command area to specify the Y-axis scale.
26. Type 2 at the command area to specify the Z-axis scale.
27. Press the ENTER key to terminate the command.

    The box is scaled nonuniformly.

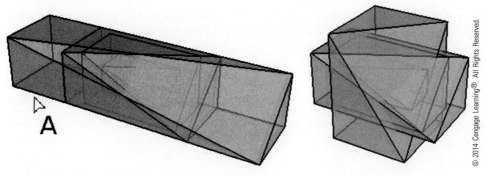

FIGURE 2–36 *Box being selected (left) and selected box scaled nonuniformly (right)*

Scale by Plane

To scale an object along a selected plane in two directions, continue with the following steps:

28. Type ScaleByPlane at the command area or click on the Scale By Plane button on the New in the V5 toolbar.
29. Select box A, shown in Figure 2–37, and press the ENTER key.
30. If copy = no, click on it at the command area to set copy option to yes.
31. Click on the Plane option at the command area to select a construction plane.
32. Click on 3point at the command area to define a construction plane.
33. Set Osnap to END so that the cursor will snap to endpoints of selected geometry.
34. Select endpoint A to define the origin.
35. Select endpoint B to define the direction of the X-axis.
36. Select endpoint C to define the XY plane orientation.
    A construction plane is defined.
37. Type 0,0 at the command area to specify the origin point.
38. Type 10,5 at the command area to specify the first reference point.
39. Type 20,5 at the command area to specify the second reference point.
40. Press the ENTER key to terminate the command.
    The box is scaled nonuniformly.
41. Do not save your file.

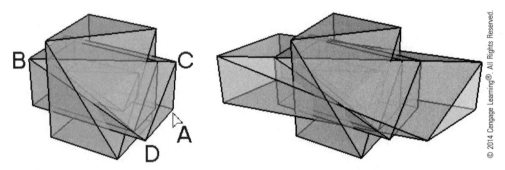

FIGURE 2–37 *Box being selected (left) and selected box scaled along a defined construction plane (right)*

Array

An array is a repetition of a selection of curves, surfaces, or polygon meshes. There are six types of arrays. The first type is the linear array, repeating objects in a single direction. The second type is the rectangular array, in which objects are repeated in X, Y, and Z directions. The third type is a polar array, in which objects are repeated around a center of array on the active construction plane. The fourth type of array repeats selected objects along a curve. The fifth type repeats selected objects along the U and V directions of a surface. The sixth type of array repeats objects along a curve residing on a surface.

Array Linear

Curves, surfaces, and polygon meshes can be repeated in a specific direction, forming a linear pattern. Perform the following steps.

1. Select File > Open and select the file ArrayLinear.3dm from the Chapter 10 folder on the Student Companion Website.
2. Select Transform > Array > Linear.
3. Select polygon mesh A, curve B, and surface C (shown in Figure 2–38) and press the ENTER key.
4. Type 0,0 at the command area to specify the first reference point.
5. Type 0,15 at the command area to specify the second reference point.
6. Do not save the file.

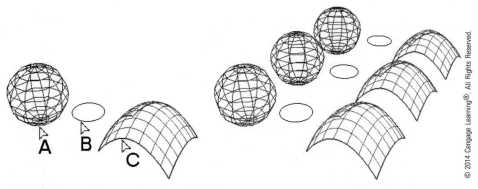

FIGURE 2–38 *Objects being selected (left) and selected objects arrayed (right)*

Rectangular Array

Curves, surfaces, and polygon meshes can be repeated in X, Y, and Z directions, forming a rectangular pattern. Perform the following steps.

1. Select File > Open and select the file ArrayRectangular.3dm from the Chapter 10 folder on the Student Companion Website.
2. Select Transform > Array > Rectangular.
3. Select polygon mesh A, curve B, and surface C (shown in Figure 2–39) and press the ENTER key.
4. Type 3 to specify the number of objects in the X direction.
5. Type 4 to specify the number of objects in the Y direction.
6. Type 2 to specify the number of objects in the Z direction.
7. Type 60 to specify the X spacing.
8. Type 40 to specify the Y spacing.
9. Type 50 to specify the Z spacing.

   Note that negative values for distances change the direction of an array. A rectangular array is constructed.

10. Do not save your file.

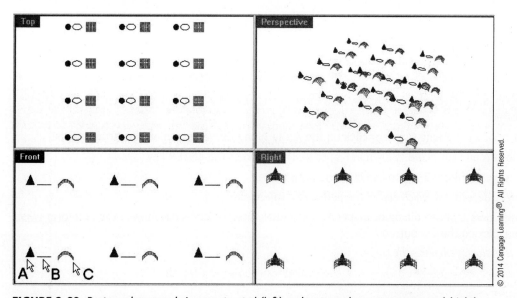

FIGURE 2–39 *Rectangular array being constructed (left) and rectangular array constructed (right)*

Polar Array

Curves, surfaces, and polygon meshes can be repeated in a circular pattern. You specify the center of the pattern, the number of repeated objects, and the angle to be filled by the repeated objects. Perform the following steps.

1. Select File > Open and select the file ArrayPolar.3dm from the Chapter 10 folder on the Student Companion Website.
2. Select Transform > Array > Polar.
3. Select polygon mesh A, curve B, and surface C (shown in Figure 2–40) and press the ENTER key.
4. Select point D (Figure 2–40) to specify the center of the array.
5. Type 4 to specify the number of elements in the array.
6. Type 90 (or press the ENTER key if the default is 360) to specify the angle to be filled.

    A polar array is constructed. Step angle is the angle between contiguous copies.

7. Do not save your file.

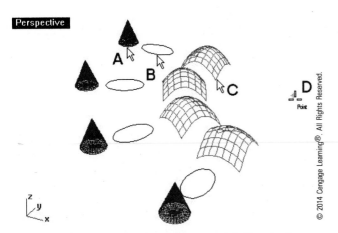

FIGURE 2–40 *Polar array being constructed (left) and polar array constructed (right)*

Array along a Curve

Apart from repeating objects in a rectangular or circular pattern, you can array selected objects along a path curve. You specify the path curve and the number of repeated objects or the distance between contiguous objects. Perform the following steps.

1. Select File > Open and select the file ArrayAlongCurve.3dm from the Chapter 10 folder on the Student Companion Website.
2. Select Transform > Array > Along Curve or click on the Array Along Curve button on the Array tool bar.
3. Select polygon mesh A and surface B (shown in Figure 2–41) and press the ENTER key.
4. Select free-form curve C (Figure 2–41) to use it as the path curve.

    The direction of array depends on which end of the path curve you select.

5. In the Array Along Curve Options dialog box, specify 3 in the Number of items box, select the Freeform twisting style, and then click on the OK button.

    The selected objects are arrayed along the curve.

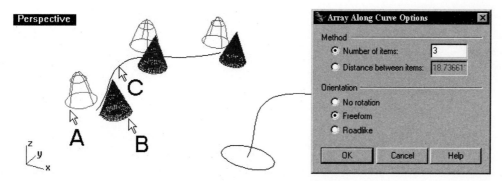

FIGURE 2–41 *Polygon mesh and surface being arrayed along a curve*
Source: Robert McNeel and Associates Rhinoceros® 5

6. Select Transform > Array > Along Curve or click on the Array Along Curve button on the Array tool bar.
7. Select curve A (shown in Figure 2–42) and press the ENTER key.
8. Select free-form curve B (shown in Figure 2–42) to use it as the path curve.
9. In the Array Along Curve Options dialog box, specify 8 in the Distance between items box, select the Roadlike style, and then click on the OK button.
10. Click on the Perspective viewport to specify a construction plane.
    The Perspective viewport has the same construction plane as the Top viewport. The selected objects are arrayed along the curve.
11. Do not save your file.

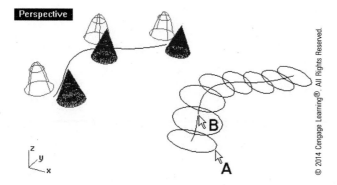

FIGURE 2–42 *Curve being arrayed along a curve*

Array along Surface's UV Directions
Objects can be arrayed along the U and V directions of a surface. Perform the following steps.

1. Select File > Open and select the file ArrayOnSurface.3dm from the Chapter 10 folder on the Student Companion Website.
2. Check the Cen box on the Osnap dialog box.
3. Select Transform > Array > Along Surface.
4. Select surface A and curve B, shown in Figure 2–43, and press the ENTER key.
5. Select center C (Figure 2–43) to specify the base point.
6. Press the ENTER key to use the CPlane's Z-axis as the reference normal.
7. Select surface D (Figure 2–43).
8. Type 3 to specify the number of elements in the U direction.
9. Type 4 to specify the number of elements in the V direction.
    An array is constructed.
10. Do not save your file.

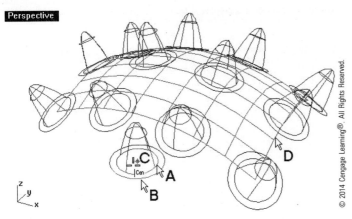

FIGURE 2–43 *Curve and surface being arrayed along the U and V directions of a surface*

Array along Curve on Surface

Apart from arraying along the U and V directions of a surface, you can also array objects on a curve residing on a surface. Perform the following steps.

1. Select File > Open and select the file ArrayCurveOnSurface.3dm from the Chapter 10 folder on the Student Companion Website.
2. Select Transform > Array > Along Curve on Surface.
3. Select surface A and curve B (shown in Figure 2–44) and press the ENTER key.
4. Select center point C (Figure 2–44) as the base point.
5. Select curve D. (See Figure 2–44.)
6. Select surface E. (See Figure 2–44.)
7. Select the Divide option at the command area.
8. Type 4 at the command area.
   The circle is arrayed along a curve on a surface.
9. Press the ENTER key to terminate the command.
10. Do not save your file.

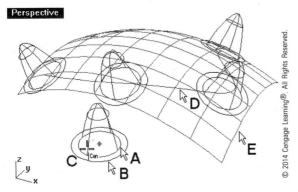

FIGURE 2–44 *Circle being arrayed along a curve on the surface*

Mirroring Objects

To mirror objects, you need to define a mirror plane, which can be specified using two methods. In the first method, the mirror plane is perpendicular to the active construction plane and you need to select only two points. In the second method, the mirror plane is 3D, and you need to specify three points to define the plane. To construct a mirror copy of selected objects, perform the following steps.

1. Select File > Open and select the file Mirror.3dm from the Chapter 2 folder on the Student Companion Website.
2. Click on the Osnap button on the Status bar to display the Osnap dialog box, if it is not already displayed.
3. Check the End box in the Osnap dialog box.
4. Select Transform > Mirror.
5. Select surface A, shown in Figure 2–45, and press the ENTER key.
6. Select points B and C, shown in Figure 2–45.

   The selected object is mirrored.

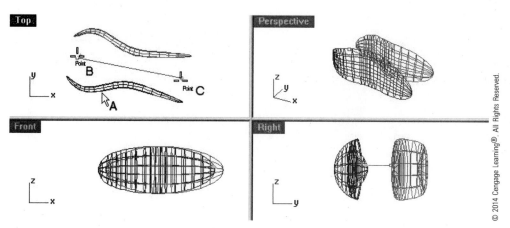

FIGURE 2–45 *Object being mirrored*

> **NOTE** If the Copy option is set to Yes, the source object is retained. Otherwise, it is deleted after being mirrored.

Undo and Redo

If you make any mistakes while working, you can use the undo command to revert to the previous state of your work. To reapply the undone commands, you use the redo command.

Number of Undos

Theoretically, you can perform undo as many times as you wish. However, there is a memory limitation because the system has to remember your work in order for the undo command to function properly. To set the minimum number of undos and the maximum memory used for storing undo information, you use the General tab of the Rhino Options dialog box.

When the memory allocated for storing undo information is used up, the undo information will be removed to hold new undo information.

Clearing Undo

To enable the redo command, recently undone work is stored in the memory. One way to free memory space is to clear the undo buffer using the ClearUndo command.

Named Position

If you are going to experiment on moving objects around and you may wish to revert objects to their original position, you may save the object's position in memory so that you can relocate them back to their original position. Perform the following steps:

1. Select File > Open and select the file AdvancedEdit.3dm from the Chapter 2 folder on the Student Companion Website.
2. Click on Named Positions button on the Move toolbar.
3. In the Named Positions pane of the Rhinoceros dialog box, click on the Save As button.
4. Click on the OK button in the Save Position dialog box.
5. Select A and B indicated in Figure 2–46 and press the ENTER key.

   Positions of the selected objects are stored in memory.

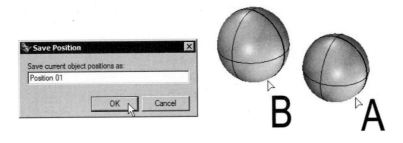
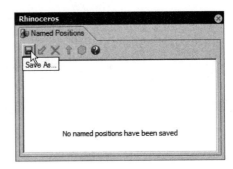

FIGURE 2–46 *Positions of objects being saved*
Source: Robert McNeel and Associates Rhinoceros® 5

Bounding Box Edit

A bounding box is the smallest rectangular box that fully encloses an object. With reference to an imaginary bounding box of a set of objects, you can collectively or individually change their size, scale them, relocate their position, and rotate them. Continue with the following steps:

6. Select Panels > Box Edit to display the Box Edit panel, if it is not already displayed.
7. Hold down the SHIFT key and select A and B indicated in Figure 2–47.
8. In the Box Edit dialog box, check the Transform objects individually and Show bounding box check boxes.

   Two individual bounding boxes are displayed, showing that the objects will be manipulated individually in terms of size, scale, position, and rotation.

9. Clear the Transform objects individually.

   The individual bounding boxes are merged into a single bounding box, indicating that objects will be manipulated collectively.

10. With reference to Figure 2–47, set the X value of the bounding box size to 100, Y scale to 1.2, Z position to 25, and X rotation to 45.
11. Click on the Apply button.

    The objects are resized, repositioned, and rotated.

12. In the Named Positions pane of the Rhinoceros dialog box, right-click and select Restore Named Position.

    The objects are relocated back to their original position as saved in memory.

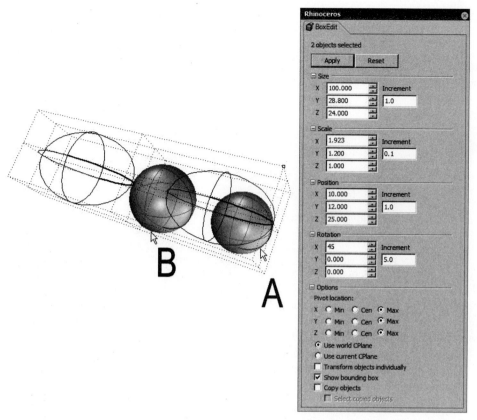

FIGURE 2–47 *From left to right: Objects being box edited, BoxEdit pane, and object box edited*
Source: Robert McNeel and Associates Rhinoceros® 5

Gumball

Gumball is a plug-in, providing an user-friendly transformation tool that enables us to translate, scale, and rotate selected objects in 3D. Continue with the following steps:

13. Turn on Gumball by clicking the Toggle Gumball button on the Gumball toolbar.
14. Hold down the SHIFT key and select A and B indicated in Figure 2–48.

    As you can see, there are three arrows, three small square boxes, and three arcs, depicting positions, scales, and rotations, which change their position as you select the objects, indicating that gumball is first applied to the first selected object and then applied to both selected objects.

15. Double-click on the arrow C and type 5 in the text box displayed to indicate a translation of 2 units in that direction.
16. Double-click on square box D and type 1.2 in the text box displayed to indicate a scale of 1.2 in that direction.
17. Double-click on arc E and type 25 in the text box to indicate a rotation of 25 degrees in that direction.

    Objects are moved, scaled, and rotated. Again, you can make use of the Name Position pane's saved position to restore the objects.

18. Do not save your file.

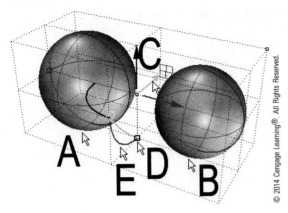

FIGURE 2–48 *Objects being moved, scaled, and rotated by using the gumball*

HISTORY MANAGER

History Manager is a modeling aid that helps keep track of the changes made to the source objects from which another object is constructed or derived. If the History Manager is turned on and an object is constructed from another object, the two objects will have a parent–child relationship. When you change the parent, the child will change accordingly. If it is not necessary to keep track of the changes, you may consider turning off the History Manager. If a file is saved with the History Manager turned on, history information will be saved as well. Because the History Manager remembers the relationship between parent and child objects, file size may become very large. Naturally, you may purge history to reduce file size.

To manage history, you can right-click on the Record History button of the status bar, as shown in Figure 2–49. The History Manager has four options: Record, Update, Lock, and Broken History Warning.

- To turn on the History Manager, click on Always Record History.
- To set the children to update automatically as the parent changes, click on Update Children.
- To lock the children so that they will not change even if the parent is modified, click on Lock Children.
- To display the Warning dialog box if history is broken due to an operation, click on History Break Warning.

FIGURE 2–49 *History Manager's options*
Source: Robert McNeel and Associates Rhinoceros® 5

To appreciate how the History Manager works, perform the following steps.

1. Select File > Open and select the file HistoryManager.3dm from the Chapter 2 folder on the Student Companion Website.
2. Click on the History Settings button on the History toolbar or type History at the command area.
3. If Record=No, select it to change it to Yes. Otherwise, proceed to the next step.
   History data will be recorded.
4. If Update=No, select it to change it to Yes. Otherwise, proceed to the next step.
   Any change made to a source object will cause an automatic update in the repeated objects.
5. Press the ENTER key to exit the command.
6. Select Transform > Copy.

7. Select curve A, surface B, and polygon mesh C, indicated in Figure 2–50, and press the ENTER key.
8. Set object snap mode to Point.
9. Select points D, E, and F and press the ENTER key.

   Point D is the point to copy from, and points E and F are the points to copy to. The curve, surface, and polygon mesh are copied.

To appreciate how the History Manager works, continue with the following steps.

10. Select Transform > Scale > Scale 3-D.
11. Select curve A, surface B, and polygon mesh C, indicated in Figure 2–50, and press the ENTER key.
12. Select point D, shown in Figure 2–50, to specify the base point.
13. If copy = yes, click on it at the command area to set it to No.
14. Type 0.5 at the command area to specify the scale factor.

    Because history update is turned on, the copied objects are also scaled.

15. Click on the History Settings button on the History toolbar or type History at the command area.
16. Select the Update option at the command area to change it to No.
17. Select Transform > Scale > Scale 3-D.
18. Select curve A, surface B, and polygon mesh C, shown in Figure 2–50, and press the ENTER key.
19. Select point D, shown in Figure 2–50, to specify the base point.
20. Type 2 at the command area to specify the scale factor.

    Because history update is recorded and turned off, the copied objects do not change at all.

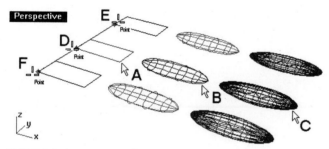

FIGURE 2–50 *Curve, surface, and polygon mesh being copied*
© 2014 Cengage Learning®. All Rights Reserved.

21. To discover the update situation, type HistoryReport at the command area.
22. Select object A, shown in Figure 2–51.

    The History Report is displayed.

23. Do not save your file.

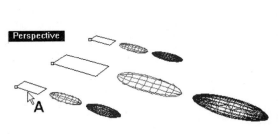
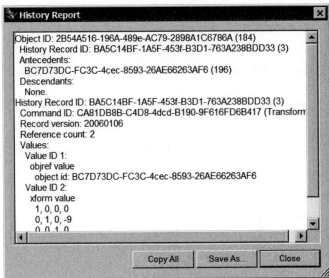

FIGURE 2-51 *History report*
Source: Robert McNeel and Associates Rhinoceros® 5

Making Symmetric Objects

You can construct symmetric curve elements by specifying a symmetry plane, and by turning History Manager beforehand, histories about symmetric and copied objects are kept in memory during the working session. Perform the following steps:

1. Select File > Open and select the file Symmetry.3dm from the Chapter 10 folder on the Student Companion Website.
2. Select Transform > Symmetry.
3. Select curve A, shown in Figure 2-52.
4. Select endpoints B and C.

   A symmetric curve is constructed.

5. Repeat the command.
6. Select surface D at edge D.
7. Select endpoints E and F.

   A symmetric surface is constructed.

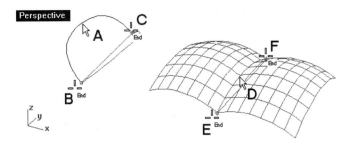

FIGURE 2-52 *Symmetric objects being constructed*
© 2014 Cengage Learning®. All Rights Reserved.

Updating History

To appreciate how History Manager helps maintain a relationship between the original objects and the symmetric objects, continue with the following steps.

8. Select Edit > Control Points > Control Points on and select curve A and surface B, shown in Figure 2-53.
9. Select Transform > Move.

10. Select control points C and D.
11. Click anywhere on the graphics area.
12. Type r0,0,10 at the command area.

    The selected control points are moved a distance of 10 units in the Z direction. Note that the symmetric curve and surface also change in shape.

13. Do not save your file.

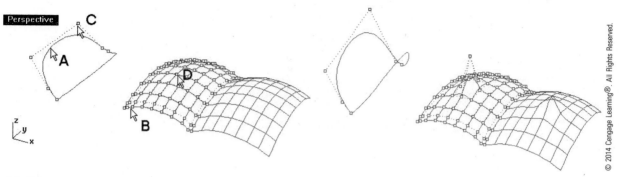

FIGURE 2–53 *Control points turned on and selected (left) and symmetric objects change together (right)*

It must be noted that history is retained only during the current working session. In other words, the next time you open a file, the history will be lost.

LAYER MANIPULATION

The term "layer" originates from manual drafting. It refers to overlay of clear transparent sheets. Layering is a management mechanism in which sets of objects are drawn on different transparent sheets. By removing or overlaying the sheets, you control which set or sets of objects are shown on a drawing.

Managing Objects in Layers

In computer-aided design, layers are not physical sheets; they are conceptual layers. You construct layers in a file and place objects on different layers. By turning layers on or off, you control the display of the objects on the screen. You can also lock a layer so that objects placed on the layer can be seen and snapped to but cannot be selected or manipulated (such as moving or erasing). In a multilayer Rhino file setup, you can move objects from one layer to another. To try editing a layer, perform the following steps.

1. Select File > Open and select the file Layer.3dm from the Chapter 2 folder on the Student Companion Website.
2. Click on the layer pane in the Rhinoceros dialog box. If the Layer pane is not there, select Edit > Layers > Edit Layers, or click on the Edit Layers button on the Standard toolbar.

In the Layers dialog box, shown in Figure 2–54, you can add new layers, delete existing layers, set visibility of layers, lock objects on layers, and define the color and material properties of objects placed on a layer as desired.

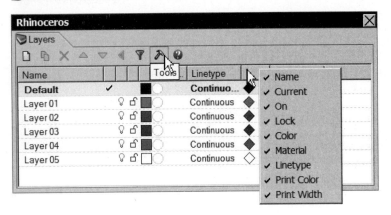

FIGURE 2–54 *Layers dialog box*
Source: Robert McNeel and Associates Rhinoceros® 5

Layers Dialog Box Column Title

The Layers dialog box has a number of columns. You may right-click over the columns to display a list of column titles. Those with checks are visible. Continue with the following steps to hide some columns.

3. Right-click over the columns of the Layers dialog box.

    A list of column titles is displayed. By clearing the check mark of a title, the column is hidden.

Setting Color

Color setting can be done in three ways in the dialog box as shown in Figure 2–55.

- The first way is to select a color from the pre-established color list.
- The second way is to use the HSB (hue, saturation, and brightness) color system.
- The third way is to use the RGB (red, green, blue) color system.

In a collaborative design environment in which a project is handled by a team, specifying color values (HSB or RGB) can accurately describe the color.

HSB Color System

In the Select Color dialog box, the HSB settings are labeled Hue, Sat, and Val. The hue represents a color ranging from red through yellow, green, and blue. Saturation describes the intensity of the hue. Brightness concerns the color's value or luminance. To select a color by using the HSB color system, you first select a color from the hue color range; you then set the saturation (intensity) and the brightness (luminance).

RGB Color System

Specifying the RGB value instructs the computer to project a color mix of red, green, and blue to each individual pixel (picture element) of the monitor.

Continue with the following steps.

4. Select a layer and then click on the Color column of the Layers dialog box.
5. In the Select Layer Color dialog box, specify a color by selecting a color from the color swatch or by inputting hue, saturation, and brightness (RGB, or red, green, and blue) values in these respective fields. (See Figure 2–55.)

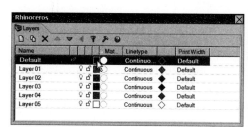
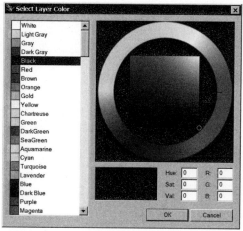

FIGURE 2–55 *Setting a layer's color*
Source: Robert McNeel and Associates Rhinoceros® 5

Setting Layer's Material Assignment

Continue with the following steps to set a layer's material.

6. Specify the material of the object (its visual appearance as rendered) by selecting the Material column of the Layers dialog box.

   This accesses the Layer Material dialog box, shown in Figure 2–56. You can simply change the color in the basic settings area to specify the rendering color. You will learn more about material assignment in Chapter 14.

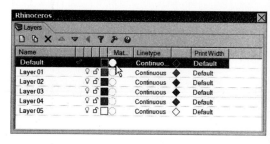
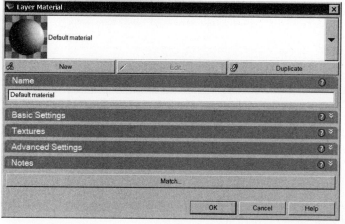

FIGURE 2–56 *Setting a layer's material*
Source: Robert McNeel and Associates Rhinoceros® 5

Changing an Object's Layer

Changing an object's layer refers to moving an object from one layer to another layer. Continue with the following steps.

7. Click on the X button at the upper-right corner of the Layers dialog box to close it.

> To display it again in the Rhinoceros dialog box, select Panels > Layers.   **NOTE**

8. Select Edit > Layers > Change Object Layer.
9. Select sphere A, shown in Figure 2–57, and press the ENTER key.
10. In the Layer for Objects dialog box, click on Layer 02 and the OK button.

    The selected object is moved to Layer 02.

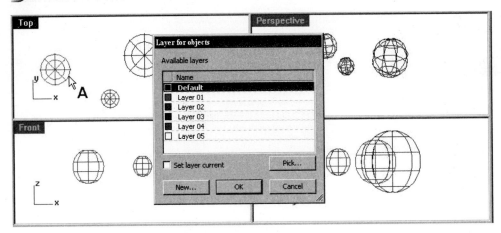

FIGURE 2–57 *Layer for Objects dialog box*
Source: Robert McNeel and Associates Rhinoceros® 5

Copying Selected Objects from One Layer to Another Layer

You can copy selected objects from one layer to another layer, as follows:

11. Select Edit > Layers > Copy Objects to Layer.
12. Select sphere A, shown in Figure 2–58, and press the ENTER key.
13. In the Layer to Copy Objects dialog box, select Layer 01 and click on the OK button.

    The selected object is copied to a new layer.

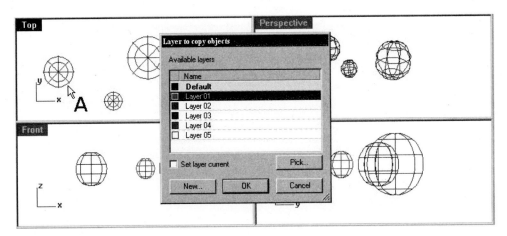

FIGURE 2–58 *Layer to Copy Objects dialog box*
Source: Robert McNeel and Associates Rhinoceros® 5

Duplicating Objects from One Layer to Another Layer

Instead of copying individual selected objects from one layer to another layer, you can copy all the objects residing in one layer to another layer. Continue with the following steps.

14. Select Edit > Layers > Duplicate Layer.
15. In the Duplicate Layer dialog box, select Layer 01, type Layer 06 in the Name of new layer to create box, and click on the OK button.

    All the objects in Layer 01 are copied to a new layer, as shown in Figure 2–59.

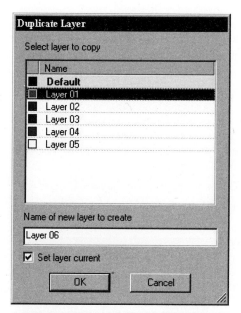

FIGURE 2–59 *Duplicate Layer dialog box*
Source: Robert McNeel and Associates Rhinoceros® 5

Ways to Turn On/Off Layers

To turn off all layers except one, continue with the following steps.

16. Select Edit > Layers > One Layer On.
17. In the Layer to Leave on dialog box, select Layer 06 and click on the OK button.
    All layers, except Layer 06, are turned off, as shown in Figure 2–60.

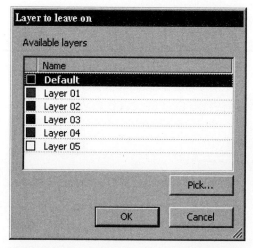

FIGURE 2–60 *Layer to Leave on dialog box*
Source: Robert McNeel and Associates Rhinoceros® 5

To turn on all layers, continue with the following steps.

18. Select Edit > Layers > All Layers On.
    All the layers are turned on. To turn off a layer on which an object is residing, continue with the following steps.
19. Select Edit > Layers > One Layer Off.
20. Select an object (any object).
    The object's layer is turned off. Naturally, the selected object becomes invisible.
21. Do not save your file.

Layer Group

You can put one or more layers under another selected layer to form a layer group. Layers that are grouped under another layer are called sublayers. Apart from being manipulated in a normal way individually, a sublayer can also be manipulated collectively by manipulating the layer that governs the sublayer. Perform the following steps.

1. Select File > Open and select the file LayerGroup.3dm from the Chapter 2 folder on the Student Companion Website.
2. Select Edit > Layers > Edit Layers.
3. Hold down the CONTROL key, select Layer 02 and Layer 03 from the Layers dialog box, and drag the layers to Layer 01.
4. Release the mouse button. The selected layers (Layer 02 and Layer 03) are grouped under Layer 01.

   Layer 02 and Layer 03 now become the sublayers of Layer 01. (See Figure 2–61.)

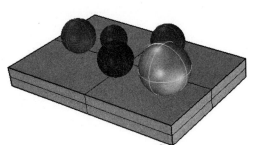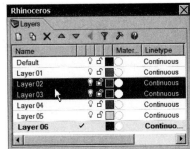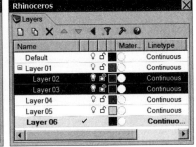

FIGURE 2–61 *From left to right: models, selecting layers in the Layers dialog box, and layers grouped*
Source: Robert McNeel and Associates Rhinoceros® 5

To appreciate how sublayers can be manipulated collectively, continue with the following steps:

5. Turn off Layer 01.

   Layer 02 and Layer 03, being grouped under Layer 01, are also turned off. (See Figure 2–62.)

6. Turn on Layer 01.

   Layer 02 and Layer 03 are also turned on.

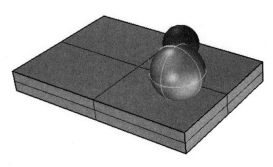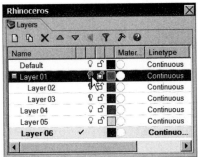

FIGURE 2–62 *Models (left) and layer group turned off (right)*
Source: Robert McNeel and Associates Rhinoceros® 5

Sublayers can be ungrouped, as follows:

7. Drag Layer 02 from its location to a location below Layer 06.

   Layer 02 is ungrouped. (See Figure 2–63.)

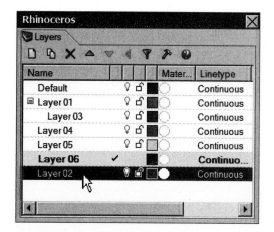

FIGURE 2–63 *Layer being ungrouped*
Source: Robert McNeel and Associates Rhinoceros® 5

8. Select Layer 02, right-click, and select New Sublayer.

   A new sublayer is constructed. (See Figure 2–64.)

9. Do not save your file.

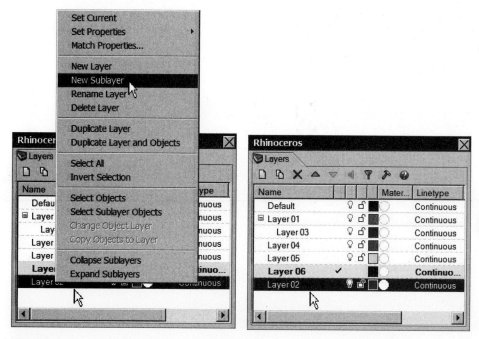

FIGURE 2–64 *New sublayer being constructed (left) and sublayer constructed (right)*
Source: Robert McNeel and Associates Rhinoceros® 5

Visibility and Locking

Other than turning off a layer to make objects residing on the layer invisible, you can hide selected objects so that they are not displayed, even if the layer on which they reside is still turned on. You can unhide a hidden object to make it visible again. You can also lock an object, whereby it is visible but cannot be modified. Perform the following steps.

1. Select File > Open and select the file Visibility.3dm from the Chapter 2 folder on the Student Companion Website.
2. Select Edit > Visibility > Hide.
3. Select objects A, B, C, D, and E, indicated in Figure 2–65, and press the ENTER key.

   The selected objects are hidden.

4. Select Edit > Visibility > Show Selected.

   In the display, the visible objects are hidden temporarily and the hidden objects are shown.

5. Select hidden objects A and B and press the ENTER key.

   The selected hidden objects are unhidden.

6. Select Edit > Visibility > Swap Hidden and Visible.

   The visible objects are hidden, and the hidden objects are unhidden.

7. Select Edit > Visibility > Show.

   All the hidden objects are displayed.

8. Select Edit > Visibility > Lock.

9. Select objects A, B, and C, shown in Figure 2–65, and press the ENTER key.

   The objects are locked. A locked object is still displayed but cannot be manipulated.

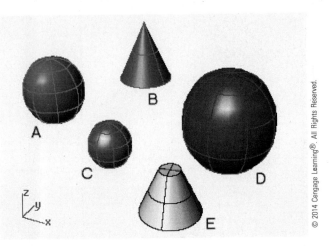

FIGURE 2–65 *Visibility*

Snapping Objects Residing on a Locked Layer

Objects residing on a locked layer can be set to be snappable, depending on the setting defined by the SnapToLocked command. To make appropriate setting, perform the following step:

10. Select Tools > Object Snap > Snap to Locked Objects and select one of the options: Enable, disable, and Toggle.

Layer State Manager

The manner in which one layer is turned on/off, made visible/invisible, and locked/unlocked can be managed by using the Layer State Manager, as follows:

1. Select File > Open and select the file LayerState.3dm from the Chapter 2 folder on the Student Companion Website.
2. Select Edit > Layers > Layer State Manager.
3. Click on the Save button on the Layer State Manager dialog box.
4. In the Save Layer State dialog box, click on the OK button.

   The current layer state is saved. Note that you may specify another layer state name.

5. Click on the Close button of the Layer State Manager to close it.
6. Turn off Layer 02, Layer 03, and Layer 04.
7. Repeat steps 3 through 5 to save the current layer state.
8. Select Layer State 01 in the Save Layer State list.

9. Click on the Restore button on the Layer State Manager dialog box.
   The saved layer state is restored. (See Figure 2–66.)
10. Do not save your file.

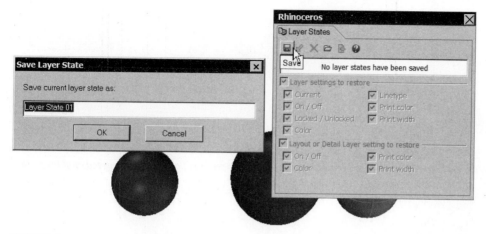

FIGURE 2–66 *Layer State Manager*
Source: Robert McNeel and Associates Rhinoceros® 5

OBJECT PROPERTIES

This section introduces how to manage object's properties by manipulating the parameters in the Properties pane of the Rhinoceros dialog box.

Viewport Properties

The Properties pane of the Rhinoceros dialog box is context sensitive, with information displayed dependent on the type of objects selected. If nothing is selected, it displays the information about the active viewport as explained in Chapter 1.

Light Object Properties

To appreciate how to manipulate object's properties via the Properties dialog box, perform the following steps:

1. Select File > Open and select the file Object Properties.3dm from the Chapter 2 folder on the Student Companion Website.
2. If the Properties pane of the Rhinoceros dialog box is not displayed, select Edit > Object Properties.
3. Click on the Properties pane on the Rhinoceros dialog box.
4. If nothing is selected, what you see in the pane are the properties about the current viewport.
5. Select light A as shown in Figure 2–67.
   Properties about the light are displayed.
6. Click on the Details button in the Properties pane, and details regarding the selected light are displayed in an Object Description dialog box.

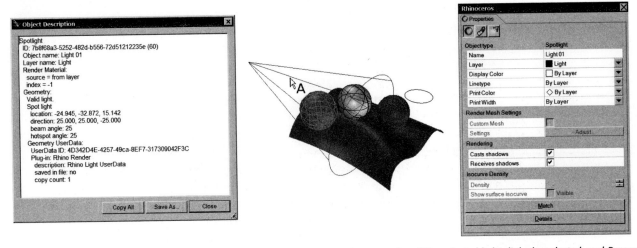

FIGURE 2–67 *From left to right: Object Properties dialog box showing the properties of the selected light, light be selected, and Properties panel*
Source: Robert McNeel and Associates Rhinoceros® 5

Geometric Object Properties

As shown in Figure 2–67, the Properties pane has three buttons, Object, Material, and Texture Mapping, enabling you to manage the selected object's basic properties (Object), material properties (Material), and rendering texture properties (Texture Mapping). As lighting, material properties, and texture mapping will be discussed in more detail in Chapter 14, now continue with the following step:

7. Close the Object Description dialog box and select circle B shown in Figure 2–68.

    Properties about the circle are displayed.

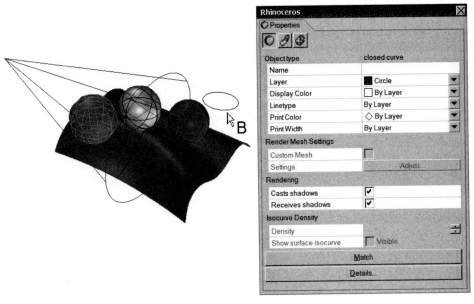

FIGURE 2–68 *Circle selected (left) and its properties displayed (right)*
Source: Robert McNeel and Associates Rhinoceros® 5

Properties Matching

To appreciate how to change an object's properties in accordance with another object's properties, continue with the following steps:

8. Select mesh C as shown in Figure 2–69 to display its properties in the Properties pane.
9. Click on the Match button of the Properties pane of the Rhinoceros dialog box.

10. Click on the OK button of the Match dialog box.
11. Select sphere D.

    The object C's properties are changed, matching those of object D.

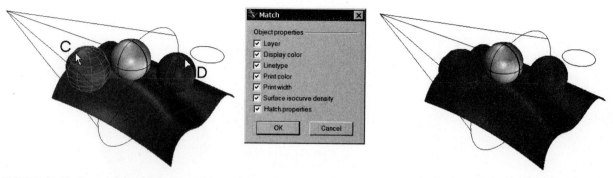

FIGURE 2–69 *From left to right: objects being selected, Match dialog box, and selected objects' properties matched*
Source: Robert McNeel and Associates Rhinoceros® 5

Isocurve Display for NURBS Surfaces

In reality, there are no curves or lines on a 3D free-form smooth surface. Therefore, you should find only edges on the surface's boundaries. However, boundary edges alone do not provide sufficient information to depict the profile and silhouette of the surface. Hence, a set of isocurves in two orthogonal directions, color shading, or a set of isocurves together with color shading is used to better illustrate a free-form object in the computer display. To avoid confusion with the X and Y axes of the coordinate system, the isocurve directions are called U and V. Figure 2–70 shows how isocurves are used to help visualize a surface.

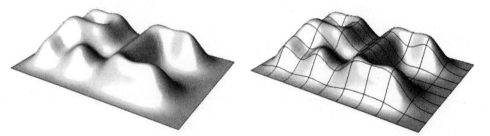

FIGURE 2–70 *A free-form smooth surface (left) and the same surface with U and V isocurves (right)*
© 2014 Cengage Learning®. All Rights Reserved.

> To select a surface, you select one of the isocurves or the boundary of the surface. Although the isocurves are not physical curves, you may still use object snap tools, such as intersection, to locate the intersections of U and V isocurves. You can set isocurve density by selecting Edit > Properties and changing the density value in the Properties dialog box. Isocurves are also known as isoparametric curves.

Isocurve Density

For a very simple surface, such as a planar surface, one or two isocurves are adequate to provide enough information on the curvature of the surface. For more complex surfaces, you need more isocurves. Although there are more isocurves to better represent the profile of the surface, selection of individual objects from a bunch of objects with high isocurve density may become difficult. Perform the following steps.

1. Select File > Open and select the file Isocurves.3dm from the Chapter 2 folder on the Student Companion Website.
2. Select Edit > Object Properties or click on the Object pane on the Rhinoceros dialog box.

3. Select surface A, shown in Figure 2–71, and press the ENTER key.
4. In the Properties dialog box, change the isocurve density to 10.

   The isocurve density of the surface is changed, as shown in Figure 2–71 (right).

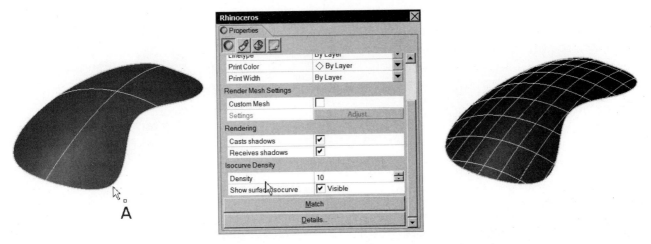

FIGURE 2–71 *From left to right: Original surface, Properties dialog box, and isocurve density changed*
Source: Robert McNeel and Associates Rhinoceros® 5

CONSTRUCTION PLANE MANIPULATION

In each viewport, there is, by default, a construction plane. If we use the pointing device to select a location in one of the viewports, we select a point on the viewport's default construction plane. In addition to selecting points, some commands, such as point and curve projection and object trimming, are construction-plane dependent. They project in a direction perpendicular to the current construction plane. Therefore, the ability to manipulate construction planes helps in constructing and manipulating objects. Construction planes can be categorized as native construction planes, universal construction planes, and mobile construction planes.

Native Construction Plane

As explained in Chapter 2, a construction plane is defined in each viewport. In a four-viewport configuration, for example, the Top and Perspective viewports have a construction plane corresponding to the X-Y plane of the World coordinates, the Front viewport has a construction plane corresponding to the Z-X plane of the World coordinates, and the Right viewport has a construction plane corresponding to the Z-Y plane of the World coordinates. Before elaborating on how construction planes can be manipulated, it must be reiterated that construction planes are not associated with the viewing direction of the viewport, although they seem to be. In other words, the construction plane in the Top viewport remains lying on the World X-Y plane, even after the view is rotated. In previous releases, Rhino provided only tools to manipulate construction planes in each individual viewport. With the advent of two additional types of construction planes, these original planes are referred to as native construction planes for easy reference.

Options of Native Construction Plane Command

The native construction command, CPlane, has twelve options: All, Curve, Elevation, Next, Object, Previous, Rotate, Surface, Through, View, World, and 3Point. You can type CPlane at the command area and then select an option, or click on a button from the toolbar, shown in Figure 2–72.

FIGURE 2–72 *Options of the CPlane command*
Source: Robert McNeel and Associates Rhinoceros® 5

All. When working with the construction plane command, if there are several viewports displayed, the active viewport when the command begins is the one that gets changed. If the All option is set to Yes, all other viewport origins will be changed.

1. Select File > Open and select the file CPlane.3dm from the Chapter 2 folder on the Student Companion Website.
2. Select View > Set CPlane > Origin.
3. If All=No, select the All option from the command area to change it to Yes and press the ENTER key to accept the default origin location. Otherwise, proceed to the next step.

Curve. The Curve option sets the construction plane perpendicular to a selected curve, as follows:

4. Repeat the command and select the Curve option.
5. Set Osnap to Mid and select curve A. (See Figure 2–73.)
6. Click on midpoint B.

   The construction plane of the Perspective viewport is changed.

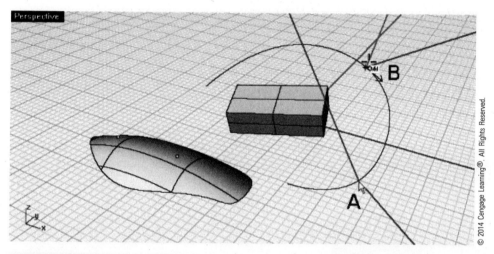

FIGURE 2–73 *A curve and its midpoint being selected to set the construction plane*

Previous and Next. If you want to revert the construction plane to its previous location, you cannot use the Undo command. Instead, you should use the Previous option. You can redo the construction plane change by using the Next option. Continue with the following steps.

7. Repeat the command and select the Previous option.

   The construction plane of the Perspective viewport is reverted, in a way similar to Undo.
8. Repeat the command and select the Next option.

   The Previous option is undone.

Object. The Object option sets the construction plane to the coordinates of a selected object, as follows:

9. Repeat the command and select the Object option.
10. Select object A, shown in Figure 2–74.

    The construction plane is set to the orientation of the selected object.

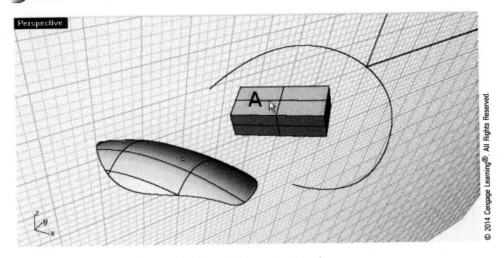

FIGURE 2–74 *An object being selected to set the construction plane*

Rotate. You can rotate a construction plane by using the Rotate option, as follows:

11. Repeat the command and select the Rotate option.
12. Select endpoint A, shown in Figure 2–75, to specify the start of the rotation axis.
13. Select endpoint B to specify the end of the rotation axis.
14. Type 45 at the command area to specify a rotation angle.

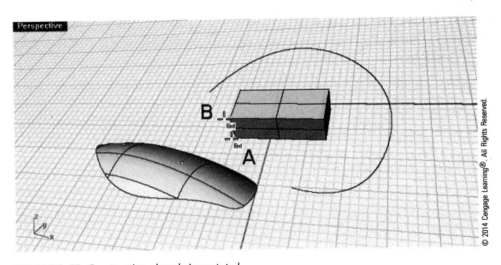

FIGURE 2–75 *Construction plane being rotated*

Surface. The Surface option sets the construction onto a selected surface, as follows:

15. Repeat the command and select the Surface option.
16. Set Osnap to Point.
17. Select surface A. (See Figure 2–76.)
18. Select point B to specify the origin.
19. Select point C to specify the X-axis direction.

    The construction plane is set to the surface.

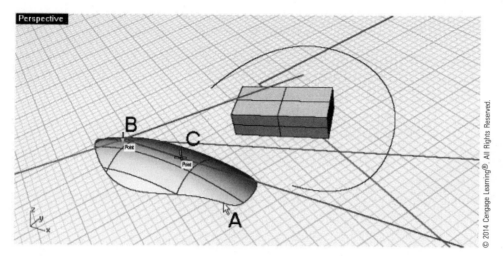

FIGURE 2–76 *Construction plane being set to the surface*

Elevation. The Elevation option changes the location of the construction in the Z direction, as follows:

20. Repeat the command and select the Elevation option.
21. Type 10 at the command area.
    The elevation of the construction plane is changed.

Through. The Through option also changes the elevation of the construction plane. It sets the construction plane to pass through a selected point, as follows:

22. Repeat the command and select the Through option.
23. Select endpoint A. (See Figure 2–77.)
    The construction plane is set.

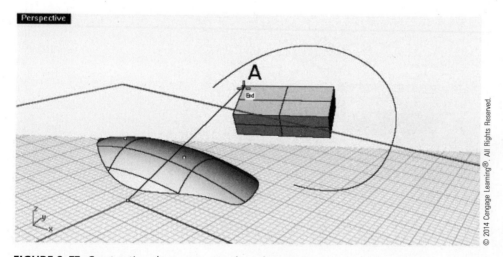

FIGURE 2–77 *Construction plane set to pass through a point*

View. You can use the View option to set the construction plane parallel to the viewport's viewing direction, as follows:

24. Repeat the command and select the View option.
    The construction plane is set. (See Figure 2–78.)

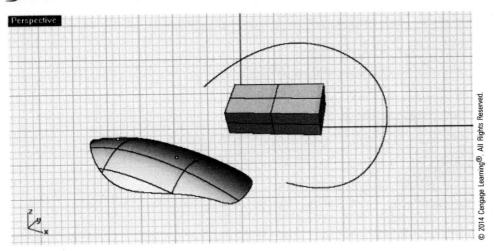

FIGURE 2–78 *Construction plane set to parallel to the viewport*

World. The World option sets the construction to one of the World's default orientations, as follows:

25. Repeat the command and select the World option.
26. Select the Top option.

    The construction plane is set to the default top orientation.

3Point. You can use three points to orient the construction plane. The first point specifies the origin, the second point specifies the X-axis direction, and the third point specifies a point on the X-Y plane, as follows:

27. Repeat the command and select the 3Point option.
28. Select endpoints A, B, and C, shown in Figure 2–79.

    The construction plane is set.

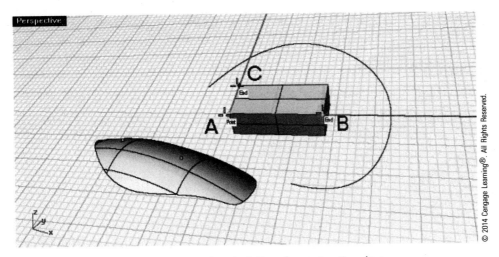

FIGURE 2–79 *Using three points to set the orientation of a construction plane*

Named Construction Plane. You can save, restore, delete, rename, and import construction plane configurations, as follows:

29. Select View > Set CPlane > Named CPlanes.
30. In the Named CPlanes dialog box, shown in Figure 2–80, click on the Import button.

31. In the Import dialog box, select the file CPlaneName.3dm from the Appendix A folder on the Student Companion Website.

    The saved construction plane configurations are imported.

32. Select 3point-a from the list and click on the Restore button.

    The construction plane is set.

33. Do not save your file.

FIGURE 2–80 *Named CPlane dialog box*
Source: Robert McNeel and Associates Rhinoceros® 5

Universal Construction Plane

The universal construction plane command provides similar functionality to the native construction plane command. In addition to changing the construction plane in the active viewport, the other viewports' construction planes will be updated to the standard top, front, and right views of the modified construction plane. To activate universal construction plane, right-click on the Set Uplane mode/Set Cplane mode button of the CPlanes toolbar, shown in Figure 2–81. Alternatively, you can select Tools > Options to open the Rhino Options dialog box, select the Modeling Aids tab, and check the Universal construction planes check box. Now perform the following steps.

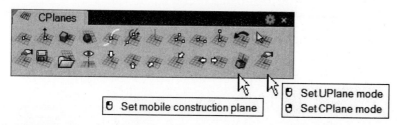

FIGURE 2–81 *CPlanes toolbar*
Source: Robert McNeel and Associates Rhinoceros® 5

1. Select File > Open and select the file UPlane.3dm from the Appendix A folder on the Student Companion Website.
2. Right-click on the Set Uplane mode/Set Cplane mode button on the CPlanes toolbar.
3. Select View > Set CPlane > 3 Points.
4. Select endpoints A, B, and C, shown in Figure 2–82.

   The construction plane of the Perspective viewport is changed, and the other viewports' viewing direction and construction planes are changed as well. (See Figure 2–83.)

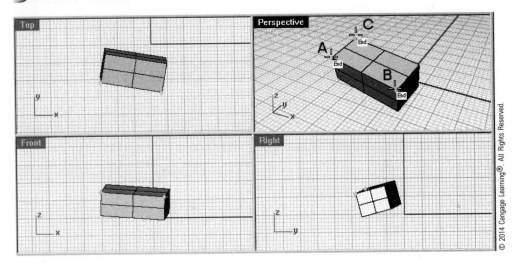

FIGURE 2–82 *Setting universal construction plane*

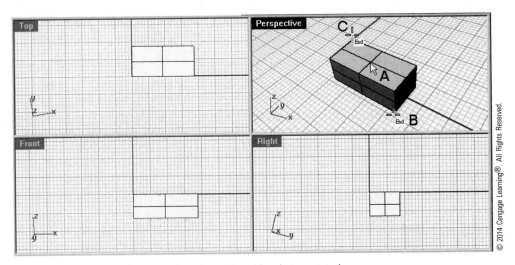

FIGURE 2–83 *Universal construction plane set and box being rotated*

Mobile Construction Plane

To attach a construction plane to an object and maintain the relationship between the object and the construction plane, you set up a mobile construction plane. Continue with the following steps.

5. Select Transform > Rotate 3D.
6. Select A, shown in Figure 2–83, and press the ENTER key.
7. Select endpoints B and C.
8. Type 15 at the command area.

    The box is rotated. Note that the construction plane configurations in the viewports do not change. (See Figure 2–84.)

9. Select View > Set CPlane > Mobile Construction Plane.
10. Select A, shown in Figure 2–84.
11. Select the Attach option.
12. Select endpoints B, C, and D.

13. If Automatic=No, select the Automatic option to change it to Yes. Otherwise, proceed to the next step.
14. Press the ENTER key.

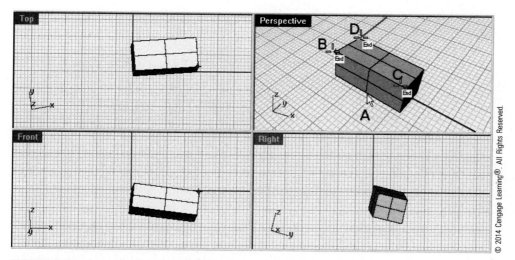

FIGURE 2–84 *Box rotated and mobile construction plane being constructed*

A universal construction plane is set and attached. (See Figure 2–85.)
15. Select Transform > Rotate 3D.
16. Select A, shown in Figure 2–85, and press the ENTER key.
17. Select points B and C.
18. Type 20 at the command area.

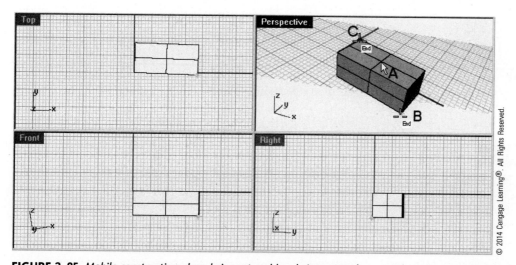

FIGURE 2–85 *Mobile construction plane being set and box being rotated*

The object is rotated, and the construction plane changes. (See Figure 2–86.)
19. Do not save your file.

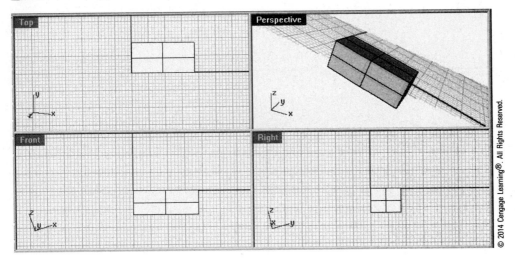

FIGURE 2–86 *Box rotated and construction planes changed as well*

CONSOLIDATION

When you use the mouse to select a point in one of the default four viewports in the graphics area, you select a point on an imaginary plane, called the construction plane, corresponding to the selected viewport. In addition to these three default construction planes, you can set up construction planes in various ways.

To input a point at the command area, you use either the construction plane coordinate system or the World coordinate system. To help construct objects in the graphics area, you can use various drawing aids. In particular, the Smart Tracking tool is very useful in terms of tracking geometric objects.

To manipulate objects that are already constructed, you have to select them by using various selection tools. Basic object manipulation methods include moving, copying, rotating, scaling, arraying, and mirroring. (More advanced manipulation tools will be discussed in Chapter 10.) To help track changes made to objects, you can turn on the History Manager.

As a surface model becomes more complicated, there will be many geometric objects. To organize these objects, you put them into layers. Objects you construct are placed on the current layer.

A smooth surface does not have any thickness or profile curves on it. In essence, you see only its boundaries and its silhouette. To enhance visualization, isocurves are placed on the surface.

REVIEW QUESTIONS

1. Explain the concepts of the construction plane.
2. Distinguish between native construction plane, universal construction plane, and mobile construction plane.
3. Describe the ways to input precise coordinate systems.
4. List the drawing aids available.
5. Give a brief account of the History Manager.
6. Outline how objects can be organized by using layers.
7. What is meant by isocurves?
8. In what ways can a surface be displayed?

CHAPTER 3

Rhinoceros NURBS Surface Modeling

INTRODUCTION

This chapter outlines the concepts of surface modeling and surface and curve continuity. In addition, this chapter divides Rhino's surfaces into four categories and delineates how these surfaces are constructed.

OBJECTIVES

After studying this chapter, you should be able to

- Think critically on how to use surface models to represent 3D objects
- Comment on various degrees of surface and curve continuity in modeling
- Plan your design by incorporating four major categories of Rhinoceros surfaces

OVERVIEW

Surface modeling involves a top-down thinking process to consider how an object can be decomposed into a number of surfaces and how these surfaces are constructed individually. After completing this analysis, you proceed to the bottom-up process of making the individual surfaces and eventually composing a model from the set of surfaces produced. To be able to start the top-down thinking process and subsequently perform the bottom-up process of building the surface model, you need to know what Rhino surfaces are available and how they can be built. To help you gain a thorough understanding in a logical way, we will divide Rhino's surface modeling commands into four main categories and discuss each one. After you gain more knowledge on curves and points construction in Chapters 4 through 6, you will further enhance your NURBS surface modeling skill later in Chapters 7, 8, and 10.

SURFACE MODELING CONCEPTS

In our daily lives, we encounter many objects of free-form shape. Some examples are the handle of a razor, the blade of an electric fan, the casing of a computer pointing device (mouse), the casing of a mobile phone, the handle of a joystick, and the body panels of an automobile. To help visualize and enhance the design of these products, you represent them in the computer using surface modeling tools. To reiterate, surface modeling is a means of representing the geometric shape of a 3D object in the computer by using a set of surfaces put together to resemble the boundary faces of the object. Each surface is a mathematical expression that represents a 3D shape with no thickness.

Design Needs

Free-form surfaces are used in objects to meet two basic design needs: aesthetic and functional.

Aesthetically, an object has to be eye-pleasing and eye-catching to attract customers. Hence, various types of free-form shapes are used in many consumer products.

Functionally, a surface needs to comply with a certain form and shape to serve specific purposes. For example, the blade of an electric fan should conform to aerodynamic requirements, the profile and silhouette of a shoe should match the shape of a human foot, and the joystick handle has to match the human hand.

SURFACE MODELING APPROACHES

Because a surface model represents an object by using a set of surfaces, surface modeling involves the construction of a set of surfaces to depict the skin of the model. Naturally, the process of making a surface model involves a top-down thinking process to decompose a model into a set of discrete surfaces and a bottom-up construction process to make the surfaces in accordance with the appearance of the object.

Top-Down Thinking

Before making the individual surfaces of a surface model, you need to go through two thinking processes. You think about what individual surfaces are required to compose the model and then about how the individual surfaces are constructed. To be capable of analyzing what surfaces are needed and how the surfaces are made, you need to acquire a thorough understanding about the repertory of surfaces available and the characteristics of these surfaces.

Thinking about the Surfaces

If you have an object to model in the computer as surfaces, you need to look at the entire object in a holistic manner to deconstruct the object into a set of surfaces that can be represented by the types of surfaces available from the surface modeling application.

To begin, you start thinking about the boundary faces of the object. With careful analysis of the shape of the object, you deconstruct the object's surface into a number of discrete regions of surfaces, with each region following a particular geometric pattern. Then you think about the shape of each individual surface and identify ways of constructing it.

After this initial thinking process, you construct the surfaces accordingly. Simply speaking, the process of surface modeling consists of two major steps: reduction of the complex 3D object into simple objects (individual surfaces) and making the simple objects to form the complex object. Figure 3–1 shows the surface model of a car body and the individual surfaces exploded.

FIGURE 3–1 *Surface model (left) and individual surfaces exploded apart (right)*

Thinking about the Points and Curves

After you deconstruct the surface of a complex 3D object into a set of surfaces, you start thinking about how to construct each of the surfaces. Among the many ways to construct an individual surface, the fundamental method is to construct 3D curves and/or points as a framework defining the profile and silhouette of the surface, and apply NURBS surface modeling tools on the framework.

To construct a surface from a framework of curves and/or points, you need to think about the curves and points and reduce the surface to its defining curves and/or points. Because the curves and points of a surface are not readily perceived, reduction of a surface into its defining curves and points requires in-depth analysis of what curves and/or points are needed, where they should be located, and what surface modeling commands are to be applied on them to obtain the required surface.

To be able to identify the locations and types of curves and/or points and the command to apply, you need to master various surface construction methods. You also need a good understanding of the shapes of surfaces generated from these methods. In any design project, you will find that the most tedious job in surface modeling is constructing curves and points, and that making surfaces from curves and points is simple. You need only use the appropriate surface construction commands. Hence, the prime focus of making individual surfaces is thinking about designing and making the curves and points. Try to determine the curves and/or points for making the surface shown in Figure 3–2.

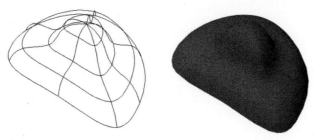

FIGURE 3–2 *Rail revolve surface*
© 2014 Cengage Learning®. All Rights Reserved.

Point and Curve Data

To guide you through various curve and surface construction techniques, some of the data for the curves in this book are provided in the tutorials. However, you must realize that this type of data is not readily available when you design a surface. You need to think about the curves and determine the data for the curves on your own. There are three approaches to constructing points and curves: using graph paper and sketches, using a digitizer, and direct construction.

Graph Paper and Sketches

One way to determine the curves is to use graph paper and sketches. After you have discerned the relationships among surface shapes, the defining curves, and the applicable surface modeling commands, you construct freehand sketches on graph paper that depict the curves. (See Figure 3–3.) Sketches can be used in two ways: extraction of coordinates from the graph and using the sketch as a background image. (See Figure 3–4.) From these sketches, you obtain the coordinates of interpolation points. Using the interpolation points, you construct preliminary curves in the computer. From the preliminary curves, you improvise and construct curves to better represent the surface according to your design intent.

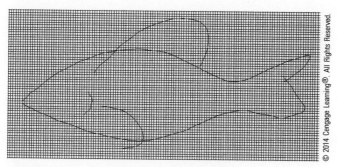

FIGURE 3–3 *Curves constructed on graph paper*

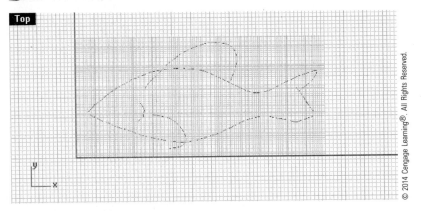

FIGURE 3–4 *Background image imported to a viewport*

Digitizing a Physical Model

Another way to define curves is to make a physical model of the object and use a digitizer to obtain the coordinates of the control points of the surface. Using the coordinates of the markings, you construct curves. Figure 3–5 shows the 3D digitizer being used to obtain coordinates of a physical model.

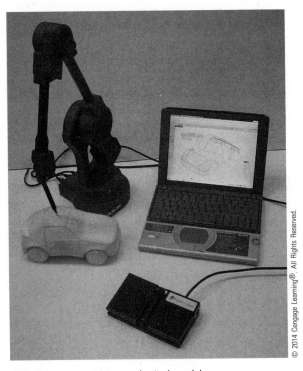

FIGURE 3–5 *Digitizing a physical model*

Direct Construction

After you master the skills of constructing curves and points in the computer, you can construct them directly from what you think they should be. Eventually, you should find this to be the best approach in design innovation and creation.

Prerequisite Knowledge and Thinking

Bearing in mind that the curves and points for the surface are implicit and imaginary, you may not find them on the object. Therefore, a good understanding of the characteristics of various surface modeling commands and the types of curves and points required for these commands is essential. Insight into the types of curves and points needed for a particular type of surface is advantageous. Figure 3–6 shows a set of curves, a surface constructed from the curves, and a rendered image of the surface.

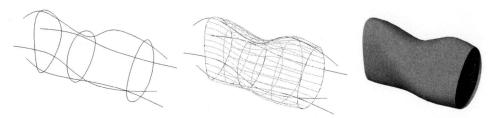

FIGURE 3–6 *Surface and its curves*
© 2014 Cengage Learning®. All Rights Reserved.

Manipulating a Surface Model

There will be times when you are not satisfied with the surface constructed from a set of curves. It is probable that the surface you construct does not match the surface you perceive in your mind. To improvise, you have two options: (1) think about the curves and points again, reconstruct or modify the curves and points, and make a new surface from the curves and points;* or (2) modify the surface by editing and transforming. Among the many methods of surface modification, one way is to manipulate control points. It is important to note that when you translate a control point on a surface, not only is the control point moved but quite a large area of the surface is affected. Figure 3–7 shows a surface modified by translating two control points.

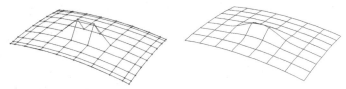

FIGURE 3–7 *Surface modified by moving its control points*
© 2014 Cengage Learning®. All Rights Reserved.

CONCEPTS OF SURFACES AND CURVES CONTINUITY

Apart from dealing with objects of very simple shape, such as a ball, surface modeling usually involves making two or more surfaces and putting them together. Before building a surface (except primitive surface) for a surface model, you may need to construct curves. In the process of making curves or surfaces, attention has to be paid to the smoothness at the joint between contiguous curves or surface elements, because it has a direct impact on the outcome of the shape of the model. A way to describe the smoothness at the junction between two contiguous curves or surfaces is called continuity. Rhino 4 uses its proprietary G-Infinity blending technology to provide five kinds of continuity: G0, G1, G2, G3, and G4.

G0 Continuity

G0 continuity is also known as positional continuity. At a G0 (positional) continuity joint, the endpoints of the curves or the edges of two surfaces simply coincide, having common control points. Their curvature radii and tangent directions at the common joint are different. Using curves as an example, Figure 3–8 illustrates the concept of G0 continuity. Both the curvature radius R_1 and tangent direction T_1 of curve A are not the same as those (R_2 and T_2) of curve B.

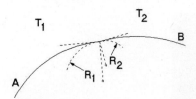

FIGURE 3–8 *G0 (positional) continuity*
© 2014 Cengage Learning®. All Rights Reserved.

---

*Rhino's History Manager, which you learned in Chapter 2, can help track changes made to the curves and modify the surface accordingly.

G1 Continuity

G1 continuity is also known as tangent continuity. In a G1 continuity joint, the tangent directions of the control points at the endpoints of the curves or the edges of the surfaces are the same. However, their curvature radii are different. Shown in Figure 3–9 is a G1 joint, where the tangent directions (T_1 and T_2) are the same, but the curvature radii (R_1 and R_2) at the joint are not the same.

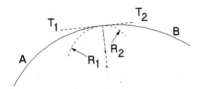

FIGURE 3–9 *G1 (tangent) continuity*
© 2014 Cengage Learning®. All Rights Reserved.

G2 Continuity

G2 continuity is also called curvature continuity. In a G2 continuity joint, the curvature radii and tangent directions of the control points at the endpoints of the curves or the edges of the contiguous surfaces are the same. Shown in Figure 3–10 is a G2 joint, where the curves' tangents (T_1 and T_2) and the curvature radii (R_1 and R_2) at the joint are the same.

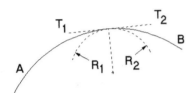

FIGURE 3–10 *G2 (curvature) continuity*
© 2014 Cengage Learning®. All Rights Reserved.

G3 Continuity

At a G3 continuity joint, the change of rate of curvature between contiguous curves/surfaces is constant, providing a much smoother joint than the G2 continuity joint.

G4 Continuity

A G4 continuity joint is even smoother because the rate of change of curvature at the common endpoint/edge of contiguous curves/surfaces is constant. A G4 joint is so smooth that it can hardly be seen as separate curves/surfaces.

Figure 3–11 shows how smoothness between contiguous surfaces is improved as the continuity grade increases.

The black and white image shown in Figure 3–11 may not be clear enough to illustrate how an increase in continuity causes corresponding increase in smoothness. You may refer to the image "continuity.tif" in the Chapter 3 folder that you downloaded from the Student Companion Website.

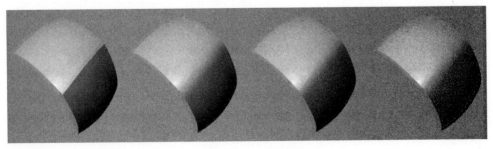

FIGURE 3–11 *From left to right: G0, G1, G2, and G4 continuity*
© 2014 Cengage Learning®. All Rights Reserved.

RHINO'S MAJOR ADVANTAGE

One major advantage of using Rhino to construct free-form surfaces is that the framework of points/curves need not be accurately defined. For example, the sets of curves in a network curve surface need not intersect exactly, and the endpoints of the cross-section curves of swept surfaces need not lie exactly on the rail curves.

In other words, Rhino is an accurate NURBS surface tool on one hand and is a very flexible tool on the other hand.

RHINO'S NURBS SURFACE

In terms of free-form surface design and manufacture, NURBS (nonuniform rational B-spline) surfaces are more significant because surfaces can be represented in a more accurate way. Therefore, we will first focus on NURBS surfaces in the early chapters of this book and discover more about polygon meshes in Chapter 9. For the sake of easy reference, we will divide Rhino's surfaces into four categories:

- The first category concerns the four general kinds of surfaces that you may find in most computer-aided design applications.
- The second category covers surfaces that are more specific to Rhinoceros.
- The third category is the planar surfaces.
- The fourth category covers surfaces derived from existing surfaces.

Concerns

There are three concerns while constructing surfaces for making a model:

1. Continuity between contiguous surfaces
2. The use of curves and surface edges as framework
3. Application of the History Manager

Continuity between Contiguous Surfaces

If surface edges are used in the construction of surfaces, continuity between the new surface to be built and the existing surface is sometimes taken into account. As mentioned in Chapter 1, whenever two or more contiguous curves/surfaces are concerned, continuity at the joint has to be considered. Rhino provides five types of continuity: G0 (positional), G1 (tangent, G2 (curvature), G3 (smooth change of curvature), and G4 (constant rate of change of curvature). Among these types, G4 continuity provides such a smooth joint between contiguous curves/surfaces that it is hardly visible.

Curves and Surface Edges

In building surfaces requiring a framework of curves, you can use existing surface edges as well as curves. Using curves, the surface constructed will have G0 continuity with the selected curve. Using surface edges, the surface constructed can have one of the five kinds of continuities with the selected surface edges.

History Manager

Quite a lot of surface construction commands explained in this chapter can be governed by the History Manager. As explained in Chapter 2, the History Manager remembers changes made to the source objects and causes corresponding changes to the objects derived from the source. Therefore, if a surface construction command is governed by the History Manager and the History Manager is turned on, changes made to the curves that are used to construct the surface during the working session cause the surface to modify automatically. However, it must be emphasized that the History Manager takes effect only in the current session and before the surface is further modified by other means.

COMMON FREE-FORM SURFACES

The first surface category covers surfaces that are common to most computer-aided design applications. There are four major kinds of free-form surfaces: extruded, revolved, swept, and lofted. In common, these surfaces are all constructed from a framework defining their cross-section profiles.

Extruded Surfaces

Basically, an extruded surface has a cross-section of uniform shape. The surface profile is constructed by translating a curve in a straight line or along a curve. Using Rhino's surface modeling tool, you can extrude a curve in six ways: Straight, Tapered, To Point, Ribbon, Along Curve, and Fin.

Extrude Straight

This option extrudes a curve along a straight line direction. If the curve is a planar curve, the direction of extrusion is perpendicular to the plane of the curve. If the curve is 3D, the direction of extrusion is perpendicular to the active construction plane. Perform the following steps.

1. Select File > Open and select the file Extrude01.3dm from the Chapter 3 folder on the Student Companion Website.
2. Select Surface > Extrude Curve > Straight.
3. Select planar curve A, indicated in Figure 3–12, and press the ENTER key.
4. Click on location B, shown in Figure 3–12.

   The planar curve is extruded in a straight line perpendicular to the plane of the curve, and an extrude straight surface is constructed.

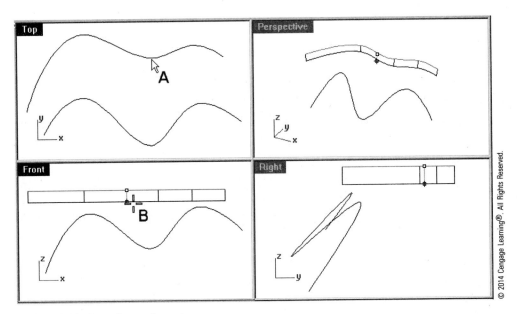

FIGURE 3–12 *Extruding a planar curve*

Continue with the following steps to extrude a 3D curve in a direction perpendicular to the active construction plane.

5. Select Surface > Extrude > Straight.
6. Select 3D curve A, indicated in Figure 3–13, and press the ENTER key.
7. Click on location B, shown in Figure 3–13.

   The 3D curve is extruded in a direction perpendicular to the construction plane from which the curve is selected and a surface is constructed. In other words, if you select the curve in a different viewport, the outcome will be different.

8. Do not save your file.

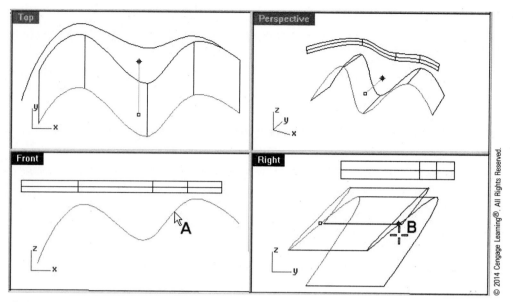

FIGURE 3–13 *Extruding a 3D curve in a direction perpendicular to the active construction plane*

Straight Extrusion with a Taper Angle

To uniformly scale up or down the cross-section profile while extruding the defining curve, you apply a taper angle. Perform the following steps.

1. Select File > Open and select the file Extrude02.3dm from the Chapter 3 folder on the Student Companion Website.
2. Select Surface > Extrude Curve > Tapered.
3. Select curve A, shown in Figure 3–14 (Top viewport), and press the ENTER key.
4. Select the DraftAngle option at the command area.
5. Type 10 to set the draft angle (taper angle) to 10 degrees.
6. Select location B, shown in Figure 3–14.

   A tapered extruded surface is constructed.
7. Do not save the file.

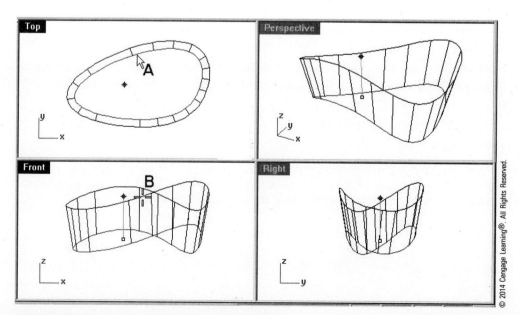

FIGURE 3–14 *Tapered extruded surface being constructed*

Extruding to a Point

An extreme case of extruding a curve with a tapered angle is to extrude it to a point. In essence, the surface profile uniformly diminishes to a single point. Perform the following steps.

1. Select File > Open and select the file Extrude03.3dm from the Chapter 3 folder on the Student Companion Website.
2. Select Surface > Extrude Curve > To Point.
3. Select curve A, shown in Figure 3–15, and press the ENTER key.
4. Select location B, shown in Figure 3–15.

    Exact location is unimportant for the purpose of this tutorial. A curve is extruded to a point, and a surface is constructed.

5. Do not save the file.

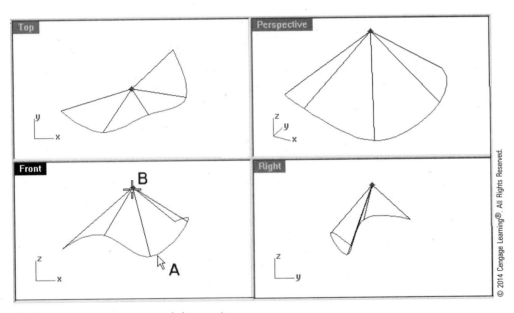

FIGURE 3–15 *Curve being extruded to a point*

Extruding along a Curve or a Subcurve

Instead of extruding a curve in a straight line, you use the curve or a portion of the curve (subcurve) as the path of extrusion. While extruding, the cross-section remains constant and parallel. Now perform the following steps.

1. Select File > Open and select the file Extrude04.3dm from the Chapter 3 folder on the Student Companion Website.
2. Select Surface > Extrude Curve > Along Curve.
3. Select curve A, shown in Figure 3–16, and press the ENTER key.
4. Click on curve B, shown in Figure 3–16.

    Curve A is extruded along curve B, and a surface is constructed.

> **NOTE** The direction of extrusion depends on which end of the path curve you select.

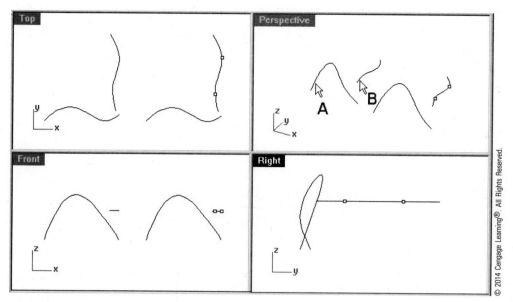

FIGURE 3-16 *Curve being extruded along a path curve*

5. Set object snap mode to Point.
6. Select Surface > Extrude Curve > Along Curve.
7. Select curve A, shown in Figure 3-17, and press the ENTER key.
8. If SubCurve option at the command area is No, click on it to change it to Yes.
9. Select curve B, shown in Figure 3-17.
10. Select location C along curve B, shown in Figure 3-17.
11. Select location D along curve B, shown in Figure 3-17.

    A curve is extruded along a path defined by two points along a path curve, and a surface is constructed, as shown in Figure 3-18.

12. Do not save your file.

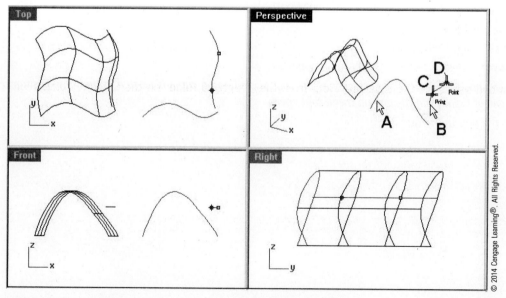

FIGURE 3-17 *A portion of a path curve being used to extrude a curve*

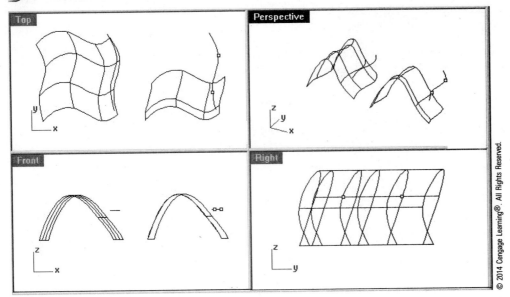

FIGURE 3-18 *Curve extruded along a portion of a curve*

Extruding a Ribbon

This is a special kind of extruded surface. It is called a ribbon because the resulting surface resembles a ribbon. In essence, the operation includes offsetting the original curve and then filling the gap between the original curve and the offset curve with a surface. Perform the following steps.

1. Select File > Open and select the file Extrude05.3dm from the Chapter 3 folder on the Student Companion Website.
2. Select Surface > Extrude Curve > Ribbon.
3. Select curve A, shown in Figure 3-19.
4. Select the Distance option at the command area.
5. Type 4 to assign the ribbon's width.
6. Click on location B, shown in Figure 3-19, in the Top viewport.

   A ribbon surface is constructed.

7. Select Surface > Extrude > Ribbon.
8. Select curve A, shown in Figure 3-20, in the Front viewport.

   Because there are two edges of the ribbon surface you just constructed and a curve is located in the same position, you need to select one of them from the pop-up menu that opens.

9. Select Curve from the pop-up menu.

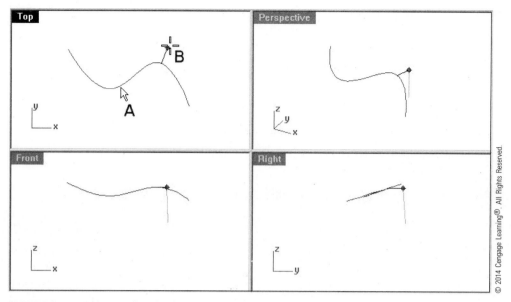

FIGURE 3–19 *Ribbon surface being constructed in a direction parallel to the active construction plane*

10. Select location C, shown in Figure 3–20, in the Front viewport.
    The second ribbon surface is constructed. (See Figure 3–21.)
11. Do not save your file.
    As you can see, the outcome depends on the active construction plane.

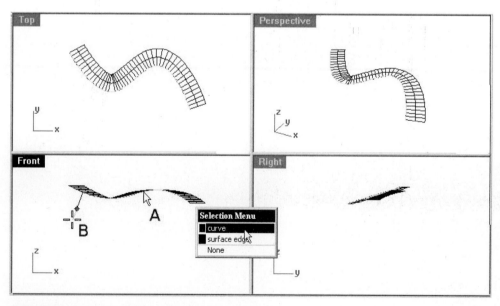

FIGURE 3–20 *Second ribbon surface being constructed*
Source: Robert McNeel and Associates Rhinoceros® 5

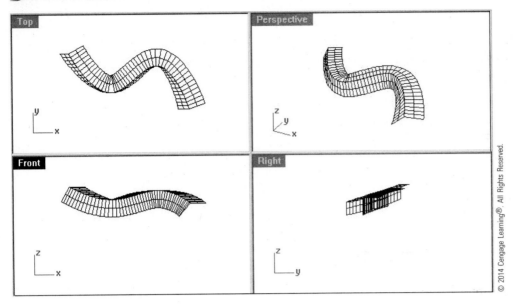

FIGURE 3-21 *Ribbon surfaces constructed*

Extruding Normal to a Surface (Making a Fin)

The Fin command produces a surface that resembles the shape of a fish fin. It extrudes a curve in a direction perpendicular to a selected surface. To construct a fin, perform the following steps.

1. Select File > Open and select the file Extrude06.3dm from the Chapter 3 folder on the Student Companion Website.
2. Select Surface > Extrude Curve > Normal to Surface.
3. Select curve A, shown in Figure 3-22.
4. Select surface B, shown in Figure 3-22.
5. Click on location C, shown in Figure 3-22.
   A fin surface is constructed.
6. Do not save your file.

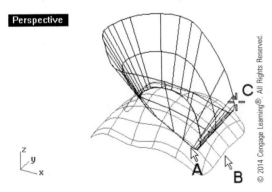

FIGURE 3-22 *Fin surface being constructed*

History Manager

Among the six ways of constructing an extruded surface, Straight, Tapered, To Point, and Along Curve are governed by the History Manager. In other words, if the History Manager is turned on prior to making these surfaces, subsequent modification of the curves causes the surface to change accordingly.

Revolved Surfaces

A revolved surface is constructed by revolving a curve around an axis. Using Rhino's tool, you can revolve a curve in two ways: revolve the curve around an axis or revolve the curve along a rail. Both commands are governed by the History Manager if it is turned on.

Revolving around an Axis

A revolved surface has a uniform cross-section. It is constructed by revolving a curve around an axis. Perform the following steps:

1. Select File > Open and select the file Revolve01.3dm from the Chapter 3 folder on the Student Companion Website.
2. Check the Point box in the Osnap dialog box.
   Point objects will be snapped onto automatically.
3. Select Surface > Revolve
4. Select curve A, shown in Figure 3–23, and press the ENTER key.
5. Select points B and C, shown in Figure 3–23.
6. Click on location D (Figure 3–23) to specify the start point of revolution.
7. Drag the cursor in a clockwise direction and click on location E, shown in Figure 3–24.
   A revolved surface is constructed.

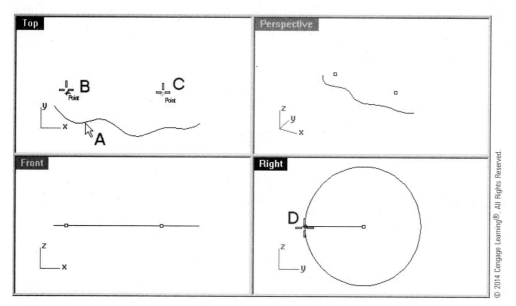

FIGURE 3–23 *Revolved surface being constructed*

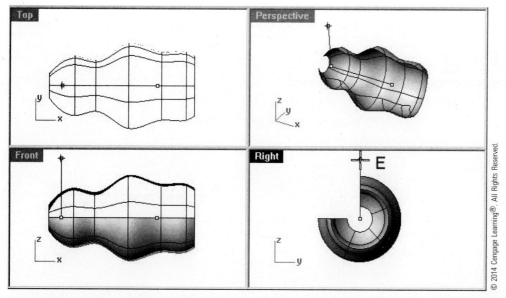

FIGURE 3–24 *Revolved surface constructed*

Revolved Surface and History Manager

The History Manager, if it is turned on, can help track changes. You may modify the curve used to make the revolved surface, and the surface will update automatically. You will learn more about curve manipulation, including modifying the shape of a curve, in Chapters 6 and 10. To appreciate how the History Manager helps track changes, continue with the following steps.

8. With reference to Figure 3-25 (left), select the curve that is used to construct the revolved surface.

    Because there are a curve and a surface edge, a pop-up menu will display, allowing you to choose the curve or the surface edge.

9. Drag the curve to a new position as shown in Figure 3-25 (center).

    The surface is modified, as shown in Figure 3-25 (right). This change is caused by a change in location between the curve and the revolve axis.

10. Do not save the file.

    Note that the History Manager will work fine as long as no further work is done on the surface. For example, if you move the surface, you may obtain unexpected result.

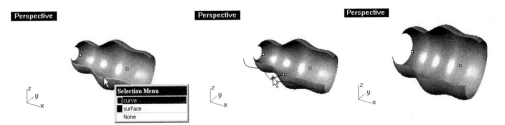

FIGURE 3–25 *From left to right: curve selected, curve dragged, and surface modified by the History Manager*
Source: Robert McNeel and Associates Rhinoceros® 5

Rail Revolve

To impose control over the surface cross-section while it is being revolved around an axis, you can add a guide rail. The angle of revolution of the surface depends on the shape of the guide rail curve and its relative location to the revolve axis. Simply speaking, if the guide rail is a closed loop curve, the rail revolve surface will revolve 360 degrees. Perform the following steps.

1. Select File > Open and select the file Revolve02.3dm from the Chapter 3 folder on the Student Companion Website.
2. Select Surface > Rail Revolve.
3. Select curve A, shown in Figure 3-26, as the profile curve.
4. Select curve B, shown in Figure 3-26, as the rail curve.
5. Select points C and D, shown in Figure 3-26, to define the axis of revolution.

    A revolved surface is constructed, as shown in Figure 3-27.

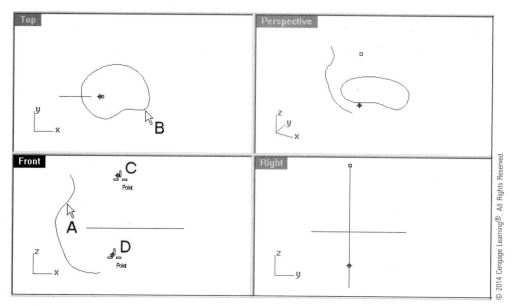

FIGURE 3–26 Rail revolve surface being constructed

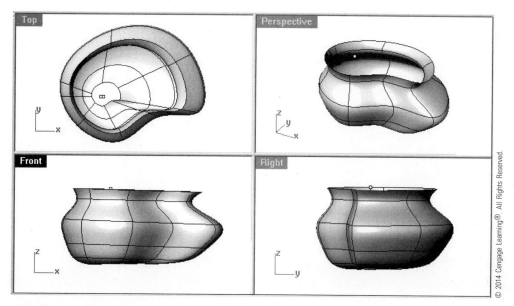

FIGURE 3–27 Rail revolve surface constructed

History Manager and Rail Revolve Surface

In Figure 3–28, two curves are involved; changes to either or both of them cause corresponding changes to the rail revolve surface. Continue with the following steps to drag the rail to a new location.

    6. Referencing Figure 3–28 (left), select the rail curve and drag it to a new location.

       The surface is modified, as shown in Figure 3–28 (right). The change in surface is caused by the change of the rail in relation to the revolve axis.

    7. Do not save the file.

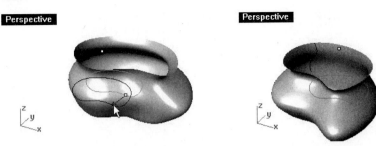

FIGURE 3–28 *Rail curve being dragged to a new location (left) and the surface modified (right)*
© 2014 Cengage Learning®. All Rights Reserved.

Swept Surfaces

A swept surface is constructed by a sweeping process in which one or more cross-section curves sweep along one or two guide rails.

Sweep 1 Rail Surface

A sweep 1 rail surface is constructed by first selecting a rail curve and then selecting one or more cross-section curves. The rail curve can be a single curve or a number of curves in a chain. If one section curve is selected, the surface profile will run along the entire length of the rail. If more than one section curve is selected, the surface file will interpolate from the first through the last section curve.

Comparing Extruding along a Curve and Sweeping a Curve along One Rail

Naturally, the simplest swept surface is constructed by sweeping a single cross-section curve along a rail curve. However, you must not confuse sweeping a single curve along a single rail with extruding a curve along a curve. As shown in Figure 3–29, the cross-section profile of an extruded along curve surface is always parallel to the original curve. As for sweeping a single profile along a single curve, the cross-section maintains a constant angle between the normal of the rail curve.

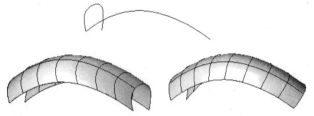

FIGURE 3–29 *Extruding along a curve (left) and sweeping a section along a rail (right)*
© 2014 Cengage Learning®. All Rights Reserved.

Perform the following steps.

1. Select File > Open and select the file Sweep101.3dm from the Chapter 3 folder on the Student Companion Website.
2. Select Surface > Sweep 1 Rail.
3. Select curve A, shown in Figure 3-30, as the rail curve; select curve B, indicated in Figure 3-30, as the cross-section curve; and press the ENTER key.
4. In the Sweep 1 Rail Options dialog box, accept the default and then click on the OK button.

   A swept surface is constructed, as shown in Figure 3–31.

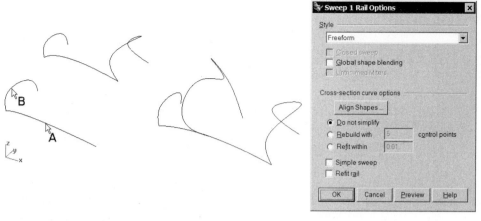

FIGURE 3–30 *A rail and a section selected (left) and Sweep 1 Rail Options dialog box (right)*
Source: Robert McNeel and Associates Rhinoceros® 5

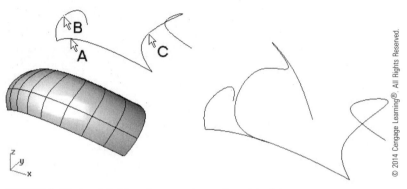

FIGURE 3–31 *A rail (left) and two sections selected (right)*

5. Referencing Figure 3–31, repeat the Sweep 1 Rail command with curve A as the rail and curves B and C as the sections.
6. Using curve A, shown in Figure 3–32, and curves B, C, and D as the sections, construct another swept surface.
   Three swept surfaces are constructed, as shown in Figure 3–33.
7. Do not save your file.

> **NOTE** When two or more cross-sections are used, the surface profile will transit gradually from the first section to the second section and then to the next cross-section.

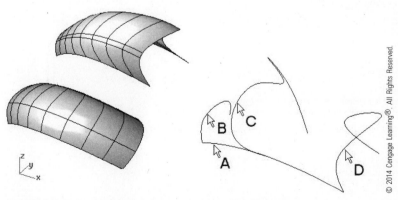

FIGURE 3–32 *A rail (left) and three sections selected (right)*

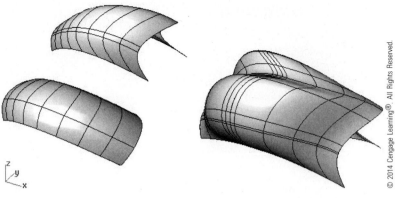

FIGURE 3-33 *Surfaces constructed by sweeping curve(s) along one rail*

Surface Continuity in Sweeping

If the rail curve used in making a sweep 1 rail surface is the edge of another surface, the swept surface that is constructed can be made to twist with the surface edge, and continuity between the swept surface and the existing surface can be enabled. Perform the following steps.

1. Select File > Open and select the file Sweep102.3dm from the Chapter 3 folder on the Student Companion Website.
2. Select Surface > Sweep 1 Rail.
3. Select edge A, shown in Figure 3-34, as the rail curve; select curves B and C, shown in Figure 3-34, as the cross-section curves; and press the ENTER key.
4. In the Style pull-down list of the Sweep 1 Rail Options dialog box (Figure 3-34), select Align with Surface, and click on the OK button.

    A swept surface that aligns with the surface whose edge is used as sweeping rail is constructed.

For comparison, the swept surface constructed from the same set of curves with free-form option is also shown in Figure 3-35.

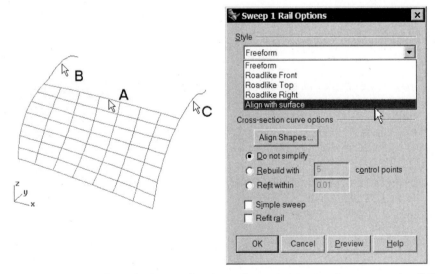

FIGURE 3-34 *Surface edge being selected as rail curve in swept surface construction (left) and Align with surface option selected (right)*
Source: Robert McNeel and Associates Rhinoceros® 5

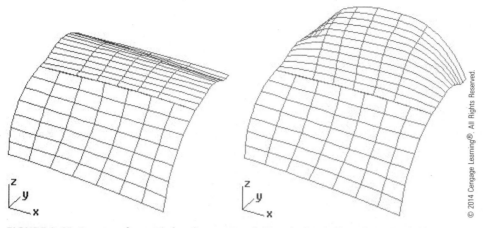

FIGURE 3–35 *Swept surface with free-form option (left) and Align with surface option (right)*

Edge Chaining for Sweeping Rail

If a rail is made up of a number of curves and/or surface edges, you do not have to first join the curves to form a single rail. Instead, you can use edge chaining option to select multiple curves and/or edges. At the command area, select the ChainEdges option at the command area and then select contiguous rail curves.

1. Select File > Open and select the file Sweep103.3dm from the Chapter 3 folder on the Student Companion Website.
2. Select Surface > Sweep 1 Rail.
3. Select the ChainEdges option at the command area.
4. Select the ChainContinuity option at the command area.
5. Select the Curvature option at the command area.
6. Select surface edges A, B, and C, indicated in Figure 3–36, and press the ENTER key.
7. Select curves D and E and press the ENTER key.
8. In the Sweep 1 Rail Options dialog box, select Align with surface from the Style pull-down list box, check the Global shape blending box, and click on the OK button.

   A swept surface using multiple edges as a single rail is constructed, as shown in Figure 3–37.

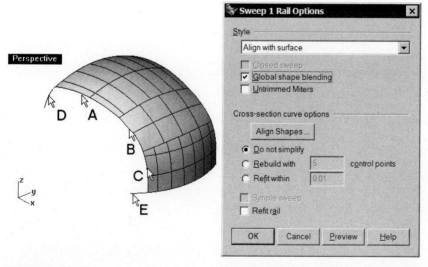

FIGURE 3–36 *Chain edges with two cross-sections selected (left) and Sweep 1 Rail Options dialog box (right)*
Source: Robert McNeel and Associates Rhinoceros® 5

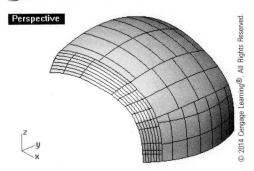

FIGURE 3–37 Swept surface constructed

Shape Alignment in Sweeping

In multiple cross-section sweep 1 rail surface construction, you can reverse the direction of selected cross-section curves. Perform the following steps.

1. Select File > Open and select the file Sweep104.3dm from the Chapter 3 folder on the Student Companion Website.
2. Select Surface > Sweep 1 Rail.
3. Select surface edge A and curves B and C, shown in Figure 3–38, and press the ENTER key.
4. Click on the Align Shapes button on the Sweep 1 Rail Options dialog box.
5. Select endpoint A, shown in Figure 3–39.
   The cross-section curve is reversed.
6. Press the ENTER key, and then click on the OK button.
   A surface is constructed.
7. Do not save the file.

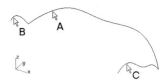

FIGURE 3–38 Curves selected for making sweep 1 rail surface
© 2014 Cengage Learning®. All Rights Reserved.

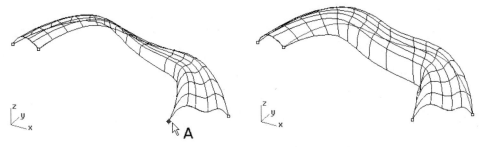

FIGURE 3–39 Cross-section curve being selected (left) and reversed (right)
© 2014 Cengage Learning®. All Rights Reserved.

Sweeping to a Point

Apart from interpolating among the cross-section curves, you can sweep a surface to a point. Perform the following steps.

1. Select File > Open and select the file Sweep105.3dm from the Chapter 3 folder on the Student Companion Website.
2. Click on Osnap button on the Status bar and set object snap mode to Point.

3. Select Surface > Sweep 1 Rail.
4. Select curves A and B, shown in Figure 3–40.
5. Select the Point option at the command area.
6. Select point C, shown in Figure 3–40, and press the ENTER key.
7. Click on the OK button in the Sweep 1 Rail Options dialog box.

   A sweeping to a point surface is constructed, as shown in Figure 3–40.
8. Do not save the file.

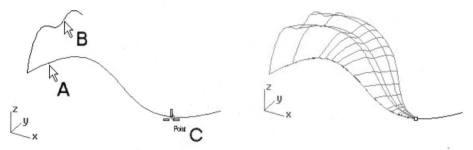

FIGURE 3–40 *Curves and point location selected (left) and sweeping to a point surface constructed (right)*
© 2014 Cengage Learning®. All Rights Reserved.

Sweep 2 Rails Surface

To add further control to the surface profile while sweeping, you use two guiding rails instead of one guiding rail. In addition to interpolating from one cross-section curve to another cross-section curve, the cross-section translates in accordance with the distance between the rails and the locations of the rails. Using Rhino's sweep 2 rails command, you select two rails and then one or more cross-section curves to construct a sweep 2 rails surface. Moreover, you can incorporate additional cross-section alignments to control how the surface is constructed. Similar to sweep 1 rail surface, you can use a series of curves and/or surface edges as sweeping rails. Perform the following steps.

1. Select File > Open and select the file Sweep201.3dm from the Chapter 3 folder on the Student Companion Website.
2. Select Surface > Sweep 2 Rails.
3. Select curves A, B, and C, shown in Figure 3–41, and press the ENTER key.

   The first two selected curves are rails, and subsequently selected curves are sections.

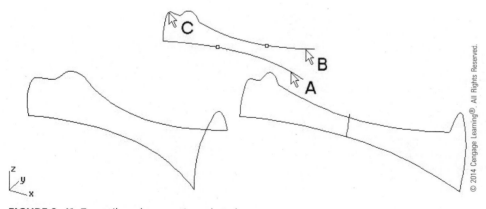

FIGURE 3–41 *Two rails and one section selected*

Adding Slash to a Sweep 2 Rails Surface

Slashes are additional cross-section alignments. To incorporate additional slash along the rails, continue with the following steps.

4. In the Sweep 2 Rail Options dialog box, click on the Add Slash button. (See Figure 3–42.)
5. Select points A and B, shown in Figure 3–42.

   Note the effect of adding a slash.
6. Press the ENTER key.
7. Click on the OK button.

   A swept surface with additional slash is constructed.

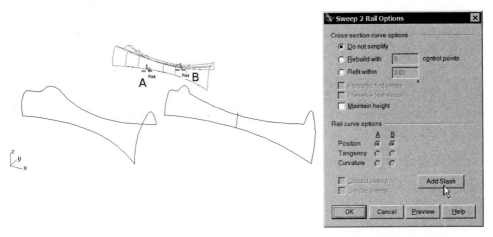

FIGURE 3–42 *Additional slash being defined (left) and Sweep 2 Rails Options dialog box (right)*
Source: Robert McNeel and Associates Rhinoceros® 5

Maintaining Height in a Sweep 2 Rails Surface

As a cross-section curve is swept along two rails, its overall cross-section is scaled in accordance with the distance between the rails. In order words, the height of cross-sections as they interpolate along the rail curves depends on the distance between the rails. To maintain the height during interpolation, you can use the Maintain Height option. Continue with the following steps.

8. Repeat the Sweep 2 Rails command.
9. Select curves A, B, C, and D, shown in Figure 3–43, and press the ENTER key.
10. Check the Maintain height button and the OK button in the Sweep 2 Rail Options dialog box.

    A swept surface with height adjustment is constructed. Note the effect of having height adjustment on the final shape of the surface.

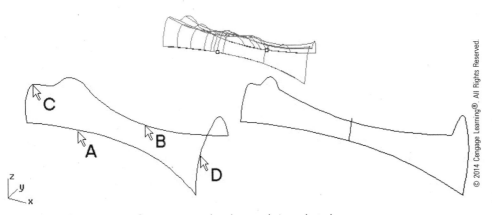

FIGURE 3–43 *A swept surface constructed and curves being selected*

11. Repeat the Sweep 2 Rails command.
12. Select curves A, B, C, D, and E, shown in Figure 3–44, and press the ENTER key.
13. Check the Maintain Height button and click on the Preview button to discover the difference between having and not having the Preview button checked.
14. Click on the OK button.

    The third swept surface is constructed. (See Figure 3–45.)

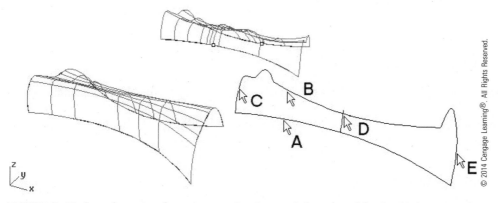

FIGURE 3–44 *Second swept surface constructed and curves being selected for the third swept surface*

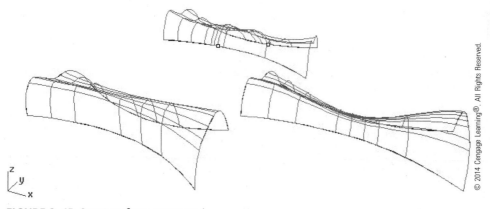

FIGURE 3–45 *Swept surfaces constructed*

Edge Continuity between Edge Rails and Sweep 2 Rails Surface

If the rails for making a sweep 2 rails surface are surface edges, edge continuity between the swept surface and the existing surfaces can be adjusted. Perform the following steps.

1. Select File > Open and select the file Sweep202.3dm from the Chapter 3 folder on the Student Companion Website.
2. Select Surface > Sweep 2 Rails.
3. Select curves A, B, C, D, and E, shown in Figure 3–46, and press the ENTER key.
4. Check the Curvature buttons in the Sweep 2 Rail Options dialog box and click on the OK button.

   A swept surface is constructed.

5. Do not save your file.

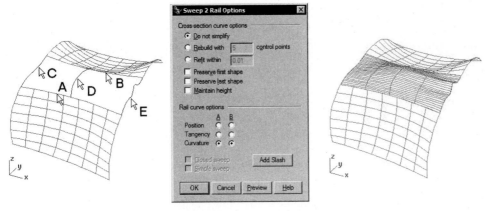

FIGURE 3–46 *From left to right: surface edges as rails, Sweep 2 Rail Options dialog box, sweep 2 rails surface constructed with surface edge matching*
Source: Robert McNeel and Associates Rhinoceros® 5

Lofted Surfaces

Lofted surfaces are constructed by interpolating one or two sets of cross-section curves. Using Rhino's tool, you can construct this type of surface in three ways: Loft, Curve Network, and Edge Curve. All three types of surfaces are governed by the History Manager, if it is turned on.

Loft Surface

Rhino's loft command uses only one set of cross-section curves. It produces a surface profile that interpolates from the first cross-section curve to the second, and to the next curve. Perform the following steps.

1. Select File > Open and select the file Loft01.3dm from the Chapter 3 folder on the Student Companion Website.
2. Select Surface > Loft.
3. Select curves A, B, and C, shown in Figure 3–47, and press the ENTER key.

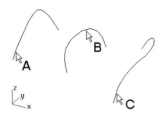

FIGURE 3–47 *Curves being selected*
© 2014 Cengage Learning®. All Rights Reserved.

Curve Alignment in Loft Surface Construction

Due to the way the curves are selected, the preview shows a bow-tie shape surface, which is obviously not what we want.

4. In the Loft Options dialog box, click on the Align Curves button and then click on endpoint A, shown in Figure 3–48.

   The curve's direction is reversed, and the curve is aligned with other curves.

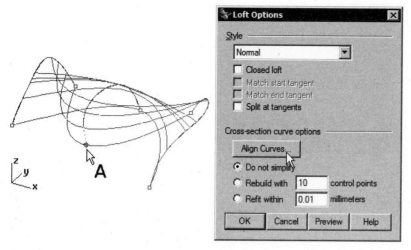

FIGURE 3–48 *Cross-section curve being aligned*
Source: Robert McNeel and Associates Rhinoceros® 5

Style in Loft Surface

To cope with downstream operations such as sheet metal fabrication, you may need to have a developable surface. A developable surface is a surface that can be developed into a flat sheet. Continue with the following steps.

   5. In the Style pull-down list box, there are Normal, Loose, Tight, Straight sections, Developable, and Uniform options. Select Developable and click on the Preview button.

      A developable surface is displayed.

> There is a slight difference between straight sections and developable in that the former produces a lofted surface with straight sections between contiguous cross-section curves, and the latter produces a surface that can be developed into a flat sheet. You may click on both options to have a preview in order to appreciate their differences.
>
> **NOTE**

   6. Select Uniform and then click on the OK button.

      A uniform loft surface is constructed.

   7. Do not save your file. (See Figure 3–49.)

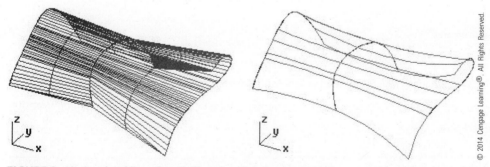

FIGURE 3–49 *Developable surface (left) and uniform surface (right)*

Tangency between Loft Surface and Edge Curves

To appreciate how surface edges can be used in making a lofted surface, perform the following steps.

   1. Select File > Open and select the file Loft02.3dm from the Chapter 3 folder on the Student Companion Website.

2. Select Surface > Loft.
3. Referencing Figure 3–50, select edge A, click on Surface edge from the pop-up menu, select curve B, and select edge C.
4. Press the ENTER key.
5. In the Loft Options dialog box, check the Match start tangent box and the Match end tangent box, and click on the OK button.

   A lofted surface with tangent matching is constructed, as shown in Figure 3–51.
6. Do not save your file.

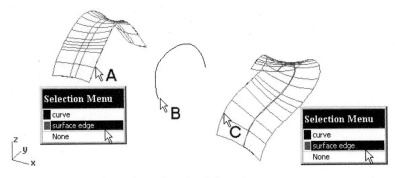

FIGURE 3–50 *Surface edges selected as loft sections*
Source: Robert McNeel and Associates Rhinoceros® 5

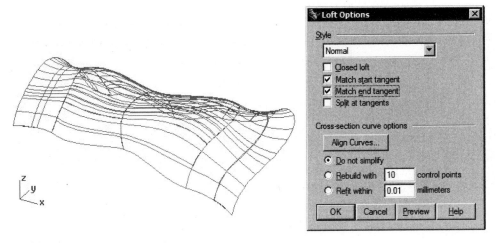

FIGURE 3–51 *Tangent matched*
Source: Robert McNeel and Associates Rhinoceros® 5

Loft to a Point

Using the point option, you can have a lofted surface terminated at a point. Perform the following steps.

1. Select File > Open and select the file Loft03.3dm from the Chapter 3 folder on the Student Companion Website.
2. Set object snap mode to Point.
3. Select Surface > Loft.
4. Select curves A and B, shown in Figure 3–52.
5. Select the Point option at the command area.
6. Select Point C, shown in Figure 3–52, and press the ENTER key.
7. Press the ENTER key in response to the prompt to adjust curve seams.
8. In the Loft Options dialog box, click on the OK button.

   A lofted surface terminating at a point is constructed.
9. Do not save your file.

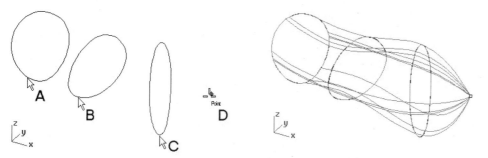

FIGURE 3–52 *Curves and points selected (left) and lofted surface terminating at a point being constructed (right)*
© 2014 Cengage Learning®. All Rights Reserved.

> In this tutorial, the curves for making the lofted surface are closed loop curves. Therefore, there is a prompt to tell you to adjust the curve seam. As for the point object, it is provided here for your convenience. You can specify a point by clicking anywhere or specifying a set of coordinates.

Curve Network Surface

Rhino's curve network surface can be regarded as a kind of lofted surface. It is constructed from two sets of cross-section curves in two orthogonal directions. The surface profile interpolates in two directions, defined by two sets of curves in two orthogonal directions.

In essence, valid curves will be automatically sorted into two sets of orthogonal curves, regardless of the sequence of selection. However, if one of the curves is invalid or ambiguously defined, the system will prompt you to select the curve sets manually. Then you should select one set of curves, press the ENTER key, continue to select the second set of orthogonal curves, and press the ENTER key again. Perform the following steps.

1. Select File > Open and select the file CurveNetwork01.3dm from the Chapter 3 folder on the Student Companion Website.
2. Select Surface > Curve Network.
3. Select all curves and press the ENTER key.

After selecting the curves, the Surface From Curve Network dialog box displays, where you specify edge tolerance and continuity of matching edges, as follows.

4. In the Surface From Curve Network dialog box, shown in Figure 3–53, accept the default and then click on the OK button.

   A surface is constructed from a curve network.

5. Do not save your file.

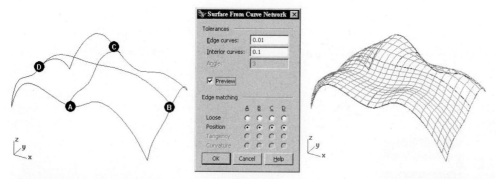

FIGURE 3–53 *From left to right: curves selected for making From curve network surface, Surface From Curve Network dialog box, and surface constructed*
Source: Robert McNeel and Associates Rhinoceros® 5

> **NOTE** Note in the Surface From Curve Network dialog box that there are options for you to decide how the edge of the surface is to match with the input curves.

Manual Sorting of U and V Curves in Network Curve Surface Construction

Perform the following steps to understand what would happen if the system fails to sort the selected curves automatically.

1. Select File > Open and select the file CurveNetwork02.3dm from the Chapter 3 folder on the Student Companion Website.
2. Select Surface > Curve Network.
3. Select all curves and press the ENTER key.

Because a curve that does not conform to the requirement for making a network curve surface is put here intentionally, the NetworkSrf Sorting Problem dialog box displays, as shown in Figure 3–54 (left).

4. Click the Yes button.
5. Select curves A, B, C, and D (Figure 3–54, middle) and press the ENTER key.
   These are the curves in the first direction.
6. Select curves E, F, G, and H (Figure 3–54, middle) and press the ENTER key.
   These are the curves in the second direction.
7. Click the OK button from the Surface From Curve Network dialog box.
   A surface is constructed.
8. Do not save your file.

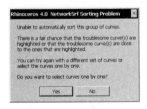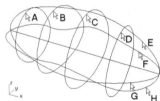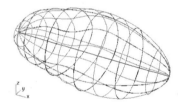

FIGURE 3–54 *From left to right: NetworkSrf Sorting Problem dialog box, curves being selected sequentially, and surface constructed*
Source: Robert McNeel and Associates Rhinoceros® 5

Edge Matching in Curve Network Surface Construction

If surface edges are used in making curve network surfaces, edge matching can be set. Perform the following steps.

1. Select File > Open and select the file CurveNetwork03.3dm from the Chapter 3 folder on the Student Companion Website.
2. Select Surface > Curve Network.
3. Referencing Figure 3–55, select surface edges A and D and curves B, C, and E. Then press the ENTER key.

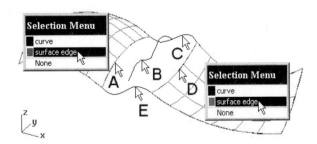

FIGURE 3–55 *Surface edges and curves selected*
Source: Robert McNeel and Associates Rhinoceros® 5

4. In the Surface From Curve Network dialog box, set the angle value to 0.

    This is the angle tolerance between the tangent directions of adjacent edges. (See Figure 3–56.)

5. Check the Curvature boxes for edges B and D.
6. Click on the OK button.

    A surface is constructed. (See Figure 3–57.)

7. Do not save the file.

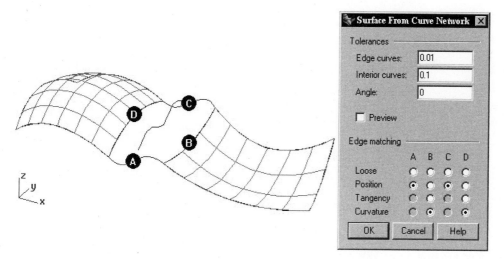

FIGURE 3–56 *Curvature continuity at edges B and D (left) and Surface From Curve Network dialog box (right)"*
Source: Robert McNeel and Associates Rhinoceros® 5

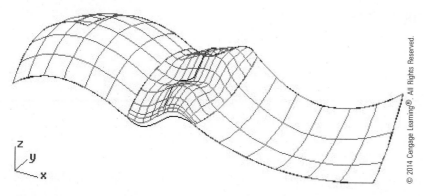

FIGURE 3–57 *Curve network surface constructed*

Edge Curves Surface

You may also regard Rhino's edge curves surface as a special kind of lofted surface. It is constructed by specifying two, three, or four edges of the surface. Perform the following steps to construct surfaces from edge curves.

1. Select File > Open and select the file EdgeCurves.3dm from the Chapter 3 folder on the Student Companion Website.
2. Select Surface > Edge Curves.
3. Select curves A and B, shown in Figure 3–58, and press the ENTER key.

    A surface is constructed from two edge curves.

4. Select Surface > Edge Curves.
5. Select curves C, D, and E, shown in Figure 3–58, and press the ENTER key.

    A surface is constructed from three edge curves.

6. Select Surface > Edge Curves.
7. Select curves F, G, H, and J, shown in Figure 3–58.
8. Do not save your file.

   Three surfaces constructed are shown in Figure 3–59.

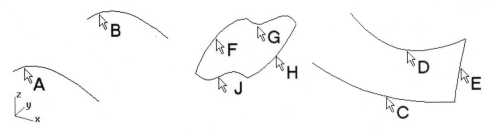

FIGURE 3–58 *Curves being selected*
© 2014 Cengage Learning®. All Rights Reserved.

FIGURE 3–59 *Surfaces constructed from two edge curves (left), three edge curves (middle), and four edge curves (right)*
© 2014 Cengage Learning®. All Rights Reserved.

OTHER KINDS OF RHINO FREE-FORM SURFACES

Apart from the four basic kinds of free-form surfaces delineated earlier, you can use Rhino tools to construct the following types of free-form surfaces: Patch, Corner Points, Point Grid, Heightfield from Image, and Drape.

Patch Surface

A patch surface provides a very flexible way to construct a surface, in particular, by filling holes in a surface model. The input data can be curves and/or points, creating a patch surface from a closed-loop curve, from point objects, from a closed-loop curve and point objects, from an open-loop curve and point objects, and from a number of curves. One major application of patch surface is to "patch" openings in existing surface models. Perform the following steps.

1. Select File > Open and select the file Patch.3dm from the Chapter 3 folder on the Student Companion Website.
2. Set the current layer to Layer01.
3. Select Surface > Patch.
4. Select surface edge curve A (Figure 3–60) and press the ENTER key.
5. In the Patch Surface Options dialog box, click on boxes Adjust Tangency and Automatic Trim, if they are not already checked.
6. Click on the OK button.

   A patch is constructed from the edge of a closed-loop surface.

7. Repeat the Patch command.
8. Select surface edge B and curve C (Figure 3–60), and press the ENTER key.
9. Click on the OK button of the Patch Surface Options dialog box.

   A patch surface is constructed from a closed-loop surface edge and a curve.

> **NOTE** The curve has to be reasonably close to the edge for the surface to be valid.

10. Repeat the Patch command.
11. Select surface edge D and points E, F, and G (Figure 3-60), and press the ENTER key.
12. Click on the OK button of the Patch Surface Options dialog box.

    A patch surface is constructed from a closed-loop surface edge and three point objects. Again, the point objects have to be reasonably close to the surface edge.

13. Repeat the Patch command.
14. Select curve H (there is a curve and surface edge) and point objects J, K, and L, and press the ENTER key.
15. Click the OK button of the Patch Surface Options dialog box.

    A patch surface is constructed from a curve and three point objects. Note the difference between using a curve and a surface edge, as shown in Figure 3-61.

16. Do not save your file.

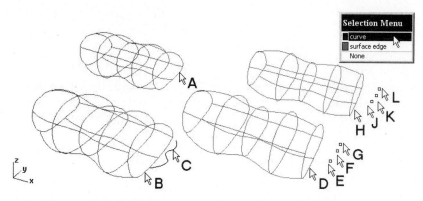

FIGURE 3-60 *Surface edges, point objects, and curves being selected*
Source: Robert McNeel and Associates Rhinoceros® 5

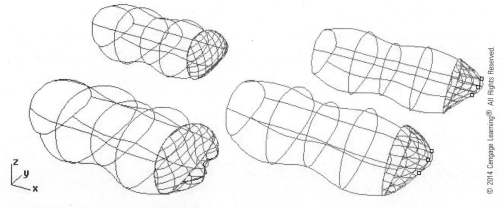

FIGURE 3-61 *Patch surfaces constructed*

Corner Points Surface

A corner points surface is a surface defined by three or four corner points. If three input points are used, a triangular planar surface is produced. If four input points are used, a quadrilateral surface is constructed. Perform the following steps:

1. Select File > Open and select the file CornerPoints.3dm from the Chapter 3 folder on the Student Companion Website.
2. Check the Point button on the Osnap dialog box.
3. Select Surface > Corner Points.
4. Select points A, B, and C, shown in Figure 3-62, and press the ENTER key.

   A corner point surface from three points is constructed.

5. Select Surface > Corner Points.
6. Select points D, E, F, and G, shown in Figure 3–62.

   A corner point surface from four points is constructed.
7. Do not save your file.

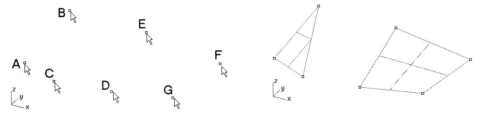

FIGURE 3–62 *Point objects (left) and corner points surfaces constructed (right)*
© 2014 Cengage Learning®. All Rights Reserved.

Point Grid Surface

A point grid surface is derived from a matrix of points arranged in rows and columns. Surface constructed using this method may use these points as interpolation points or control points. Using these points as interpolation points, the surface profile will lie on the points. Using them as control points, they become the control points of the surface. If you use them as control points, you need to specify the degree of polynomial as well. The concept of polynomial degree and control points will be explained in Chapter 4. Perform the following steps.

1. Select File > Open and select the file PointGrid.3dm from the Chapter 3 folder on the Student Companion Website.
2. Check the Point box on the Osnap dialog box.
3. Select Surface > Point Grid. (This command interpolates a surface from input point objects.)
4. Type 4 at the command area twice to specify the number of points in rows and columns.
5. Select points A1, A2, A3, A4, B1, B2, B3, B4, C1, C2, C3, C4, D1, D2, D3, and D4, shown in Figure 3–63.
6. Type SrfControlPtGrid at the command area. (This command uses input point objects as control points to build a surface.)
7. Press the ENTER key twice to accept the number of rows and columns, which should be 4 for both.
8. Select points E1, E2, E3, E4, F1, F2, F3, F4, G1, G2, G3, G4, H1, H2, H3, and H4, shown in Figure 3–63.
9. Do not save your file.

Surfaces constructed are shown in Figure 3–64.

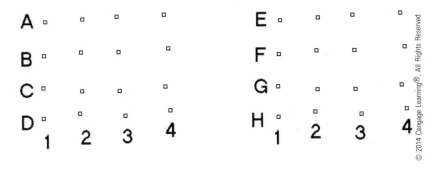

FIGURE 3–63 *Points and surfaces being constructed*

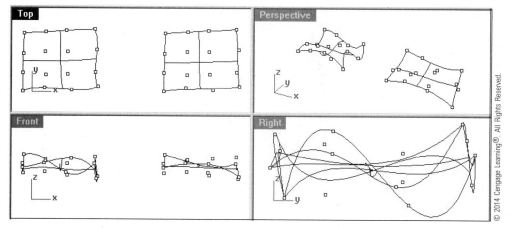

FIGURE 3–64 *Point grid surface*

Heightfield from Image Surface

A heightfield from image surface is an interesting way to construct a surface. It uses the color values of a bitmap image to define a matrix of points. Similar to a point grid surface, you can use these points as interpolation points on the surface or control points for the surface. To construct a heightfield from image surface, you select an image, indicate the location of the image on the construction plane, and then specify the number of sample points in two orthogonal directions and the height value of the sample points.

Basically, the bitmap can be color or black and white. However, it is more appropriate to use a black-and-white image because only the brightness value is considered when a surface is derived. If you use a black-and-white image, you can better perceive the outcome before making the surface. Naturally, you need a bitmap to make a surface of this type. Figure 3–65 shows digital black-and-white images being taken via digital camera. Perform the following steps.

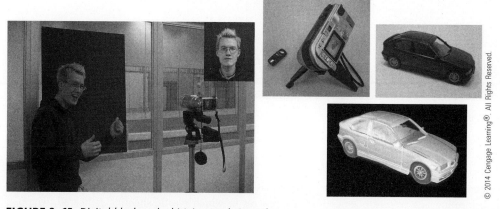

FIGURE 3–65 *Digital black-and-white images being taken*

1. Start a new file. Use the Small Objects, Millimeters template.
2. Maximize the Top viewport.
3. Select Surface > Heightfield from Image.
4. In the Select Bitmap dialog box, select the image file Car.tga from the Chapter 3 folder on the Student Companion Website.
5. Pick two points A and B in the Top viewport to indicate the size of the surface to be derived from the bitmap, as shown in Figure 3–66.
6. In the Heightfield dialog box, set the number of sample points to 200 times 200, set Height to 0.5, select Interpolate surface through samples, and then click on the OK button.

   A surface is constructed from the bitmap image, as shown in Figure 3–67. The surface is complete.
7. Do not save your file.

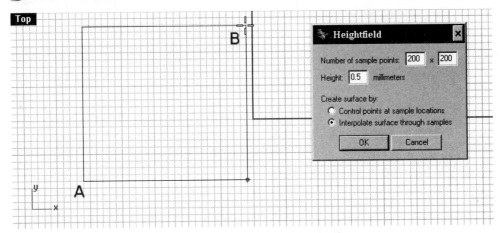

FIGURE 3–66 *Heightfield surface being constructed*
Source: Robert McNeel and Associates Rhinoceros® 5

FIGURE 3–67 *Heightfield surface constructed*

Drape Surface

Making a draped surface is analogous to wrapping a rectangular piece of plastic sheeting on a set of 3D objects in a way similar to vacuum forming. Vacuum forming is a type of plastics-forming process in which plastic sheeting is heated, wrapped onto a 3D object (the mold), and a vacuum is applied to deform the sheet. Figure 3–68 shows a vacuum-form machine. Perform the following steps.

1. Select File > Open and select the file Drape.3dm from the Chapter 3 folder on the Student Companion Website.
2. Select Surface > Drape.
3. Select locations A and B, shown in Figure 3–69.

    A draped surface is constructed, as shown in Figure 3–70.

4. Do not save your file.

FIGURE 3–68 *Vacuum-forming machine*

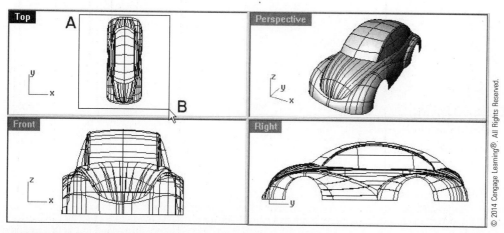

FIGURE 3–69 *Draped surface being constructed*

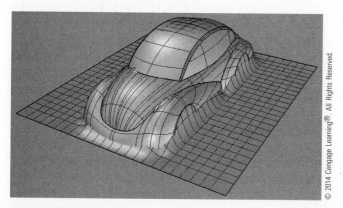

FIGURE 3–70 *Draped surface constructed*

PLANAR SURFACES

A planar surface is a flat surface, which is rectangular in shape. However, you may also construct a planar surface with irregular boundary edges.

Rectangular Planar Surfaces

Using Rhino's tool, you can construct planar rectangular surfaces in five ways: Corner-to-Corner, 3 Points, Vertical, Through Points, and Cutting Plane.

Corner-to-Corner Rectangular Surface

A corner-to-corner surface is a rectangular planar surface constructed by specifying two diagonal points of the surface. Perform the following steps.

1. Select File > Open and select the file Rectangle01.3dm from the Chapter 3 folder on the Student Companion Website.
2. Check the Point box on the Osnap dialog box.
3. Select Surface > Plane > Corner to Corner.
4. Select Deformable option at the command area.

> **NOTE** Deformable refers to how the surface can be deformed by manipulating its control points. In essence, it concerns the degree of polynomial of the surface produced. By default, a planar rectangular surface is degree 2. A deformable rectangular surface has a degree of 3 or above.

5. Select points A and B, shown in Figure 3–71.

   A rectangular planar surface is constructed.

3 Points Rectangular Surface

A 3 points surface is a rectangular planar surface constructed by specifying two endpoints of an edge and a point along the opposite edge of the surface. Continue with the following steps.

6. Select Surface > Plane > 3 Points.
7. Select points C, D, and E, shown in Figure 3–71.

   A rectangular planar surface is constructed.

Vertical Rectangular Surface

A vertical surface is a rectangular planar surface constructed perpendicular to a construction plane on which you specify two endpoints of an edge. Perform the following steps.

8. Select Surface > Plane > Vertical.
9. Select points F, G, and H, shown in Figure 3–71.

   A vertical planar surface is constructed.

10. Do not save your file.

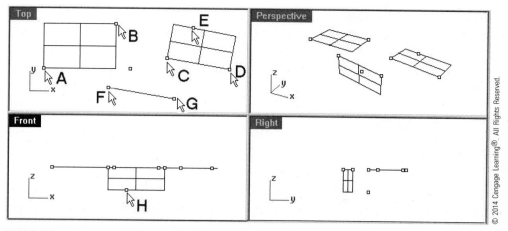

FIGURE 3–71 Rectangular planar surfaces constructed

Through Points Rectangular Surface

A through points surface is a planar rectangular surface constructed by interpolating among a set of points. Perform the following steps.

1. Select File > Open and select the file Rectangle02.3dm from the Chapter 3 folder on the Student Companion Website.
2. Select Surface > Plane > Through Points.
3. Select all the point objects and press the ENTER key.

    A planar surface is fitted through the points, as shown in Figure 3–72.
4. Do not save the file.

Cutting Plane Rectangular Surface

A cutting plane rectangle is a planar rectangular surface constructed by defining a section plane across a set of surfaces. Perform the following steps.

1. Select File > Open and select the file Rectangle03.3dm from the Chapter 3 folder on the Student Companion Website.
2. Select Surface > Plane > Cutting Plane.
3. Select surface A and curves B and C, shown in Figure 3–73, and press the ENTER key.
4. Select locations D and E, F, and G, and H and J.

    Exact location is unimportant for the purpose of this tutorial.
5. Press the ENTER key.

    Three rectangular planar surfaces that cut through the selected objects are constructed.
6. Do not save your file.

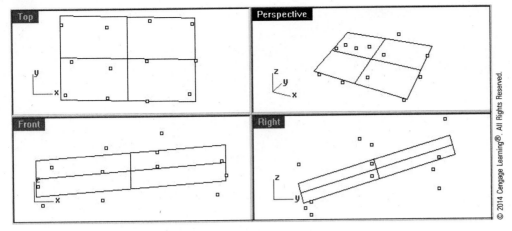

FIGURE 3-72 Planar surface fitted through a set of points

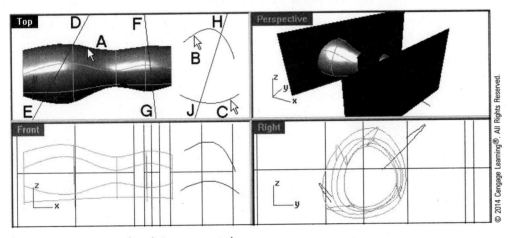

FIGURE 3-73 Cutting plane being constructed

Picture Frame Rectangular Surface

A picture frame rectangular surface is a rectangular surface with a specified bitmap image attached as texture. You can view the surface's mapped texture by setting the display to Rendered. Now perform the following steps:

1. Select File > Open and select the file PictureFrame.3dm from the Chapter 3 folder on the Student Companion Website.
2. Click on the Picture Frame button on the Surface Creation toolbar.
3. Select the image file "Bubble Car.tif" from the Chapter 3 folder on the Student Companion Website.
4. Click on point A, shown in Figure 3-74, and then point B.

    Note that the second selected point serves only to specify the width and orientation of the rectangular surface, because the height of the rectangle is governed by the bitmap's aspect ratio.

5. Do not save your file.

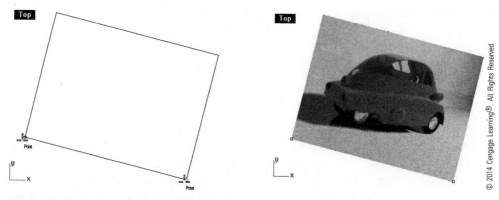

FIGURE 3–74 Points being selected (left) and picture frame constructed (right)

Planar Surface with Irregular Boundary Edges

A planar surface with irregular boundary edges is constructed from one or more planar curves. In essence, it is a trimmed surface. Rhino constructs a rectangular planar surface that is large enough to encompass the selected curves and then uses the curves to trim the surface. You will learn more about trimming a surface in Chapter 7. Perform the following steps.

1. Select File > Open and select the file PlanarTrim.3dm from the Chapter 3 folder on the Student Companion Website.
2. Select Surface > Planar Curve.
3. Select curves A, B, and C, shown in Figure 3–75, and press the ENTER key.
   Two trimmed planar surfaces are constructed.
4. Do not save the file.

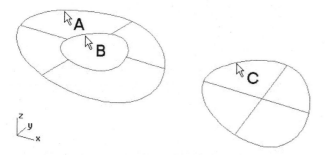

FIGURE 3–75 Trimmed planar surface being constructed (from a planar closed-loop curve)
© 2014 Cengage Learning®. All Rights Reserved.

DERIVED SURFACES

You can derive two kinds of surfaces from existing surfaces, obtaining a new surface by offsetting an existing surface and obtaining a 2D flat pattern of an existing 3D surface.

Offset Surfaces

Offset surfaces, which are used frequently in product styling, resemble in shape the existing surfaces and at a distance from them. There are two kinds of offset surfaces: uniform offset surface and variable offset surface.

Uniform Offset Surface/Polysurface

A polysurface is a set of surfaces joined together. (You will learn more about polysurfaces in Chapters 7 and 8.) To construct a surface or polysurface that resembles in shape to an existing surface or a polysurface and at a uniform distance from that surface or polysurface, you construct an offset surface/polysurface. Every point on the offset surface is equal in distance from the original surface. However, before you construct an offset surface/polysurface, you must note that offsetting can be done only if the resulting surfaces' radius of curvature at any one location is greater than zero. Perform the following steps to appreciate how offset surfaces can be made.

1. Select File > Open and select the file OffsetPolySurface.3dm from the Chapter 3 folder on the Student Companion Website.
2. Select Surface > Offset Surface.
3. Select A, shown in Figure 3–76.

    Note the tracking lines that indicate the normal direction. If you do not find them on your screen, set the color of the Tracking Lines option to black on the Appearance > Color tab of the Rhino Options dialog box.

    Among the five options available, the FlipAll option enables you to flip the offset direction and the Solid option enables you to construct a solid object. Rhino's solid modeling method will be explained in Chapter 8.

4. If the arrow direction is not the same as that shown in Figure 3–76 (left), select the FlipAll option from the command area.
5. Click on the Distance option and set the offset distance to 8.
6. Press the ENTER key.

    You will notice that offsetting cannot be done. It is because the distance set is so large that it causes the radius of curvature of a certain portion of the surface to reduce to zero or negative.

7. Repeat the command on the polysurface B again.
8. Set offset distance to 4 and select the Solid option to change it to Yes.

    An offset polysurface is constructed.

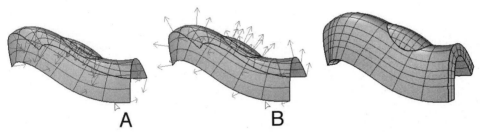

FIGURE 3–76 *From left to right: polysurface being offset inward, polysurface being offset outward, and solid constructed*
© 2014 Cengage Learning®. All Rights Reserved.

Variable Offset Surface

To meet certain aesthetic requirements, you may wish to have an offset surface with variable offsetting distances. However, you can offset only one surface element at a time, not the set of surfaces (polysurface) collectively. Continue with the following steps to construct a variable offset surface.

9. Undo the last command or open the OffsetSurface.3dm from the Chapter 3 folder on the Student Companion Website again.
10. Select Surface > Variable Offset Surface.
11. Select surface A, shown in Figure 3–77.
12. Click on points B, C, D, and E one by one and drag to new locations.
13. Press the ENTER key.

    A variable distance offset surface is constructed.

14. Do not save your file.

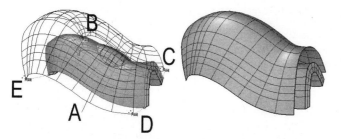

FIGURE 3–77 *Variable distance offset surface being constructed (left) and variable distance offset surface constructed (right)*
© 2014 Cengage Learning®. All Rights Reserved.

Surfaces Interpolating between Two Surfaces

In order to meet some aesthetic requirements in design, you may need to interpolate a series of surfaces between two existing surfaces. Now perform the following steps:

1. Select File > Open and select the file Tween Surface.3dm from the Chapter 3 folder on the Student Companion Website.
2. Select Surface > Tween Surfaces.
3. Select surfaces A and B, indicated in Figure 3–78.
4. At the command area, set NumberOfSurfaces=3.
5. Press the ENTER key.

   Three surfaces interpolating between the two selected surfaces are constructed.
6. Do not save your file.

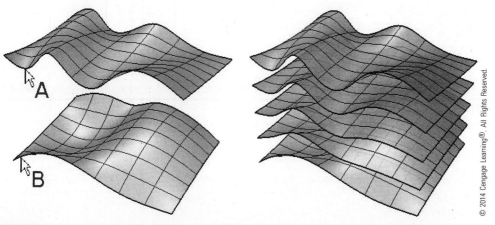

FIGURE 3–78 *Start and end surfaces selected (left) and interpolated surfaces constructed (right)*

2D Flat Pattern (Surface Development)

A way to produce a sheet metal product with a free-form shape is to first cut out a shape from a flat 2D sheet and then deform it by one of the many manufacturing methods, such as pressing. To figure out how the 2D sheet should look like, you need to construct a 2D pattern, or a surface development.

Unrolling Surfaces

The process of unrolling is useful in the sheet metal working industry in that a 3D sheet metal object is first unrolled into a 2D surface for subsequent forming operations. Basically, only single-curve surface can be unrolled into a 2D surface with an acceptable accuracy. Now perform the following steps.

1. Select File > Open and select the file Develop.3dm from the Chapter 3 folder on the Student Companion Website.

2. Select Surface > Unroll Developable Srf.
3. Select surface A, indicated in Figure 3-79, and then press the ENTER key.
   The surface is unrolled, as shown in Figure 3-79.
4. Do not save your file.

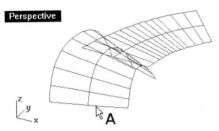

FIGURE 3-79 *Surface being unrolled and unrolled surface*
© 2014 Cengage Learning®. All Rights Reserved.

Smashing

To obtain better approximation of the flat pattern of a compound curve surface or a polysurface, you should use the Smash command instead. However, it must be noted that this command is useful for only the surface that is not extremely curved. Now perform the following steps.

1. Select File > Open and select the file Smash.3dm from the Chapter 3 folder on the Student Companion Website.
2. Select Surface > Smash.
3. If Label=No, select it to change it to Yes.
4. Select surface A, shown in Figure 3-80.
5. At the command area, click on LinearDirection.
6. Click on V and press the ENTER key.
7. Select curve B and press the ENTER key.

   A smashed surface is constructed. If you apply this command on a polysurface, you will obtain a separate smashed surface for each individual surface element of the polysurface.

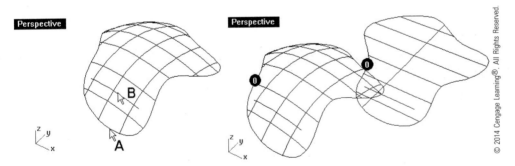

FIGURE 3-80 *Free-form surface (left) and smashed surface (right)*

8. Repeat the command and set Linear Direction to U.
   The smashed surface will have a different orientation, and the smashed line's length will differ slightly.
9. Do not save your file.

Flattening Curve Edges

Apart from smashing, it is possible to construct a 2D curve by projecting the edges of a developable surface through flattening. Again, the outcome is only an approximation. Perform the following steps:

1. Select File > Open and select the file Flatten.3dm from the Chapter 3 folder on the Student Companion Website.

2. Right-click on the Unroll Developable Surface/Flatten Surface button of the Surface Tools toolbar.
3. Select edges A and B (Figure 3-81).
4. Press the ENTER key to accept the default sample spacing.

   A set of 2D curves is constructed.
5. Do not save your file.

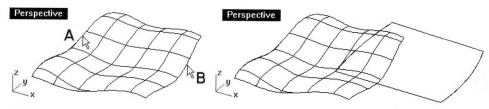

FIGURE 3-81 *Developable surface (left) and edges flattened (right)*
© 2014 Cengage Learning®. All Rights Reserved.

Squishing and Squishing back

The outcome of squishing is quite similar to smashing in that an approximated flat pattern of a free-form surface can be obtained. However, with squishing, you can construct objects on the squished flat surface and squish the newly constructed object back to the original surface. Now perform the following steps:

1. Select File > Open and select the file Squish.3dm from the Chapter 3 folder on the Student Companion Website.
2. Type Squish at the command area.
3. Select A, indicated in Figure 3-82.

   An approximated flat pattern of the selected surface is constructed.

Now you will construct a circle at the intersection point of the squished surface's isocurves. Before you can do that, you need to extract the isocurves. (Manipulation of curves and construction of curves from existing objects will be discussed in more detail in later chapters.)

4. Enable object snap and click on the Mid and Inter buttons of the Object Snap (Onap) toolbar.
5. Select Curve > Curve From Objects > Extract Isocurve.
6. Click on the intersection point C of surface B in Figure 3-82.
7. Press the ENTER key.

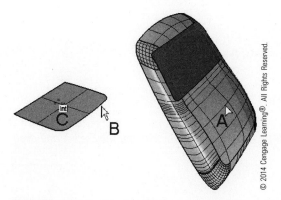

FIGURE 3-82 *Squished surface constructed (left) and isocurves being extracted (right)*

8. Select Curve > Circle > Center, Radius.
9. Click on midpoint A, shown in Figure 3-83.

10. Type 20 at the command area.

    A circle is constructed, as shown in Figure 3–83.

11. Type Squishback at the command area.

12. Select squished surface B and the circle C, shown in Figure 3–83, and press the ENTER key.

    The circle is transferred to the original free-form surface at the location corresponding to the location of the circle you constructed on the squished surface.

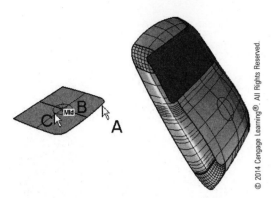

FIGURE 3–83 *Circle constructed on the squished surface (left) and transformed to the original surface (right)*

LIGHTWEIGHT EXTRUSION OBJECTS

In order to reduce memory requirement for construction of models with a large number of extrusion surfaces, Rhino 5 introduces a new kind of data structure called extrusion surface, which requires less memory and less computer resources.

You can set the default of whether to construct extrusion objects or a polysurface as well as converting an extrusion object to an ordinary NURBS surface, as follows:

1. Select File > Open and select the file Convert.3dm from the Chapter 3 folder on the Student Companion Website.
2. Type UseExtrusion at the command area.
3. Select Extrusion option at the command area.
4. Select Surface > Extrude Curve > Straight.

    Now the default is lightweight extrusion object.

5. Select curve A, indicated in Figure 3–84, and press the ENTER key.
6. Click on B to indicate the distance of extrusion. Exact location and whether extrusion is both side or not are unimportant here.

    A lightweight extrusion object is constructed.

If for any reason you need to convert it back to a NURBS surface, continue with the following steps.

7. Type ConvertExtrusion at the command area.
8. Click on surface C, indicated in Figure 3–84, and press the ENTER key.
9. Set DeleteInput=No and press the ENTER key.

    The lightweight extrusion object is retained, and an ordinary NURBS surface is constructed. You may select and drag them apart.

10. Do not save your file.

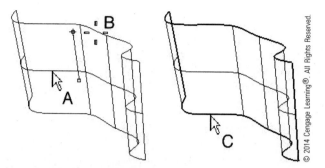

FIGURE 3–84 *Curve being extruded to a lightweight extrusion object (left) and the lightweight extrusion object being converted to an ordinary NURBS surface (right)*

CONSOLIDATION

Construction of a surface model to represent a 3D free-form object involves a top-down thinking process to think about how to decompose the object into manageable pieces of individual surfaces that can be constructed in the computer, how to construct these individual surfaces, and how to treat the edges between contiguous surfaces.

In order to be able to construct these individual surfaces, you need to know the repertoire of surfaces that you can construct in Rhino. For the sake of easy reference, we divide Rhino's NURBS surfaces logically into four categories.

The first surface category concerns the four major types of surfaces that are, in essence, common to most computer-aided design applications. They are extruded surfaces, revolved surfaces, swept surfaces, and lofted surfaces. Among them, extruded surface is further divided into six types (straight, tapered, to point, along curve, ribbon, and fin); revolved surface into two types (normal revolve and revolve rail); swept surface into two types (sweep along one rail and sweep along two rails); and lofted surface into three types (loft, edge curve, and curve network). The second surface category includes Patch, Corner Points, Point Grid, Heightfield from Image, and Drape. The third surface category is planar surfaces. The fourth category is surfaces derived from existing surfaces. You can derive an offset surface, which offsets uniformly or nonuniformly. You can derive a flat pattern if the surface is a developable surface or has an approximated development.

In this chapter, you have examined various kinds of surfaces and learned how some of them are constructed from a set of curves and/or points. Because the curves and points are already given, constructing the surfaces is simple. However, making the curves and points can be a tedious job. Therefore, you will learn various techniques in point and curve construction in Chapters 4 through 6. However, if you are already familiar with Rhinoceros's curve construction and manipulation tools, you may proceed to Chapter 7 to continue with ways to manipulate NURBS surfaces.

REVIEW QUESTIONS

1. Explain key concepts of surface modeling.
2. What kinds of curve and surface continuity can be achieved in Rhino?
3. List the four major kinds of Rhino surfaces that are common to most computer-aided design applications.
4. Apart from the free-form surfaces common to most computer-aided design systems, what other kinds of free-form surfaces can you construct using Rhino?
5. How many ways can a planar surface be constructed? What are they?
6. List the two kinds of derived surfaces.

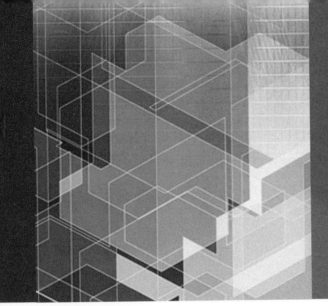

CHAPTER 4

Free-Form NURBS Curves and Point Objects

INTRODUCTION

This chapter explains the concepts of NURBS curves and illustrates various methods of constructing and manipulating free-form curves and point objects.

OBJECTIVES

After studying this chapter, you should be able to

- Construct free-form curves and point objects for building surface models

OVERVIEW

After learning how free-form surfaces can be constructed from a framework of points and/or curves in Chapter 3, this chapter and Chapters 5 and 6 provide you with a solid understanding of curve and point construction. In this chapter, you will learn how to use Rhino as a tool for constructing and manipulating free-form curves and point objects. In Chapter 5, you will learn how to construct curves of regular pattern. Chapter 6 provides more information about Rhino curve tools. After equipping yourself with skills in 3D curve manipulation, in Chapters 7 and 8 you will continue to perform various methods of surface and solid construction and manipulation.

CURVES FOR SURFACE MODELING

Although the use of curves alone in 3D computer modeling is diminishing, you still need to learn how to construct 3D curves because they are required in the construction of surfaces and solids, particularly in free-form surface modeling. As explained in Chapter 3, surface modeling requires prior thinking about what surfaces are needed to represent an object as well as thinking about the curves for building the surfaces. Because there are many ways to construct a surface, you need to know various ways of constructing curves for surface modeling. Figure 4–1 shows four 3D curves and a surface constructed from the curves.

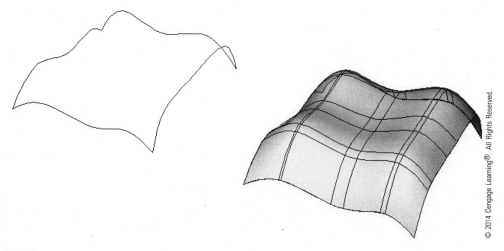

FIGURE 4–1 *Curves (left) and the surface constructed from them (right)*

Of the many types of curves you can use to construct 3D surfaces and solids, the spline is the most important to the designer because it enables you to construct complex free-form shapes.

Spline

Simply speaking, a spline is a free-form curve. Although the focus of this book is on the application of Rhino's tools in design and not on mathematics, you still need to understand some of the characteristics in order to prepare yourself for making 3D free-form surfaces. Mathematically, a spline is a way of defining a free-form curve by specifying two endpoints and two or more tangent vectors that control the profile of the curve. There are many mathematical methods of defining a spline. Some of the most popular methods are outlined in the sections that follow.

Polynomial Spline

A polynomial spline is depicted by a polynomial equation in the computer. It is a set of contiguous spline segments. At the endpoints of each spline segment there is a tangent vector having a direction and magnitude, as shown in Figure 4–2. (You will learn how to manipulate these tangents later in this chapter.) The effect of the tangent defines the curvature of the segment. Because the tangent vector is described by a polynomial equation, the spline is called a polynomial spline.

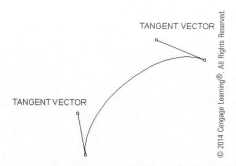

FIGURE 4–2 *Polynomial spline segment and its tangent vectors*

How the tangent vector affects the shape of the spline segment is determined by the degree of the polynomial equation. Simply speaking, the degree of polynomial refers to the power of the x, y, and z values of the polynomial equation describing the curve. Therefore, the x, y, and z values of the equation depicting a degree 2 curve have a power of 2 and those of a degree 3 curve have a power of 3. As designers using Rhino to construct a free-form curve, the exact mathematics is unimportant to us. However, you might need to know that a degree 1 curve is a rigid curve, like a straight line. (You can use the analogy of a steel rod so rigid that it cannot be bent.) Another important issue you need to realize is that the higher the polynomial degree, the more flexible is the curve. For example, a degree 3 curve is analogous to a bamboo, and an even higher degree curve is analogous to a very thin flexible steel wire.

B-Spline

A B-spline is a kind of multisegment spline curve. A connection point between two contiguous spline segments is called a knot. Being an extension of the polynomial spline, each spline segment is formed from the weighted sum of four local polynomial basis functions. Hence, the spline is called a basis spline (or B-spline). Four control points lying outside the spline control the shape of a B-spline segment, as shown in Figure 4–3. Movement of a control point affects only four segments of the curve.

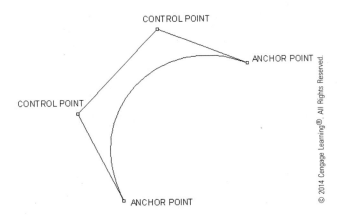

FIGURE 4–3 *B-spline segment and its control points*

Nonuniform Rational B-Spline

This is the kind of spline that is used in most surface modeling tools. Mathematically, a nonuniform rational B-spline (NURBS) curve is a derivative of the B-spline curve. It has two additional characteristics: It is nonuniform and rational. Unlike the uniform B-spline, in which a uniform parameter domain defines each spline segment, the parameter domain of a nonuniform B-spline need not be uniform. Because of the nonuniform characteristic, different levels of continuity between the spline segments can be attained, unequal spacing between the knot points is allowed, and the spline can more accurately interpolate among a set of given points. With a rational form, the curve can better represent conic shapes. To the designer, one major advantage of using NURBS mathematics is that you can trim a curve at any point and the curve will retain its original shape.

RHINO CURVES

Common to most other contemporary surface modeling tools, Rhino uses NURBS mathematics to define curves and surfaces. In essence, a NURBS curve is a set of connected spline segments, and the joint between two contiguous spline segments is a knot. The degree of the polynomial equation, the control point location, and the weight of the control points determine the shape of each spline segment.

For ease of classification, we will divide Rhino's curves into two major types: free-form curves and curves with regular pattern such as line, arc, and circle. In this chapter, we will focus on Rhino's free-form curves and then proceed to curves of regular pattern in the next chapter. Because all Rhino curves are NURBS curves, methods used to manipulate a free-form curve delineated in this chapter also apply to other types of curves.

To help you gain a deep understanding of NURBS curve manipulation and be confident in using it to design free-form surfaces, we will examine various ways to construct a free-form NURBS curve and methods to modify it.

Basic Curve Construction

When making surfaces, you select curves and points that are already constructed. But you need to specify a location to construct individual point objects and basic curves. To specify a location, you use your pointing device to select a point in one of the viewports, or input a set of coordinates at the command area. To help construct 3D points, you can use the elevator mode (holding down the CONTROL key to select two locations from two viewports to compose a 3D point) and the planar mode (checking the Planar button on the Status bar).

As explained in Chapter 2, each viewport includes a construction plane. Selecting a point in one of the viewports specifies a location on the corresponding construction plane. To key in a set of coordinates, you use either the construction plane coordinate system or the World coordinate system. The X and Y directions of the construction plane coordinates correspond to the X (red in color, by default) and Y (green in color, by default) axes of the construction plane. Therefore, typing the same construction plane coordinates (other than the origin) at the command area results in different locations in 3D space, depending on where the cursor is placed, for example, on the Top viewport as opposed to the Front viewport. In the following exercises, you will use the pointing device to select points in viewports.

Interpolated Curve and Control Point Curve

Using Rhino, there are many ways to construct a free-form curve. Two basic ways are specifying a set of points through which the curve has to interpolate and setting the control points of the curve.

Interpolated Curve

To construct an interpolated curve, you specify a number of points through which the curve has to interpolate. Practically, you may regard this type of curve as a best-fit curve through the selected points. As a result, this method is considered to be more user-friendly because you can better perceive the shape of the curve while constructing it. Perform the following steps.

1. Select File > Open and select the file FreeFormCurve01.3dm from the Chapter 4 folder that you downloaded from the Student Companion Website.

In the file, you will find four-point objects already constructed for you. Now you will construct a free-form curve passing through these points.

2. Check the Osnap button on the Status bar.
3. In the Osnap dialog box, check the Point box.
4. Select Curve > Free-Form > Interpolate Points.
5. Select points A, B, C, and D, shown in Figure 4–4, and press the ENTER key to terminate the command.
   An interpolated curve is constructed passing through the selected points.
6. Do not close the file.

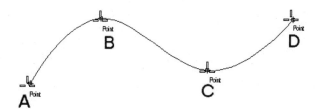

FIGURE 4–4 *Interpolated curve constructed*
© 2014 Cengage Learning®. All Rights Reserved.

> **NOTE** You may go back to Chapter 2 to learn how to input 3D points by holding down the CONTROL key and clicking on two adjacent viewports.

Control Point Curve

The control point curve is more mathematically based. It is constructed by specifying the locations of the control points of the curve. You use this method to construct a curve when you know only the locations of the control points rather than the points through which the curve passes.

Continue with the following steps to construct a control point curve using the point objects as control point locations.

7. Set current layer to Layer01.
8. Select Curve > Free-Form > Control Points.
9. Select points A, B, C, and D, shown in Figure 4–5, and then press the ENTER key to terminate the command.
   A control point curve is constructed. Note the difference in shape between the interpolated curve and the control point curve in that the resulting control point curve does not pass through input points.

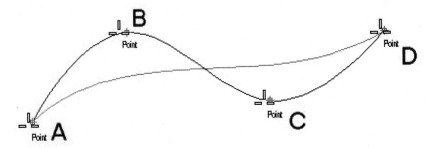

FIGURE 4–5 *Control point curve constructed*
© 2014 Cengage Learning®. All Rights Reserved.

Comparing Interpolated Curve and Control Point Curve

Regardless of the method you use to construct a curve, a set of control points governs the shape of the curve. To appreciate the difference between the interpolated curve and the control point curve in terms of the number of control points and their locations, let us turn on their control points, as follows.

10. Select Edit > Control Points > Control Points On.
11. Select the free-form curves and press the ENTER key.

    The control points are turned on, as shown in Figure 4–6.

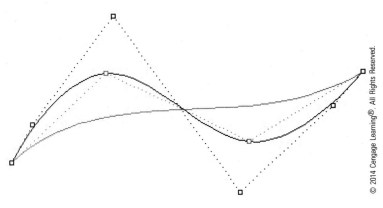

FIGURE 4–6 *Control points turned on*

The control point curve (curve constructed by specifying four control points) has control points coincident with the point objects. The interpolated curve (curve constructed by specifying four interpolated points) has six control points, with the curve showing more segments. Excepting the endpoints, both curves do not pass through the control point locations. Turn off the control points, as follows.

12. Select Edit > Control Points > Control Points Off.
13. Do not save your file.

Control Point Location

Normally, control points lie outside the curve. In an open-loop curve, only the first and last control points coincide with the endpoints of the curve. These endpoints are also called anchor points. In a closed-loop curve, all control points lie outside the curve, an example of which is shown in Figure 4–7 (right).

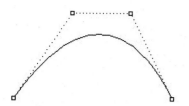
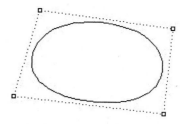

FIGURE 4–7 *Control point locations of an open-loop curve (left) and a closed-loop curve (right)*
© 2014 Cengage Learning®. All Rights Reserved.

Polynomial Degree Concepts

As stated earlier, a NURBS curve is a kind of polynomial spline, and the degree of a polynomial equation has a direct impact on the complexity of the shape of the curve. A higher polynomial degree curve has more control points than one with a lower polynomial degree. Basically, a line is a degree 1 NURBS curve, an arc is a degree 2 NURBS curve, and free-form curve is a NURBS curve of degree 3 or above.

Changing the Polynomial Degree of a Curve

One of the ways to modify a curve is to raise or reduce its polynomial degree. Simply raising the degree of a polynomial spline curve does not change the curve's shape, but it does increase the number of control points. In turn, more control points enable you to modify the curve to create a more complex shape. (You will learn how to modify the shape of a curve by manipulating its control points in the next paragraph.) On the other hand, reducing the degree decreases the number of control points and hence simplifies the shape of the curve. Perform the following steps.

1. Select File > Open and select the file FreeFormCurve02.3dm from the Chapter 4 folder on the Student Companion Website.

   Here, you have three identical curves. Their degree of polynomial is 3.

2. Select Edit > Control Points > Control Points On.
3. Select the free-form curves A, B, and C, shown in Figure 4–8, and press the ENTER key.

   The control points are turned on.

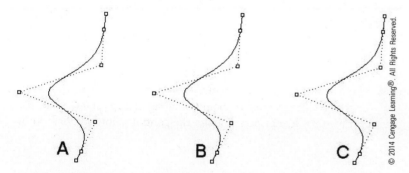

FIGURE 4–8 *Control points turned on*

4. Select Edit > Change Degree.
5. Select curve A (Figure 4–9) and press the ENTER key.
6. Type 2 at the command prompt to change the degree of polynomial to 2.
7. Repeat the Change Degree command.
8. Select curve B (Figure 4–9) and press the ENTER key.
9. Type 5 at the command prompt to change the degree of polynomial to 5.

   As can be seen, the number of control points in curve A is reduced and in curve B is increased. In regard to the shape of the curve, curve A is simplified but curve B remains unchanged.

10. Do not save your file.

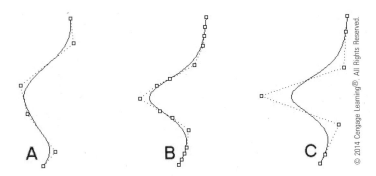

FIGURE 4-9 *Degree of polynomial being changed*

Control Point Manipulation

Understanding that the degree of polynomial has a direct influence on the number of control points, you will now learn how to manipulate control points to modify a curve. You can change the location and the weight of the control points. You can also delete control points to simplify the shape of the curve.

Moving Control Points

A way to change the shape of a curve is to move the location of the curve's control points. You may select the control point, hold down the mouse button, and drag it to a new position. To move the control point in a more precise way, you can use the nudge keys, which is the combination of the ALT key and the arrow keys. To move the control point in three directions independent of the current construction plane, you can use the buttons on the Organic toolbar. Perform the following steps.

1. Select File > Open and select the file FreeFormCurve03.3dm from the Chapter 4 folder on the Student Companion Website.
2. Turn on the control points of the curve.
3. Click on control point A, indicated in Figure 4-10, and drag it to point B.

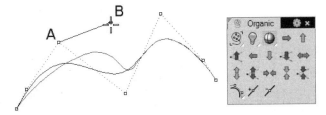

FIGURE 4-10 *Dragging a control point to change the shape of the curve (left) and Organic toolbar (right)*
Source: Robert McNeel and Associates Rhinoceros® 5

> **NOTE** Two other ways by which you can move the control points are holding down the ALT key and then pressing the arrow keys, and using the buttons on the Organic toolbar.

Dragging Orientation

Apart from using the Organic toolbar, you can use the DragMode command as explained in Chapter 2 to change the drag orientation.

Deleting Control Points

By deleting one or more control points of a curve, you simplify its shape. Continue with the following steps.

4. Select control point A, shown in Figure 4–11.
5. Press the DELETE key.

The selected control point is deleted, and the curve is simplified, in terms of the number of control points.

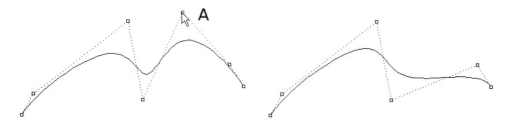

FIGURE 4–11 *Control point to be deleted (left) and control point deleted (right)*
© 2014 Cengage Learning®. All Rights Reserved.

> In addition to turning on a control point, selecting a control point, and pressing the DELETE key to remove a control point, you can click on the Remove a control point button on the Point Edit toolbar and then select a curve to remove unwanted control points.

NOTE

Inserting Control Points

You can add control points to a curve, as follows:

6. Select Edit Control Points > Insert Control Point.
7. Select curve A and then location B, shown in Figure 4–12.

A control point is added to the curve.

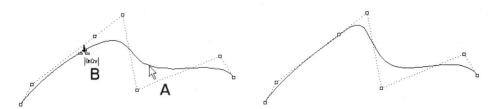

FIGURE 4–12 *Control point being added (left) and control point added (right)*
© 2014 Cengage Learning®. All Rights Reserved.

Adjusting Control Point Weight

The weight of control points has a significant effect on the shape of a spline segment. You can regard the weight of a control point as a pulling force that pulls the spline curve toward the control point. The higher the weight, the closer the curve will be pulled to the control point, an example of which is shown in Figure 4–13 (right). To edit the weight of a control point, continue with the following steps.

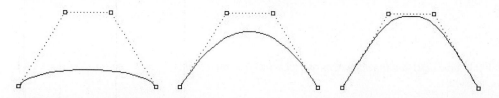

FIGURE 4–13 *Control points with different weights (from left to right: weight 0.1, 5, and 7)*
© 2014 Cengage Learning®. All Rights Reserved.

8. Select Edit > Control Points > Edit Weight.
9. Select control point A (you can select multiple points) and press the ENTER key.
10. In the Set Control Point Weight dialog box, shown in Figure 4–14, set the weight of the selected point(s) to 0.2, and then click on the OK button.

    The curve is pulled more closely to the selected control point.
11. Do not save your file.

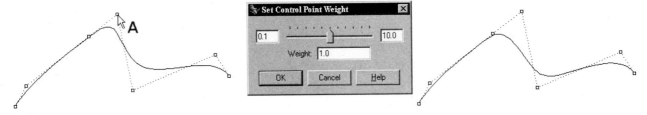

FIGURE 4–14 *Point weight being modified in the Set Control Point Weight dialog box from its default value of 1 (left) to 0.2 (right)*
Source: Robert McNeel and Associates Rhinoceros® 5

Spline Segments

Because a NURBS curve is a polynomial spline with a number of segments connected end-to-end at the segments' endpoints, called knots, the number of segments has a direct impact on the number of control points. A curve with more segments has more control points than one with fewer segments. A way to modify the number of segments of a curve is to add or remove the knots of the curve. Adding knots to a NURBS curve increases the number of spline segments without changing the shape of the curve. However, a curve with more segments has more control points. Subsequently, you can modify the curve to create a more complex shape.

Inserting Knot Points

A knot is the junction between two contiguous spline segments. One way to increase the complexity of the shape of a curve is to increase the number of spline segments by inserting knots along the curve. In the following exercise, you will add a knot to the curve. Adding knots to a curve increases the number of spline segments. As a result, there will be more control points. However, the shape of the curve will not change until you manipulate the control points. Perform the following steps.

1. Select File > Open and select the file FreeFormCurve04.3dm from the Chapter 4 folder on the Student Companion Website.
2. Turn on the control points of the curve.
3. Select Edit > Control Points > Insert Knot.
4. Select curve A, shown in Figure 4–15.
5. Select location B, shown in Figure 4–15, along the curve where you want to insert a knot.

   Note that you can insert multiple knot points.
6. Press the ENTER key to terminate the command.

   A knot is inserted.

FIGURE 4–15 *Knot being inserted in the curve (left) and the number of control points increased (right)*
© 2014 Cengage Learning®. All Rights Reserved.

Comparing Figure 4–15 (left) with Figure 4–15 (right), you will find that there is an increase in the number of control points. But there is no change in the shape of the curve. Depending on the exact location where you insert the knot, Figure 4–15 (right) may not look the same as that displayed in your screen.

Kink Point

A spline curve is analogous to bamboo: You can bend bamboo to a great degree and still retain a very smooth curve. If we continue to bend the bamboo further, it will fracture at a certain point. The fracture point is equivalent to a kink in a spline curve.

Mathematically, a kink point is a special type of knot on a curve in which the tangent directions of the contiguous spline segments are not the same. A kink occurs when you join two curves with different tangent directions, or when you explicitly add a kink to a curve. You add a kink point to a curve as follows.

7. Select Edit > Point Editing > Insert Kink.
8. Select curve A, shown in Figure 4–16 (left).
9. Select location B, shown in Figure 4–16 (left), along the curve where you want to insert a kink.
   Note that you can insert multiple kink points.
10. Press the ENTER key to terminate the command.
    A kink is inserted.
11. Select kink point C, shown in Figure 4–16 (right); hold down the left mouse button; and drag the mouse to a new location to see the effect of a kink.

FIGURE 4–16 *Kink point being inserted (left) and inserted kink point being moved (right)*
© 2014 Cengage Learning®. All Rights Reserved.

Removal of Knot and Kink Points

To remove a kink point or to reduce the complexity of a curve, we remove the kink points or knots along the curve. Because removing knots on a curve reduces the complexity of the curve, the curve's shape will change. To remove a kink point or other type of knot from a curve, perform the following steps.

12. Select Edit > Control Points > Remove Knot.
13. Select curve A, shown in Figure 4–17 (left).
14. Select knots B and C, shown in Figure 4–17 (left), and press the ENTER key.
    A knot of the curve is removed, as shown in Figure 4–17 (right).
15. Do not save your file.

FIGURE 4–17 *Knots being selected (left) and knot removed (right)*
© 2014 Cengage Learning®. All Rights Reserved.

After you remove a knot from a curve, the number of control points decreases. As a result, the shape of the curve is simplified and changed. Note that Figure 4–17 (right) may not be the same as that shown in your screen because the exact locations of the inserted kink and knot points may be different.

Multiple Knots at a Location

To discover if there are any multiple knots at the same location on a curve, you may type RemoveMultiKnot at the command area or click on the Remove Multiple Knots from Surface or Curve button on the Surface Tool toolbar.

Note that this command is also applicable to removing multiple knots on a surface.

SPLINE SEGMENT, POLYNOMIAL DEGREE, AND CONTROL POINT

Because the number of control points of a NURBS curve depends on both the number of spline segments and the polynomial degree, you may increase the segment number or the polynomial degree to increase the number of control points. For better results, it is recommended that you keep the polynomial degree to 3 and increase or decrease the number of segments to modify the number of control points.

Periodic Curve

A periodic curve is a special kind of curve in that it is a closed-loop curve with no kink point. Figure 4–18 shows a closed-loop curve with a kink and a periodic curve. To make a curve periodic, perform the following steps.

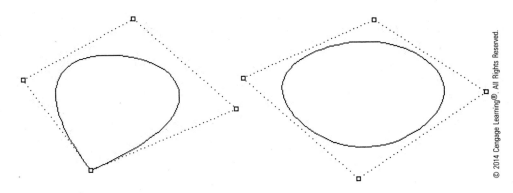

FIGURE 4–18 *Closed-loop curve with a kink (not a periodic curve) (left) and a periodic curve (right)*

1. Select File > Open and select the file FreeFormCurve05.3dm from the Chapter 4 folder on the Student Companion Website.
2. Turn on the control points so that you can see the effect of making the curve periodic.
3. Select Edit > Make Periodic.
4. Select curve A, shown in Figure 4–19, and press the ENTER key.
5. Select the Yes option at the command area to delete the original curve.

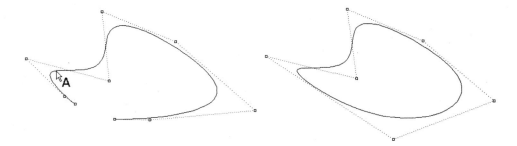

FIGURE 4–19 *Original curve (left) and periodic curve (right)*
© 2014 Cengage Learning®. All Rights Reserved.

> **NOTE** The same command can be applied to a surface to change it to a *periodic surface*. Details will be discussed in Chapter 7.

Adjusting the Seam of a Closed Curve

Although a periodic curve is a closed curve, it contains a seam point that is both the start and endpoints of the curve. If you construct a surface from one or more periodic curves, the location of the seam point may affect the final outcome of the surface. Therefore, you may want to adjust the seam point location, as follows.

6. Select Curve > Curve Edit Tools > Adjust Closed Curve Seam.
7. Select curve A (Figure 4–20) and press the ENTER key.

8. Click on the seam point and drag it along the curve.
9. Press the ENTER key.
   The seam point location is changed.
10. Do not save your file.

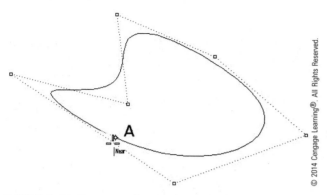

FIGURE 4–20 *Seam point location being changed*

Edit Point

Because control points are normally not lying along the curve, you may find it difficult to make a curve pass through designated locations by manipulating the control point. To add flexibility in modifying a NURBS curve, Rhino enables you to use edit points along the curve. Edit points are independent of degree, control points, and knots of a curve.

Edit Point Manipulation

Edit points lie along a curve. To change the shape of a curve, you display the edit points, select them, and drag them to new locations. To turn on the edit points of the curve and manipulate the points to modify the curve, perform the following steps.

1. Select File > Open and select the file FreeFormCurve06.3dm from the Chapter 4 folder on the Student Companion Website.
2. Set object snap mode to Point.
3. Select Edit > Control Points > Show Edit Points.
4. Select curve A, shown in Figure 4–21, and press the ENTER key.
5. Select edit point B (Figure 4–21), hold down the left mouse button, and drag the mouse to point C (Figure 4–21).
   The curve is modified.
6. Do not save the file.

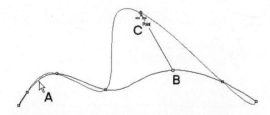

FIGURE 4–21 *Selected edit point being dragged*
© 2014 Cengage Learning®. All Rights Reserved.

POINT EDITING

The shape of a curve is controlled by the location and weight of control points. Except for the two anchor points at the endpoints of a curve, all the control points of the curve lie outside the curve, so it may be more predictable to manipulate the edit points along the curve. Apart from manipulating the control points or the edit points, you can add or remove knots from a curve and add kink points to a curve.

Handlebar Editor

Edit points enable you to modify a curve only in a limited scope. To change the tangent direction and the location of any point along a curve, you use the handlebar. The handlebar is a special type of editing tool consisting of a point on the curve and two tangent lines. Moving the central point of the handlebar changes the location of a point along the curve. Selecting and dragging the endpoints of the handlebar changes the weight and tangent direction of the curve at a selected location. Continue with the following steps to modify the curve by using the Handlebar Editor.

7. Select Edit > Point Editing > Handlebar Editor.
8. Select the curve, shown in Figure 4–22.

    The handlebar is displayed at the point where you selected the curve.

9. Click on an endpoint of the handlebar and drag, as shown in Figure 4–22.
10. Press the ENTER key.

    The curve is modified.

Using a handlebar to edit the shape of a curve, the curve profile changes, but the location of the selected point at the handlebar does not change.

11. Do not save your file.

FIGURE 4–22 *Handlebar being displayed (left) and handlebar being manipulated (right)*
© 2014 Cengage Learning®. All Rights Reserved.

Handle Curve

Having learnt how to use the handlebar to modify a curve, we will now introduce the use of handlebar to construct a curve, as follows.

1. Start a new file. Use the Small Objects, Millimeters template.
2. Maximize the Top viewport.
3. Select Curve > Free-Form > Handle Curve.
4. Click on location A, shown in Figure 4–23, to specify the first curve point location.
5. Click on location B (Figure 4–23) to specify the first handle location.

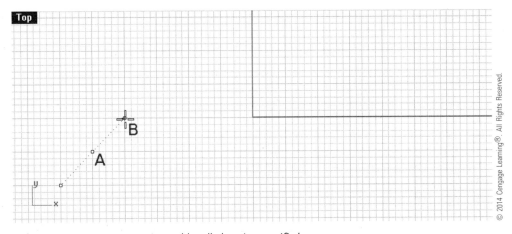

FIGURE 4–23 *First curve point and handle location specified*

6. Click on location C, shown in Figure 4–24, to specify the first curve point location.
7. Click on location D (Figure 4–24) to specify the first handle location.

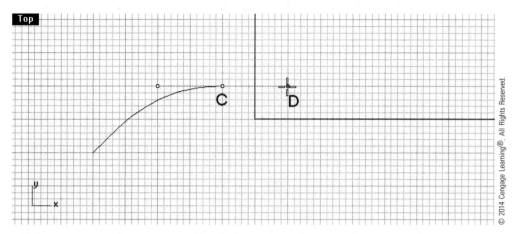

FIGURE 4–24 *Second curve point and handle location specified*

8. Click on location E, shown in Figure 4–25, to specify the first curve point location.
9. Click on location F (Figure 4–25) to specify the first handle location.
10. Press the ENTER key.

    A curve is constructed. You may turn its control points on to discover how complex the curve is.

11. Do not save your file.

If the CONTROL key is pressed while placing a handle point, the previous point will be moved to a new location.

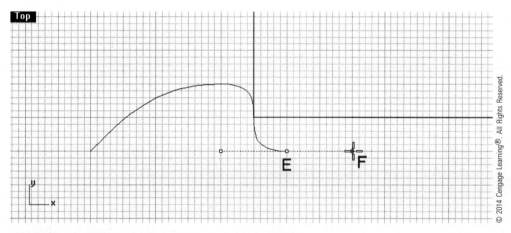

FIGURE 4–25 *Third curve point and handle location specified*

Sketch Curve

Apart from specifying the interpolated points or the control points and using the handlebar, another way to construct a free-form sketch is to use sketching technique. With sketching technique, you use the mouse as a drafting pen to construct a curve. Perform the following steps to construct a sketch curve.

1. Start a new file. Use the Small Objects, Millimeters template.
2. Maximize the Top viewport.
3. Select Curve > Free-Form > Sketch.
4. Select location A, shown in Figure 4–26.

5. Hold down the left mouse button and drag the cursor to location B.
6. Release the mouse button.
7. Press the ENTER key.

   A free-form sketch curve is constructed. Unlike the other three methods (specifying the control points, specifying the edit points, and using the handlebar), this method of constructing a curve lets you use your mouse as an electronic drawing pen. It is particularly useful if you embed a background bitmap in the viewport as a reference.

8. Do not save your file.

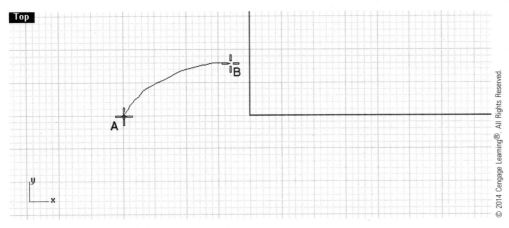

FIGURE 4–26 *Sketch curve being constructed*

Zooming 1 to 1 Calibration

While you are constructing free-form sketches, you may better realize the actual size of your sketch by setting the zoom scale as follows.

1. Select View > Zoom > Zoom 1 to 1 Calibration.
2. Take out a ruler to measure the length of the horizontal blue bar of the Zoom 1 to 1 Calibration dialog box, shown in Figure 4–27.
3. Input the measured distance at the Length of bar box.
4. Click on the OK button.

   The zoom scale is set.

5. To zoom the display, select View > Zoom > Zoom 1 to 1.

 NOTE You may apply a scale factor while inputting the measure *distance. This way, you can have a display that is zoomed to a known scale factor.*

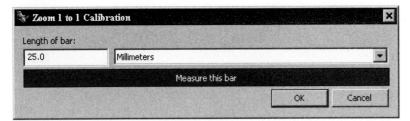

FIGURE 4–27 *Zoom 1 to 1 Calibration dialog box*
Source: Robert McNeel and Associates Rhinoceros® 5

Tracing Background Images

Knowing that the fundamental issue of curve construction is to specify a location, you need to be able to use the pointing device to select a point or input the coordinates at the command area. To help determine the locations, you might sketch your design idea on graph paper. From the graph paper, you extract point locations. Another way of using the sketch is to scan it as a digital image and insert it in the viewport as a background image. If the image is properly scaled, you can directly sketch in the viewport. When a background image is selected, point objects will be displayed at the four corners. You can click on these corner points to manipulate the image. To try this way of sketching, perform the following steps.

1. Obtain a piece of graph paper.
2. Use a pencil to construct a freehand sketch depicting the curve you want to construct in the computer.
3. Scan the sketch as a digital image.
4. After scanning, use an image-processing application to trim away the unwanted portion of the image and write down the dimensions of the grid lines in the image. For example, the bitmap image shown in Figure 4–28 is 30-mm wide and 35-mm tall.

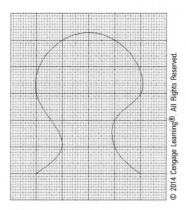

FIGURE 4–28 Bitmap image

Incorporate the digital image as a background image in the Front viewport by continuing with the following steps.

5. Start a new file. Use the Small Objects, Millimeters template.
6. Maximize the Top viewport.
7. Check the Snap button on the Status bar to turn on snap mode.
8. Select View > Background Bitmap > Place.
9. In the Open Bitmap dialog box, select the bitmap file Background Image.tga from the Chapter 4 folder on the Student Companion Website and then click on the Open button.
10. Select a point on the Front viewport, as shown in Figure 4–29.
11. Select a point 30 mm to the right of the first selected point, because the width of the bitmap used in this example is 30-mm wide.
12. Select Curve > Free-Form > Sketch.

Alternatively, you can use the Curve from the Control Points option or the Interpolated Curve option to trace the images.

13. Hold down the left mouse button to sketch a free-form curve.
14. Select View > Background Bitmap > Hide.
15. Turn off grid display.

    A sketch is constructed, and the bitmap is hidden. You should note that the freehand sketch constructed in this way is not smooth enough, and you may need to use the editing tools and transformation tools to modify it.

16. Do not save your file.

> **NOTE**
>
> If you save your file, the bitmap will also be saved in the Rhino file. Therefore, it is not necessary to save the bitmap file separately. Because the background image is saved in the Rhino file, you may save it back to your memory device as an individual bitmap file.
>
> Although the bitmap file is saved in the Rhino file, there is still a link with the original bitmap file. Therefore, you may modify the source bitmap file and use the Refresh option of the BackgroundBitmap command to update the embedded bitmap.

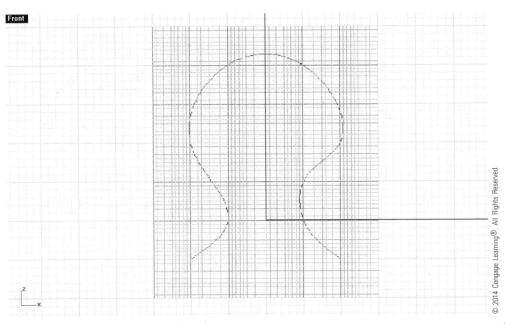

FIGURE 4–29 *Image placed on the background of Front viewport*

Other Ways to Construct Free-Form Curves

There are six more ways to construct a free-form curve. You can construct an interpolated curve on a surface, sketch a curve on a surface, sketch a curve on a polygonal mesh object, fit a curve from a series of point objects, fit a curve from a polyline, and construct a shortest curve between two points on a surface.

Interpolated Curve on a Surface

You can construct an interpolated curve by selecting point locations on a selected surface, producing a curve on the surface. Perform the following steps.

1. Select File > Open and select the file FreeFormCurve07.3dm from the Chapter 4 folder on the Student Companion Website.
2. Select Curve > Free-Form > Interpolate on Surface
3. Select surface A, shown in Figure 4–30.
4. Select locations B, C, and D, shown in Figure 4–30, and then press the ENTER key.

   A free-form curve is constructed on the surface.

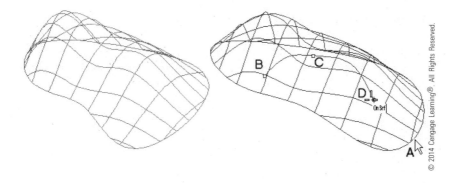

FIGURE 4–30 *Free-form curve being constructed on a surface*

Free-Form Sketch Curve on a Surface

In addition to sketching on a construction plane, you can sketch on a surface. Continue with the following steps.

5. Select Curve > Free-Form > Sketch on Surface.
6. Select surface A, shown in Figure 4–31.
7. Select location B (Figure 4–31), hold down the mouse button, and drag to location C (Figure 4–31).
8. Release the mouse button when finished.
9. Press the ENTER key.

    A free-form sketch curve is constructed on the surface.

10. Do not save your file.

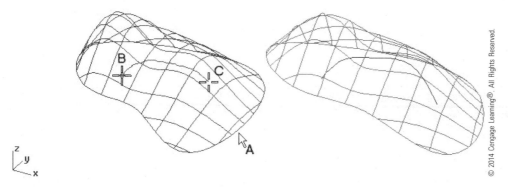

FIGURE 4–31 *Free-form sketch curve being constructed on a surface (left) and free-form sketch curve constructed (right)*

Free-Form Sketch Curve on a Polygon Mesh

In addition to sketching curves on NURBS surfaces, you can sketch a curve on a polygon mesh, as follows. (To reiterate, a polygon mesh is an approximation of a smooth surface via a set of planar polygons.)

1. Select File > Open and select the file FreeFormCurve08.3dm from the Chapter 4 folder on the Student Companion Website.
2. Select Curve > Free-Form > Sketch on Polygon Mesh.
3. Select location A, shown in Figure 4–32, on the polygon mesh hold down the mouse button; and drag it to location B (Figure 4–32).
4. Release the mouse button and press the ENTER key.

    A free-form sketch curve is constructed on the polygon mesh.

5. Do not save your file.

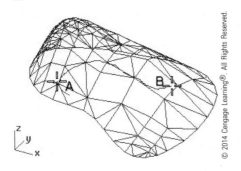

FIGURE 4–32 *Free-form sketch curve being constructed on a polygon mesh*

Fitting a Curve through a Set of Points

If you already have constructed a set of point objects, you can select all of them collectively in one operation to construct a free-form curve, using the points as interpolated points or control points. Perform the following steps.

1. Select File > Open and select the file FreeFormCurve09.3dm from the Chapter 4 folder on the Student Companion Website.
2. Select Curve > Free-Form > Fit to Points.
3. Select all the point objects and press the ENTER key.

    A preview of the curve is shown. (Note that the setting of the curve depends on previous settings made to this command. The preview here shows a curve of degree 3 interpolating through the selected points.)

4. If the Degree option is not 3 or above, select it from the command area and then type 3 to set the degree of polynomial to 3.
5. Select the CurveType option at the command area.
6. If CurveType=ControlPoint, select the Interpolated option at the command area to change to an interpolated curve. Otherwise, proceed to the next step.
7. Press the ENTER key.

    A curve is constructed. (See Figure 4–33.)

8. Do not save your file.

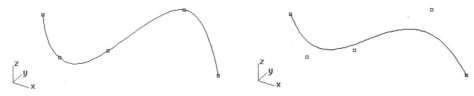

FIGURE 4–33 *Interpolated curve constructed from a set of point objects (left) and control point curve constructed from a set of point objects (right)*
© 2014 Cengage Learning®. All Rights Reserved.

Fitting a Curve through the Vertices or the Control Points of a Polyline

Other than using point objects, you can use the vertices of a polyline as the interpolated points or control points for making a curve. (Ways to construct a polyline will be discussed in the next chapter.) Perform the following steps.

1. Select File > Open and select the file FreeFormCurve10.3dm from the Chapter 4 folder on the Student Companion Website.
2. Select Curve > Free-Form > Fit to Polyline.
3. Select polyline A, shown in Figure 4–34, and press the ENTER key.
4. If Degree=3, proceed to the next step. Otherwise, select the Degree option at the command area and then type 3 at the command area.

5. If CurveType=ControlPoint, select the CurveType option at the command area and then select the Interpolated option at the command area. Otherwise, proceed to the next step.
6. Press the ENTER key.

   A curve is constructed.
7. Do not save your file.

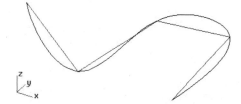

FIGURE 4–34 *Control point curve constructed from the vertices of a polyline (left) and interpolated curve constructed from the vertices of a polyline (right)*
© 2014 Cengage Learning®. All Rights Reserved.

Fitting a Curve through the Control Points of a Surface

A surface has a matrix of control points in two orthogonal directions. You can use these control points as follows:

1. Select File > Open and select the file FreeFormCurve11.3dm from the Chapter 4 folder on the Student Companion Website.
2. Type CurveThroughSrfControlPt at the command area.
3. Select surface A, shown in Figure 4–35.

   The control points of the surface are displayed.
4. Select the Direction option at the command area.
5. Select Both.
6. Select CurveType option at the command area.
7. Select Interpolated.
8. Select control point B and press the ENTER key.

   Two curves are constructed in two directions passing through the selected control point of the surface.
9. Do not save your file.

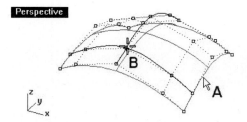

FIGURE 4–35 *Curves constructed through the control points of a surface*
© 2014 Cengage Learning®. All Rights Reserved.

Shortest Curve between Two Points on a Surface

You can construct a free-form curve on a surface to depict the shortest path between two designated points. Perform the following steps.

1. Select File > Open and select the file ShortPath.3dm from the Chapter 4 folder on the Student Companion Website.
2. Click on Geodesic curve button on the Curve from the Objects toolbar or type ShortPath at the command area.
3. Select surface A, shown in Figure 4–36.

4. Set object snap mode to point.
5. Select points A and B.

    A free-form curve is constructed.

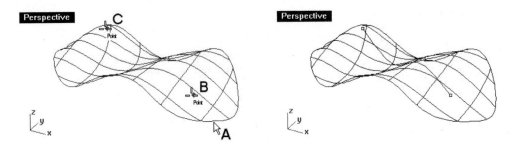

FIGURE 4–36 *Surface and points being selected (left) and curve constructed (right)*
© 2014 Cengage Learning®. All Rights Reserved.

POINTS AND POINT CLOUDS

A point is a node. A point cloud is a set of points. Using point objects, you specify locations in 3D space, define vertices and definition points of curves, and portray locations on a surface.

Multiple Points

You can construct point objects by specifying their locations. Typically, you can select a location on one of the construction planes, key in the coordinates of a point, or use a digitizer to directly input the coordinates of locations from a real physical object. Perform the following steps.

1. Start a new file. Use the Small Objects, Millimeters template.
2. Double-click on the Top viewport title to maximize the viewport.
3. Select Curve > Point Object > Multiple Points.
4. Select locations A, B, C, and D, as shown in Figure 4–37.

    The exact location of points in this exercise is unimportant.

5. Press the ENTER key to terminate the command.

    Four-point objects are constructed.

6. Do not save your file.

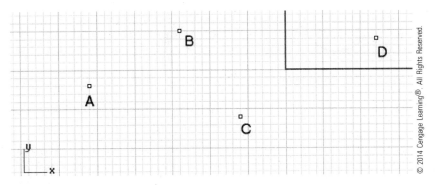

FIGURE 4–37 *Four-point objects*

Points near a Curve and along a Curve

If you already have a curve, you can construct a point on the curve that is closest to a target object, points at the endpoints of the curve, a set of points along the curve at specified spacing, and a set of points along the curve at regular intervals. Perform the following steps.

1. Select File > Open and select the file Point01.3dm from the Chapter 4 folder on the Student Companion Website.

2. Set object snap mode to Quad. (Quad means quadrant locations of circles and arcs.)
3. Select Curve > Point Object > Closest Point.
4. Select curve A, shown in Figure 4–38, and press the ENTER key.
5. Select quadrant location B, shown in Figure 4–38.

   A point closest to the selected location is constructed along the curve.

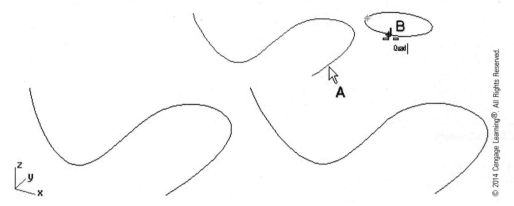

FIGURE 4–38 *Point on a curve closest to the quadrant point of a circle being constructed*

6. Select Curve > Point Object > Mark Curve Start.
7. Select curve A, shown in Figure 4–39, and press the ENTER key.
8. Select Curve > Point Object > Mark Curve End.
9. Select curve A, shown in Figure 4–39, and press the ENTER key.

   Points are constructed at the start point and endpoint of the curve.

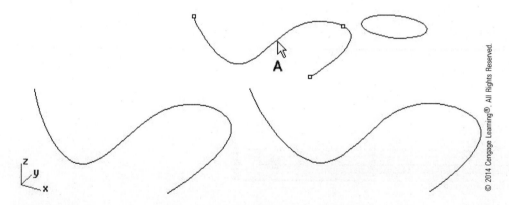

FIGURE 4–39 *Points at the start point and endpoint of a curve constructed*

To construct a set of point objects along a curve at regular intervals, perform the following steps.

10. Select Curve > Divide Curve by > Length of Segments.
11. Select curve A, shown in Figure 4–40, and press the ENTER key.
12. Type 14 at the command area to specify a length.

    A set of points at 14-unit intervals is constructed along the curve, with the first point object constructed at the start point of the curve.

13. Select Curve > Divide Curve by > Number of Segments.
14. If MarkEnds=No, select the MarkEnds option at the command area to change to MarkEnds=Yes.
15. Select curve B, shown in Figure 4–40, and press the ENTER key.

16. Type 7 at the command area to specify six segments.

    Eight points are constructed, as shown in Figure 4–40.

17. Do not save your file.

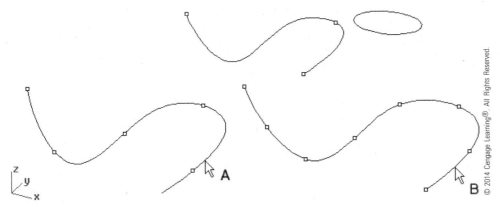

FIGURE 4–40 *Points at 14-unit intervals and points equally spaced along the curve constructed*

Focal Points

You can construct point objects at the focal points of parabola, hyperbola, and ellipse. Perform the following steps.

1. Select File > Open and select the file Focal Points.3dm from the Chapter 4 folder on the Student Companion Website.
2. Select Curve > Point Object > Mark Foci.
3. Select parabola A, hyperbola B, and ellipse C in Figure 4–41 and press the ENTER key.

    Focal points are constructed.

4. Do not save your file.

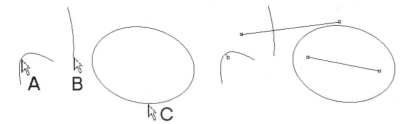

FIGURE 4–41 *Curves selected (left) and focal points constructed (right)*
© 2014 Cengage Learning®. All Rights Reserved.

Drape Points

If you already have a surface or a polygon mesh, you can construct a matrix of point objects on them. Using these point objects, you can proceed to constructing new curves for making additional surfaces. Perform the following steps.

1. Select File > Open and select the file Point02.3dm from the Chapter 4 folder on the Student Companion Website.
2. Select Curve > Point Object > Drape Points.
3. Select points A and B, shown in Figure 4–42.

    A set of points is constructed.

4. Turn off layer Surface.
5. Do not save or close the file.

    You will continue to work on the point objects.

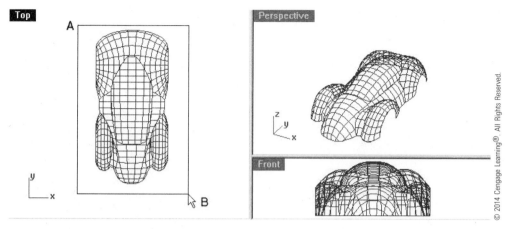

FIGURE 4–42 Drape points being constructed

Point Cloud Construction

A point cloud is a set of point objects grouped together. Point clouds are typical products of digitizing. Using point clouds, you can construct curves for making surfaces. To understand the use of point clouds, continue with the following steps.

6. Select Curve > Point Cloud > Create Point Cloud.
7. Select locations A and B, shown in Figure 4–43, and press the ENTER key.

   A point cloud is constructed from a set of points.

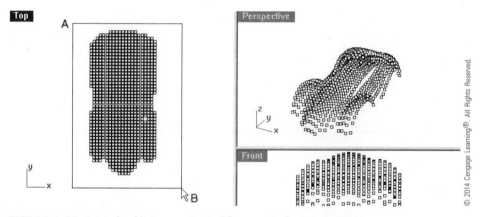

FIGURE 4–43 Point cloud being constructed from a set of selected points

8. To set the point cloud style, select Curve > Point Cloud > Point Cloud Style, and then select a style from the options available.

> You can add point objects to a point cloud by selecting *Curve > Point Cloud > Add Points* and remove point objects from a point cloud by selecting *Curve > Point Cloud > Remove Points*. **NOTE**

Curve Construction from Point Cloud

A way to reconstruct a surface from a point cloud is to construct sections across the point cloud. Continue with the following steps.

9. Set object snap mode to Point.
10. Select Curve > Point Cloud > Point Cloud Section.
11. Select the point cloud and press the ENTER key.

12. In the Point Cloud Section Options dialog box (Figure 4-44), clear the Create polylines, if it is checked, and click on the OK button.

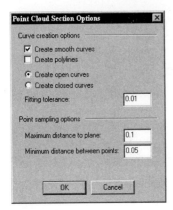

FIGURE 4-44 *Point Cloud Section Options dialog box*
Source: Robert McNeel and Associates Rhinoceros® 5

13. Select locations A and B, shown in Figure 4-45, to describe a section line across the point cloud.

If the section line is located too far from the points, a prompt "Too few points within minimum distance of requested section plane" will be displayed at the command area. If you see this message, try again.

14. Continue to select pairs of start and endpoints.
15. Press the ENTER key when finished.

    Point cloud curves are constructed.

16. Do not save your file.

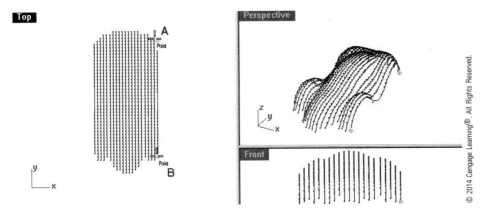

FIGURE 4-45 *Point cloud curves*

In the previous steps, constructing a point cloud from the surfaces and constructing point cloud curves serve to demonstrate how drape points, point clouds, and point cloud curves are constructed. In reality, point clouds are usually obtained by digitizing a real physical model for the purpose of reconstructing the surfaces of the model in the computer.

Draft Angle Point

To help define a parting line along an object for splitting into two in mould making, you can use the DraftAnglePoint command to construct a series of points on the surface. By constructing a curve to interpolate along these points, you will get a parting line.

1. Select File > Open and select the file Draft Angle Point.3dm from the Chapter 4 folder on the Student Companion Website.

2. Type DraftAnglePoint at the command area.
3. Continuously click on the surface to obtain a series of point objects.
4. Turn off layer Surface to appreciate the outcome (Figure 4–46).
5. Do not save your file.

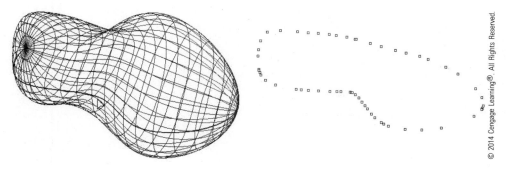

FIGURE 4–46 *Surface object (left) and draft angle points constructed (right)*

CONSOLIDATION

Rhino, like most modeling applications, uses NURBS (nonuniform rational basis spline) curves to construct free-form surfaces. Simply speaking, a NURBS curve is a set of polynomial spline segments. The shape of each spline segment in a curve is influenced by the degree of polynomial equation used to define the curve. For most designers, it is not necessary to know about the mathematics related to such equations. However, understanding the relationship between the degree of polynomial and the shape of a curve is an advantage.

A line is degree 1, a circle is degree 2, and a free-form curve is degree 3 or above. The higher the degree, the more control points exist in a spline segment. For a curve of degree 3, each spline segment has four control points. Between two contiguous segments of a spline curve is a knot. Adding or removing knots increases or reduces the number of segments in a curve. This, in turn, has an influence on the total number of control points in a curve.

A special type of knot on a curve is called a kink, which is a junction between two contiguous segments where the tangent directions and radii of curvature are not congruent. A closed curve without any kink is called a periodic curve. The overall shape of a NURBS curve is affected by the number of segments in the curve, the degree of polynomial equation, the location of control points, and the weight of the control points.

Points are as important as curves in making a framework for constructing surfaces. Point objects can be constructed from scratch or by digitizing.

REVIEW QUESTIONS

1. State the characteristics and key features of a NURBS curve.
2. Illustrate the methods of editing a curve by manipulating its control points, edit points, knots, and degree of polynomial.
3. What is a point cloud, and how is it used?

CHAPTER 5

Curves of Regular Pattern

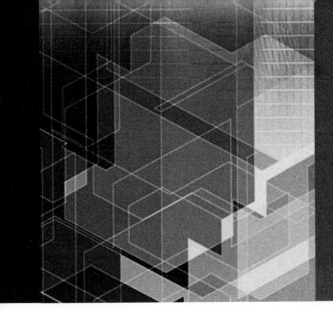

INTRODUCTION

This chapter delineates various kinds of curves of regular pattern that you may use in surface modeling.

OBJECTIVES

After studying this chapter, you should be able to

- Construct curves of regular pattern in 3D surface modeling

OVERVIEW

In addition to using free-form curves in your design, you may also need to deploy curves with regular pattern, such as lines, arcs, circles, and polygons. Using Rhino, you can construct twelve kinds of regular curves, categorized in terms of their degree of polynomial as follows.

- Degree 1 curves are line, polyline, rectangle, and polygon.
- Degree 2 curves are circle, arc, ellipse, parabola, hyperbola, and conic.
- Degree 3 curves are helix and spiral.

All Rhino's curves are NURBS curves, so you can modify them by increasing or reducing the degree of polynomial, manipulating the control points or edit points, adding or deleting control points, and adding or removing knots and kink points. You can also use the Handlebar Editor.

LINE

A straight line is a degree 1 curve. You can construct it in thirteen ways, as follows.

Single Line and Line Segments

You can construct a single line segment by specifying two endpoints or construct a series of individual line segments connected end to end by specifying their adjoining endpoints. Note that the line segments constructed are separate, individual line segments. Perform the following steps.

1. Select File > Open and select the file Line01.3dm from the Chapter 5 folder that you downloaded from the Student Companion Website.

2. Set object snap mode to Point.
3. Select Curve > Line > Single Line.
4. Select points A and B, shown in Figure 5–1.
5. Select Curve > Line > Line Segments.
6. Select points C, D, E, and F, shown in Figure 5–1, and then press the ENTER key.

   A series of line segments is constructed.
7. Do not save your file.

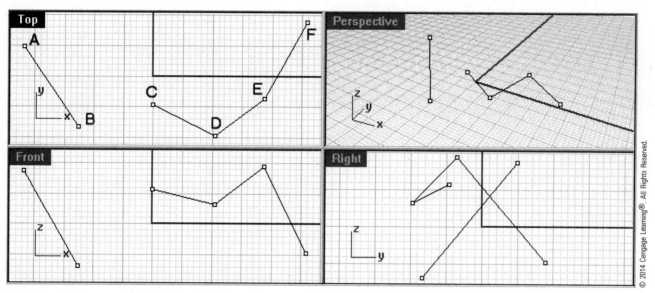

FIGURE 5–1 *Single line and line segment constructed*

Line Perpendicular/Tangent to a Curve and Line Perpendicular/Tangent to Two Curves

You may construct a line perpendicular/tangent to a curve from a selected point and construct a line perpendicular/tangent to two curves. Depending on the location of the selected point and the shape of the curve (for a line perpendicular/tangent to a curve from a point) or the shape of the curves (for a line perpendicular/tangent to two curves), there may be no solutions at all or more than one solution. Perform the following steps.

1. Select File > Open and select the file Line02.3dm from the Chapter 5 folder on the Student Companion Website.
2. Set object snap mode to Point, if it is not already set.
3. Select Curve > Line > Perpendicular from Curve.
4. Select point A and then curve B, shown in Figure 5–2.

   Exact location of curve B is unimportant. A line perpendicular to a curve from a point is constructed.

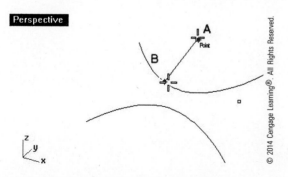

FIGURE 5–2 *Line through a point and perpendicular to a curve being constructed*

5. Select Curve > Line > Perpendicular to 2 Curves.
6. Select curve A and then curve B, shown in Figure 5–3.

   A line perpendicular to two curves is constructed.

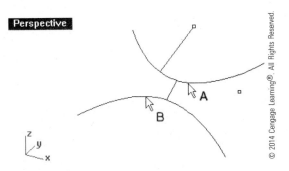

FIGURE 5–3 *Line perpendicular to two curves being constructed*

7. Select Curve > Line > Tangent from Curve.
8. Select curve B and then point A, shown in Figure 5–4.

   A line tangent to a curve from a point is constructed.

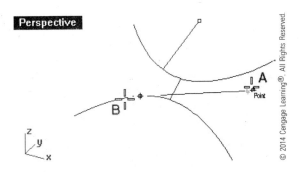

FIGURE 5–4 *Line through a point and tangent to a curve being constructed*

9. Select Curve > Line > Tangent to 2 Curves.
10. Select curves A and B, shown in Figure 5–5.
11. Do not save your file.

    A line tangent to two curves is constructed.

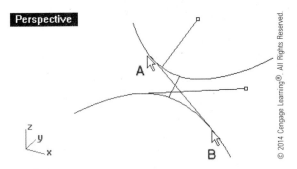

FIGURE 5–5 *Line tangent to two curves being constructed*

Line Tangent to a Curve and Perpendicular to Another Curve

You may construct a line that is tangent to a curve and also perpendicular to another curve. Likewise, there may be no solutions or more than one solution to the problem. Perform the following steps.

1. Select File > Open and select the file Line03.3dm from the Chapter 5 folder on the Student Companion Website.
2. Select Curve > Line > Tangent, Perpendicular.
3. Select curve A where the tangent point will be placed, as shown in Figure 5–6.
4. Select curve B where the perpendicular point will be placed, as shown in Figure 5–6.

   A line tangent to a curve and perpendicular to another curve is constructed.

5. Do not save your file.

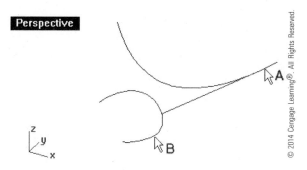

FIGURE 5–6 *Line tangent to a curve and perpendicular to another curve being constructed*

Line at an Angle, Line Bisecting Two Lines, and Four-Point Line

You can construct a line at an angle to a selected reference line, construct a line bisecting two nonparallel lines, and construct a line in a direction defined by two points with the endpoints nearest to two selected points (i.e., a four-point line). Perform the following steps.

1. Select File > Open and select the file Line04.3dm from the Chapter 5 folder on the Student Companion Website.
2. Set object snap to End and Point.
3. Select Curve > Line > Angled.
4. Select endpoints A and B, shown in Figure 5–7.
5. Type 45 to specify an angle.
6. Select point C, shown in Figure 5–7.

   A line at 45 degrees to the selected line is constructed. It has an endpoint closest to the selected reference point. The angle is measured on a plane parallel to the active construction plane. Here, the Perspective viewport's construction plane is the same as that of the Top viewport.

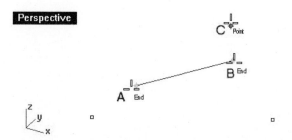

FIGURE 5–7 *Line segment at an angle to a reference line being constructed*
© 2014 Cengage Learning®. All Rights Reserved.

7. Select Curve > Line > Bisector.
8. Select endpoints A and B, shown in Figure 5–8, to specify the first reference line (start of the bisector line and start of the angle to be bisected).
9. Select endpoints A and C, shown in Figure 5–8, to indicate the end of the angle to be bisected.
10. Select point D, shown in Figure 5–8.

    A line bisector is constructed. Its endpoint is closest to the selected reference point D.

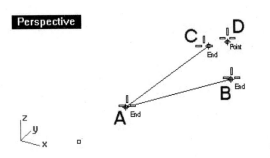

FIGURE 5–8 *Bisector line being constructed*
© 2014 Cengage Learning®. All Rights Reserved.

11. Select Curve > Line > From 4 Points.
12. Select points A and B, shown in Figure 5–9, to indicate a direction reference.
13. Select endpoints C and D, shown in Figure 5–9.

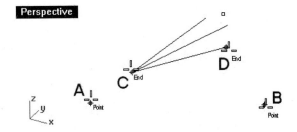

FIGURE 5–9 *A four-point line being constructed*
© 2014 Cengage Learning®. All Rights Reserved.

A line lying on the reference direction line and with its endpoints closest to two reference points is constructed. (See Figure 5–10.)

14. Do not save your file.

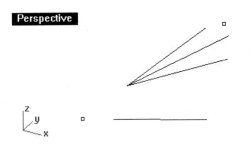

FIGURE 5–10 *A four-point line constructed*
© 2014 Cengage Learning®. All Rights Reserved.

Line Normal to a Surface and Line Vertical to a Construction Plane

You can construct a line normal to a surface, and you can construct a line perpendicular to a construction plane. Perform the following steps.

1. Select File > Open and select the file Line05.3dm from the Chapter 5 folder on the Student Companion Website.
2. Set object snap mode to Int (intersection) and Point.
3. Select Curve > Line > Normal to Surface.
4. Select surface A, shown in Figure 5–11.
5. Select location B (isocurve's intersection), shown in Figure 5–11, to indicate the start point of the line on the surface.
6. Select reference point C, shown in Figure 5–11.

   A line normal to the surface is constructed. Its endpoint is closer to the selected reference point.

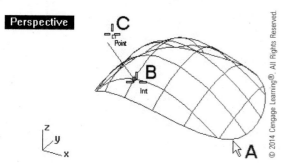

FIGURE 5–11 *Normal line constructed*

7. Select Curve > Line > Vertical to CPlane.
8. Select endpoint A to specify the start point of the line, and select endpoint B to specify the reference point.

   This creates a line perpendicular to the construction plane, as shown in Figure 5–12.

   Height is defined by the vertical distance between the first selected point and the second selected point.

9. Do not save your file.

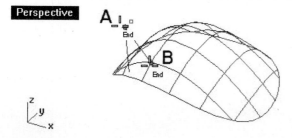

FIGURE 5–12 *Line perpendicular to the construction plane being constructed*
© 2014 Cengage Learning®. All Rights Reserved.

Line Passing through a Set of Points

You can fit a line through a set of point objects and/or control points. Perform the following steps:

1. Select File > Open and select the file Line06.3dm from the Chapter 5 folder on the Student Companion Website.
2. Click on the Line through points button on the Lines toolbar.
3. Referencing Figure 5–13, click on location A and drag to location B to select all the point objects, and press the ENTER key.

   A line is fitted.

4. Do not save your file.

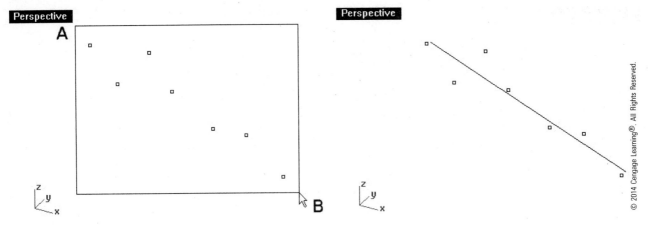

FIGURE 5–13 *Point objects selected (left) and line fitted (right)*

Drawing a Line with a Digitizer

If you have already attached a digitizer to your computer, you can use the DigLine command in conjunction with the digitizer to construct a line perpendicular to a surface.

POLYLINE

A polyline is a series of contiguous line and/or arc segments joined together at their endpoints. In essence, you can explode a polyline to obtain a set of line and/or arc segments. Conversely, you can join a set of contiguous line and/or arc segments to become a polyline. You can construct a polyline in much the same way as constructing a series of line segments. You can also fit a polyline to a set of points, similar to fitting a curve to a set of points. Moreover, you can construct a polyline on a polygon mesh. To construct a polyline, perform the following steps.

1. Select File > Open and select the file Polyline01.3dm from the Chapter 5 folder on the Student Companion Website.
2. Check the Osnap pane on the Status bar to display the Osnap dialog box.
3. In the Osnap dialog box, check the Point box to establish object snap mode.
4. Select Curve > Polyline > Polyline.
5. Select point A, shown in Figure 5–14.
6. At the command area, if mode=arc, select the mode option to change to mode=line. Otherwise, proceed to the next step.
7. Select point B, shown in Figure 5–14.
8. Select the mode option at the command area to change to mode=arc.
9. Select point C, shown in Figure 5–14.
10. Select the mode option at the command area to change to mode=line.
11. Select point D, shown in Figure 5–14.
12. Select the Close option at the command area.

    A closed polyline with line and arc segments is constructed.

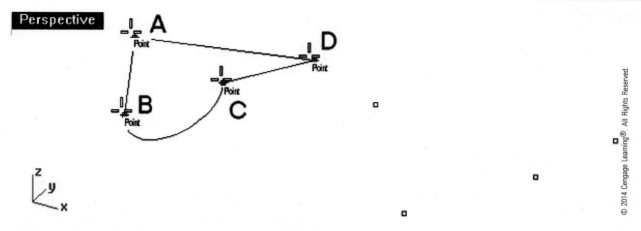

FIGURE 5–14 *A closed polyline with line and arc segments being constructed*

Polyline through a Set of Points

You can construct a polyline to pass through a set of points. Perform the following steps.

13. Set the current layer to Layer02 and turn off Layer01.
14. Select Curve > Polyline > Polyline Through Points.
15. Click on A and drag to location B (Figure 5–15) to select four-point objects, and press the ENTER key.

    Because this command is the same as Curve > Free-Form > Fit To Points, the degree of polynomial setting will determine the outcome.

16. If Degree option is not equal to 1, select it from the command area and then type 1 at the command area to set the degree of polynomial to 1. Otherwise, proceed to the next step.

    As will be explained in next chapter, the line segment's degree of polynomial is 1.

17. Press the ENTER key.

    A polyline is constructed.

18. Do not save your file.

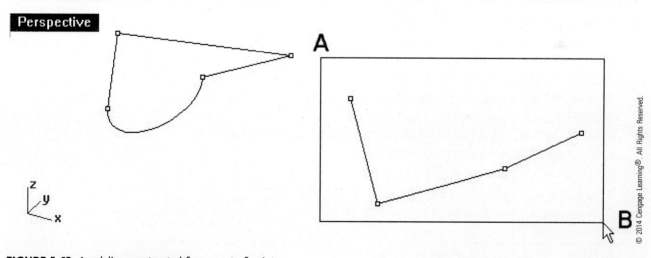

FIGURE 5–15 *A polyline constructed from a set of points*

Polyline on a Polygon Mesh

To construct a polyline on a polygon mesh, perform the following steps.

1. Select File > Open and select the file Polyline02.3dm from the Chapter 5 folder on the Student Companion Website.
2. Set object snap mode to Point.

3. Select Curve > Polyline > On Mesh.
4. Select polygon mesh A, shown in Figure 5–16.
5. Select vertices B, C, and D, shown in Figure 5–16, and then press the ENTER key.

   A polyline is constructed on a polygon mesh.
6. Do not save your file.

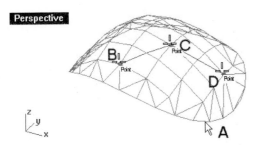

FIGURE 5–16 *Polyline constructed on a polygon mesh*
© 2014 Cengage Learning®. All Rights Reserved.

Connected Line Segments and Polyline

At a glance, the connected line segments and the polyline seem to be similar. In fact, the polyline is a single object and the connected line segments are separate line segments with their endpoints coincident with each other. To separate the polyline into individual line segments, you explode it by selecting Edit > Explode. To convert a set of connected line segments to a single polyline, you join the line segments by selecting Edit > Join.

RECTANGLE

Using Rhino, you can construct three types of rectangles. The first, consisting of four joined line segments, is a degree 1 curve. The second, consisting of four line segments and four arcs (filleted corners) joined, is a degree 2 curve. The third, consisting of four joined conic curves, is also a degree 2 curve. If you explode a rectangle, the first rectangle type will decompose into four separate line segments, the second will decompose into line and arc segments, and the third will decompose into four conic curves.

Four Ways to Construct a Regular Rectangle

You can construct any type of rectangle in four ways: specifying its diagonal corners (Corner to Corner), specifying the center and a corner (Center, Corner), specifying an edge and a point on the opposite edge (Three Points), and specifying an edge and its height from the construction plane (Vertical). Perform the following steps.

1. Select File > Open and select the file Rectangle01.3dm from the Chapter 5 folder on the Student Companion Website.
2. Check the Point box in the Osnap dialog box to establish object snap mode.
3. Select Curve > Rectangle > Corner to Corner.
4. Select points A and B, shown in Figure 5–17.
5. Select Curve > Rectangle > Center, Corner.
6. Select points C and D, shown in Figure 5–17.
7. Select Curve > Rectangle > 3 Points.
8. Select points E, F, and G, shown in Figure 5–17.
9. Select Curve > Rectangle > Vertical.
10. Select points H, J, and K, shown in Figure 5–17.
11. Do not save your file.

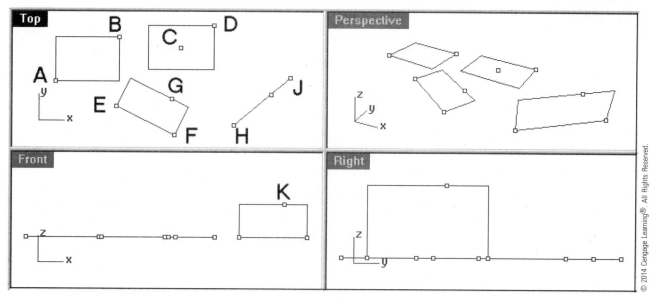

FIGURE 5–17 *Four rectangles constructed*

Rounded Rectangle and Conic Rectangle

However you construct a rectangle, you can have a filleted rectangle or conic rectangle instead of a rectangle with four line segments. Perform the following steps.

1. Select File > Open and select the file Rectangle02.3dm from the Chapter 5 folder on the Student Companion Website.
2. Select Curve > Rectangle > Corner to Corner and select the Rounded option at the command area.
3. Select points A and B, shown in Figure 5–18.
4. At the command area, if Corner=Conic, select the Corner option to change it to Arc. Otherwise, proceed to the next step.
5. Type 4 at the command area to specify the radius.

   A rectangle with circular arc corners is constructed.
6. Select Curve > Rectangle > Corner to Corner and select the Rounded option at the command area.
7. At the command area, set Corner=Conic.
8. Select locations C and D (Figure 5–18).
9. Select location E (Figure 5–18).

   A conic rectangle is constructed.
10. Do not save your file.

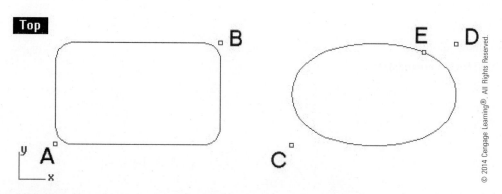

FIGURE 5–18 *Filleted rectangle (left) and conic rectangle (right) constructed*

POLYGON

A polygon's polynomial degree is 1. It is a regular polygon in shape, consisting of a set of lines of equal length joined together. You can construct two types of polygons, regular and star-shaped, in several ways: specifying the number of sides, the center, and a corner; specifying the center and a midpoint of an edge; or specifying two endpoints of an edge. To construct a star, you specify the number of sides, corner of the star, and the second radius of the star. In addition to constructing a polygon or star on the current construction plane, you can construct it around a curve.

Polygon and Polygon around a Curve

To construct a regular polygon, star polygon, regular polygon around a curve, and star polygon around a curve, perform the following steps.

1. Select File > Open and select the file Polygon.3dm from the Chapter 5 folder on the Student Companion Website.
2. Check the Point and End boxes in the Osnap dialog box to establish object snap mode.
3. Select Curve > Polygon > Center, Radius.
4. Select the NumSides option at the command area.
5. Type 7 to specify the number of sides.
6. Select point A, shown Figure 5–19, to specify the center.
7. Select point B, shown in Figure 5–19, to specify the radius.

   A polygon specified by its center and the radius of inscribed circle is constructed. Radius of the inscribed circle is depicted by the distance between A and B.

8. Repeat the Polygon command and select Circumscribed option at the command area.
9. Select points C and D, shown in Figure 5–19. A polygon specified by its center and the radius of circumscribed circle is constructed.

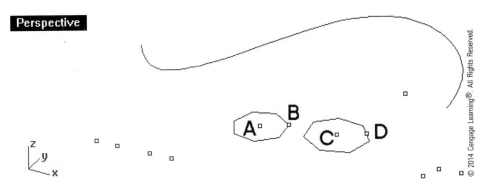

FIGURE 5–19 *Inscribed polygon (left) and circumscribed polygon (right)*

10. Select Curve > Polygon > By Edge.
11. Select points A and B, shown in Figure 5–20.
12. Repeat the Polygon command.
13. Select Vertical option at the command area.
14. Select points C and D, shown in Figure 5–20.

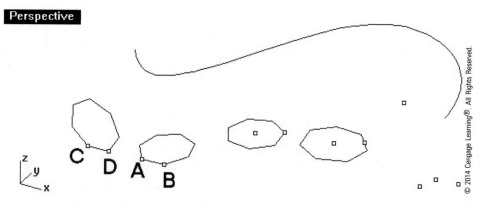

FIGURE 5–20 *Edge polygon and vertical polygon constructed*

15. Select Curve > Polygon > Star.
16. Select point A, shown Figure 5–21, to indicate the center.
17. If you wish to change the number of sides, select the NumSides option at the command area and then type a value.
18. Select point B, shown in Figure 5–21, to indicate a corner point.
19. Select point C, shown in Figure 5–21, to indicate the second radius.

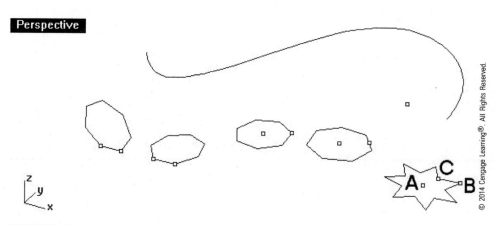

FIGURE 5–21 *Star polygon constructed*

20. Select Curve > Polygon > CENTER, Radius.
21. Select the AroundCurve option at the command area.
22. Select curve A, shown in Figure 5–22.
23. Select endpoint B, shown in Figure 5–22, to indicate the corner location of the polygon.
24. Type 12 to specify the radius.
25. Select reference point C.

A polygon is constructed around the selected curve. One of the corners is closest to the selected reference point.

26. Do not save your file.

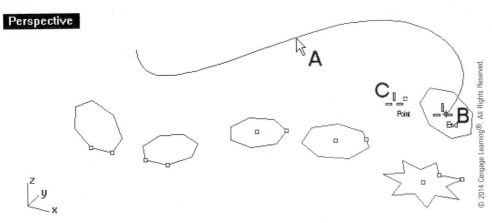

FIGURE 5-22 *Polygon around a curve constructed*

CIRCLE

Using Rhino, you can construct two kinds of circular objects: true circle and deformable circle. A true circle is a degree 2 curve with four kink points at the quadrant positions. A deformable circle is a degree 3 periodic curve in the shape of a circle. You can construct a circle in many ways.

Center and Radius, Diameter, Three Points, and Multipoints

Perform the following steps to construct a circle by specifying its center and radius, its diameter, and three points on the circumference, and by fitting to a set of points.

1. Select File > Open and select the file Circle01.3dm from the Chapter 5 folder on the Student Companion Website.
2. Select Curve > Circle > Center, Radius.
3. Select points A and B, shown in Figure 5-23, to specify the center and a point on the circumference, respectively.

    A circle with its center at point A and its circumference on point B is constructed.

4. Select Curve > Circle > Diameter.
5. Select points C and D, shown in Figure 5-23.

    A circle with its diameter defined by points C and D is constructed.

6. Select Curve > Circle > 3 Points.
7. Select points E, F, and G, shown in Figure 5-23, to specify the center and a point on the circumference.

    A circle passing three points is constructed.

8. Select Curve > Circle > Fit Points.
9. Click on location H and drag J, shown in Figure 5-23.
10. Press the ENTER key.

    A circle fitting to the selected points is constructed.

11. Do not save your file.

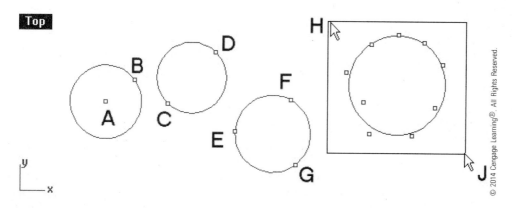

FIGURE 5–23 Circles constructed

Tangent Circle

You can construct a circle tangent to two curves or three curves. Perform the following steps.

1. Select File > Open and select the file Circle02.3dm from the Chapter 5 folder on the Student Companion Website.
2. Select Curve > Circle > Tangent, Tangent, Radius.
3. Select curves A and B, shown in Figure 5–24.
4. Type 25 at the command area to specify the radius.
5. Select Curve > Circle > Tangent to Curves.
6. Select curves A, B, and C, shown in Figure 5–24.
7. Do not save your file.

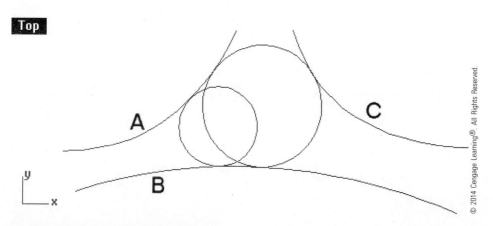

FIGURE 5–24 Tangent circles constructed

Vertical Circle and Circle around a Curve

You can construct a circle vertical to a construction plane and a circle around a curve. Perform the following steps.

1. Select File > Open and select the file Circle03.3dm from the Chapter 5 folder on the Student Companion Website.
2. Select Curve > Circle > Center, Radius, and select Vertical option at the command area.
3. Select points A and B, shown in Figure 5–25, to specify the center and the radius, respectively.
4. Select Curve > Circle > Center, Radius, and select AroundCurve option at the command area.
5. Select curve C, shown in Figure 5–25.
6. Select point D, shown in Figure 5–25, to indicate the center location.
7. Type 40 at the command area to specify the radius.
8. Do not save your file.

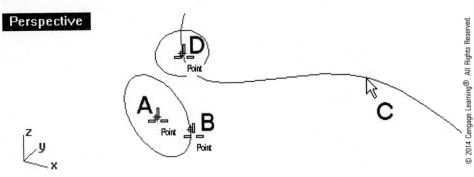

FIGURE 5–25 *Vertical circle (left) and circle around a curve (right) constructed*

Deformable Circle

As mentioned, a circle is a degree 2 curve. If you turn on its control points, you will find that it has four kink points. (Although it is a closed-loop curve, it is not a periodic curve because it has kink points.) However, you can use the deformable option to construct a deformable circle of degree 3 or above and without a kink point. In essence, a deformable circle is not a true circle. Instead, it is a free-form curve in the shape of a circle. Perform the following steps.

1. Select File > Open and select the file Circle04.3dm from the Chapter 5 folder on the Student Companion Website.
2. Select Curve > Circle > Center, Radius.
3. Select points A and B, shown in Figure 5–26, to specify the center and a point on the circumference, respectively.
4. Select Curve > Circle > Center, Radius.
5. Select the Deformable option at the command area.
6. Select points C and D, shown in Figure 5–26, to specify the center and a point on the circumference, respectively.

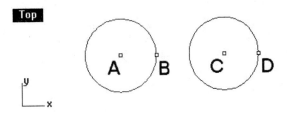

FIGURE 5–26 *True circle (left) and deformable circle (right) being constructed*
© 2014 Cengage Learning®. All Rights Reserved.

7. Select Edit > Control Point > Control Points On.
8. Select circles A and B, shown in Figure 5–27, and press the ENTER key.

    The control points are turned on. As shown, the true circle consists of four arc segments and four kink points, and the deformable circle is a free-form curve without any kink point.

9. Do not save your file.

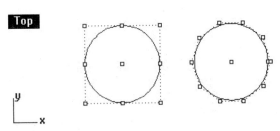

FIGURE 5–27 *True circle (left) and deformable circle (right) being constructed*
© 2014 Cengage Learning®. All Rights Reserved.

ARC

An arc is a degree 2 curve. There are seven ways to construct an arc, by specifying

- The center point, the start point, and the included angle
- The start point, the endpoint, and the tangent direction at the start point
- The start point, the endpoint, and a point along the arc
- The start point, a point along the arc, and the endpoint
- The start point, the endpoint, and the radius of the arc
- Two curves that the arc will be tangent to and the radius of the arc
- Three curves that the arc will be tangent to

Perform the following steps to construct five arcs.

1. Select File > Open and select the file Arc01.3dm from the Chapter 5 folder on the Student Companion Website.
2. Select Curve > Arc > Center, Start, Angle.
3. Select points A and B, shown in Figure 5–28, to specify the center and the start point of the arc, respectively.
4. Type 45 at the command area to specify the angle.

   An arc specified by the center, start point, and arc angle is constructed.
5. Select Curve > Arc > Start, End, Point.
6. Select points, C, D, and E, shown in Figure 5–28.

   An arc specified by the start point, the endpoint, and a point on the arc is constructed.
7. Select Curve > Arc > Start, End, Direction.
8. Select points F, G, and H, shown in Figure 5–28, to specify the start point, the endpoint, and the direction of the arc at the start point, respectively.

   The third arc is constructed.
9. Select Curve > Arc > Start, End, Radius.
10. Select points J, K, and L, shown in Figure 5–28.

    J and K are the start and endpoints of the arc, and JL defines the radius of the arc.
11. Select Curve > Arc > Start, Point, End.
12. Select points M, N, and P, shown in Figure 5–28.

    The fifth arc is constructed.
13. Do not save your file.

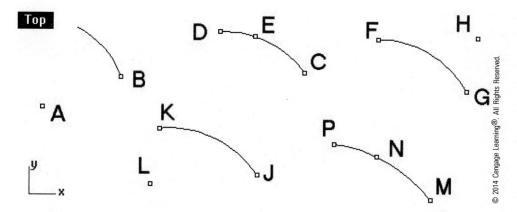

FIGURE 5–28 *Various ways of arc construction*

Perform the following steps to construct two more arcs.

1. Select File > Open and select the file Arc02.3dm from the Chapter 5 folder on the Student Companion Website.
2. Select Curve > Arc > Tangent, Tangent, Radius.

3. Select curves A and B, shown in Figure 5-29.
4. Type 5 to specify the radius of the arc.
5. Because there are two possible solutions, click on location C.

   An arc tangent to two curves is constructed.
6. Select Curve > Arc > Tangent to Curves.
7. Select curves D, E, and F, shown in Figure 5-29.

   By moving the cursor around, you may find that there are six possible arcs.
8. Click on location G.

   Another arc is constructed.
9. Do not save your file.

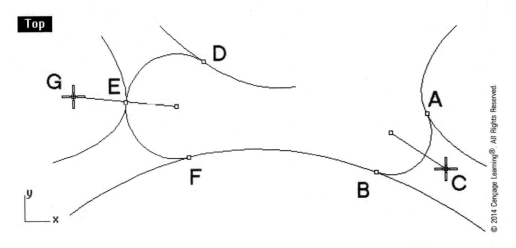

FIGURE 5-29 *Arc tangent to three curves (left) and arc tangent to two curves (right) constructed*

Modifying Radius

You can modify the radius of a circle or an arc. Continue with the following steps.

1. Select File > Open and select the file Modify Radius.3dm from the Chapter 5 folder on the Student Companion Website.
2. Type ModifyRadius at the command area.
3. Select circle A and arc B in Figure 5-30 and press the ENTER key.
4. Type 5 at the command area.

   The radii are changed.
5. Do not save your file.

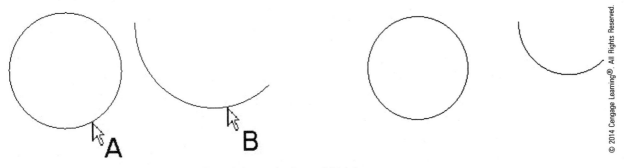

FIGURE 5-30 *Circle and arc selected (left) and their radii changed (right)*

ELLIPSE

Similar to a circle, a true ellipse is a degree 2 curve with four kink points. However, you can also construct a deformable ellipse, which, in essence, is a degree 3 curve in the shape of an ellipse.

From Center, Diameter, from Foci, and Ellipse by Corners

You can construct an ellipse by specifying its center and its axes, by specifying one of its diameters and the second axis, and by specifying its foci and a point on the circumference of the ellipse. Perform the following steps by specifying two corners of a rectangle circumscribing the ellipse.

1. Select File > Open and select the file Ellipse01.3dm from the Chapter 5 folder on the Student Companion Website.
2. Select Curve > Ellipse > From Center.
3. In the command area, select Deformable.
   If you do not select this option, you will get a true ellipse in degree 2 polynomial and has four kink points.
4. Select points A, B, and C, shown in Figure 5-31, to specify the center and two axes, respectively.
5. Select Curve > Ellipse > Diameter.
6. Select points D and E to specify the diameter, and select point F to specify a point on the ellipse, as shown in Figure 5-31.
7. Select Curve > Ellipse > From Foci.
8. Select points G and H to specify the foci, and select point J to specify a point on the ellipse, as shown in Figure 5-31.
9. Select Curve > Ellipse > From Center, and select the Corner option at the command area.
10. Select points K and L to describe a rectangular region in which the ellipse is constructed, as shown in Figure 5-31.
11. Do not save your file.

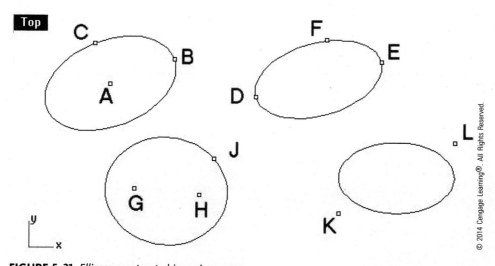

FIGURE 5–31 *Ellipses constructed in various ways*

Vertical Ellipse and Ellipse around a Curve

You can construct an ellipse vertical to a construction plane and an ellipse around a curve. Perform the following steps.

1. Select File > Open and select the file Ellipse02.3dm from the Chapter 5 folder on the Student Companion Website.
2. Select Curve > Ellipse > Diameter.
3. Select Vertical option at the command area.
4. Select points A and B in the Top viewport to specify the diameter, and select point C in the Front viewport to specify a point on the ellipse, as shown in Figure 5-32.

5. Select Curve > Ellipse > From Center and select AroundCurve option at the command area.
6. Select endpoint D (Figure 5-32) along the curve to indicate the center on the curve.
7. Select point E (Figure 5-32) to indicate a radius (distance between D and E).
8. Select point C, shown in Figure 5-32, to indicate the other radius (distance from the center to the selected point).
   An ellipse is constructed around a curve.
9. Do not save your file.

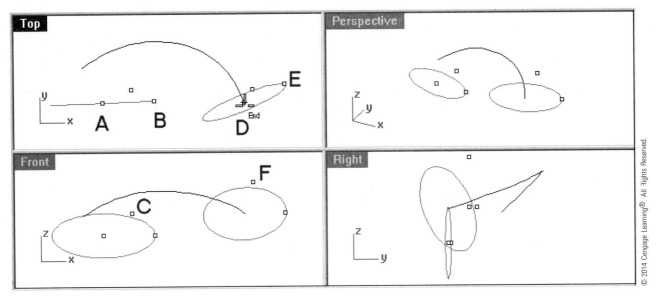

FIGURE 5–32 *Ellipse vertical to the construction plane and ellipse around a curve being constructed*

Deformable Ellipse

A true ellipse consists of four elliptical arcs joined at four kink points. A deformable ellipse is a free-form curve of degree 3 or above in the shape of an ellipse. Figure 5–33 shows the control points of a true ellipse and a deformable ellipse.

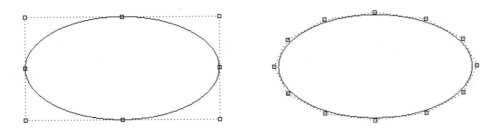

FIGURE 5–33 *True ellipse (left) and deformable ellipse (right)*
© 2014 Cengage Learning®. All Rights Reserved.

PARABOLA

A parabola is degree 2 curve. You can construct a parabola by specifying its focus and direction, and by specifying its vertex and focus. Perform the following steps.

1. Select File > Open and select the file Parabola.3dm from the Chapter 5 folder on the Student Companion Website.
2. Select Curve > Parabola > Focus, Direction.
3. Select points A and B, shown in Figure 5-34, to specify the focus point and the direction, respectively.
4. Select point C, shown in Figure 5-34, to specify one of the endpoints of the parabola.
5. Select Curve > Parabola > Vertex, Focus.
6. Select points D and E, shown in Figure 5-34, to specify the focus point and the direction, respectively.

7. Select point F, shown in Figure 5–34, to specify a reference point.
   A parabola is constructed.
8. Do not save your file.

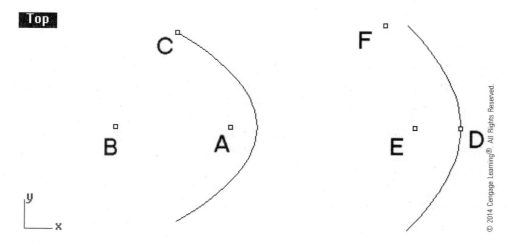

FIGURE 5–34 *Parabola curves*

HYPERBOLA

A hyperbola is also a degree 2 curve. Perform the following steps.

1. Select File > Open and select the file Hyperbola.3dm from the Chapter 5 folder on the Student Companion Website.
2. Set object snap mode to Point.
3. Select Curve > Hyperbola > Center, Coefficient and select the FromCoefficient option at the command area.
4. Select points A, B, and C, shown in Figure 5–35.
5. Select Curve > Hyperbola > From Foci and select the FromFoci option at the command area.
6. Select points D, E, and F, shown in Figure 5–35.
7. Select Curve > Hyperbola > Vertex, Focus and select the FromVertex option at the command area.
8. Select points G, H, and J, shown in Figure 5–35.
   Three hyperbola curves are constructed.
9. Do not save your file.

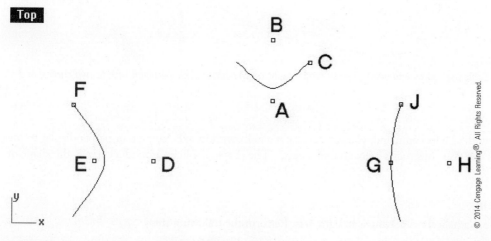

FIGURE 5–35 *Hyperbola curves being constructed*

CONIC

A conic is also a degree 2 curve. You can construct it in several ways. You can construct a conic by specifying its end points, its apex, and a point on the conic. You can also construct a conic that is perpendicular to a selected curve and a conic that is tangent to two curves. Perform the following steps.

1. Select File > Open and select the file Conic.3dm from the Chapter 5 folder on the Student Companion Website.
2. Set object snap mode to End and Point.
3. Select Curve > Conic.
4. Select points A and B (Figure 5–36) to indicate the end points of the conic.
5. Select point C (Figure 5–36) to indicate the apex location.
6. Select location D (Figure 5–36) to indicate a point on the conic curve.
7. Select Curve > Conic and select Perpendicular option at the command area.
8. Click on end point E (Figure 5–36).
9. Select the Perpendicular option at the command area.
10. Click on end point F (Figure 5–36).
11. Select point G.

    A conic perpendicular to two curves and passing through a reference point is constructed.

12. Select Curve > Conic and select Tangent at the command area.
13. Click on end point H (Figure 5–36) to specify the tangent end point.
14. Select the Tangent option at the command area.
15. Select end point J (Figure 5–36) to specify the other end point.
16. Select point K (Figure 5–36).

    A conic tangent to two curves and passing through a reference point is constructed.

17. Do not save your file.

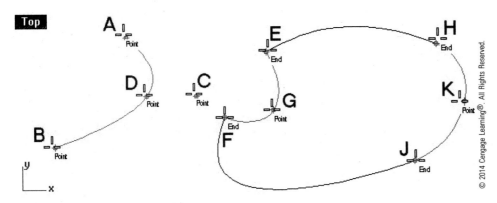

FIGURE 5–36 *From left to right: Conic defined by end points, vertex, and a point on the curve; conic perpendicular to two curves; and conic tangent to two curves*

HELIX

A helix is a degree 3 curve. Using the Helix command, you can construct a helix with its axis lying on or perpendicular to the current construction plane. You can also construct a helix around a curve. The shape of a helix around a curve is the shape of the selected curve.

Helix with a Straight Axis

The simplest kind of helix curve is a helical coil having a straight axis. Perform the following steps.

1. Select File > Open and select the file Helix.3dm from the Chapter 5 folder on the Student Companion Website.

2. Select Curve > Helix.
3. Select points A and B, shown in Figure 5-37, in the Top viewport to specify the axis end points.
4. Select the Turns option at the command area and type 3 at the command area to specify the number of coils.
5. Select point C (Figure 5-37) in the Top viewport to specify the radius of the helix.

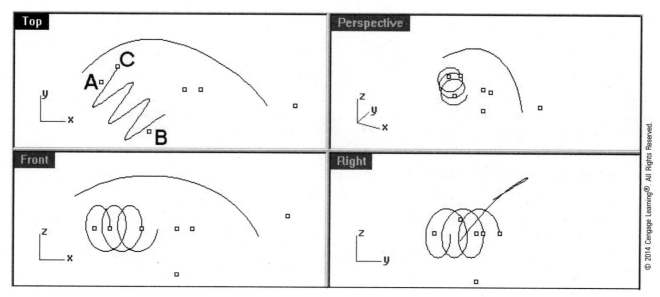

FIGURE 5-37 *Helix lying on the active construction plane constructed*

Vertical Helix

A vertical helix has its axis vertical to the construction plane. Continue with the following steps:

6. Select Curve > Helix and select Vertical option at the command area.
7. Select point D, shown in Figure 5-38, in the Top viewport to specify an axis end point.
8. Select point E (Figure 5-38) in the Front viewport to specify the other axis end point.
9. Select the Turns option at the command area, and type 4 at the command area to specify the number of coils.
10. Select point F (Figure 5-38) in the Top viewport to specify the radius.

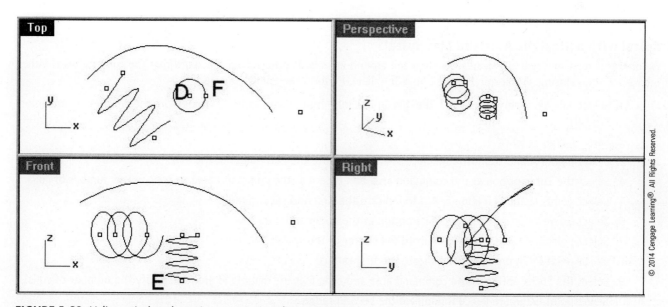

FIGURE 5-38 *Helix vertical to the active construction plane constructed*

Helix around a Curve

A helix around a curve has its axis following the flow of a selected curve. Continue with the following steps.

11. Select Curve > Helix.
12. Select the AroundCurve option at the command area.
13. Select the free-form curve G, shown in Figure 5–39.
14. Type 5 at the command area to set the helix radius.
15. Select the Turns option at the command area and type 6 at the command area to specify the number of coils.
16. Select point H (Figure 5–39) to specify a reference.
17. Do not save your file.

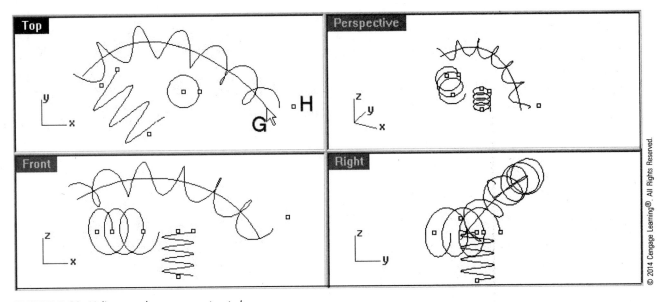

FIGURE 5–39 *Helix around a curve constructed*

SPIRAL

A spiral is a degree 3 curve. You can use the Spiral command to construct a 3D spiral or a flat spiral. You can also construct a 3D spiral around a curve.

Spiral with a Straight Axis and Flat Spiral

A spiral is similar to a helix in terms of being a coil around an axis. It differs from a helix in that the coil's radius is either increasing or decreasing. A flat spiral is a coil on a 2D plane. Perform the following steps.

1. Select File > Open and select the file Spiral.3dm from the Chapter 5 folder on the Student Companion Website.
2. Select Curve > Spiral.
3. Select points A and B, shown in Figure 5–40, to indicate the end points of the axis.
4. Select the Turns option at the command area and type 4 at the command area to specify the number of coils.
5. Select points C and D (Figure 5–40) to indicate the first and second radii.
6. Select Curve > Spiral and select Flat option at the command area.
7. Select point E (Figure 5–40) to indicate the center of the spiral.
8. Select point F (Figure 5–40) to indicate the first radius.
9. Select the Turns option at the command area and type 5 at the command area to specify the number of coils.
10. Select point G (Figure 5–40) to indicate the second radius.

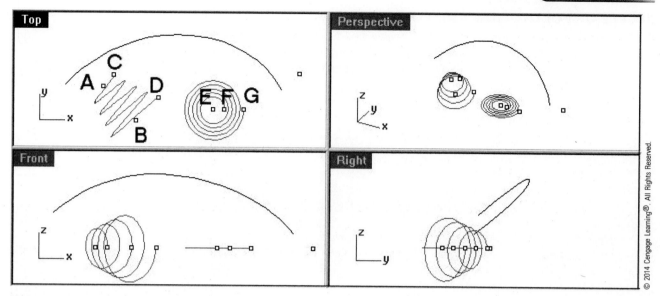

FIGURE 5–40 *A straight spiral and flat spiral constructed*

Spiral around a Curve

You can construct a spiral around a curve, as follows:

11. Select Curve > Spiral.
12. Select the AroundCurve option at the command area.
13. Select curve H, shown in Figure 5–41.
14. Select the Turns option at the command area, and type 6 to specify the number of coils.
15. Type 5 at the command area to specify the first radius.
16. Select point J (Figure 5–41) to specify a reference point.
17. Type 10 at the command area to specify the second radius.

    A spiral around the conic curve is constructed.

18. Do not save your file.

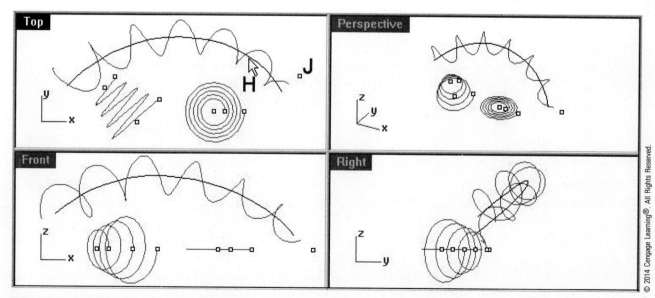

FIGURE 5–41 *Spiral around a curve constructed*

DEGREE OF POLYNOMIAL AND POINT EDITING

Because Rhino uses NURBS mathematics to define curves, the curve manipulating tools delineated in the previous chapter also apply to curves of regular pattern explained above. Naturally, you may change the degree of polynomial of the curves, manipulate the control points and edit points, and use the Handlebar Editor.

Because increasing the degree of polynomial will not change the shape of the curve, it is advisable to change the degree of polynomial of the curves to 3 after they are completed. To avoid having kink points in the curve, you should use deformable circle and deformable ellipse unless true circle and true ellipse are required in your design.

CONSOLIDATION

Apart from free-form curves of irregular pattern that you learned in the previous chapter, you can use curves of regular pattern in surface construction. These curves are line, polyline, rectangle, polygon, circle, arc, ellipse, parabola, hyperbola, conic, helix, and spiral.

Regular curves may be classified by their degree of polynomial. Line, polyline, rectangle, and polygon are degree 1 curves. True circle, arc, true ellipse, parabola, hyperbola, and conic are degree 2 curves. In particular, true circle and true ellipse have kink points at their quadrant locations. The deformable circle, deformable ellipse, helix, and spiral are degree 3 curves of regular pattern. To produce a free-form curve without a kink point in the shape of a circle or in the shape of an ellipse, you construct a deformable circle or deformable ellipse.

Because Rhino uses NURBS mathematics to define curves and surfaces, you can use all the editing tools delineated in Chapter 4 on curves of regular shapes. You can change the degree of polynomial, add or remove knots and kink points, adjust control point location and point weight, and add or remove control points. Naturally, you can use the Handlebar Editor as well.

REVIEW QUESTIONS

1. Give a brief account of the types of curves you can construct using Rhino.
2. Classify the curves in accordance with their degree of polynomial.
3. Differentiate between deformable circle and deformable ellipse with true circle and true ellipse.

CHAPTER 6

Curve Manipulation

INTRODUCTION

This chapter explores various curve manipulation methods, including manipulation of a curve's length, treatment of two or more curves, refinement of curve profile, and deriving points and curves from existing objects.

OBJECTIVES

After studying this chapter, you should be able to

- Manipulate curves and points in the context of surface modeling

OVERVIEW

After learning how to construct free-form curves and curves of regular pattern in Chapters 4 and 5, respectively, this chapter addresses various curve manipulation tools, categorized into four groups. The first group deals with changing the length of a curve; the second group concerns the treatment of two or more separate curves; the third group explains ways to refine a curve; and the fourth group deals with deriving curves and point objects from existing objects.

MANIPULATING A CURVE'S LENGTH

To meet design need, you may extend a curve to add extra length to it, trim a curve to remove unwanted portions, split a curve into two curves, and delete a portion of it, without changing the shape of the existing portion of the curve. If you have an open polyline, you can close it by adding a line segment.

Continuing an Existing Curve

You can increase the length of an existing curve by continuing with more control points or interpolate points. Perform the following steps.

1. Select File > Open and select the file Continue Curve.3dm from the Chapter 6 folder that you downloaded from the Student Companion Website.
2. Select Curve > Free-Form > Continue Control Point Curve.
3. Click on A indicated in Figure 6–1.
4. Check the Point box in the Osnap dialog box.

5. Click on points B, C, and D and press the ENTER key.

   The curve is lengthened by adding more control points to it.

6. Select Curve > Free-Form > Continue Interpolated Curve.
7. Click on E indicated in Figure 6–1.
8. Click on points F, G, and H and press the ENTER key.

   The curve is lengthened by adding more interpolate points to it.

9. Do not save your file.

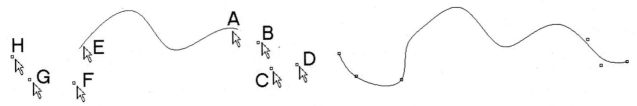

FIGURE 6–1 *Original curve (left) and curve increased in length (right)*
© 2014 Cengage Learning®. All Rights Reserved.

Extending the Length of a Curve

You can extend a curve in several ways: by extending it to a boundary curve, dragging an end of the curve to a new position, adding a straight line segment to the curve, adding an arc segment to the curve, or extending it to meet the edge of a surface.

Extending to a Boundary

The simplest way of extending a curve is to extend it to a boundary. However, if the endpoint of the curve is too far from the boundary curve, a solution may not be available, in which case you will need to try other methods of extending the curve. Perform the following steps.

1. Select File > Open and select the file Extend01.3dm from the Chapter 6 folder on the Student Companion Website.
2. Select Curve > Extend Curve > Extend Curve.
3. Select curve A, shown in Figure 6–2, and press the ENTER key.
4. Select the Type option at the command area.

NOTE Four options are available: Natural, Line, Arc, and Smooth. Obviously, the Line and Arc options add a line or arc segment to the existing curve, and the Smooth option adds a free-form curve segment. The Natural option is quite special: If the original curve is a line, it adds a line segment; if the original curve is an arc, it adds an arc segment; and if the original curve is a free-form curve, it adds a free-form curve segment.

5. Select the Line option at the command area.
6. Select curve B to extend it to A.
7. Press the ENTER key to terminate the command.

   A curve is extended to a boundary curve.

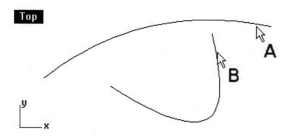

FIGURE 6–2 *Curve being extended to a boundary curve*
© 2014 Cengage Learning®. All Rights Reserved.

Extend by Dragging

This method enables you to determine the endpoint location of the extended curve by dragging it to a new location. Continue with the following steps.

    8. Repeat the command, press the ENTER key to extend dynamically, and select the Type = smooth option.
    9. Select curve A (shown in Figure 6–3) and then select location B.

        For the purpose of this tutorial, the exact location is unimportant. The selected curve is extended by adding a spline segment at the endpoint.

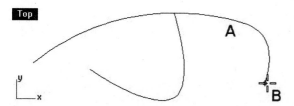

FIGURE 6–3 *Curve being extended by adding a spline segment*
© 2014 Cengage Learning®. All Rights Reserved.

Adding a Line Segment

This method enables you to add a line segment to the curve and determine the endpoint of the line segment. Continue with the following steps.

    10. Select Curve > Extend Curve > By Line.
    11. Select curve A (shown in Figure 6–4) and then select location B.

        The selected curve is extended by attaching to it a tangent line segment.

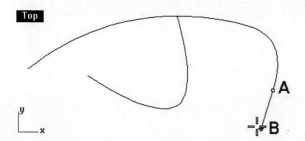

FIGURE 6–4 *Curve being extended by adding a tangent line segment*
© 2014 Cengage Learning®. All Rights Reserved.

Adding an Arc Segment with a Specified Endpoint

This method enables you to add an arc segment to the curve and determine the endpoint of the arc segment. Continue with the following steps.

12. Select Curve > Extend Curve > By Arc to Point.
13. Select curve A (shown in Figure 6–5) and then select location B.

    The selected curve is extended by attaching to it a tangent arc segment defined by the arc's endpoint.

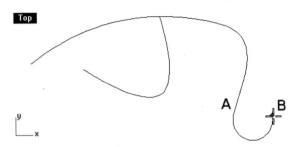

FIGURE 6–5 *Curve being extended by adding a tangent arc segment with a specified end point*
© 2014 Cengage Learning®. All Rights Reserved.

Adding an Arc Segment with Equal Radius of Curvature

This method enables you to add an arc segment having the same radius of curvature at the endpoint of the original curve.

14. Select Curve > Extend Curve > By Arc.
15. Select curve A (shown in Figure 6–6) and then select location B.

    The selected curve is extended by attaching to it a tangent arc segment with a radius equal to the radius of curvature at the endpoint of the curve.

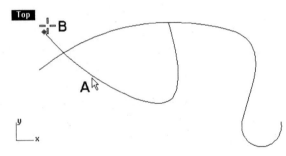

FIGURE 6–6 *Curve being extended by adding a tangent arc segment with equal radius of curvature*
© 2014 Cengage Learning®. All Rights Reserved.

Adding an Arc Segment with Specified Radius

This method enables you to add an arc segment by specifying the radius and endpoint location of the arc. Continue with the following steps.

16. Select Curve > Extend Curve > By Arc with Center.
17. Select curve A (shown in Figure 6–7).
18. Select location B (Figure 6–7) to specify the radius of the arc.
19. Select location C (Figure 6–7) to specify the endpoint of the arc.

    The selected curve is extended by attaching to it a tangent arc.

20. Do not save your file.

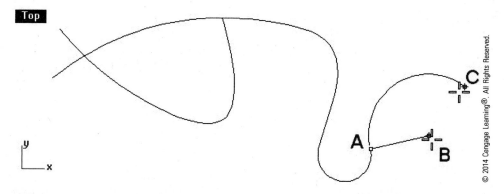

FIGURE 6–7 *Curve being extended by adding a tangent arc segment of designated radius*

Extending Curve on a Surface

To extend a curve on a surface to the surface's boundary edge, perform the following steps.

1. Select File > Open and select the file Extend02.3dm from the Chapter 6 folder on the Student Companion Website.
2. Select Curve > Extend Curve > Curve on Surface, or click on the Extend Curve on Surface button.
3. Select curve A (shown in Figure 6–8) and press the ENTER key.
4. Select surface B.

    The curve on the surface is extended to the boundary edges of the surface.

5. Do not save your file.

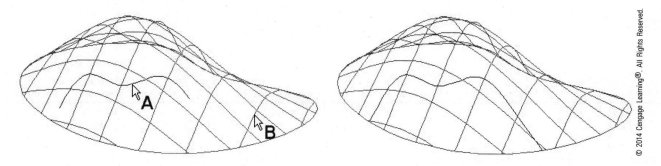

FIGURE 6–8 *Curve being extended (left) and curve extended to the edge of the surface (right)*

Trimming a Curve

Unwanted portions of a curve can be trimmed away. After trimming, the shape of the remaining portion of the curve is unchanged. Perform the following steps.

1. Select File > Open and select the file TrimSplit.3dm from the Chapter 6 folder on the Student Companion Website.
2. Select Edit > Trim.
3. Select curve A, shown in Figure 6–9, and press the ENTER key.

    This is the cutting edge.

4. Select curve B (Figure 6–9).

    This is the portion to be trimmed away.

5. Press the ENTER key to terminate the command.

    The curve is trimmed.

FIGURE 6-9 *Curve being trimmed (left) and curve trimmed (right)*
© 2014 Cengage Learning®. All Rights Reserved.

Splitting a Curve into Two Curves

You can split a curve into two curves and keep the shape of the split curves unchanged. Continue with the following steps.

6. Select Edit > Split.
7. Select curve A, shown in Figure 6-10, and press the ENTER key.
   This is the curve to be split.
8. Select curve B (Figure 6-10).
   This is the cutting object.
9. Press the ENTER key to terminate the command.
   Because curve B intersects curve A at two locations, curve A is split into three curves.
10. Do not save your file.

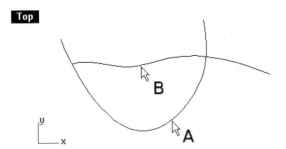

FIGURE 6-10 *Curve being split*
© 2014 Cengage Learning®. All Rights Reserved.

Deleting a Portion of a Curve

A portion of a curve can be deleted. As a result, two curves are derived from a single curve. Perform the following steps.

1. Select File > Open and select the file DelSubCurve.3dm from the Chapter 6 folder on the Student Companion Website.
2. Set object snap mode to Point.
3. Select Curve > Curve Edit Tools > Delete Subcurve.
4. Select curve A and then points B and C (shown in Figure 6-11).
   The portion of the curve between B and C is deleted.
5. Do not save your file.

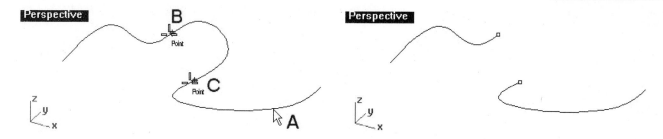

FIGURE 6–11 *Original curve (left) and a portion deleted (right)*
© 2014 Cengage Learning®. All Rights Reserved.

Closing an Open Polyline

An open polyline consists of line segments only and can be closed by adding a line segment to it, as follows.

1. Select File > Open and select the file ClosePolyline.3dm from the Chapter 6 folder on the Student Companion Website.
2. Select Curve > Curve Edit Tools > Close Curve.
3. Select polycurve A, shown in Figure 6–12, and press the ENTER key.
   A segment is added, and the polycurve is closed.
4. Do not save your file.

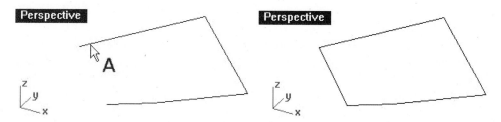

FIGURE 6–12 *Open loop polycurve (left) and polycurve closed (right)*
© 2014 Cengage Learning®. All Rights Reserved.

TREATING TWO OR MORE SEPARATE CURVES

Chamfering, filleting, and blending are three basic ways to modify two separate curves to become a set of contiguous curves. Between the curves, chamfering adds a bevel edge, which, in essence, is a straight line; filleting adds a tangent arc; and blending adds a smooth curve. As may be necessary, both chamfering and filleting extend or trim the existing curves to meet the bevelled edge or filleted arc. In accordance with your design needs, you may have to join two or more curves into one, or explode a joined curve into individual curve components.

As explained in Chapter 3, with two contiguous curves meeting each other endpoint to endpoint, the continuity between them can be described as G0, G1, G2, G3, or G4. Typically, chamfered edge has G0 continuity; filleted edge has G1 continuity; and blended edge can have G2, G3, or G4 continuity.

In addition to chamfering, filleting, and blending, you can connect two coplanar curves to become a joined curve, modify a curve to match another curve, or modify two curves to match each other. If curves intersect each other and there are closed regions between them, boundaries of such regions can be constructed using curve Boolean operations.

Chamfering

Chamfering joins two selected curves with a bevelled line. In essence, chamfering extends or trims the existing curves and inserts a line segment between them. While chamfering, you need to specify the distances of the endpoints of the inserted line segment measured from the imaginary intersection point of the two curves. Perform the following steps.

1. Select File > Open and select the file Joint.3dm from the Chapter 6 folder on the Student Companion Website.
2. Select Curve > Chamfer Curves.
3. At the command area, set Trim = Yes if Trim = No.
4. Select the Distance option at the command area.

5. Type 10 to set the first distance of the line segment from the imaginary intersection point of the two curves.
6. Type 5 to set the second distance.

> **NOTE** By default, the second distance is equal to the first distance. Therefore, you may simply press the ENTER key to accept, if the second distance equals the first distance.

7. Select curves A and B, shown in Figure 6–13.

    Note that the first selected edge corresponds to the first chamfer distance. A chamfer curve is constructed.

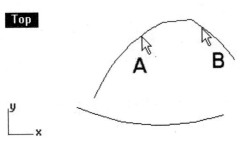

FIGURE 6–13 *Chamfer curve constructed*
© 2014 Cengage Learning®. All Rights Reserved.

Filleting

Filleting also joins two selected curves. It extends or trims the existing curves and adds a tangent arc segment between them. A filleted curve gives a G1 continuity. Continue with the following steps.

8. Select Curve > Fillet Curves.
9. Select the Radius option at the command area.
10. Type 3 to specify a radius of 3 mm.
11. Select curves A and B, shown in Figure 6–14.

    The selected curves are trimmed/extended, and a fillet curve is constructed between them.

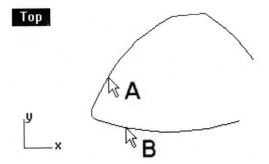

FIGURE 6–14 *Fillet curve constructed*
© 2014 Cengage Learning®. All Rights Reserved.

Filleting Vertices of a Polyline

If you have a polyline or a set of curves joined together end to end, you may fillet all of its corners in a single operation, as follows.

12. Maximize the Front Viewport.
13. Select Curve > Fillet Corners.
14. Select curve A, shown in Figure 6–15, and press the ENTER key.
15. Type 5 at the command area to specify the fillet radius.

In essence, all the corners of a polyline will be filleted. However, segments too short for the specified fillet radius will not be filleted. Here, a corner is not filleted for this reason.

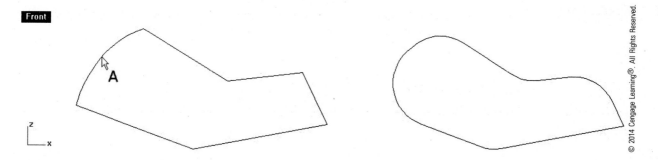

FIGURE 6–15 *Polyline (left) and five corners of the polyline filleted (right)*

Blending

To construct a curve to bridge two edges/curves with a continuity of G1, you construct a simple blend curve, as follows.

16. Maximize the Top viewport.
17. Select Curve > Quick Curve Blend.
18. At the command area, if Continuity is not equal to Curvature, click on it and select Curvature.
19. Select curves A and B, shown in Figure 6–16.

    A continuous curvature curve is constructed.

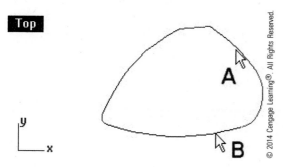

FIGURE 6–16 *Blend curve constructed*

To construct a perpendicular and angular blend curve, continue with the following steps.

20. Maximize the Perspective viewport.
21. Set object snap mode to Point.
22. Select Curve > Quick Curve Blend and click on the Perpendicular option at the command area.
23. Select edge A, shown in Figure 6–17, and then click on point B.
24. Click on the Perpendicular option at the command area.
25. Select edge C, shown in Figure 6–17, and then click on point D.

    A perpendicular blend curve is constructed.

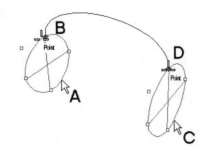

FIGURE 6–17 *Perpendicular blend curve constructed*
© 2014 Cengage Learning®. All Rights Reserved.

26. Select Curve > Blend Curve.
27. Select the At Angle option at the command area.
28. Select edge A, shown in Figure 6–18, and then click on points B and C.
29. Select the At Angle option at the command area.
30. Select edge D, shown in Figure 6–18, and then click on points E and F.

    An angular blend curve is constructed.

31. Do not save your file.

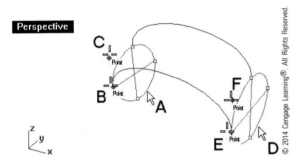

FIGURE 6–18 *Angular blend curve constructed*

Adjustable Curve Blend

To achieve a continuity beyond G2 and impose additional control on the shape while constructing a blend curve, you use the BlendCrv command. Perform the following steps.

1. Select File > Open and select the file AdjustableBlend.3dm from the Chapter 6 folder on the Student Companion Website.
2. Select Curve > Blend Curves > Adjustable Curve Blend.
3. Select curves A and B, shown in Figure 6–19.
4. In the Adjust Curve Blend dialog box, check the G4 buttons and click on the Show Curvature box.
5. In the Curvature Graph dialog box, click on the Curve hair button and then select a color from the Select Color dialog box to choose a color for the curvature hairs that appear on the curves, then click on the Surface hair button, and select a color from the Select Color dialog box to choose a color for the surface hairs.
6. Click on the Display scale and Density arrow buttons to adjust the curvature scale and density as may be necessary.
7. Click on the X mark of the dialog box to close it.
8. Select the control points and drag them to new locations to experience how the blend curve's shape is changed.
9. Press the ENTER key.

    A blend curve is constructed.

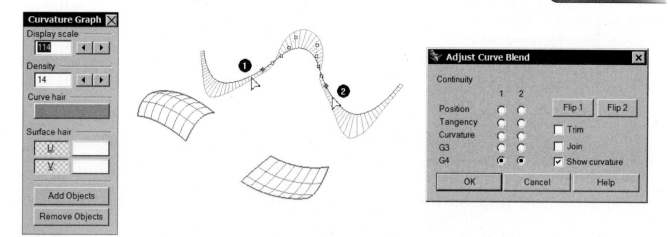

FIGURE 6–19 *From left to right: Curvature Graph dialog box, adjustable blend curve being constructed between curves, and Adjust Curve Blend dialog box*
Source: Robert McNeel and Associates Rhinoceros® 5

In addition to blending between two selected curves, you can construct blend curve between two surfaces along selected locations on the surfaces' edges. Continue with the following steps.

10. Repeat the command.
11. Select the Edges option at the command area.
12. Select edges A and B, shown in Figure 6–20.
13. To adjust the location of the curve, click on the endpoint of the curve and drag it to a new position.
14. To adjust the shape of the curve, click on the control points and drag them to a new location.
15. Press the ENTER key.

    A curve is constructed.

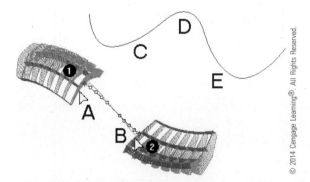

FIGURE 6–20 *Adjustable blend curve being constructed along the surfaces' edges*

Joining Contiguous Curves

To reflect our design intent, it is sometimes easier to construct a number of contiguous curves and then join them to form a single curve. Perform the following steps.

16. Select Edit > Join.
17. Select curves C, D, and E (Figure 6–20) in either clockwise or counter-clockwise direction and press the ENTER key.

    The curves are joined.

Exploding a Joined Curve

To revert a joined curve into separate, individual curves, you can explode it. Continue with the following steps.

18. Select the joined curve D (shown in Figure 6–20) and press the ENTER key.
19. Do not save your file.

Two-Arc Fillet

Instead of using a single arc to bridge the gap between two curves by filleting, you can use the ArcBlend command to insert two arcs. Although the outcome is aesthetically better than a single fillet arc, the continuity between the curves is still G1. Now perform the following steps.

1. Select File > Open and select the file Two Arc Fillet.3dm from the Chapter 6 folder on the Student Companion Website.
2. Select Curve > Blend Curves > Arc Blend.
3. Select A and B indicated in Figure 6–21.

   A two-arc fillet is constructed.
4. Click on control point C and drag to appreciate how to adjust the radii of the arcs.
5. Press the ENTER key if you are satisfied with the shape of the arcs.
6. Do not save your file.

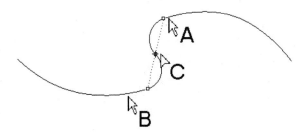

FIGURE 6–21 *Two-arc fillet being constructed*
© 2014 Cengage Learning®. All Rights Reserved.

Connecting Two Coplanar Curves to Meet

A simple way to connect two coplanar curves to become a joined curve is to add line segments to their ends for them to meet. If the curves are arcs, you may optionally add arc segments instead of line segments. Perform the following steps.

1. Select File > Open and select the file Connect.3dm from the Chapter 6 folder on the Student Companion Website.
2. Select Curve > Connect Curves.
3. If Join = No, select it at the command area to change it to Yes.
4. If ExtendArcsBy = Line, select it at the command area to change it to Arc. Otherwise, proceed to the next step.
5. Select curve A, shown in Figure 6–22.
6. Select ExtendArcsBy = Arc at the command area to change it to Line.
7. Select curve B, shown in Figure 6–22. Curve A (extended by an arc segment) and curve B (extended by a line segment) are joined.
8. Do not save your file.

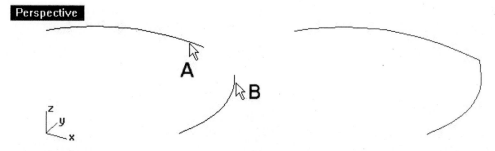

FIGURE 6–22 *Curves selected (left) and curves connected (right)*
© 2014 Cengage Learning®. All Rights Reserved.

Curve Matching

If we have two separate curves, we can modify the shape of one curve to make it match the other curve. Unlike blending, in which the original curves are not changed and a new curve is inserted, matching changes the shape of one of the curves. Between the curves, you have G2 continuity. Perform the following steps.

1. Select File > Open and select the file Match1.3dm from the Chapter 6 folder on the Student Companion Website.
2. Select Curve > Curve Edit Tools > Match.
3. Select curve A near its left end (shown in Figure 6–23).
4. Select curve B near its right end (shown in Figure 6–23).

   This way the left end of curve A will match to the right end of curve B.

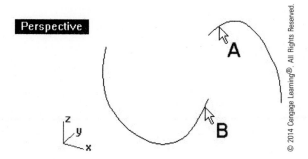

FIGURE 6–23 *Free-form curves to be matched*

The Match Curve dialog box includes several buttons. The Position, Tangency, and Curvature buttons concern the continuity between the two curves after they are matched. If the Average curves button is not selected, only the first selected curve will be modified. If the Preserve other end button is selected, the shape of the other end of the modified curve(s) will be preserved as far as possible. The Join and Merge buttons enable the curves to be joined or merged. A join curve consists of two separate segments that can be exploded. A merge curve becomes a single curve and cannot be exploded.

5. Try out various options in the Match Curve dialog box, shown in Figure 6–24.
6. Check the Curvature and Average curves, and then click on the OK button.

   Both curves are modified to match each other.

7. Do not save your file.

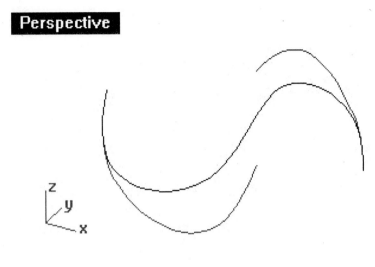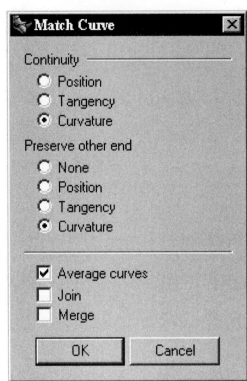

FIGURE 6–24 *Preview of matched curve (left) and Match Curve dialog box (right)*
Source: Robert McNeel and Associates Rhinoceros® 5

Matching Two Endpoints of a Curve

Two endpoints of an open curve can be matched to form a periodic curve, as follows.

1. Select File > Open and select the file Match2.3dm from the Chapter 6 folder on the Student Companion Website.
2. Select Curve > Curve Edit Tools > Match.
3. Select curve A near one end (shown in Figure 6–25).
4. Select curve B near the other end (shown in Figure 6–25).
5. In the Match Curve dialog box, click on the OK button.

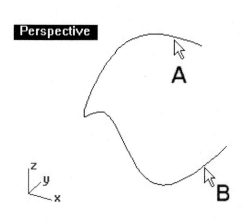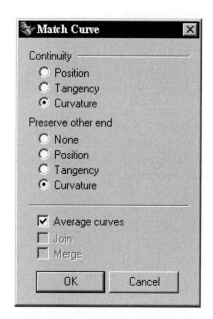

FIGURE 6–25 *From left to right: open curve, Match Curve dialog box, and preview of the matched curve*
Source: Robert McNeel and Associates Rhinoceros® 5

Curve Boolean

A quick way to construct closed-loop curves from two or more curves is to use Boolean operation. Perform the following steps.

1. Select File > Open and select the file CurveBoolean.3dm from the Chapter 6 folder on the Student Companion Website.
2. Select Curve > Curve Edit Tools > Curve Boolean.
3. Select curves A, B, C, and D, shown in Figure 6–26, and press the ENTER key.
4. At the command area, set DeleteInput = All and CombineRegions = Yes.
5. Click on location E, shown in Figure 6–26.
   A region is highlighted.
6. Click on location F, shown in Figure 6–26.
   Another region is highlighted.
7. Click on location G, shown in Figure 6–26.
   A third region is highlighted and is combined with the second region.
8. Press the ENTER key.
   Two closed-loop curves are constructed from the input curves.
9. Do not save your file.

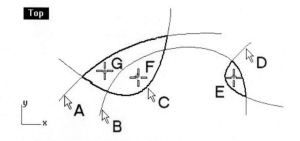

FIGURE 6–26 *Closed-loop curves being constructed*
© 2014 Cengage Learning®. All Rights Reserved.

CURVE REFINEMENT METHODS

A curve's shape can be refined in many ways: rebuilding by specifying its degree of polynomial and number of control points, rebuilding interactively, adjusting its end bulge without changing the tangent direction at its endpoints, soft editing, modifying while keeping the curve's length unchanged, moving individual segments if the curve is a polycurve, replacing a portion of the curve with a line segment, fairing, simplifying, refitting, and converting the curve into line and arc segments.

Rebuilding a Curve

After you have finalized the shape of a curve (via various editing methods, such as those you performed previously), you might need to refine the curve by rebuilding it with a different number of control points and a different degree of polynomial. Rebuilding always removes kinks; it is a common technique used by designers to both simplify curves and remove potential problems with kinks. If you reduce the number of control points, the curve simplifies and the shape changes. If you increase the number of control points, the shape will not change until next time you manipulate the control points. Perform the following steps.

1. Select File > Open and select the file Rebuild.3dm from the Chapter 6 folder on the Student Companion Website.
2. Select Curve > Control Points > Control Points On.
3. Select curve A, shown in Figure 6–27, and press the ENTER key.
4. Select Edit > Rebuild.
5. Select curve A, shown in Figure 6–27, and press the ENTER key.
6. In the Rebuild Curve dialog box, set the number of control points to 8. Leave the degree unchanged.

   The default value shown may not be the same as yours. This command always remembers the previous settings when last used.

7. Click on the Preview button.

   The maximum deviation of the rebuilt curve from the original curve is highlighted on the curve, and the deviation value is displayed in the dialog box.

8. If you are satisfied with the change, click on the OK button.

   The curve is rebuilt.

FIGURE 6–27 *From left to right: original curve, Rebuild Curve dialog box, and rebuilt curve*
Source: Robert McNeel and Associates Rhinoceros® 5

Rebuilding a Curve Interactively

To rebuild a curve interactively, continue as follows.

9. Type RebuildCrvNonUniform at the command area or click on the Rebuild Curve Non Uniform button on the Curve Tools toolbar.
10. Select curve A, shown in Figure 6–28, and then press the ENTER key.

    Note that which end you select on the curve has an effect on the outcome.

11. At the command area, set MaxPointCount=7.
12. Press the ENTER key.

    The curve is rebuilt.

13. Do not save your file.

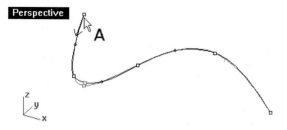

FIGURE 6–28 *Rebuilding a curve interactively*
© 2014 Cengage Learning®. All Rights Reserved.

Adjusting a Curve's End Bulge

If you want to modify the shape of a curve and yet maintain the tangent directions at the endpoints, you adjust the curve's end bulge. Continue with the following steps.

1. Select File > Open and select the file EndBulge.3dm from the Chapter 6 folder on the Student Companion Website.
2. Select Edit > Adjust End Bulge.
3. Select curve A, shown in Figure 6–29.
4. Select and drag the control points B and C one by one to appreciate the effect of modification.
5. Press the ENTER key when finished.

   The end bulge is modified.
6. Do not save your file.

   As shown, the two control points are restricted to translate along a straight line. As a result, the tangency at the endpoints of the curve is maintained even though the curve's shape is modified.

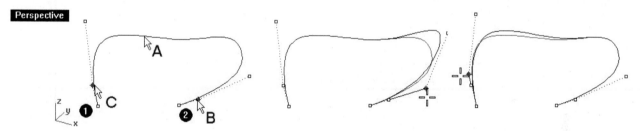

FIGURE 6–29 *From left to right: original curve, control point B being dragged, and control point C being dragged*
© 2014 Cengage Learning®. All Rights Reserved.

Soft Editing a Curve

A way to modify the shape of a curve is to soft edit it by picking a point on the curve and repositioning it to a new location. The term "soft edit" probably refers to the phenomenon that the curve, during soft editing, behaves like a soft object. To soft edit a curve, perform the following steps.

1. Select File > Open and select the file SoftEdit.Crv3dm from the Chapter 6 folder on the Student Companion Website.
2. Select Curve > Curve Edit Tools > Soft Edit.
3. Select curve A, shown in Figure 6–30.
4. Type mid at the command area and pick midpoint B as the base point.
5. At the command area, se FixEnds = Yes and set Distance = 20. This is the distance along the curve that behaves as "soft."
6. Select point C. The curve is copied, and the copy is soft edited.
   Press the ENTER key.

To appreciate the effect of the distance option, you may repeat soft editing the original curve with a distance of 50.

7. Do not save your file.

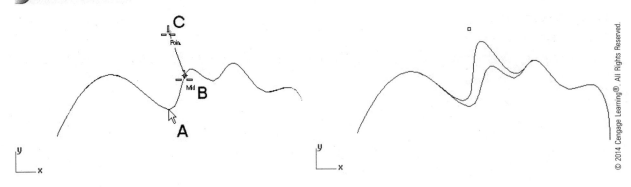

FIGURE 6–30 *Original curve (left) and curve copied and soft edited (right)*

Fixed Length Curve Edit

This process modifies the shape of a curve interactively without changing the length of the curve. Perform the following steps.

1. Select File > Open and select the file FixedLengthEdit.Crv3dm from the Chapter 6 folder on the Student Companion Website.
2. Click on the Edit curve with fixed length button on the Move toolbar.
3. At the command area, set Copy = Yes.
4. Select curve A, shown in Figure 6–31.
5. Check the Mid and Point boxes on the Object Snap (Osnap) toolbar.
6. Select midpoint B and then point C.
   The curve is modified, but the length is kept unchanged.
7. Do not save your file.

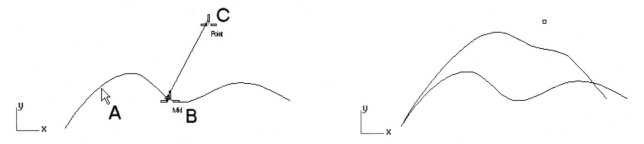

FIGURE 6–31 *Original curve (left) and curve being modified (right)*
© 2014 Cengage Learning®. All Rights Reserved.

Moving Segments of a Polyline or Polycurve

When two or more line segments are joined together, you obtain a polyline. Naturally, you can construct a polyline by using the Polyline command. When two or more curve segments are joined together, you obtain a polycurve. Individual segment of a polyline or polycurve can be moved, as follows.

1. Select File > Open and select the file MoveSegment.3dm from the Chapter 6 folder on the Student Companion Website.
2. Set object snap to point.
3. Select Curve > Curve Edit Tools > Move Curve Segment.
4. At the command area, set End = No and Copy = No.
5. Click on segment A, shown in Figure 6–32.
6. Select point B and then point C.
   The selected polyline segment is moved.

7. Repeat the command.
8. Click on segment D, shown in Figure 6–32.

   This is a polycurve constructed by joining three curves. Segment D is the central segment of the polycurve.

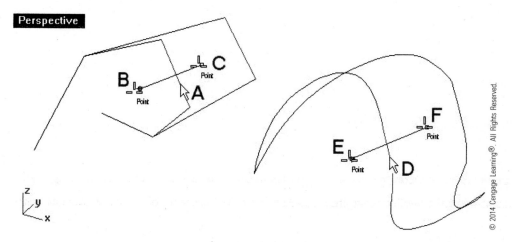

FIGURE 6–32 *Segment being moved in a polyline (left) and polycurve (right)*

9. Select point E and then point F.

   The selected curve segment is moved.
10. Repeat the command.
11. Select the End option at the command area to change it to Yes.
12. Click on vertex A, shown in Figure 6–33.
13. Select point B.

    The polyline segment's endpoint is moved.
14. Repeat the command.
15. Click on joint C, shown in Figure 6–33.
16. Select point D and then point E.

    The curve segment's endpoint is moved.
17. Do not save your file.

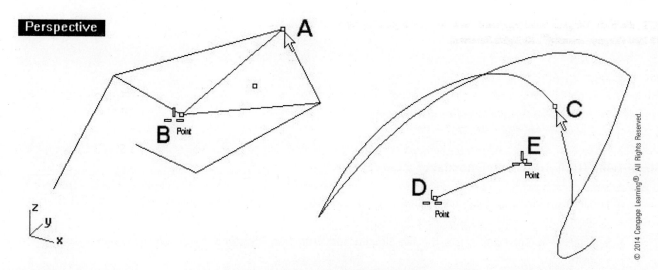

FIGURE 6–33 *Segment end point being moved in a polyline (left) and polycurve (right)*

Replacing a Portion of a Curve with a Line Segment

You can replace a portion of a curve with a line segment, as follows.

1. Select File > Open and select the file InsertLineSegment.3dm from the Chapter 6 folder on the Student Companion Website.
2. Click on the Insert Line into Curve button on the Curve Tools toolbar.
3. Select curve A and then points B and C, shown in Figure 6–34.

   The portion BC is replaced by a line segment.
4. Do not save your file.

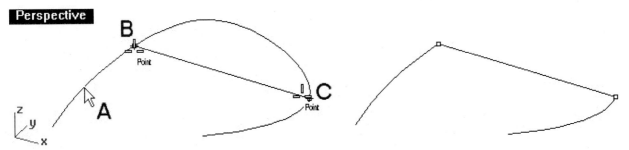

FIGURE 6–34 *Curves selected (left) and curves connected (right)*
© 2014 Cengage Learning®. All Rights Reserved.

Fairing a Curve

If you have a sketch curve or a digitized curve with large curvature deviation, you can smooth it. The process to remove large curvature variation is called fairing. Perform the following steps.

1. Select File > Open and select the file Fair.3dm from the Chapter 6 folder on the Student Companion Website.
2. Select Curve > Curve Edit Tools > Fair.
3. Select curve A (shown in Figure 6–35) and press the ENTER key.
4. Type 0.1 at the command line area to specify tolerance.

   The curve is faired.
5. Do not save your file.

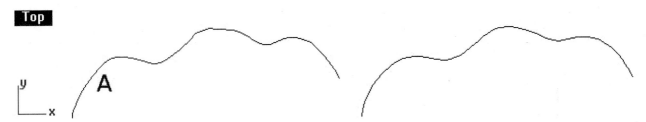

FIGURE 6–35 *Curve being faired (left) and faired curve (right)*
© 2014 Cengage Learning®. All Rights Reserved.

Simplifying (Unifying the Polynomial of a Curve)

If you extend a curve by adding a line segment or an arc segment to it, the degree of polynomial of the added segment may not be congruent with the curve itself. The process to replace such line or arc segments in a curve with a true NURBS curve is called simplifying. In the following exercise, you will add an arc segment to a curve and then simplify it.

1. Select File > Open and select the file Simplify.3dm from the Chapter 6 folder on the Student Companion Website.
2. Select Curve > Extend Curve > By Line.

3. Select curve A (shown in Figure 6–36).
4. Select location B (Figure 6–36) to indicate the endpoint of the added line segment.

   Note that for the purpose of this tutorial, the extended length is unimportant. A line segment is added.

Now you will simplify the curve, as follows.

5. Select Curve > Curve Edit Tools > Simply Lines and Arcs.
6. Select curve A (shown in Figure 6–36) and press the ENTER key.

   The extended line segment of the curve is now replaced with a true NURBS segment. You will not find much change on the shape of the curve because the curve takes on the shape of the line.

7. Do not save your file.

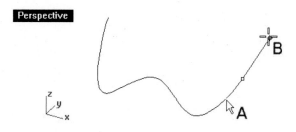

FIGURE 6–36 *Curve being simplified*
© 2014 Cengage Learning®. All Rights Reserved.

Refitting a Curve

If we want to build a smooth surface, we may need to refit a curve, as follows.

1. Select File > Open and select the file Refit.3dm from the Chapter 6 folder on the Student Companion Website.
2. Select Curve > Curve Edit Tools > Refit to Tolerance.
3. Select polyline A, shown in Figure 6–37, and press the ENTER key.
4. At the command area, set DeleteInput = No, OutputLayer = Current, and AngleTolerance = 5.
5. Type 1 at the command area to specify the fitting tolerance and press the ENTER key.

   A new curve is constructed by refitting the original curve in accordance with the input parameters. You may repeat the command and try out with a larger or smaller tolerance to see the differences.

6. Select Edit > Control Points > Control Points On and select both curves to discover the difference between the original curve and the fitted curve.
7. Do not save your file.

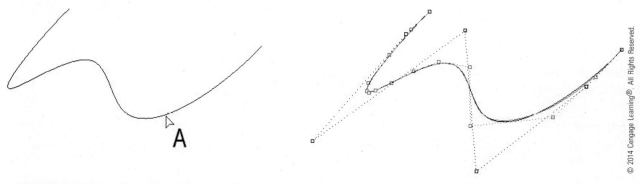

FIGURE 6–37 *Polyline (left) and refitted curve (right)*

Converting a Free-Form Curve to Line/Arc Segments

Contrary to refitting a polyline to a curve, you can convert a free-form curve to a polyline. Such interconversion is basically required when you import data from a system that outputs polylines and export to a system that reads polylines. If you are going to export curves to another computerized system that does not support splines, you may consider converting the splines to lines and arcs. If your curve to be simplified is composed of line and arc segments, you should use the SimplifyInput option to make sure that lines and arcs are converted accurately. Perform the following steps to convert a curve to a polyline and a curve to a set of joined arcs.

1. Select File > Open and select the file Convert.3dm from the Chapter 6 folder on the Student Companion Website.
2. Select Curve > Convert > Curve to Lines.
3. Select curve A, shown in Figure 6–38, and press the ENTER key.
4. At the command area, set AngleTolerance = 25 and Tolerance = 5.
5. Press the ENTER key.

    The curve is converted to a polyline. Smaller tolerance results in a higher number of line segments in the polyline.

FIGURE 6–38 *A free-form curve being converted to a polyline*
© 2014 Cengage Learning®. All Rights Reserved.

6. Maximize the Right viewport.
7. Select Curve > Convert > Curve to Arcs.
8. Select curve A, shown in Figure 6–39, and then press the ENTER key.
9. At the command area, set AngleTolerance = 15 and Tolerance = 5.
10. Press the ENTER key.

    The curve is converted to a set of connected arcs.

11. Do not save your file.

FIGURE 6–39 *A free-form curve being converted to a set of connected arcs*
© 2014 Cengage Learning®. All Rights Reserved.

Arc Curve

To construct a curve to interpolate among a set of points, and then convert the curve into a number of arc segments in one single operation, perform the following steps.

1. Select File > Open and select the file ArcCurve.3dm from the Chapter 6 folder on the Student Companion Website.
2. Select the Arc Through Points button on the Arc toolbar.
3. Drag from A to B (Figure 6–40) to describe a rectangular area to select all the point objects.

4. Press the ENTER key.

   A curve is constructed and converted to a set of arc segments.

5. Do not save your file.

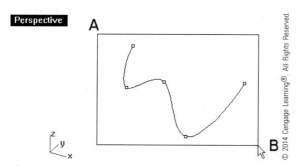

FIGURE 6–40 *Curve constructed from point objects and converted to arc segments*

CURVES AND POINTS FROM EXISTING OBJECTS

Although it is logical to construct curves and/or points before creating surfaces and polysurfaces, detailing a 3D free-form model often requires intertwined construction of curves, points, and surfaces. Therefore, you may need to construct points and curves from existing surfaces, polysurfaces, and polygon meshes. Using these points and curves, you can develop and refine your design.

Offsetting a Curve

You can offset a curve to derive a new curve resembling the existing curve and at a specified distance from it. The offset distance is measured on a plane parallel to the active construction plane. Perform the following steps.

1. Select File > Open and select the file Offset01.3dm from the Chapter 6 folder on the Student Companion Website.
2. Select Curve > Offset Curve.
3. Select the Distance option at the command area.
4. Type 4 to set offset distance.
5. Select curve A, shown in Figure 6–41.
6. Click on location B (Figure 6–41) to indicate the direction of offset.

   An offset curve is constructed.

7. Repeat the command.
8. Select curve C and click on location D (Figure 6–41).

   Another offset curve is constructed.

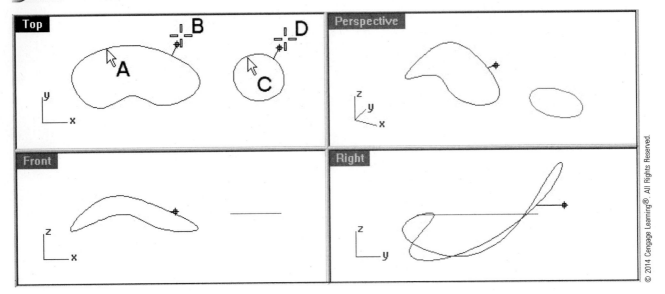

FIGURE 6–41 *Offset curves being constructed*

To appreciate how the outcome of offset is affected by the active construction plane, continue with the following steps.

9. Repeat the command.
10. Select curve A and click on location B, shown in Figure 6-42.
11. Repeat the command.
12. Select curve C and click on location D (Figure 6-42).
13. Do not save your file.

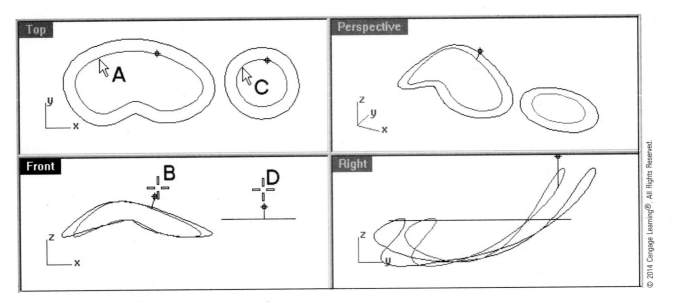

FIGURE 6–42 *Second offset curve being constructed*

Offsetting a Curve on a Surface

To offset a curve residing on a surface, perform the following steps.

1. Select File > Open and select the file Offset02.3dm from the Chapter 6 folder on the Student Companion Website.
2. Click on the Offset curve on surface button on the Curve Tools toolbar.

3. Select curve A, shown in Figure 6-43.
4. Select surface B, shown in Figure 6-43.
5. Type 3 and press the ENTER key.
   An offset curve is constructed.

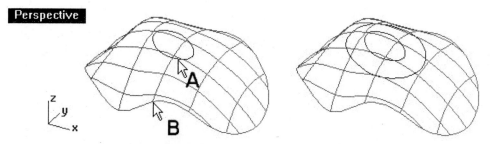

FIGURE 6-43 *Curve on a surface (left) and curve offset on a surface (right)*
© 2014 Cengage Learning®. All Rights Reserved.

Offsetting a Curve in a Direction Normal to a Surface

Continue with the following steps to offset a curve normal to a surface.

6. Select Curve > Offset Normal to Surface.
7. Select curve A, shown in Figure 6-44.
8. Select surface B.
9. Type -10 at the command area to specify the height.
   An offset curve is constructed.
10. Do not save your file.

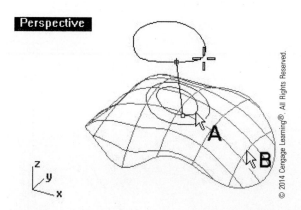

FIGURE 6-44 *Curve offset normal to a surface being constructed*

Curves Interpolating between Two Curves

You can construct a set of curves between two selected curves, as follows.

1. Select File > Open and select the file MeanCurve.3dm from the Chapter 6 folder on the Student Companion Website.
2. Select Curve > Mean Curve.
3. Select curves A and B, shown in Figure 6-45.
   A mean curve is constructed.

To construct a series of curves between two selected curves, continue with the following steps.

4. Select Curve > Tween Curves.
5. Select curves A and B, shown in Figure 6–45.
6. Type 4 at the command area.
   Two curves are constructed.
7. Do not save your file.

FIGURE 6–45 *From left to right: two curves selected (left), mean curve constructed (center), and tween curves constructed (right)* © 2014 Cengage Learning®. All Rights Reserved.

Construction of a 3D Curve from Two Planar Curves

This method constructs a 3D curve. If you already have an idea of the shape of a 3D curve in two orthographic drawing views, you can first construct two planar curves residing on two adjacent viewports, and then derive the 3D curve from the planar curves. Perform the following steps to construct a 3D curve from two planar curves.

1. Select File > Open and select the file CurveFrom2views.3dm from the Chapter 6 folder on the Student Companion Website.
2. Select Curve > Curve from 2 Curves.
3. Select curves A and B. A 3D curve is derived, as shown in Figure 6–46.
4. Do not save your file.

 NOTE For better visualization of the 2D planar curves prior to constructing the 3D curve, you may synchronize the viewports.

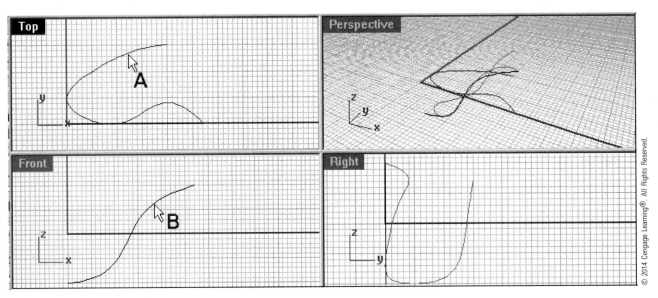

FIGURE 6–46 *3D curve derived from two planar curves*

Construction of Cross-Section Profile Curves

A cross-section profile is a closed-loop, free-form curve interpolating the intersection points between a set of longitudinal curves and a specified section plane across the curves. The cross-section curves, together with the longitudinal curves, form the U and V curves for making a surface from a network of curves. Perform the following steps to construct cross-section profile curves.

1. Select File > Open and select the file CSecProfile.3dm from the Chapter 6 folder on the Student Companion Website.
2. Select Curve > Cross Section Profiles.
3. Select, in clockwise or counter-wise direction sequentially but not randomly, the profile curves A, B, C, and D, shown in Figure 6–47, and press the ENTER key.
4. Select locations E and F (Figure 6–47).

   A cross-section curve is constructed.

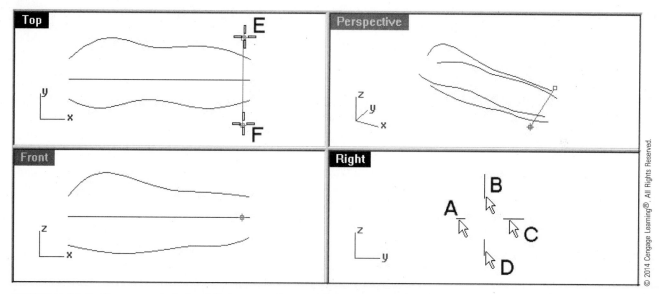

FIGURE 6–47 *Profile curves selected and first cross-section curve being constructed*

5. Select locations A and B (shown in Figure 6–48), locations C and D, locations E and F, locations G and H, and locations J and K.

   Five more cross-section curves are constructed.

6. Press the ENTER key to terminate the command.
7. Do not save your file.

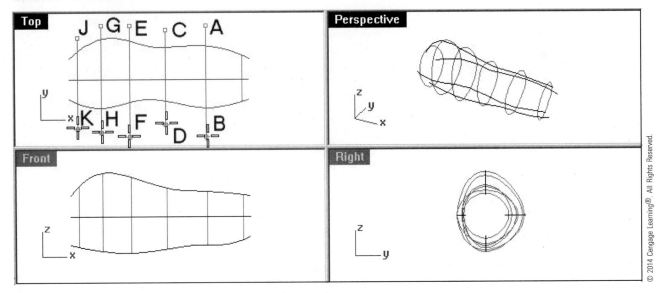

FIGURE 6-48 *Five more cross-section curves constructed*

Projecting and Pulling a Curve to a Surface

Construction of 3D curves on a surface can be made easy by first constructing a 2D curve on one of the construction planes and then projecting it or pulling it toward the surface. The direction of projection is perpendicular to the construction plane, and the direction of pull is normal to the surface. To experiment with this functionality, perform the following steps.

1. Select File > Open and select the file ProjectPull.3dm from the Chapter 6 folder on the Student Companion Website.
2. Click on the Top viewport to set it as the current viewport.
3. Select Curve > Curve From Objects > Project.
4. Select curve A (Figure 6-49) and press the ENTER key.
5. Select surface B (Figure 6-49) and press the ENTER key.

    A curve is projected. If the History Manager is turned on prior to projecting, manipulation of the source curve causes corresponding changes to the projected curve.

6. Click on the source curve A and drag it to a new location.

    The projected curve will also move.

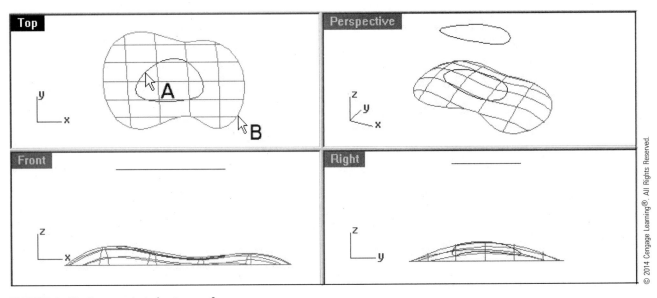

FIGURE 6-49 *Curve projected onto a surface*

7. Select Curve > Curve From Objects > Pullback.
8. Select curve A (Figure 6–50) and press the ENTER key.
9. Select surface B (Figure 6–50) and press the ENTER key.
10. Compare the result of pulling to the result of projection.
11. Do not save your file.

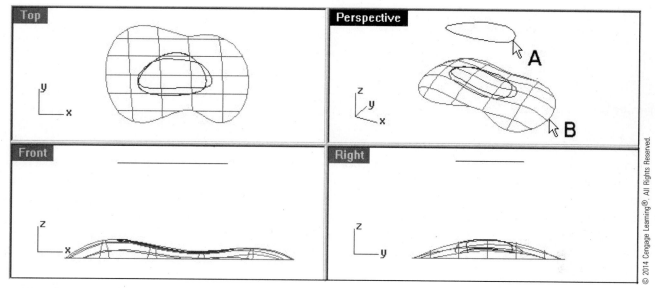

FIGURE 6–50 *Curve pulled to a surface*

Projecting and Pulling a Curve to a Polygon Mesh

The same commands described in the previous section can be used to project/pull curves onto a polygon mesh. (Polygon mesh will be discussed in Chapter 9.) Because a polygon mesh is faceted, what you will get is a zigzag curve. Perform the following steps.

1. Select File > Open and select the file ProjectPullMesh.3dm from the Chapter 6 folder on the Student Companion Website.
2. Select Curve > Curve From Objects > Project.
3. Select curve A, shown in Figure 6–51, and press the ENTER key.
4. Select polygon mesh B and press the ENTER key.
   The curve is projected onto the polygon mesh.
5. Select Curve > Curve From Objects > Pullback.
6. Select curve C, shown in Figure 6–51, and press the ENTER key.
7. Select polygon mesh D and press the ENTER key.
   The curve is projected onto the polygon mesh.
8. Do not save your file.

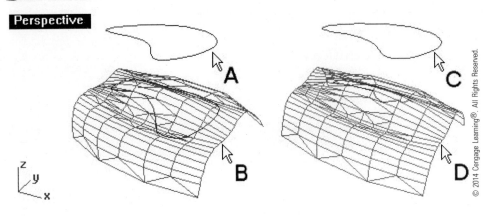

FIGURE 6–51 *Curves projected and pulled to a polygon mesh*

Duplication of Surface Edge and Border

You can construct a 3D curve from an edge or the border of a surface. Superficially, an edge seems to be analogous to a border. However, an open surface can have multiple edges but only a single border. For example, a rectangular surface has four edges but just one continuous border. Perform the following steps to duplicate a surface's edge and border.

1. Select File > Open and select the file EdgeBorder.3dm from the Chapter 6 folder on the Student Companion Website.
2. Select Curve > Curve From Objects > Duplicate Edge.
3. Select edge A (shown in Figure 6–52) and press the ENTER key.
   An edge of the surface is duplicated.
4. Select Curve > Curve From Objects > Duplicate Border.
5. Select surface B (Figure 6–52) and press the ENTER key.
   The borders of the surface are duplicated.

To appreciate that an edge is duplicated, hide the surface, as follows.

6. Press the ESC key to ensure that nothing is selected.
7. Select Edit > Visibility > Hide.
8. Select surfaces B and C (Figure 6–52) and press the ENTER key.
   The surface is hidden. (See Figure 6–53.)
9. Do not save your file.

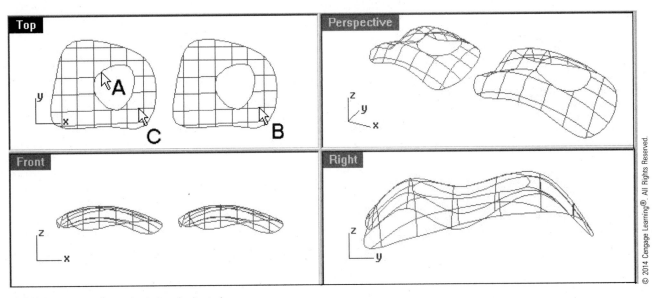

FIGURE 6–52 *Surface edge being duplicated*

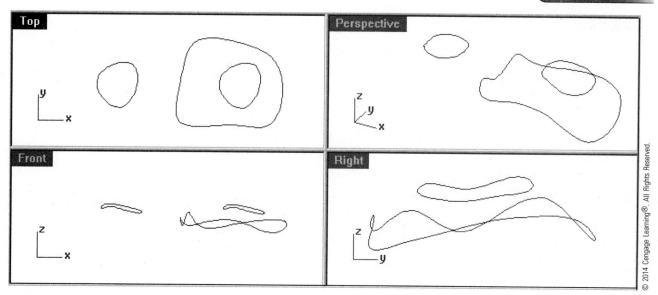

FIGURE 6–53 *Surface border being duplicated*

Duplicating Border of Individual Surfaces of a Polysurface

To extract the border of a surface of a polysurface, perform the following steps.

1. Select File > Open and select the file Border.3dm from the Chapter 6 folder on the Student Companion Website.
2. Select Curve > Curve From Objects > Duplicate Face Border.
3. Select face A of the polysurface, shown in Figure 6–54, and press the ENTER key. The border of a face of the polysurface is extracted.
4. Press the ESC key to ensure that nothing is selected.
5. Select Edit > Visibility > Hide.
6. Select polysurface A (Figure 6–54) and press the ENTER key.
7. Do not save your file.

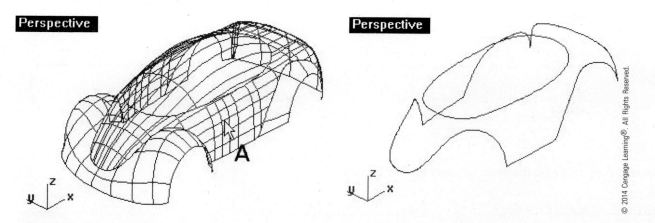

FIGURE 6–54 *Surface border of a polysurface being extracted (left) and border extracted and polysurface hidden (right)*

Duplicating Mesh Edge

Like the edges of a surface, the edges of a meshed object are not curves. If you want to make use of a meshed object's boundary edge for other drafting/construction purpose, you have to duplicate it as curves independent of the meshed object.

1. Select File > Open and select the file DuplicateBoundary.3dm from the Chapter 6 folder on the Student Companion Website.
2. Select Mesh > Mesh Edit Tools > Extract > Hole Boundary.
3. Select edge A, shown in Figure 6–55.
   The edge boundary is duplicated.
4. Do not save your file.

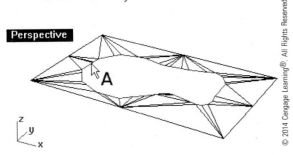

FIGURE 6–55 *Edge boundary of a polygon mesh being duplicated*

Duplicating a Mesh's Edge

You can duplicate a polygon mesh's edge in the form of a polyline, as follows.

1. Select File > Open and select the file DupMeshEdge.3dm from the Chapter 6 folder on the Student Companion Website.
2. Select Curve > Curve From Objects > Duplicate Mesh Edge.
3. Select edge A (Figure 6–56) and press the ENTER key.
   The selected edge is duplicated.
4. Repeat the command, select edge B, and press the ENTER key.
   Another edge is duplicated.
5. To appreciate the effect of edge duplication, turn off Layer Mesh. (See Figure 6–56.)
6. Do not save your file.

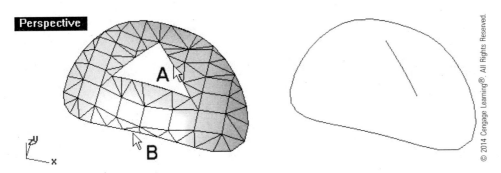

FIGURE 6–56 *Edge being duplicated (left) and edge duplicated (right)*

Extracting Mesh Face Edge

You can extract mesh face edges to obtain a sketch element. In essence, extraction of edges does not make any changes to the mesh; it simply constructs curve elements from selected edges of mesh faces. Perform the following steps.

1. Select File > Open and select the file ExtractMeshEdge.3dm from the Chapter 6 folder on the Student Companion Website.
2. Click on the Extract Mesh Edges button on the Extract Mesh toolbar.
3. Select mesh A, shown in Figure 6–57.

4. In the Extract Edges dialog box, check the Unwelded box and click on the OK button.
5. Hide mesh A.
6. Do not save your file.

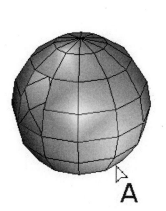
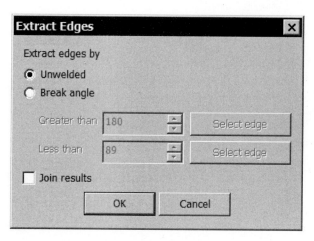

FIGURE 6–57 *From left to right: original mesh object, Extract Edges dialog box, and edges extracted*
Source: Robert McNeel and Associates Rhinoceros® 5

Constructing 2D Outlines of Surface and Polygon Mesh

An outline of selected surfaces and polygon meshes can be obtained by projecting the boundary at an angle perpendicular to the viewing direction of the active viewport. Perform the following steps.

1. Select File > Open and select the file MeshOutline.3dm from the Chapter 6 folder on the Student Companion Website.
2. Select Curve > Curve From Objects > Mesh Outline.
3. Select surface A and polygon mesh B in the Top viewport and press the ENTER key.
   Boundaries are projected onto the construction plane of the Top viewport. (See Figure 6–58.)

> **NOTE** This command is viewport dependent.

4. Do not save your file.

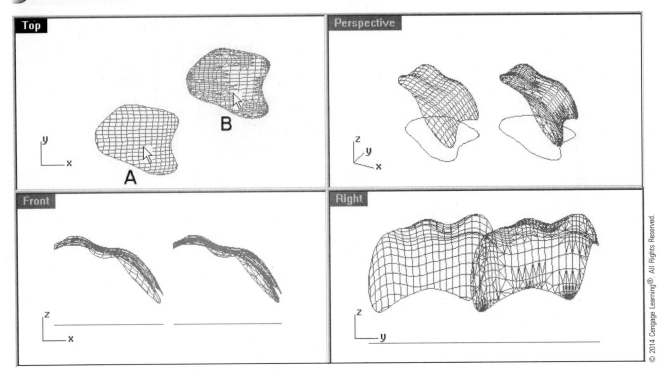

FIGURE 6-58 *Outline projected*

Intersection of Surfaces/Curves

Intersection of a surface and a curve produces a point object, and two intersected surfaces produce a curve. Turn on the History Manager to enable history, and perform the following steps.

1. Select File > Open and select the file Intersect.3dm from the Chapter 6 folder on the Student Companion Website.
2. Select Curve > Curve From Objects > Intersection.
3. Select curve A and surfaces B and C (shown in Figure 6–59), and press the ENTER key.

   A curve is constructed at the intersection of surfaces B and C, and two points are constructed at the intersection of surface C and curve A.
4. Hide surfaces B and C and curve A to see the result, shown in Figure 6–59.
5. Do not save your file.

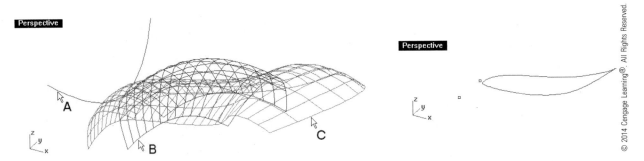

FIGURE 6-59 *Intersection objects of surfaces and curves being constructed (left) and intersected points and curve (right)*

Intersection of Polygon Meshes

You can construct a curve from the intersection of two polygon meshes by using the MeshIntersect command. Perform the following steps.

1. Select File > Open and select the file MeshIntersect.3dm from the Chapter 6 folder on the Student Companion Website.
2. Click on the Mesh Intersect button on the Mesh Tools toolbar.
3. Select polygon meshes A and B, shown in Figure 6–60.

   An intersection curve is constructed.
4. Hide meshes A and B.
5. Do not save your file.

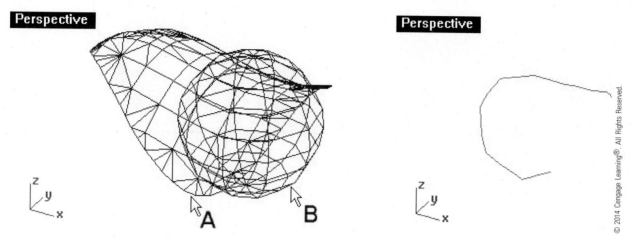

FIGURE 6–60 *Polygon meshes (left) and intersection curve constructed (right)*

Contour Lines and Section Lines of Surfaces

Contour lines and sections are both curves derived on a surface, and they are section lines cutting across a surface. Contour lines typically form a set of section lines spaced at a regular interval. To construct a set of contour lines across an ellipsoid surface, perform the following steps. Note that these commands also work on polygon meshes (you will learn about polygon meshes in Chapter 9). You can construct a NURBS surface from a set of contour lines, and you can construct sections through a polygon mesh. Perform the following steps.

1. Select File > Open and select the file ContourSection.3dm from the Chapter 6 folder on the Student Companion Website.
2. Select Curve > Curve From Objects > Contour.
3. Select surface A, shown in Figure 6–61, and press the ENTER key.
4. Select points B and C (Figure 6–61) to indicate the contour plane base point and contour plane direction.
5. Type 3 at the command area to specify the between-contour lines.

   Contour lines are constructed.

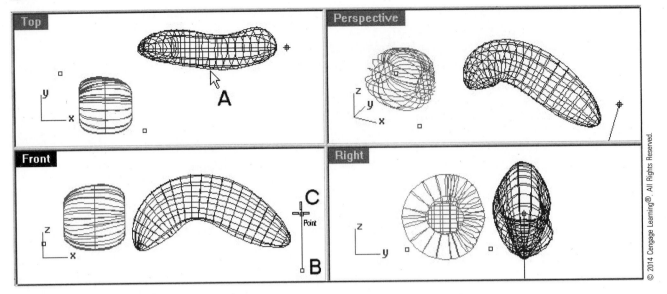

FIGURE 6–61 Contour lines being constructed

To construct sections, continue with the following steps.

6. Select Curve > Curve From Objects > Section.
7. Select surface A, shown in Figure 6–62, and press the ENTER key.
8. Select points B and C (Figure 6–62) to define a section plane.
9. A section curve is constructed.
10. Press the ENTER key to terminate the command.

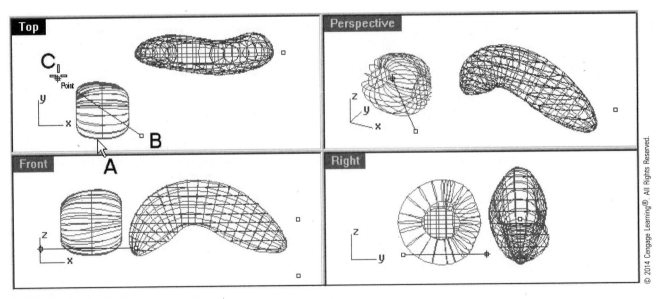

FIGURE 6–62 Section line being constructed

11. Turn off Layer01 to see the result. (See Figure 6–63.)
12. Do not save your file.

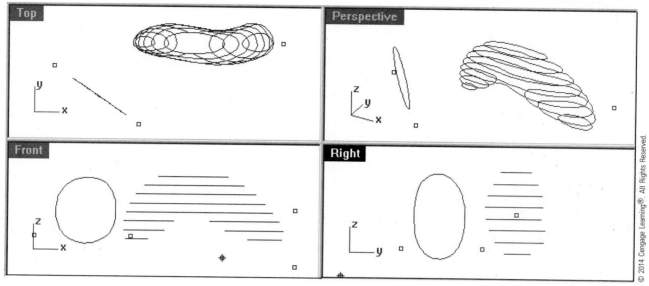

FIGURE 6–63 *Contour lines and section line constructed*

Silhouette of a Surface in a Viewport

The shape of the silhouette of a free-form surface resembles the edge of the shadow of the object projected in the viewing direction of the viewport. You see a different silhouette of the same object at different viewing angles. Perform the following steps.

1. Select File > Open and select the file Silhouette.3dm from the Chapter 6 folder on the Student Companion Website.
2. Select Curve > Curve From Objects > Silhouette.
3. Select surface A, shown in Figure 6–64, and press the ENTER key.

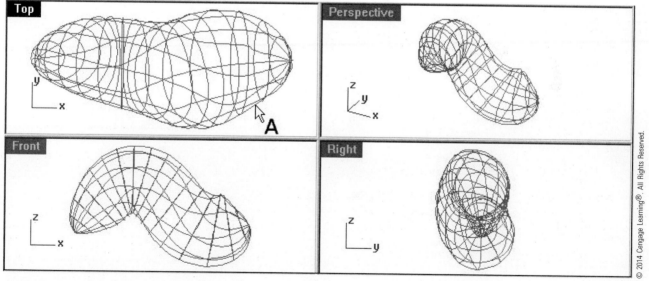

FIGURE 6–64 *Silhouette being constructed*

> **NOTE**
> This command is view dependent. The outcome will be quite different if you click on the Front viewport or the Right viewport.

4. Press the ESC key to ensure that nothing is selected.
5. Hide the surface to see the result, shown in Figure 6–65.
6. Do not save your file.

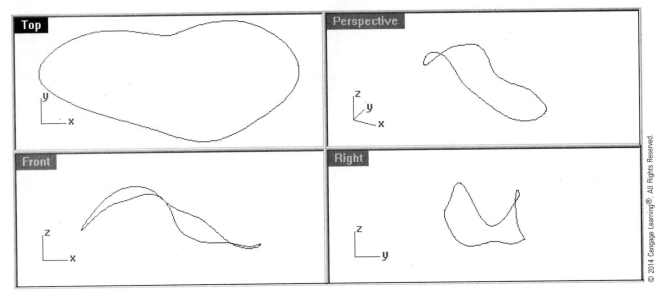

FIGURE 6–65 *Silhouette constructed and surface hidden*

Extracting Isocurves at Designated Locations of a Surface

Isocurve lines are directionally U and V lines displayed along a surface to help you visualize the profile and curvature of a surface. They are visual aids and not curves on the surface. To extract an isocurve at designated locations on the surface, perform the following steps.

1. Select File > Open and select the file Extract.3dm from the Chapter 6 folder on the Student Companion Website.
2. Select Curve > Curve From Objects > Extract Isocurve.
3. Select surface A, shown in Figure 6–66.
4. Click on point B, shown in Figure 6–66.

> **NOTE** You can use the INT osnap to snap to isocurve intersections.

5. Select the Toggle option at the command area.
6. Click on point B again (Figure 6–66).
7. Press the ENTER key.
   Two isocurve lines are constructed.

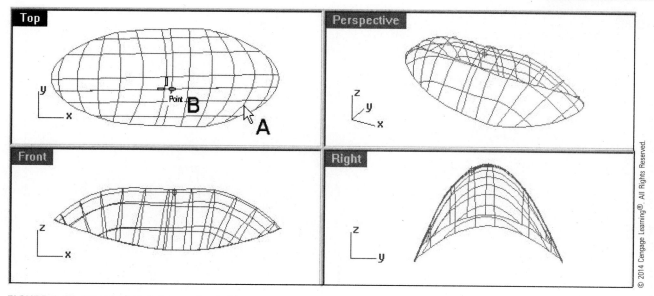

FIGURE 6–66 *Isocurve lines being constructed*

8. Hide the surface.

Extracting a Wireframe from Isocurves of a Surface

You can extract a wireframe from the isocurves of a surface for further development of your model. Continue with the following steps.

9. Select Curve > Curve From Objects > Extract Wireframe.
10. Select surface A (shown in Figure 6–66) and press the ENTER key.

    A wireframe is extracted from borders and isocurves of the surface.

Because the extracted curves and the isocurve lines of the surface lie in the same location, you will not find any visual difference after you construct the curves. To see the difference, you can hide the surface and shade the viewport, as follows:

11. Press the ESC key to ensure that nothing is selected.
12. Right-click on the Perspective viewport's label and select Shaded Display.

Extracting Control Points

Like curves, NURBS surfaces also have control points that govern their shape. To extract the control points of a surface and a curve, perform the following steps.

13. Unhide the surface and set the display to wireframe mode.
14. Select Curve > Curve From Objects > Extract Points.
15. Select surface A and curve B, shown in Figure 6–67, and press the ENTER key.

    Points are constructed at the control point locations of the surface and curve.

16. Do not save your file.

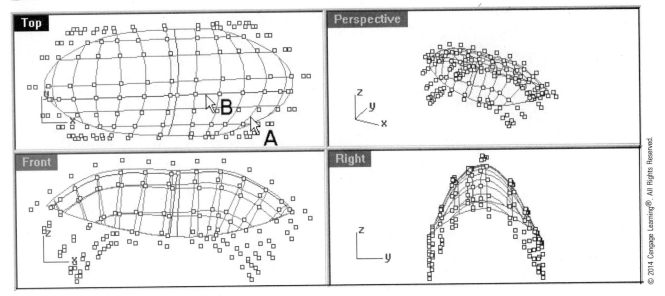

FIGURE 6–67 *Point objects constructed at the control point locations of the curve and surface*

Extracting Subcurve of a Polycurve

When two or more curves are joined together, a polycurve is formed. To extract a curve element of a polycurve, perform the following steps.

1. Select File > Open and select the file SubCurve.3dm from the Chapter 6 folder on the Student Companion Website.
2. Select Curve > Curve From Objects > Extract Curve.
3. Select curve A, shown in Figure 6–68.
4. Click on location B.
5. Select the Copy option at the command area.
6. Press the ENTER key. A subcurve is extracted.
7. Click on the extracted curve and drag it to location C to appreciate the effect.
8. Do not save your file.

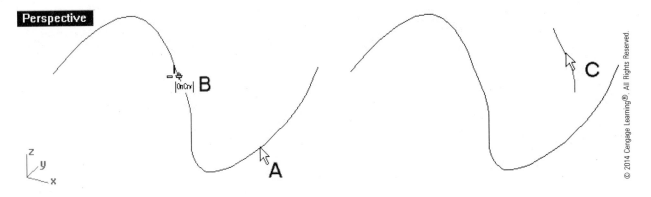

FIGURE 6–68 *Polycurve (left) and subcurve extracted and dragged (right)*

Creating Reference U and V Curves and Applying Planar Curves on Surfaces

To place a set of planar curves on a surface accurately with reference to the U and V orientation of the surface, you first construct a set of rectangular reference UV frame curves on the X-Y plane, construct a curve on the X-Y plane with reference to the reference UV curves, and apply the curves to the surface. Perform the following steps.

1. Select File > Open and select the file ApplyUV.3dm from the Chapter 6 folder on the Student Companion Website.
2. Select Curve > Curve From Objects > Create UV Curves.
3. Select surface A, shown in Figure 6–69.

   A reference rectangle representing the U and V orientations of the selected surface is constructed on the X-Y plane.

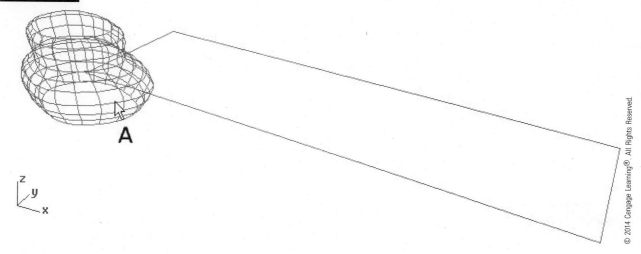

FIGURE 6–69 *Reference UV curves*

4. Referencing Figure 6–70, construct curve A relative to the reference UV curves.

   The exact shape of the curve is unimportant for this tutorial, as far as the curve is constructed within the reference rectangle.
5. Select Curve > From Objects > Apply UV Curves.
6. Select curves A and B, shown in Figure 6–70, and press the ENTER key.

> It is important to select the reference curves as well as the curve to be applied.   **NOTE**

7. Select surface C (Figure 6–70).

   Curves A and B are mapped onto the surface, as shown in Figure 6–70. In essence, you may not need to apply the reference curve B.
8. Do not save your file.

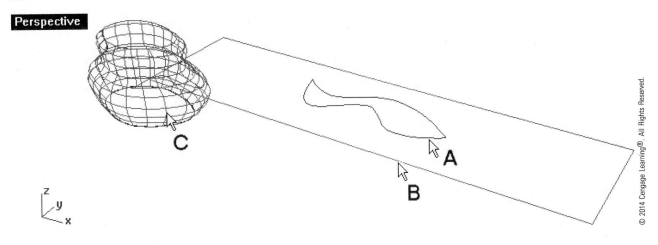

FIGURE 6–70 *Curve applied to the surface*

Comparing Create and Apply UV Curves to Squish and Squish Back

Superficially, the command CreateUVCurves and ApplyCrv commands look similar to the Squish and SquishBack commands delineated in Chapter 3. However, it must be noted that CreateUVCurves and ApplyCrv commands are more appropriate to untrimmed surfaces and Squish and Squishback commands are more accurate on trimmed surfaces.

You may experience their differences by using the CreateUVCurves and ApplyCrv commands on the example shown in Chapter 3 and use the Squish and Squishback commands on the above example.

CONSOLIDATION

The basic way to manipulate a curve is to extend it (if it is not long enough), trim away the unwanted part (if necessary), and split it into two so that you get two contiguous curves with a smooth transition between them. You can also remove a portion of the curve to break it into two curves. You can also add a line segment to an open polyline to close it.

In between two curves, you can add a chamfer, a fillet, or a blend curve. You can also add line or arc segments to two coplanar curves for them to meet. Smoothness of the chamfer joint and connected joint are G0; fillet joint is G1; and blended joint can be G2, G3, or G4. Naturally, you can join two or more contiguous curves into one and explode a joined curve into its individual curve elements. If two or more curves intersect to form regions among them, you can extract boundaries of such regions.

There are many ways to refine a curve. You can rebuild it by specifying the degree of polynomial degree and the number of control points. As an alternative to blending two separate curves, which adds a curve between them, you can match a curve to another curve to obtain G2 continuity. To change the shape of a curve and yet maintain the tangent direction at the endpoints, you adjust end bulge. You can move individual segments of a polycurve, and you can replace a portion of a curve with a line segment. You can fair a curve to reduce large curvature deviation, simplify a curve consisting of segments of varying degree of polynomial, and refit an imported polyline to a free-form curve. If you are going to export free-form curves to other applications that do not support splines, you convert curves to line and arc segments.

From existing curves, you can derive curves and point objects. You can offset a curve to obtain another curve of similar shape at a specific distance from the source curve. Between two curves, you can construct a mean curve. To help produce a 3D curve, you can first construct two planar curves in two viewports to depict two orthographic views of the curve, and then derive a 3D curve from them. You can construct a set of cross-section curves from a set of longitudinal curves. As a result, you get a set of U and V curves, with the original longitudinal curves as the U curve and the cross-section curves as the V curves.

To ease the task of sketching or interpolating on a surface or polygon mesh, you can first construct a planar curve and then project or pull it onto a surface or a polygon mesh to obtain a curve on the surface or the polygon mesh. If edges and borders of a surface or polygon mesh are required for the development of some other geometric objects, you extract them. Curves and points produced from the intersection of two surfaces/two polygon meshes and intersection of a surface/polygon mesh and a curve are sometimes required for further design work. From an existing surface, you can construct contour curves, section curves, and silhouettes, and can extract isocurves, control points, and isocurve wireframe. You can also extract a portion of a curve.

To map a curve on a surface precisely, you can first create the UV curves of the surface on a plane, construct the curve to be mapped, and then apply the UV curve and the curve to be mapped to the surface.

REVIEW QUESTIONS

1. In what ways can a curve's length be changed?
2. Outline the ways to treat two separate curves.
3. Illustrate how a curve's polynomial degree and fit tolerance can be modified.
4. Depict the methods of deriving curves from existing curves.

CHAPTER 7

NURBS Surface Manipulation

INTRODUCTION

After discussing how various kinds of surfaces are constructed in Chapter 3 and how curves and points are constructed and manipulated in Chapters 4 through 6, this chapter examines various surface manipulation methods.

OBJECTIVES

After studying this chapter, you should be able to

- Master various surface manipulation techniques and use them in design and construction of 3D surface models

OVERVIEW

To help you master surface modeling techniques systematically, this chapter divides surface manipulation methods into four sections. The first section deals with surface boundary manipulation, the second section explains ways to treat two or more separate surfaces, the third section concerns surface refinement, and the final section details surface edge treatment.

SURFACE BOUNDARY MANIPULATION

This section begins by explaining the concepts of surface trimming and delineating ways to trim a surface. It covers why a trimmed surface can be untrimmed, how the trimmed boundary can be detached, how a portion of the trimmed edge can be deleted, and how computer memory space can be saved by shrinking a trimmed surface. Following trimming and untrimming, this section describes how a surface can be split or, if there is kink along the surface, divided along creases. This section also depicts how an untrimmed surface can be extended to increase its size; extended perpendicularly to another surface, splitting it, and joining with it; and extended to become a periodic surface. Finally, this section explains one of the options of making a sweep surface from a kinked rail, in which the resulting surface can be two trimmed surfaces joined together.

Surface Trimming Concepts

To produce a smooth surface, it is necessary to use smooth defining curves and smooth boundary lines. However, most of the surfaces used to compose a design do not necessarily have smooth boundaries, although they have smooth profiles. If you use the irregular boundary edges to construct the surface directly, you get a surface with many sudden changes in curvature; the surface will not be smooth at all. To obtain a smooth surface with irregular edges, you build a larger surface from smooth wires, and then trim the smooth surface with the irregular edges. The resulting surface is a trimmed surface.

Proper Way to Construct a Smooth Surface Profile

Figure 7–1 (left) shows the surface model of an automobile body panel. This is a smooth surface, but its boundary is irregular. Given this problem, you might intuitively use the boundary wires that you see as the defining wires to construct the surface model. If you did, you would probably get an irregular surface like the one shown in Figure 7–1 (right).

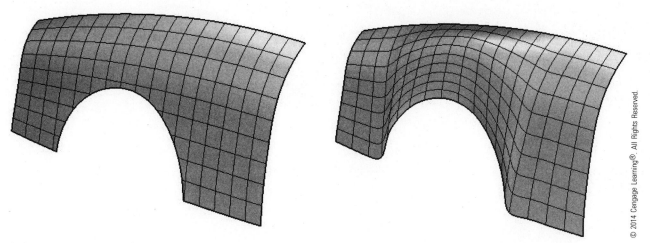

FIGURE 7–1 *Smooth surface with irregular boundaries (left) and irregular surface defined by irregular boundaries (right)*

Obviously, the surface shown in Figure 7–1 (right) is not the one we want. What has happened? The answer is that a set of irregular wires will generate an irregular surface. Unless the boundary lines are smooth wires, they cannot be used as defining wires for the surface. To obtain a smooth surface with an irregular boundary, you have to perform two steps.

You use a set of smooth wires to produce a smooth surface that is much larger than the required surface. This is called the base surface. You then use the irregular boundary wire to trim the smooth surface. In the computer, a resulting trimmed surface consists of the original untrimmed surface (base surface) and the trim boundaries (trimmed edges). Although both of these are saved in the database, only the boundary and the remaining part of the trimmed surface are displayed. As a result, we obtain a smooth free-form surface with irregular boundaries. This is called a trimmed surface.

To produce a free-form surface that is large enough for subsequent trimming, define a set of wires that encompass the required surface. To make such wires, you need to be able to visualize the defining wires that are outside the required surface. In Figure 7–2, the construction of the smooth automobile body panel starts from a set of smooth wires. From the smooth wires, a smooth surface that is much larger than the required surface is made. To obtain the required surface, an irregular boundary is used to trim the large smooth surface.

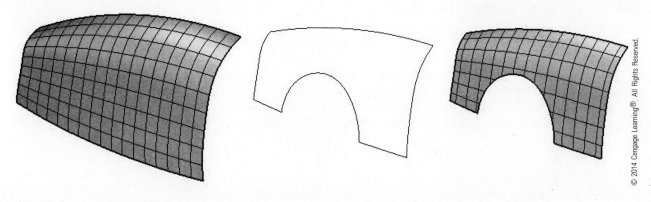

FIGURE 7–2 *From left to right: the untrimmed surface, the trimming boundary, and trimmed surface*

Methods to Trim a Surface

Basically, there are two methods to trim a surface or polysurface. (A polysurface is a set of contiguous surfaces joined together.) The first method is to construct a curve and use the curve to trim the surface. If the curve is not lying on the surface, the direction of trimming will be vertical to the active construction plane, as shown in Figure 7–3. Another way to trim a surface is to construct two intersecting surfaces and trim the unwanted portions of the surfaces away to form a sharp edge at the intersection, as shown in Figure 7–4. In both cases, the cutting object has to be long enough to cut through the surface.

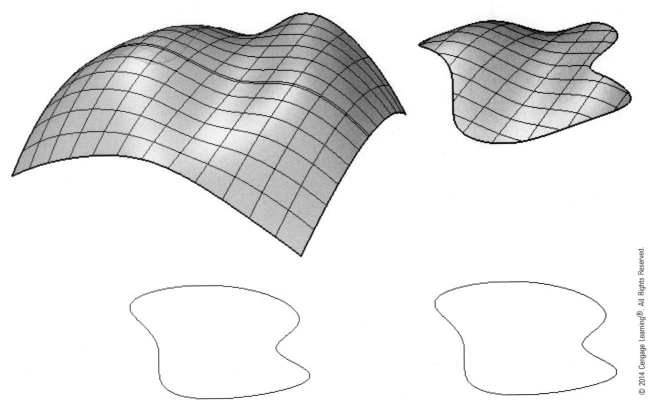

FIGURE 7–3 *Surface and wire (left) and surface trimmed by projecting the wire in a direction perpendicular to the construction plane (right)*

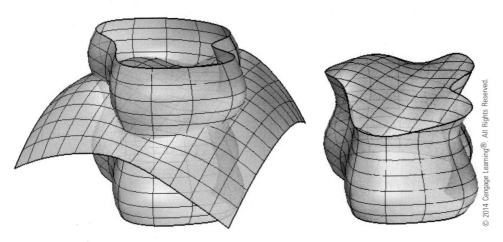

FIGURE 7–4 *Two intersecting surfaces (left) and trimmed surfaces (right)*

A surface can be cut only by a surface or curve that is long enough to pass through the surface. Perform the following steps.

1. Select File > Open and select the file TrimSurface01.3dm from the Chapter 7 folder that you downloaded from the Student Companion Website.
2. Select Edit > Trim.
3. Select curve A and surface B, shown in Figure 7-5, as cutting objects and press the ENTER key.
4. Select locations C and D, shown in Figure 7-5, to indicate the portion to be trimmed away and then press the ENTER key.

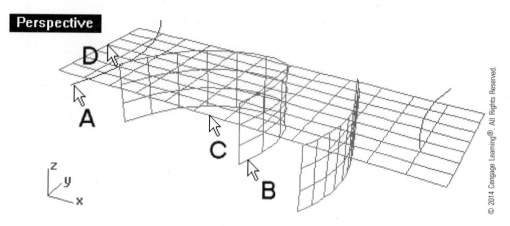

FIGURE 7-5 *Surfaces being trimmed*

5. Select Edit > Trim.
6. Select surface A and curve B, shown in Figure 7-6, as cutting objects and press the ENTER key.
7. Select location C, shown in Figure 7-6, to indicate the portion to be trimmed away and then press the ENTER key.
   The surface is not trimmed, because the cutting objects are not long enough to cut through the target surface.
8. Do not save your file.

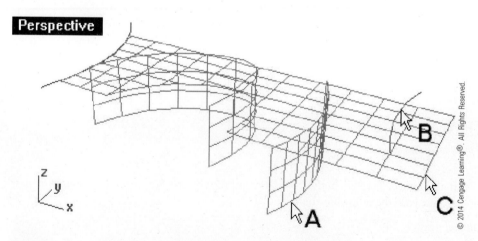

FIGURE 7-6 *Cutting objects not long enough to cut through the surface*

If two surfaces of different sizes intersect each other, the larger surface can trim the smaller one but not vice versa. Perform the following steps.

1. Select File > Open and select the file TrimSurface02.3dm from the Chapter 7 folder on the Student Companion Website.
2. Select Edit > Trim.

3. Select surfaces A and B (Figure 7–7) as cutting objects and press the ENTER key.
4. Select locations C and D, shown in Figure 7–7, to indicate the portion to be trimmed away and then press the ENTER key.

   Note that surface B (the narrower surface) is trimmed by surface A (the wider surface), but not vice versa.

5. Do not save your file.

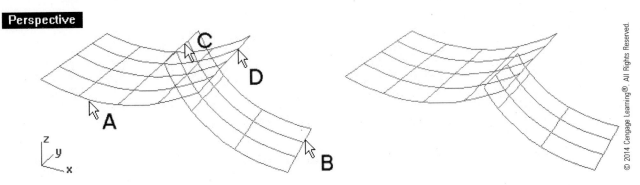

FIGURE 7–7 *Surfaces being used as cutting objects to trim each other (left) and outcome (right)*

To have two surfaces trimming each other, you may need to first trim their edges. Perform the following steps.

1. Select File > Open and select the file TrimSurface03.3dm from the Chapter 7 folder on the Student Companion Website.
2. Select Edit > Trim.
3. Select surfaces A and B (Figure 7–8) as cutting objects and press the ENTER key.
4. Select locations C and D, shown in Figure 7–8, to indicate the portion to be trimmed away and then press the ENTER key.

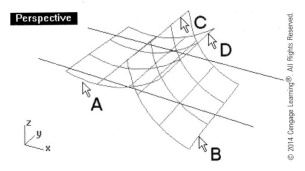

FIGURE 7–8 *Surfaces being trimmed by a curve and surface*

You will find that both surfaces are not trimmed, because both A and B (cutting objects) are not wide enough to pass through the objects to be trimmed. Continue with the following steps.

5. Maximize the Top viewport.
6. Select Edit > Trim.
7. Select curves A and B (Figure 7–9) as cutting objects and press the ENTER key.
8. Select locations C, D, E, and F, shown in Figure 7–9, to indicate the portion to be trimmed away, and then press the ENTER key.

   The surfaces are trimmed by the curves. Now their widths are the same.

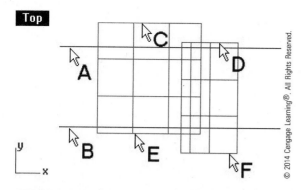

FIGURE 7-9 Surfaces being trimmed by curves

9. Maximize the Perspective viewport.
10. Select Edit > Trim.
11. Select surfaces A and B (shown in Figure 7-10) as cutting objects and press the ENTER key.
12. Select locations C and D (shown in Figure 7-10) to indicate the portion to be trimmed away and then press the ENTER key.

    Both surfaces are trimmed.

13. Do not save your file.

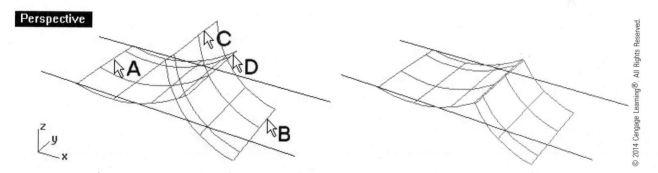

FIGURE 7-10 Surfaces trimming each other (left) and surfaces trimmed (right)

Untrimming a Trimmed Surface

In the course of refining your design, you may need to revert a trimmed surface back to its original untrimmed state. Because a trimmed surface in the computer consists of the original base surface and the trimmed edge, you can remove its trimmed boundary to change it back to its untrimmed state. This is called untrimming. Perform the following steps.

1. Select File > Open and select the file Trimboundary01.3dm from the Chapter 7 folder on the Student Companion Website.

    In this file, there are four identical trimmed surfaces to be used for various operations and comparison.

2. Select Surface > Surface Edit Tools > Untrim.
3. Select edge A, as shown in Figure 7-11.

    The surface is untrimmed. If there is more than one trim boundary in a trimmed surface, you can use the All option to untrim all the edges in one operation. When the All option is selected and an edge is selected, all edge trims will be removed. If a hole is selected, all hole trims will be removed.

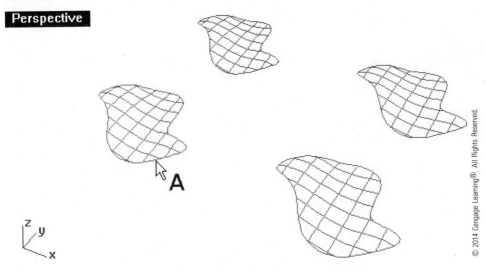

FIGURE 7–11 *Four identical trimmed surfaces and one surface being untrimmed*

Detaching a Trimmed Boundary from a Trimmed Surface

While untrimming, you have the option of retaining the trimmed boundary as a set of curves. This is called detaching. To untrim a trimmed surface and retain the trimmed boundary curve, continue with the following steps.

4. Surface > Surface Edit Tools > Detach Trim.
5. Select edge A, shown in Figure 7–12.

    The surface is untrimmed, and the trimmed boundary is retained.

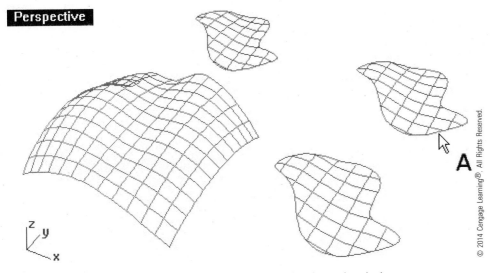

FIGURE 7–12 *A surface untrimmed and the surface's boundary being detached*

Shrinking a Trimmed Surface

To reduce the memory requirement to store a trimmed surface, you reduce the original untrimmed surface to its minimum size. This process is called shrinking. In essence, a trimmed surface retains its original smooth defining boundaries in the database while possessing a new trimmed boundary. The original surface is called the base surface, and the trimmed boundary is called the trim edge. Sometimes you might use a base surface that is much larger than required. If so, unnecessary memory space is wasted to store the unwanted part of the base surface. To reduce the memory used, you truncate the base surface of a trimmed surface. Continue with the following steps.

6. Select Surface > Surface Edit Tools > Shrink Trimmed Surface.
7. Select edge A, shown in Figure 7-13, and press the ENTER key.

   At a glance, both surfaces (the shrunk and the original) are the same. However, their base surfaces are different in size.

To appreciate the effect of shrinking, continue with the following steps.

8. Surface > Surface Edit Tools > Detach Trim.
9. Select edge A, as shown in Figure 7-13.

   The surface is untrimmed, and the boundaries are retained. (See Figure 7-14.)

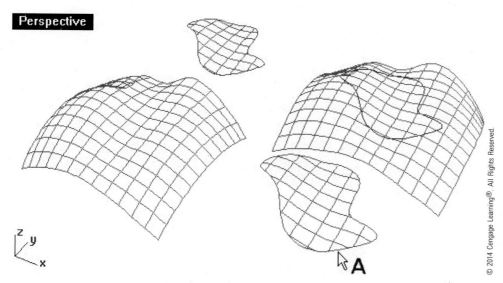

FIGURE 7-13 *A surface untrimmed with boundary detached (left) and a surface being shrunk (right)*

10. Do not save your file.

    Note that shrinking a trimmed surface also makes subsequent fillet, blend, and sweep commands work better. Apart from a reduction in memory space, the topology of the control points of a shrunk, trimmed surface is also different from that of the same trimmed surface before it is shrunk.

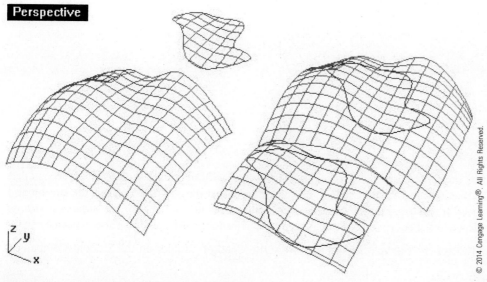

FIGURE 7-14 *Shrunk surface boundaries detached*

Shrinking to Boundary Edge

To further reduce the size of the base surface of a trimmed surface, you may shrink the base surface to the trimming edge. However, this may cause problems later, when the surface boundaries are used for other inputs. Perform the following steps.

1. Select File > Open and select the file Trimboundary02.3dm from the Chapter 7 folder on the Student Companion Website.

   In this file, there are four identical trimmed surfaces to be used for various operations and comparison.

2. Right-click on the Shrink trimmed surface/Shrink trimmed surface to edge button on the Surface Tools toolbar.
3. Select surface A, shown in Figure 7–15, and press the ENTER key.
4. Select Surface > Surface Edit Tools > Shrink Trimmed Surface.
5. Select surface B, shown in Figure 7–15, and press the ENTER key.
6. Surface > Surface Edit Tools > Detach Trim.
7. Select edges A and B, shown in Figure 7–15, and press the ENTER key.

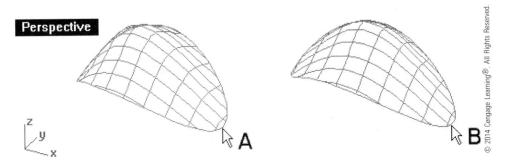

FIGURE 7–15 *Two trimmed surfaces*

   The surfaces are untrimmed, and the trimmed boundaries are detached. As shown in Figure 7–16, the surface that is shrunk to the trimmed edge is smaller.

8. Do not save your file.

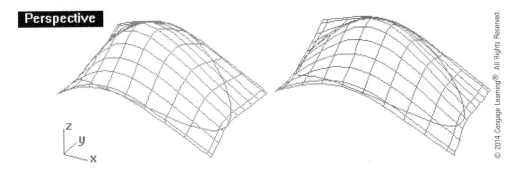

FIGURE 7–16 *Surface shrunk to the trimmed edge (left) and shrunk surface (right)*

Splitting One Surface into Two

Similar to the trimming process, you can split a surface into two contiguous surfaces with smooth transition. The splitting operation is very similar to trimming in that you need a cutting object. The difference is that the cutting object cuts the target object in two. Cutting objects can be curves or surfaces. A point to note when using curves not residing on the surfaces as cutting objects is that cutting is viewport dependent. Perform the following steps to split a surface using another surface and a curve.

1. Select File > Open and select the file SplitSurface.3dm from the Chapter 7 folder on the Student Companion Website.
2. Select Edit > Split.

3. Select surface A, shown in Figure 7–17, as the object to be split and then press the ENTER key.
4. Select surface B and curve C, shown in Figure 7–17, as splitting objects and then press the ENTER key.

   The surface is split into three surfaces.
5. Do not save your file.

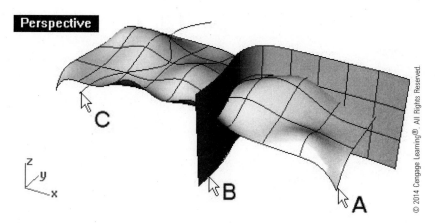

FIGURE 7–17 *Surfaces being split*

Splitting along U and V Lines

A surface can be broken into two surfaces along a selected U or V line. After the break, the surfaces still maintain the original continuity, and the profiles and silhouettes of the broken surfaces will be the same as those of the original surface. Perform the following steps.

1. Select File > Open and select the file BreakSurface.3dm from the Chapter 7 folder on the Student Companion Website.
2. Set object snap mode to Point objects.
3. Surface > Surface Edit Tools > Split at Isocurve.
4. Select surface A (shown in Figure 7–18).
5. At the command area, set Direction=Both.
6. Click on point B (shown in Figure 7–18) and press the ENTER key.

   The surface is split into four surfaces.
7. Do not save your file.

   Note that you may use the INT object snap tool to snap to the isocurve intersection to split the surface along the isocurves.

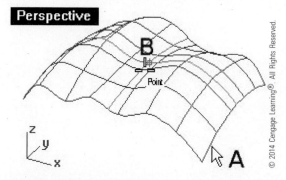

FIGURE 7–18 *Surface being split along its isocurve*

Dividing a Surface along Its Creases

Creases are kinky edges on a single surface. Depending on the prior setting of CreaseSplitting, no or yes, you can have a single surface with kinky edges or a polysurface composing of two or more surfaces joining together with sharp edges. If you already have a kinky surface, you can divide it into a number of contiguous surfaces at the kinky edges. Now perform the following steps.

1. Select File > Open and select the file Loft with Creases.3dm from the Chapter 7 folder on the Student Companion Website.
2. Type CreaseSplitting at the command area.
3. At the command area, set SplitAlongCreases=No.
4. Select Surface > Loft.
5. Select curves A and B, indicated in Figure 7–19, and press the ENTER key.
6. In the Loft Options dialog box, select Tight from the Style pull-down list box and click on the OK button.
   A single surface with creases is constructed.

To divide the creased surface along the creases, continue with the following steps.

7. Select Surface > Surface Edit Tools > Divide Surface on Creases.
8. Select surface C, indicated in Figure 7–19, and press the ENTER key.
   The creased single surface is divided into surface elements along the creases. These elements are joined together. To separate them, use the Explode command.
9. Do not save your file.

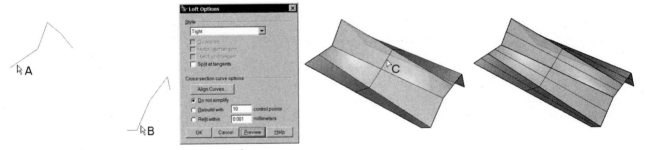

FIGURE 7–19 *From left to right: kinkycurves, Loft Options dialog box, creased single surface, and surface divided along creases*
Source: Robert McNeel and Associates Rhinoceros® 5

Extending the Untrimmed Edge of a Surface

Contrary to trimming a surface to reduce its size, you can extend a surface to increase its size. Perform the following steps.

1. Select File > Open and select the file Extend.3dm from the Chapter 7 folder on the Student Companion Website.
2. Select Surface > Extend Surface.
3. At the command area, set Type=Smooth.
4. Select edge A (shown in Figure 7–20).
5. Type 8 at the command area to set the extension distance.
   The surface is extended.
6. Do not save your file.

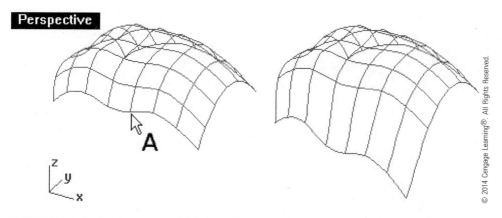

FIGURE 7–20 *Surface being extended (left) and extended surface (right)*

Extending the Trimmed Edge of a Surface

Previously, trimmed surfaces cannot be extended, but now the same command used to extend an untrimmed surface can also work on a trimmed surface. Perform the following steps.

1. Select File > Open and select the file ExtendTrimmedEdge.3dm from the Chapter 7 folder on the Student Companion Website.

   There are two identical trimmed surfaces in this file: one for extending and one for reference.

2. Select Surface > Extend Surface.
3. Select the Type=Smooth option at the command area to change it to Line.
4. Select surface edge A, shown in Figure 7–21.
5. Type 5 at the command area to specify the extended factor and press the ENTER key.

   The trimmed edge of the surface is extended.

6. Select Surface > Surface Edit Tool > Untrim.
7. Select surface B, indicated in Figure 7–21, and press the ENTER key.

   The surface is untrimmed. See Figure 7–22 to compare the results.

8. Do not save your file.

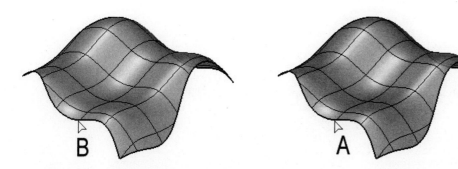

FIGURE 7–21 *Two identical trimmed surfaces*

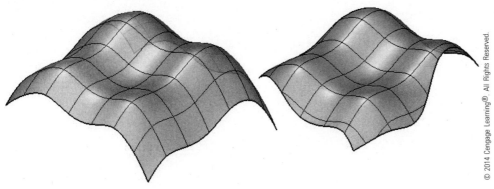

FIGURE 7–22 *Surface untrimmed (left) and trimmed edge extended (right)*

Extending a Surface to Make It Periodic

A periodic surface is a smooth, closed-loop surface. By making a surface periodic, the surface is extended to form a closed loop. If the surface is a trimmed surface, it will be untrimmed automatically before extending to form a closed loop. To appreciate how to make a surface periodic, perform the following steps.

1. Select File > Open and select the file Periodic.3dm from the Chapter 7 folder on the Student Companion Website.
2. Select Edit > Make Periodic.
3. Select edge A, shown in Figure 7–23, and press the ENTER key.
4. If the default is Yes for delete input, press the ENTER key. Otherwise, select Yes option from the command area.

   The surface is changed to a periodic surface.
5. Repeat the command.
6. Select edge B (Figure 7–23) and press the ENTER key.
7. Press the ENTER key again.

   The second surface is made periodic. (See Figure 7–24.)
8. Do not save your file.

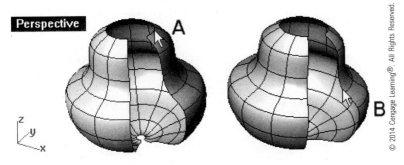

FIGURE 7–23 *Surfaces being changed to periodic surfaces*

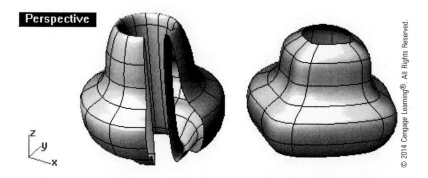

FIGURE 7–24 *Surfaces made periodic*

Surface Extension in Making a Sweep Surface from a Kinked Rail

Extending what you have learned about sweep surface construction in Chapter 3, if the rail for making a sweep surface has a kink in it, the resulting surface will be broken into two surfaces and joined together as a polysurface. Depending on the options selected while making the surface, you may have trimmed surfaces or mitered surfaces. Perform the following steps.

1. Select File > Open and select the file KinkRail.3dm from the Chapter 7 folder on the Student Companion Website.
2. Select Surface > Sweep 1 Rail.
3. Select curves A and B (Figure 7–25) and press the ENTER key.
4. Click on the OK button.

   A sweep surface consisting of two trimmed surfaces joined together is constructed.

5. Repeat the command.
6. Select curves C and D (Figure 7–25) and press the ENTER key.
7. Check the Untrimmed Miters button and click on the OK button on the Sweep 1 Rail Options dialog box.

   A sweep surface consisting of two mitered surfaces joined together is constructed.

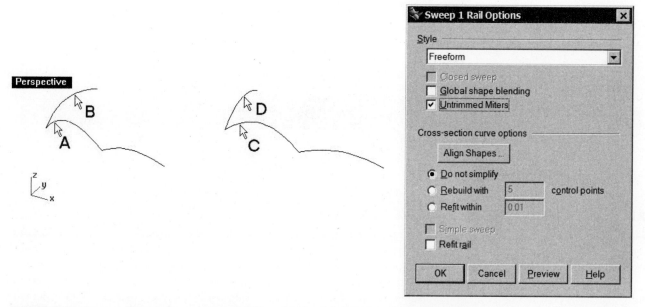

FIGURE 7–25 *Sweep surfaces being constructed (left) and Sweep 1 Rail Options dialog box (right)*
Source: Robert McNeel and Associates Rhinoceros® 5

8. Select Edit > Explode.
9. Select A and B (Figure 7–26) and press the ENTER key.
   The surfaces are broken into four surfaces.
10. Select Surface > Surface Edit Tools > Untrim.
11. Select edge A (Figure 7–26).
    There are two surface edges. Select either one of them.
12. Repeat the command.
13. Select edge A (Figure 7–26), the remaining edge.
14. Repeat the command two more times to untrim the edges at B (Figure 7–26).
    The surfaces are untrimmed. Compare the result. (See Figure 7–27.)

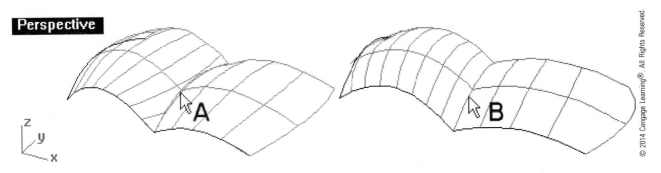

FIGURE 7–26 *Surfaces being exploded and untrimmed*

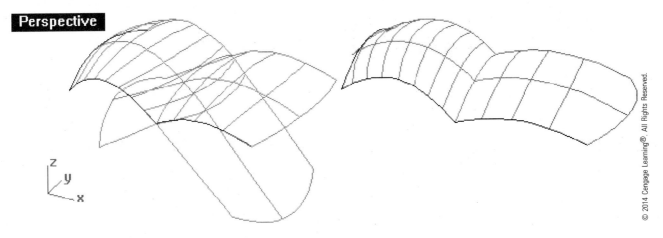

FIGURE 7–27 *Surfaces exploded and untrimmed*

TREATING TWO OR MORE SEPARATE SURFACES

In composing a surface model consisting of more than one surface, you can treat the joint between two consecutive surfaces by simply extending and trimming. Alternatively, you can insert a chamfer surface, a fillet surface, or a blended surface between them. The chamfer distance, fillet radius, or blend curvature along the joint can be constant or variable. The inserted chamfer, fillet, or blended surface, together with the original surfaces, is a typical example of five kinds of continuity between contiguous surfaces, with G0 (position) for chamfer surface; G1 (tangency) for fillet surface; and G2 (curvature), G3, and G4 for blended surface.

Continuity between Contiguous Surfaces

To reiterate, a G0 (position) joint simply has edges of two contiguous surfaces meeting each other. Both the tangency direction and the radius of curvature at the joint are different. In a G1 (tangency) joint, the tangency directions of the edges of two joined surface edges are the same, but the radii of curvature are different. In a G2 (curvature) joint, both the tangency directions and the radii of curvature of the edges are the same. For G3 continuity, the rate of change of curvature at the joint is constant. The smoothest joint is G4, in which the rate of change of curvature is constant. The rendered image shown in Figure 7–28 provides an idea about the differences in continuity between contiguous surfaces.

FIGURE 7–28 *From left to right: G0 (position), G1 (tangency), and G4 continuity*

Instead of inserting a surface, you can change the shape of one surface for it to match with the edge of the other one. To modify the shape of a surface, and yet maintain its degree of continuity with its contiguous surface, you modify its end bulge.

To handle two or more surfaces collectively, you join them into a single polysurface. The opposite of joining is called exploding. If the surface boundaries between the contiguous surfaces are untrimmed edges, you may, instead of joining, merge them into a single surface. However, it must be noted that two surfaces that are merged together cannot be exploded into two surfaces.

Chamfering, Filleting, and Blending

Depending on the styling requirement, you may insert a chamfer, fillet, or blended surface between two consecutive surfaces.

Constructing a Chamfer Surface

A chamfered surface is a flat surface that forms a bevelled edge between the two original surfaces. Because edges of the chamfered face and the contiguous surfaces simply join together without any tangent relationship or congruent radius of curvature, the joint has G0 continuity. Preferably, the original surfaces have to intersect each other. However, if they are not intersecting, the gap between them must be narrower than the bevelled edge. While chamfering, some of the original surfaces are trimmed away to have the chamfer surface fitted in between. As there are four possible chamfered surfaces on any given pair of intersecting surfaces, the location at which you select the original surface determines the location of the chamfered surface. Perform the following steps.

1. Select File > Open and select the file ChamferFillet.3dm from the Chapter 7 folder on the Student Companion Website.
2. Select Surface > Chamfer Surfaces.
3. At the command area, select the Distances option.

4. Type 5 to set the first chamfer distance.

   The first chamfer distance applies to the first selected surface.

5. Type 4 to set the second chamfer distance.

> **NOTE** By default, the second chamfer distance is set to be equal to the first chamfer distance. Therefore, you can simply press the ENTER key to accept

6. Select surfaces A and B (shown in Figure 7-29).

   A chamfer surface is constructed.

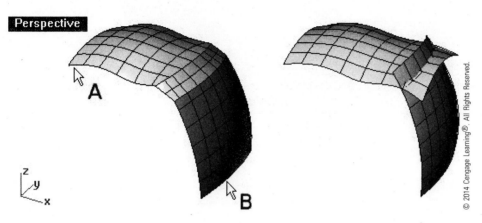

FIGURE 7-29 *Chamfer surface constructed*

Remember, the location where you select the surfaces affects the outcome, and, if there is a gap between the surfaces to be chamfered and the chamfer distance is too small in comparison to the gap's width, chamfer may not be successful.

Constructing a Fillet Surface

A filleted surface has an arc-shaped cross-section. You may consider a filleted surface as being derived by rolling a ball between two surfaces. Because the joints between the added fillet surface and the original surface have only tangent relationship without having the same radius of curvature, filleted surface and contiguous surfaces have G1 continuity. Like chamfering, the surfaces preferably have to intersect each other. If the surfaces in question are not intersecting, the widest gap between them must not be so wide that the rolling ball falls away from the surfaces. While filleting, you choose to trim the surfaces so that the filleted surface and the trimmed surfaces form a continuous surface profile or to keep the original surfaces intact. To construct a filleted surface, you specify the radius of the fillet and select two nonparallel surfaces. Like chamfering, any pair of intersecting surfaces has four possible filleted surfaces. Continue with the following steps.

7. Select Surface > Fillet Surfaces.
8. Select the Radius option from the command area.
9. Type 7 at the command area.
10. Select surfaces A and B (shown in Figure 7-30).

    A filleted surface is constructed.

11. Do not save your file.

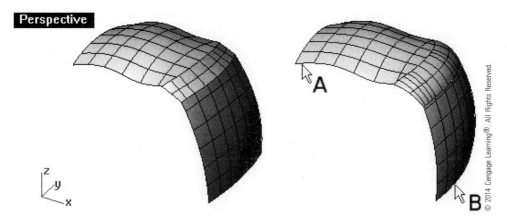

FIGURE 7–30 *Filleted surface constructed*

Constructing Blended Surfaces

A blended surface produces a smooth transition between the two original surfaces, and the junction between the blended surface and contiguous surfaces has G2, G3, or G4 continuity. Perform the following steps.

1. Select File > Open and select the file Blend.3dm from the Chapter 7 folder on the Student Companion Website.
2. Select Surface > Blend Surface.

While constructing a blended surface, you can select multiple contiguous surface edges, adjust end bulge, and set blend surface height.

3. Select edge A, shown in Figure 7–31, and press the ENTER key to select the first set of edges.
4. Select edge B, shown in Figure 7–31, and press the ENTER key to select the second set of edges.
5. In the Adjust Surface Blend dialog box, check the G4 boxes for edges 1 and 2.
6. Click on the Same height and Preview button.
7. Experience the effect of the control point distribution in relation to the slider bars' location.
8. Clear the Same height shapes button and click on the OK button.

    A blended surface is constructed.

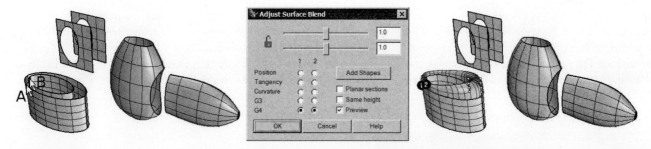

FIGURE 7–31 *From left to right: surface edges selected and curve seam displayed, Adjust Surface Blend dialog box, and preview of the blended surface*
Source: Robert McNeel and Associates Rhinoceros® 5

9. Repeat the command.
10. Select surface edge A, shown in Figure 7–32, and press the ENTER key.
11. Select surface edge B, shown in Figure 7–32, and press the ENTER key.
12. Press the ENTER key to accept the default seam location.

13. In the Adjust Surface Blend dialog box, click on the Same height button and set the slider bars' value to 0.8.
14. Click on the OK button. Another blended surface is constructed.

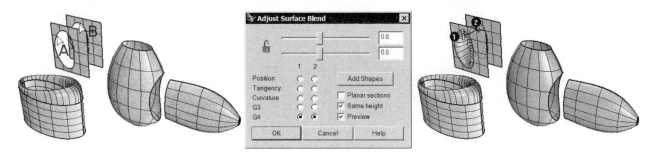

FIGURE 7–32 *From left to right: two surface edges selected, Adjust Surface Blend dialog box, and preview of the blended surface*
Source: Robert McNeel and Associates Rhinoceros® 5

In the following steps, you are going to construct a blended surface to bridge two sets of edges.

15. Repeat the command again.
16. At the command area, set AutoChain=Yes.

    Because there is a seam at the openings of the two original surfaces, each surface has two edges to blend.

17. Select chained surface edge A, shown in Figure 7–33, and press the ENTER key.
18. Select chained surface edge B, shown in Figure 7–33, and press the ENTER key.
19. In the Adjust Surface Blend dialog box, set the slider bars' value to 0.5.
20. Click on the OK button.

    A blended surface is constructed. A rendered image of the blended objects is shown in Figure 7–34.

21. Do not save your file.

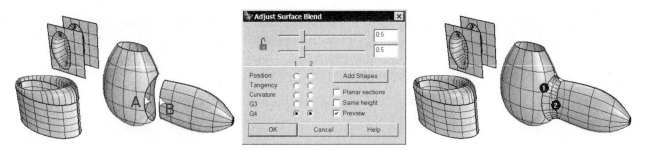

FIGURE 7–33 *From left to right: two sets of edges selected, Adjust Surface Blend dialog box, and preview of blended surface*
Source: Robert McNeel and Associates Rhinoceros® 5

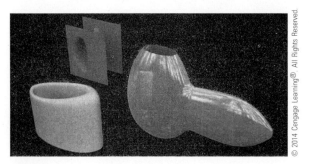

FIGURE 7–34 *Rendered image of the blended objects*

Variable Chamfer, Fillet, and Blended Surfaces

If you have two intersecting surfaces, you can construct a variable distance chamfer, variable radius fillet, or variable curvature blended surface between them and appropriately trim the original surfaces. Perform the following steps.

1. Select File > Open and select the file Variable.3dm from the Chapter 7 folder on the Student Companion Website.
2. Select Surface > Variable Fillet/Blend/Chamfer > Variable Chamfer Surfaces.
3. Type 6 to set the first chamfer distance.
4. Select surface A, indicated in Figure 7–35.
5. Type 8 to set the second chamfer distance.
6. Select surface B, indicated in Figure 7–35.
7. At the command area, set TrimAndJoin=Yes and RailType=DistBetweenRails.
8. Click on the Mid button of the Osnap dialog box.
9. Select the AddHandle option.
10. Type 10 to specify the new chamfer distance.
11. Click on midpoint C indicated in Figure 7–35.
12. Press the ENTER key twice.

    A variable chamfer is constructed.

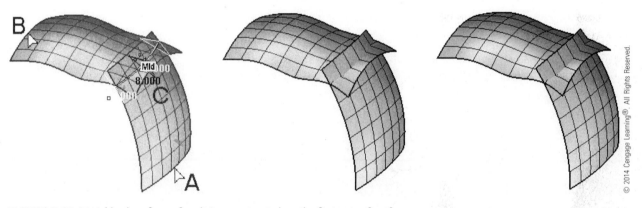

FIGURE 7–35 *Variable chamfer surface being constructed on the first pair of surfaces*

13. Select Surface > Variable Fillet/Blend/Chamfer > Variable Fillet Surfaces.
14. Type 5 at the command area to specify the first fillet radius.
15. Select surface A, shown in Figure 7–36.
16. Type 7 at the command area to specify the second fillet radius.
17. Select surface B, shown in Figure 7–36.
18. At the command area, set TrimAndJoin=Yes and RailType=RollingBall.
19. Select the AddHandle option.
20. Type 10 at the command area to specify the radius of the new handle.
21. Click on midpoint C, shown in Figure 7–36.
22. Press the ENTER key twice.

    A variable fillet is constructed.

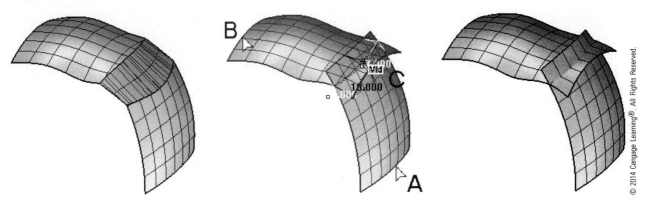

FIGURE 7–36 *Variable chamfer surface constructed on the first pair of surfaces, and variable fillet surface being constructed on the second pair of surfaces*

23. Select Surface > Variable Fillet/Blend/Chamfer > Variable Blend Surfaces.
24. Type 7 at the command area to specify the first blend radius. (Because the blended surface is not a fillet, the radius here is only indicative.)
25. Select surface A, shown in Figure 7–37.
26. Type 9 at the command area to specify the second blend radius.
27. Select surface B, shown in Figure 7–37.
28. At the command area, set TrimAndJoin=Yes and RailType=RollingBall.
29. Select the AddHandle option.
30. Type 11 at the command area to specify the radius of the new handle.
31. Click on midpoint C, shown in Figure 7–37.
32. Press the ENTER key twice.

    A variable blend is constructed. (See Figure 7–38 for a rendered image.)

33. Do not save your file.

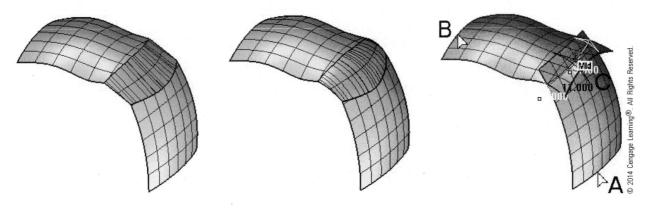

FIGURE 7–37 *Variable radius fillet surface constructed on the second pair of surfaces and variable blend surface being constructed on the third pair of surfaces*

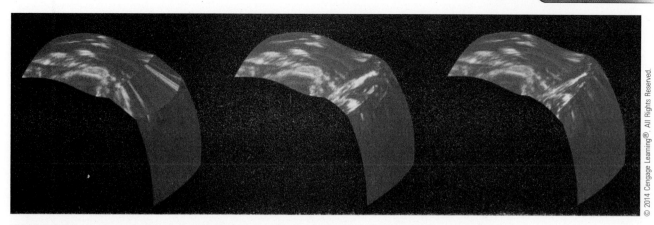

FIGURE 7–38 Rendered image of the three pairs of surfaces: variable chamfer, fillet, and blend surfaces

Connecting Two Surfaces

Connecting concerns extending or trimming two surfaces for them to meet at a sharp edge, as follows.

1. Select File > Open and select the file ConnectSrf.3dm from the Chapter 7 folder on the Student Companion Website.
2. Select Surface > Connect Surfaces.
3. Select surfaces A and B, shown in Figure 7–39.
4. Because the surfaces are separated, you have to select edges A and B.
   The surfaces are extended to meet.
5. Repeat the command.
6. Select surfaces C and D, shown in Figure 7–39. Because the surfaces are intersecting each other, they are trimmed and meet at a sharp edge.
7. Do not save your file.

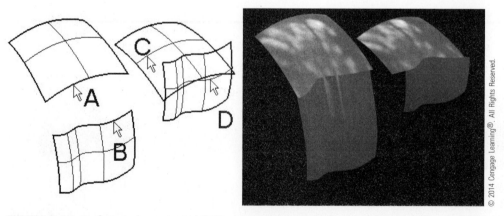

FIGURE 7–39 Surfaces to be connected (left) and rendered image of the connected surfaces (right)

Matching an Untrimmed Surface Edge with Another Surface Edge

Matching concerns extending an untrimmed edge of a surface and modifying its shape for it to match the edge (either trimmed or untrimmed) of another surface, to yield one of the three types of continuity (G0, G1, or G2) between the surfaces. Perform the following steps.

1. Select File > Open and select the file MatchSurface.3dm from the Chapter 7 folder on the Student Companion Website.
2. Select Surface > Surface Edit Tools > Match.

3. Select edge A, shown in Figure 7–40, as the untrimmed edge to change.
4. Select edge B, shown in Figure 7–40, as the segment to match.

   Note that where you pick the surfaces is important in terms of eliminating twisted surfaces.
5. Press the ENTER key.
6. In the Match Surface dialog box, check the Curvature boxes in the Continuity and Preserve other end areas, the Average surfaces box, and the Automatic box in the Isocurve direction adjustment area.
7. Click on the OK button.

   A surface's edge is changed to match the other, shown in Figure 7–40.

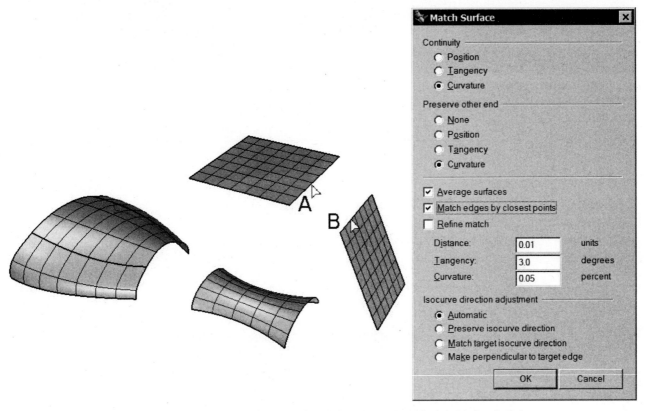

FIGURE 7–40 *Surface being modified to match the other surface (left) and Match Surface dialog box (right)*
Source: Robert McNeel and Associates Rhinoceros® 5

8. Repeat the command.
9. Select the MultipleMatches option at the command area.
10. Select edge A, shown in Figure 7–41.
11. Select the ChainEdges option at the command area.
12. Select edges B, C, and D, shown in Figure 7–41, and press the ENTER key.
13. In the Match Surface dialog box, check the Curvature boxes in the Continuity and Preserve other end areas, the Match edges by closest points box, and the Automatic box in the Isocurve direction adjustment area.
14. Click on the OK button.

    A match surface is constructed. (See Figure 7–42.)
15. Do not save your file.

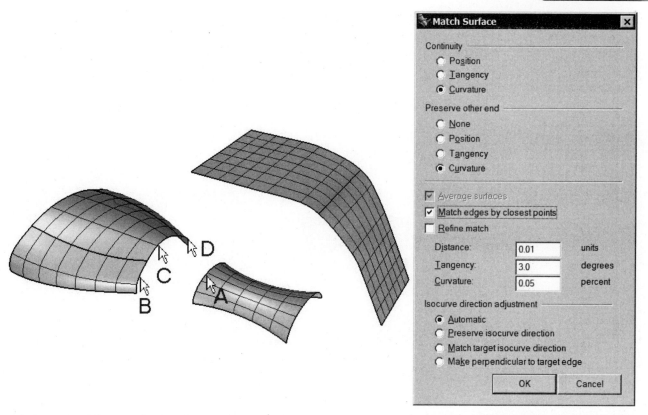

FIGURE 7–41 *A planar surface matched to another planar surface and multiple edges selected for matching (left) and Match Surface dialog box (right)*
Source: Robert McNeel and Associates Rhinoceros® 5

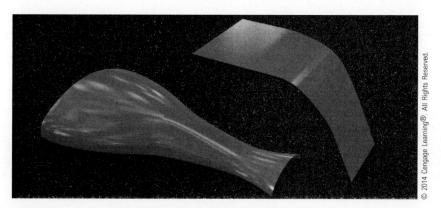

FIGURE 7–42 *Rendered image of the matched surfaces*

Adjusting a Surface's End Bulge

If you have two contiguous surfaces and want to modify the shape of one of the surfaces without altering the continuity condition between the surfaces, you adjust the end bulge. Perform the following steps.

1. Select File > Open and select the file Endbulge.3dm from the Chapter 7 folder on the Student Companion Website.
2. Select Surface > Surface Edit Tools > Adjust End Bulge.
3. Select surface edge A, shown in Figure 7–43.

    Because there are two contiguous surfaces sharing one common edge, select the surface edge in red.

4. Click on point B to specify the point to edit.
5. Click on points C and D to specify the start and end of the region to edit.
6. Drag points A and B, shown in Figure 7–44, to modify the shape of the surface.
7. Press the ENTER key when done.

   The surface is modified.

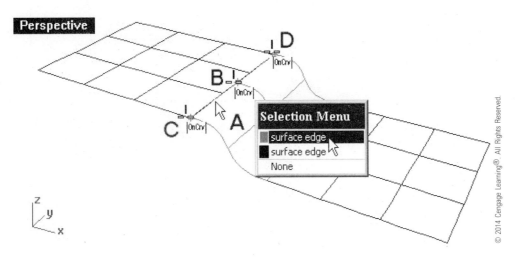

FIGURE 7–43 *Surface edge, edit point, and edit region selected*

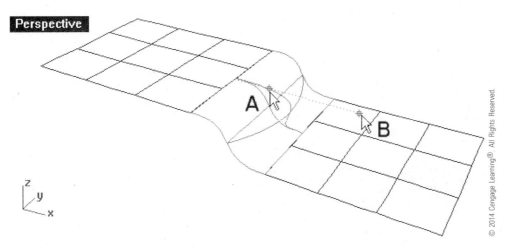

FIGURE 7–44 *Control point being manipulated*

Joining Contiguous Surfaces to Form a Polysurface

By joining two or more contiguous surfaces, you get a polysurface. It must be emphasized that the surfaces will not join unless they meet edge to edge within tolerance. Continue with the following steps.

8. Select Edit > Join.
9. Select the surface, shown in Figure 7–44, and press the ENTER key.

   The surfaces are joined to become a polysurface. Note in the command area the report stating the result of joining.

Exploding Polysurfaces to Individual Surfaces

Contrary to joining, exploding a polysurface renders a set of individual surfaces. Continue with the following steps to explode a polysurface to revert it into a set of individual surfaces.

10. Select Edit > Explode.
11. Select the polysurface that you just constructed and press the ENTER key.
    The polysurface is exploded.
12. Do not save your file.
    In Chapter 8, you will learn how to extract individual surfaces from a polysurface without exploding it.

Non-Manifold Merge

If more than two surfaces share a single edge, that object is regarded as a non-manifold object. To merge these edges, perform the following steps.

1. Select File > Open and select the file NonManifold.3dm from the Chapter 7 folder on the Student Companion Website.
2. Type NonManifoldMerge at the command area.
3. Select surfaces A, B, and C, shown in Figure 7–45, and press the ENTER key.
   Three surfaces sharing a common edge are merged into a single polysurface.
4. Repeat the command.
5. Select surfaces D and E, shown in Figure 7–45, and press the ENTER key.
   Two surfaces sharing a common edge are merged into a single polysurface.
6. Do not save your file.

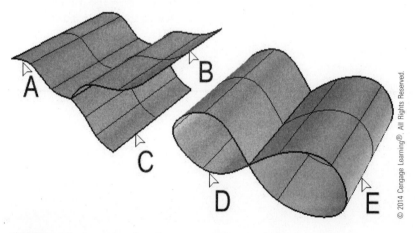

FIGURE 7–45 *Two sets of surfaces sharing common edges*

Merging Contiguous Untrimmed Surfaces to Form a Single Surface

Contiguous surfaces sharing a common untrimmed edge can be merged to become a single surface. Note that unlike joining, merging results in a single surface, not a polysurface. Therefore, you cannot explode it into the surfaces you merged. Perform the following steps.

1. Select File > Open and select the file MergeSurface.3dm from the Chapter 7 folder on the Student Companion Website.
2. Select Surface > Surface Edit Tools > Merge.
3. Select surfaces A and B, shown in Figure 7–46. The surfaces are merged into one. If the surfaces failed to merge, repeat the command, select Tolerance option from the command area, specify a larger joining tolerance value, and select the surfaces again.

4. Repeat the command.
5. Select B and C (Figure 7–46).

   The selected untrimmed surfaces are merged into a single surface.
6. Do not save your file.

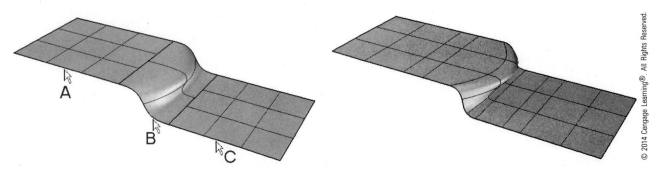

FIGURE 7–46 *Surfaces being merged (left) and merged surface (right)*

SURFACE PROFILE MANIPULATION

This section deals with how the normal direction of a surface can be flipped, ways to modify the profile of a surface, rebuilding and reparameterizing a surface, and changing its degree of polynomial.

Flip Normal Direction

A surface has no thickness. To represent a 3D object in a computer, you need a number of surfaces. For the computerized downstream manufacturing operations to recognize which side of the surface represents a void and which side represents a volume, a normal vector is used. The direction of normal is determined by the direction of the curves, the curve patterns, and also the sequence of selection when you construct a surface from the curves. This sounds too complicated for us to memorize. Hence, you may simply disregard the normal direction when you first construct the surface and flip the normal in a later stage. In Rhinoceros, the normal direction of a surface is depicted by a line normal to and at the corner of the surface. Perform the following steps.

1. Select File > Open and select the file Normal.3dm from the Chapter 7 folder on the Student Companion Website.
2. Select Analyze > Direction.
3. Select surface A, shown in Figure 7–47, and press the ENTER key.
4. Select the Flip option from the command area to flip the normal direction.
5. Press the ENTER key to terminate the command.
6. Do not save your file.

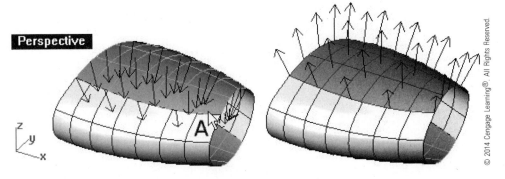

FIGURE 7–47 *Normal directions displayed (left) and flipped (right)*

The U and V isocurves on a surface also have directions. Together with normal, a surface has three directions. Using this command, you can reverse the U and V directions, swap the U with V, and change the normal direction.

> **NOTE**
> The color of the normal directions can be changed by modifying the tracking color in the Colors tab of the Rhino Options dialog box.

Editing Surface Profile

To modify the shape of a surface, you can manipulate the locations of its control points by using the nudge keys, using the MoveUVN dialog box, or using the organic toolbar; adjusting a control point's weight; using the handlebar; adding or removing knots; adding or removing control points; and soft editing the surface.

Control Point Manipulation

One method of modifying the shape of a surface is to edit its control points. There are four ways to move a control point, as follows.

- Select the control point, hold down the mouse button, and drag it to a new position.
- Select the control point and use the nudge keys. (By default, the nudge keys are the ALT key plus the arrow keys.)
- Use the MoveUVN command.
- Use any transform command.

Drag Mode and Using the Nudge Keys

As explained in Chapter 2, you can set the drag mode to one of five options. By holding down the ALT key, you can press the arrow key to move selected objects vertically or horizontally. To modify a control point of a surface using the nudge keys, perform the following steps.

1. Select File > Open and select the file SurfaceKnotPoint.3dm from the Chapter 7 folder on the Student Companion Website.
2. Select Edit > Control Points > Control Points On.
3. Select surface A, shown in Figure 7–48, and press the ENTER key.
4. Select control point B, shown in Figure 7–48, in the Top viewport.
5. Hold down the ALT key and press the Up arrow key.

    The selected control point is moved in a horizontal plane parallel to the current construction plane.

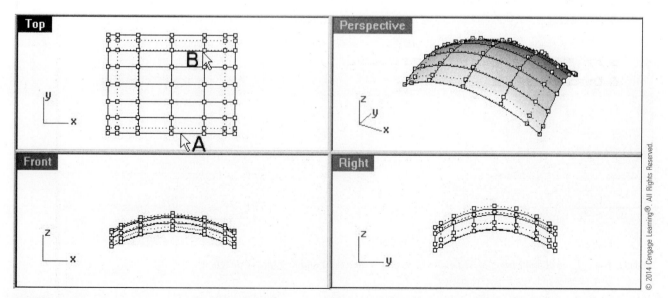

FIGURE 7–48 *Control points of a surface being modified*

Move along U, V, and N Directions

As explained previously, U, V, and N are three directions on a surface. The U and V depict the two isocurve directions, and the N depicts the normal direction. Continue with the following steps to use the MoveUVN command to move the control point's location in these three directions.

6. Maximize the Perspective viewport.
7. Select Transform > Move UVN.
8. Select control point A, shown in Figure 7–49.
9. In the MoveUVN dialog box, drag the N slider bar to move the control point in a direction normal to the surface, or drag the U and V bars to move the control point along the U and V directions of the surface.
10. Click on the X icon of the MoveUVN dialog box to close it.

NOTE You may leave the dialog box on all the time.

Using the Organic Toolbar

To manipulate objects and, in particular, control points of a surface along the X, Y, and Z directions, you can use the buttons on the Organic Toolbar. Continue with the following steps.

11. Click on the Red, Green, or Black Arrow button on the Organic toolbar to move the control point in X, Y, or Z direction. (See Figure 7–49.)

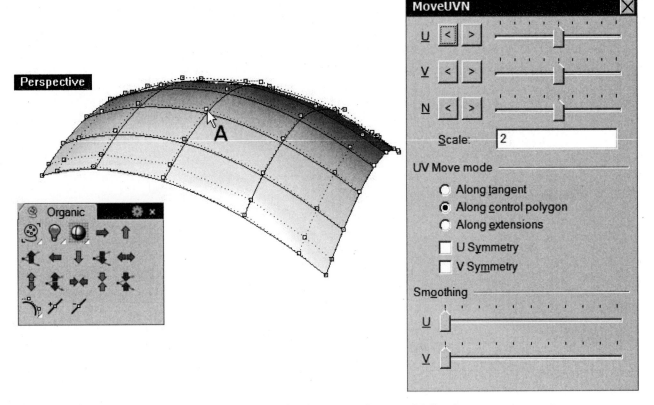

FIGURE 7–49 *Control point being moved along U, V, and N directions and X, Y, and Z directions*
Source: Robert McNeel and Associates Rhinoceros® 5

Adjusting Control Point Weight

Like the control points of a spline, the higher the weight, the closer the surface will be pulled to the control point. Continue with the following steps to change the control point's weight.

12. Select Edit > Control Points > Edit Weight.
13. Select control point A (shown in Figure 7–50) and press the ENTER key.
14. In the Set Control Point Weight dialog box, drag and move the slider bar to increase or decrease the weight of the selected control point.
15. Click on the OK button.
    The surface is modified.
16. Do not save your file.

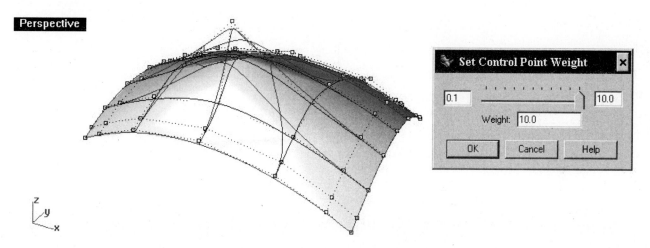

FIGURE 7–50 *Control point's weight being modified (left) and Set Control Point Weight dialog box (right)*
Source: Robert McNeel and Associates Rhinoceros® 5

Using the Handlebar Editor

The Handlebar Editor enables you to modify the shape of a surface by adjusting the tangency of the selected locations on the surface. To facilitate selection of specific locations on a surface, you may increase the isocurve density. To reiterate, isocurves are aids to help you visualize and select a surface. Changing the density does not affect the number of control points and the degree of polynomial. To edit a surface using the handlebar, perform the following steps.

1. Open the file SurfaceKnotPoint.3dm from the Chapter 7 folder on the Student Companion Website again.
2. Select Edit > Control Points > Handlebar Editor.
3. Select surface A and then location B (shown in Figure 7–51).
4. Select and drag one of the five handles of the handlebar to modify the shape of the surface at the selected location, as shown in Figure 7–51. Press the ENTER key when finished.
5. Do not save your file.

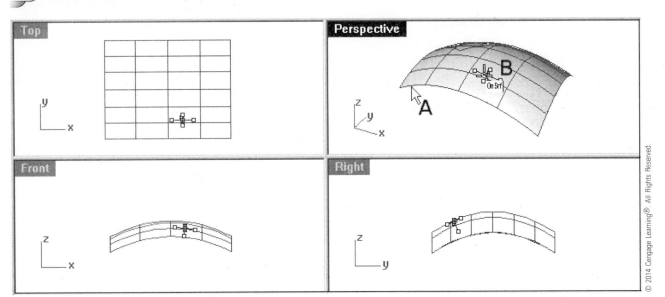

FIGURE 7–51 *Handlebar editor being activated at the selected location of a surface*

Adding and Removing Knots

The number of knot points of a surface has a significant effect on how the surface will change in shape if a control point is moved. Adding knots increases the number of control points. The surface profile will not change until the next time you manipulate its control points. However, removal of a knot point simplifies the surface and causes the surface profile to change. Perform the following steps to add knots to a surface.

1. Open the file SurfaceKnotPoint.3dm from the Chapter 7 folder on the Student Companion Website again.
2. Select Edit > Control Points > Control Points On.
3. Select surface A (shown in Figure 7–52) and press the ENTER key.
4. Select Edit > Control Points > Insert Knot.
5. Select surface A, shown in Figure 7–52.
6. Select the Direction option at the command area.
7. Select the Both option.

   A knot is added in both U and V directions.

8. Select a location on the surface to add a knot point.
9. Continue to add as many knot points as may be required. Press the ENTER key when finished.

   The number of control points is increased accordingly.

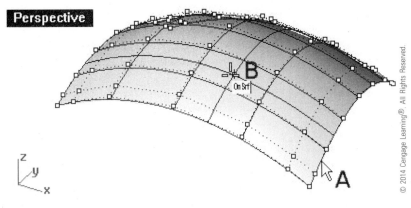

FIGURE 7–52 *Knot point being added to a surface*

To remove a knot from the surface, continue with the following steps.

10. Select Edit > Control Points > Remove Knot.
11. Select surface A (shown in Figure 7-53).
12. If Direction=V, select it to change to Direction=U. Otherwise, proceed to the next step.
13. Select location B and press the ENTER key when finished.

    The selected knot in the U direction is removed.

14. Do not save your file.

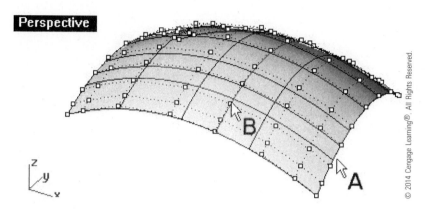

FIGURE 7-53 *Knot point being removed*

Control Point Insertion and Removal

Apart from inserting and removing knots, you can insert and remove control points of a surface. The surface can be a trimmed surface or an untrimmed surface. Perform the following steps.

1. Open the file SurfaceControlPoint.3dm from the Chapter 7 folder on the Student Companion Website.
2. Click on the Insert a control point button on the Point Edit toolbar.
3. Select surface A, shown in Figure 7-54.
4. If Direction=U, select it at the command area to change to Direction=V. Otherwise, proceed to the next step.
5. Click on location B. A control point is added in the V direction.

    A control point is added in the V direction.

6. Click on the Remove a control point button on the Point Edit toolbar.
7. Select surface A, shown in Figure 7-55.
8. If Direction=U, select it at the command area to change to Direction=V. Otherwise, proceed to the next step.
9. Click on location B.

    A control point in the V direction is removed.

10. Do not save your file.

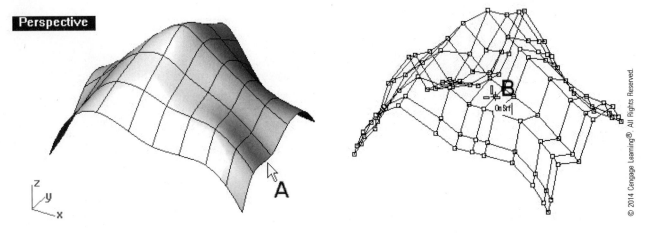

FIGURE 7–54 *Surface being selected (left) and a control point being added (right)*

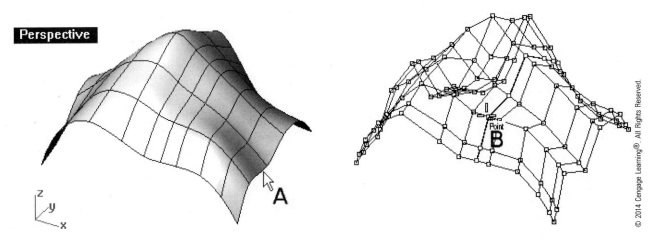

FIGURE 7–55 *Surface being selected (left) and a control point being removed (right)*

Hiding Control Points and Control Polygon

Control points and control polygon of surfaces that are behind other objects can be hidden or displayed, as follows.

1. Open the file CullControlPolygon.3dm from the Chapter 7 folder on the Student Companion Website.
2. Turn on the control points of the surfaces, shown in Figure 7–56 (left).
3. Click on the Cull control polygon backfaces button on the Point Edit toolbar.
   Control points and control polygon behind other objects are hidden. (See Figure 7–56, right).
4. Do not save your file.

This command also works on polygon meshes that will be explained in Chapter 9.

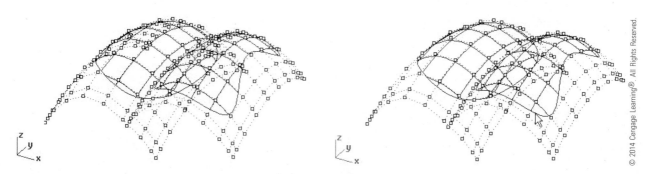

FIGURE 7–56 *Control points and control polygon shown (left) and culled (right)*

Multiple Knots

To find out if there are any multiple knots situated at a single location and to remove them, you can use the RemoveMultiKnot command.

Soft Editing a Surface

Soft editing a surface changes the shape of a surface by picking a point on the surface and moving it to a new location. Perform the following steps.

1. Select File > Open and select the file SoftEditSrf.3dm from the Chapter 7 folder on the Student Companion Website.
2. Select Surface > Surface Edit Tools > Soft Edit.
3. Select the U Distance and V Distance options one by one at the command area, and set their values to 5.
4. If FixEdges=No, select it at the command area to change it to Yes.
5. Select surface A, shown in Figure 7–57.
6. Select a point on the surface. Exact location is unimportant for this tutorial.
7. Hold down the CONTROL key and click on location C.
8. With the CONTROL key still held down, click on location D.
9. Press the ENTER key.

    The surface is soft edited. (See Figure 7–58.)

10. Do not save your file.

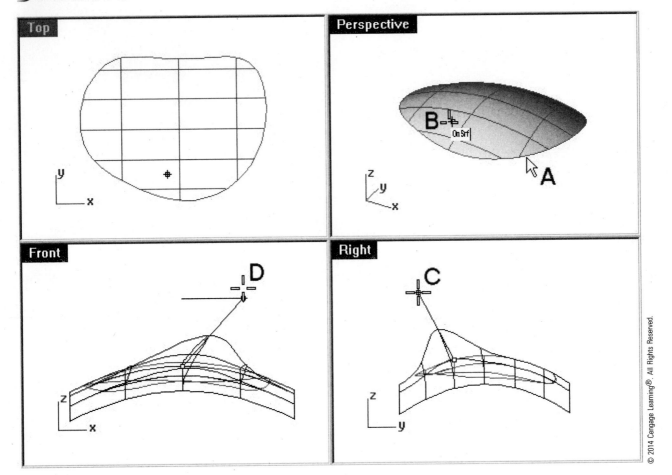

FIGURE 7–57 *Soft editing a surface*

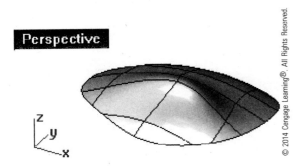

FIGURE 7–58 *Surface's profile changed*

Rebuilding

A way to improve the smoothness of a surface is to rebuild it. In rebuilding a surface, you can specify the number of control points and/or the degree of polynomials along the U and V directions of the surface. To rebuild a surface, perform the following steps.

1. Select File > Open and select the file SurfaceRebuild.3dm from the Chapter 7 folder on the Student Companion Website.
2. Select Edit > Rebuild.
3. Select surface A (shown in Figure 7–59) and press the ENTER key.

4. In the Rebuild Surface dialog box, set the U and V point count to 6 and degree of polynomial to 4, and then click on the OK button.

   The surface is rebuilt, as shown in Figure 7–59.

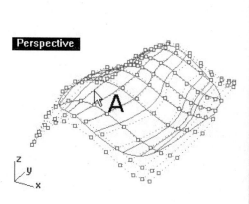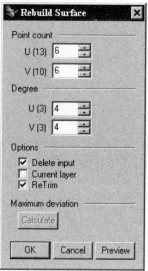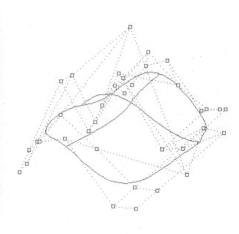

FIGURE 7–59 *Surface being rebuilt (left) and rebuilt surface (right)*
Source: Robert McNeel and Associates Rhinoceros® 5

Rebuilding a Surface in U and V Directions Independently

To rebuild a surface's U and V directions independently, continue with the following steps.

5. Type RebuildUV at the command area or click on the Rebuild Surface UV button on the New in V5 toolbar.
6. Select surface A, indicated in Figure 7–59, and press the ENTER key.
7. At the command area, set Direction=U.
8. Press the ENTER key.

   The surface is rebuilt in the U direction.

Make Uniform

To unify a surface's U or V independently without changing its degree of polynomial or number of control points, continue with the following steps.

9. Type MakeUniformUV at the command area.
10. Select surface A, indicated in Figure 7–59, and press the ENTER key.
11. At the command area, set Direction=U and press the ENTER key.

    The surface is made uniform in the U direction.

To unify a curve or a surface without changing its degree of polynomial or number of control points, continue with the following steps.

12. Click on the Make Uniform button on the Curve Tools toolbar.
13. Select surface A, indicated in Figure 7–59, and press the ENTER key.

    The surface is made uniform.

14. Do not save your file.

Reparameterizing Curves and Surfaces

Parameterization concerns recalculating the parameter space of a curve or a surface. This calculation process causes the parameter space of the objects to be approximately the same size as the object's original 3D geometry. Although there is not much change in the shape of the object, reparameterizing has two major advantages: Texture mapping will become more accurate, and trimming can be done more properly. To reparameterize a surface, perform the following steps.

1. Open any file with a surface model to be reparameterized.
2. Type Reparameterize at the command area.
3. Select the surface and/or curve to be reparameterized and press the ENTER key.
4. Select the Automatic option at the command area.

Refitting and Changing Polynomial Degree

You can refit a surface from an existing surface in accordance with a specified tolerance. Perform the following steps.

1. Select File > Open and select the file SurfacePolynomial.3dm from the Chapter 7 folder on the Student Companion Website.
2. Select Surface > Surface Edit Tool > Refit to Tolerance.
3. Select A (shown in Figure 7–60) and press the ENTER key.
4. Type 2 at the command area to specify the tolerance.

    The surface is refitted in accordance with a tolerance of 2 units, as shown in Figure 7–60 (middle).

If you increase the degree of polynomial, the surface's shape will not change. However, if you reduce the degree of polynomial, the surface's shape is simplified. To change the polynomial degree of a surface along its U and V directions, continue with the following steps.

5. Select Edit > Change Degree.
6. Select B (shown in Figure 7–60) and press the ENTER key.
7. Type 1 at the command area to set the degree of polynomial of the surface in the U direction.
8. Type 4 at the command area to set the degree of polynomial of the surface in the V direction.

    The surface's degree of polynomial is changed, as shown in Figure 7–60 (right).

9. Do not save your file.

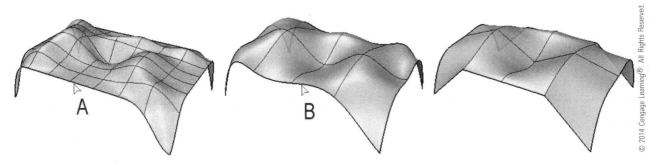

FIGURE 7–60 *From left to right: original surface, refitted surface, and surface with the degree of polynomial changed*

SURFACE EDGE MANIPULATION

There are five ways to manipulate the edges of a surface. You can show selected edges to distinguish them from the isocurves. You can split an edge into two so that you can use a portion of the edge in some other operations. Contrary to splitting, you can merge split edges. You can join edges of two contiguous surfaces, and you can rebuild an edge.

Show Edges

In the computer display, both surface edges and isocurves are displayed as curves. To see clearly the edges of a surface, you highlight them, as follows.

1. Select File > Open and select the file Edge.3dm from the Chapter 7 folder on the Student Companion Website.
2. Select Surface > Edge Tools > Show Edges.
3. Select the surface A, shown in Figure 7-61, and press the ENTER key.
4. In the Edge Analysis dialog box, click on the All Edges box.

    This way, the seam edge is also highlighted.

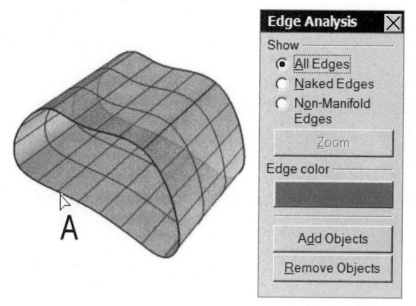

FIGURE 7-61 *All edges highlighted (left) and Edge Analysis dialog box (right)*
Source: Robert McNeel and Associates Rhinoceros® 5

5. If you wish to change the color of the highlighted edges, click on the Edge color button and select a color.
6. If you want to display edges of some other objects, click on the Add Objects button and select additional objects.
7. Click on the Remove Objects button and select objects to deselect them.
8. Click on the Naked Edges box in the Edge Analysis dialog box.

    The seam edge is not highlighted.

Zoom Naked Edge

To take a closer look at the naked edge, you can use the ZoomNaked command by clicking on the Zoom button of the Edge Analysis dialog box or typing ZoomNaked at the command area.

9. Close the Edge Analysis dialog box.

Splitting an Edge

By splitting an edge of a surface into two, you can make use of one of the edges to further construct surfaces. Continue with the following steps to split an edge.

10. Select Surface > Edge Tools > Split Edge.
11. Select edge A (shown in Figure 7-62).
12. Select location B (shown in Figure 7-62). Exact location is unimportant in this tutorial.

    The selected edge is split.

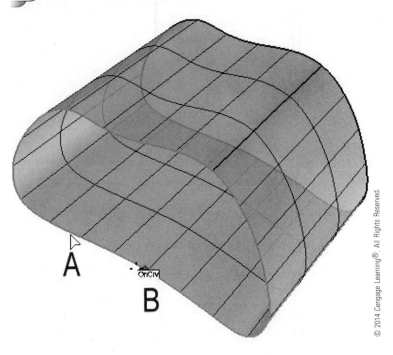

FIGURE 7–62 *Edge being split*

To appreciate how a portion of an edge can be used, continue with the following steps.

13. Select Surface > Extrude > Straight.
14. Select edge A (shown in Figure 7–63) and press the ENTER key.
15. Select location B (Figure 7–63) to indicate the extrusion height. Exact location of B is unimportant here.
    A split edge is extruded.

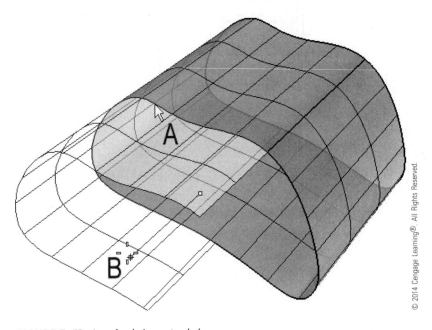

FIGURE 7–63 *An edge being extruded*

Merging Edges

To see how surfaces separated a small distance apart can be joined, you need to move the extruded surface, as follows.

16. Move the extruded surface A, as shown in Figure 7–64.
17. Select Surface > Edge Tools > Merge Edge. To merge all edges at a location, select Surface > Edge Tools > Merge All Edges.
18. Select split edge B (shown in Figure 7–64).
19. Select all in the pop-up dialog box.

   The split edges are merged into a single edge.
20. Select Edit > Explode.
21. Select A, shown in Figure 7–64, and press the ENTER key.

   The extruded object is exploded to an extruded surface.

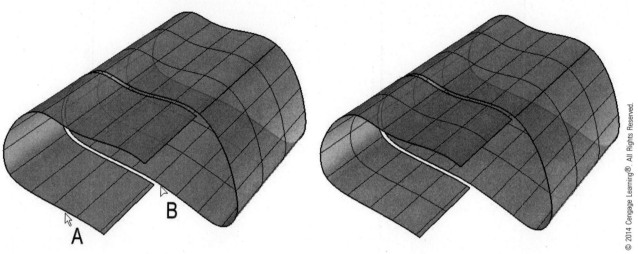

FIGURE 7–64 *Extrude surface moved apart and split edges being merged (left) and extruded object exploded to become an extruded surface (right)*

Joining Edges

Naked edges of two surfaces, although separated by a small distance, can be merged to become a single edge. Joining naked edges repairs any small gaps between contiguous surfaces. The display of the edges is forced closed. The geometry does not change. This command may work for models that are for rendering, but generally is not good enough for manufacturing. To join the naked edges of two surfaces, perform the following steps.

22. Select Analyze > Edge Tools > Join 2 Naked Edges.
23. Select edges A and B (shown in Figure 7–65).

   A dialog box informing you of the tolerances between the edges is displayed.
24. In the Edge Joining dialog box, click on the OK button.

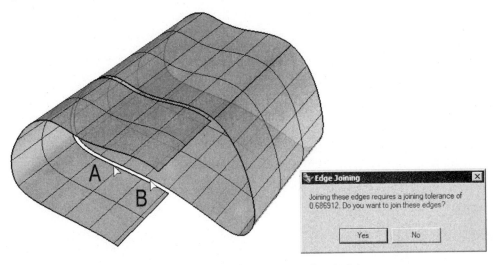

FIGURE 7–65 *Edges being joined (left) and Edge Joining dialog box (right)*
Source: Robert McNeel and Associates Rhinoceros® 5

Rebuilding Edges

To restore the original edge after you have joined edges, you explode the joined surface into individual surfaces and rebuild the edges, as follows.

25. Select Edit > Explode.
26. Select the surface and press the ENTER key.
27. Select Surface > Edge Tools > Rebuild Edges.
28. Select edge A (shown in Figure 7–66).
    Note that this command is also one of the first things to apply if your surface shows as a bad object.
29. Do not save your file.

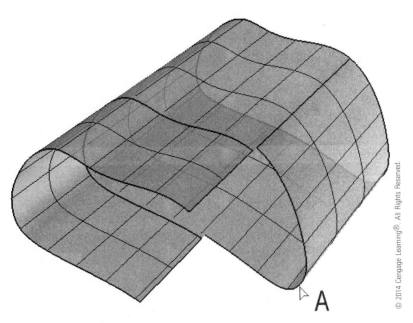

FIGURE 7–66 *Polysurface exploded and edges being restored*

CONSOLIDATION

Because smooth surfaces are constructed from smooth curves, smooth surfaces with an irregular boundary must be trimmed. To maintain the surface's profile, a trimmed surface retains its original base surface as well as the trimmed boundaries in the database. As a result, you can untrim a surface, detach trimmed boundary, or delete a trimmed edge. To reduce the memory needed to save a trimmed surface, you can shrink the base surface to reduce its size or further shrink to the trimmed edge. To obtain two surfaces with smooth continuity, you can split a surface by using a cutting object or by splitting it along its isocurves. Normally, possible kink lines in a surface automatically cause the surface to be broken into a number of surfaces. If a surface with kink lines is imported, you can break it into a set of surfaces along the creases. If a surface is too small in size, you can extend its untrimmed edges. Contrary to trimming, you can extend an untrimmed surface to increase its size, extend and form a periodic surface, and extend to a surface, splitting it and joining with it.

Apart from trimming and extending two consecutive surfaces to bridge them together, you can insert a constant or variable chamfer, fillet, or blended surface between them. These three kinds of surfaces, together with the source surfaces, are typical examples of five kinds of continuity: G0, G1, G2, G3, and G4. Another way of treating two separate surfaces is to modify one of them to match the other. To modify a surface without altering the tangency direction at its edges, you adjust its end bulge. Two or more contiguous surfaces connected at their edges can be joined to become a polysurface. If the connected edges are untrimmed, you can merge them to become a single surface. The reverse of joining surfaces to a polysurface is to explode it to become individual surfaces. Merged surfaces cannot be exploded.

Surface profiles can be edited in several ways. You can change the directions of a surface in terms of its U and V isocurves and its normal direction. To alter the shape of a surface, you can translate its control points in U, V, and N directions; adjust the weighting of the control points; use handlebar editor; add/remove knot points; and add/remove control points. Furthermore, you can rebuild a surface by specifying the point counts and degree of polynomial in the U and V directions. To refine a surface without changing its degree of polynomial and number of control points, you reparameterize it.

To manipulate the edges of a surface, you display its edges, split an edge, merge a split edge, join edges, and rebuild edges.

REVIEW QUESTIONS

1. Outline the ways to modify the boundaries of a surface.
2. How can you treat the junction between two separate surfaces?
3. Differentiate between joining and merging two surfaces.
4. List the ways you can refine or modify a surface's profile.
5. What are the ways you can manipulate the edge of a surface?

CHAPTER 8

Rhinoceros Polysurfaces and Solids

INTRODUCTION

This chapter details Rhino's polysurface and solid modeling methods.

OBJECTIVES

After studying this chapter, you should be able to

- Comment on Rhino's solid modeling method
- Differentiate between a polysurface and a Rhino solid
- Construct and detail Rhinoceros solids

OVERVIEW

As a NURBS surface modeling tool, Rhino represents a solid in the computer by using a closed-loop surface (such as a sphere or ellipsoid) or a polysurface (such as a box or a cylinder) that encloses a volume without an opening or gap. This chapter will describe various ways of using Rhinoceros to construct, manipulate, detail, and modify solid objects in the computer.

RHINO'S SOLID MODELING METHOD

There are many ways to represent a solid in the computer. Rhino uses a closed-loop surface or a closed polysurface without openings or gaps to depict a solid. This modeling method is particularly useful for constructing free-form objects because you can first decompose a complex free-form object into individual free-form surfaces and then construct individual surfaces one by one. Naturally, prior to composing a solid from the surfaces, you may need to trim the surfaces for them to enclose a volume without openings or intersections. To help you learn about using Rhino as a tool in solid modeling, Rhino's solid modeling tools are categorized into five groups.

- The first group covers ways to construct solids of regular geometric shapes.
- The second group explains how free-form solids can be composed.
- The third group covers how two or more Rhino solid objects can be combined.

- The fourth group covers ways to detail a solid object.
- The fifth group deals with various editing methods.

The fourth and fifth groups apply to both polysurfaces and solids.

Subobject Selection

An individual surface in a polysurface or a solid is a subobject. If you need to select these individual surfaces, you can hold down the CONTROL and SHIFT keys simultaneously when clicking on the object.

SOLIDS OF REGULAR GEOMETRIC SHAPES

Solid objects of regular geometric shapes range from simple 3D boxes (six planar surfaces joined together) to 3D text objects (combination of planar and curved surfaces). Among them, some are single closed-loop surfaces (sphere, ellipsoid, and torus), and some are polysurfaces joined together. Figure 8–1 shows the box, sphere, cylinder, cone, truncated cone, pyramid, ellipsoid, paraboloid, tube, pipe, torus, and text shapes.

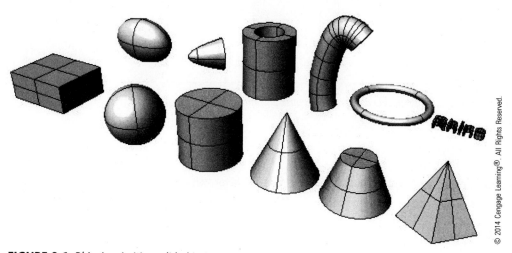

FIGURE 8–1 *Rhino's primitive solid objects*

Box

A box is a polysurface with six planar surfaces that are mutually perpendicular to one another. You can construct a box using one of five methods: specifying two corners of the box's base and the box's height; specifying the box's diagonal; using three points to specify the box's base and specifying the box's height; specifying two points in a viewport and one point in another viewport to define the box's base and specifying the box's height; and specifying the center point of the base, a corner point, and the height.

Specifying Corners and Height

Perform the following steps to construct a box by specifying corners of the base and the height.

1. Select File > Open and select the file Box.3dm from the Chapter 8 folder that you downloaded from the Student Companion Website.
2. Set object snap mode to Point.
3. Select Solid > Box > Corner to Corner, Height.
4. Select points A, B, and C (shown in Figure 8–2).

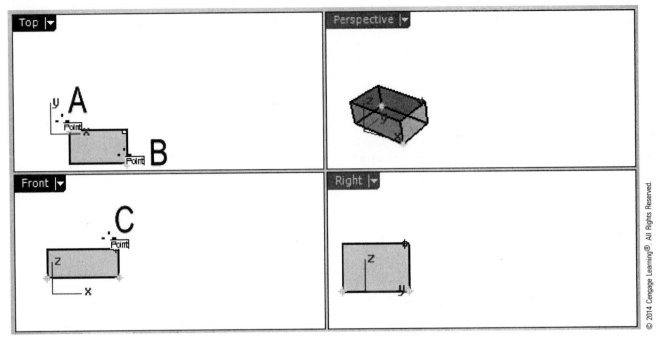

FIGURE 8–2 *Solid box being constructed by specifying two corners of the base and the height*

Specifying Diagonal

Continue with the following steps to construct a box by specifying its diagonal points.

  5. Turn off Layer 00 and turn on Layer 02. Do not change the current layer, which should be Layer 01.
  6. Select Solid > Box > Diagonal.
  7. Select points A and B, shown in Figure 8–3.

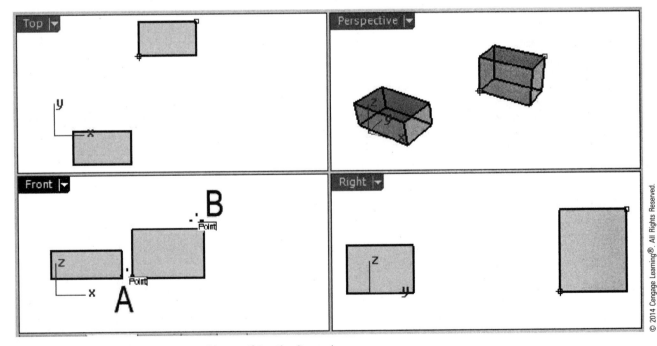

FIGURE 8–3 *Solid box being constructed by specifying the diagonal*

Specifying Three Points and Height

Continue with the following steps to construct a box by using three points to specify the base and specifying the height.

8. Turn off Layer 02 and turn on Layer 03. Again, keep the Layer 01 current.
9. Select Solid > Box > 3 points, height.
10. Select points A, B, C, and D, shown in Figure 8–4.

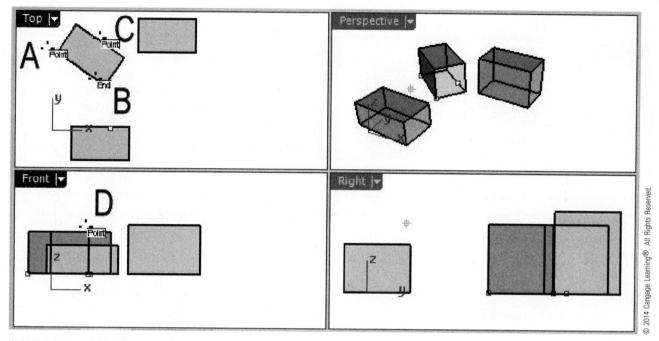

FIGURE 8–4 *Solid box being constructed by using three points to define the base and specifying the height*

Box with a Base Vertical to the Construction Plane

Continue with the following steps to construct a box vertical to the active construction plane.

11. Turn off Layer 03 and turn on Layer 04.
12. Select Solid > Box > Vertical Base.
13. Select points A, B, C, and D, shown in Figure 8–5.

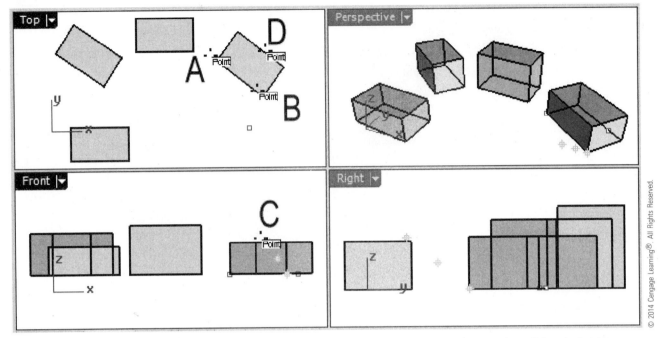

FIGURE 8–5 *Solid box being constructed by constructing the box base vertical to a selected viewport and specifying the height*

Specifying Center

Continue with the following steps to construct a box by specifying the center of the base, a corner, and the height.

14. Turn off Layer 04 and turn on Layer 05.
15. Select Solid > Box > Center of Base, Corner, Height.
16. Select points A, B, and C, shown in Figure 8–6.
17. Do not save your file.

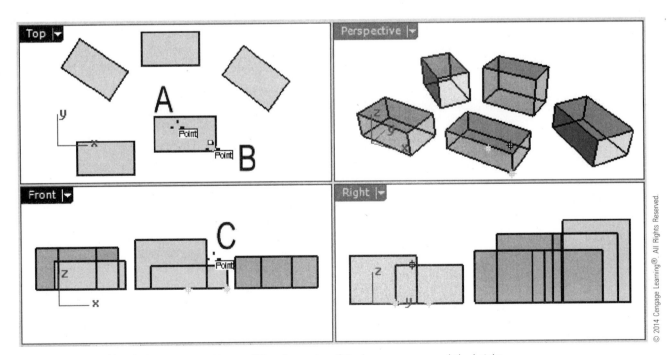

FIGURE 8–6 *Solid box being constructed by specifying the center of the base, a corner, and the height*

Sphere

A sphere is a single closed-loop surface. You can use one of seven methods to construct a sphere: specifying the center and a point on the surface; specifying the diameter; specifying three points on the surface of the sphere; specifying three tangent curves; selecting a curve and specifying the center along the curve and the radius; specifying four points on the surface of the sphere; and selecting a set of points.

Specifying Center and a Point on the Surface

Perform the following steps to construct a sphere by specifying the center and a point on the sphere's surface.

1. Select File > Open and select the file Sphere.3dm from the Chapter 8 folder on the Student Companion Website.
2. Select Solid > Sphere > Center, Radius.
3. Select points A and B, shown in Figure 8–7.

Specifying Diameter

Continue with the following steps to construct a sphere by specifying its diameter.

4. Select Solid > Sphere > 2 Points.
5. Select points C and D, shown in Figure 8–7.

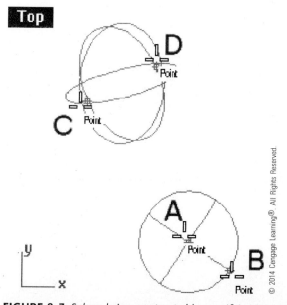

FIGURE 8–7 *Sphere being constructed by specifying its center and a point on the surface and by specifying the diameter*

Specifying Three Points

Continue with the following steps to construct a sphere by specifying three points on the surface.

6. Turn off Layer 00 and turn on Layer 02. Keep Layer 01 current.
7. Select Solid > Sphere > 3 Points.
8. Select points A, B, and C, shown in Figure 8–8.

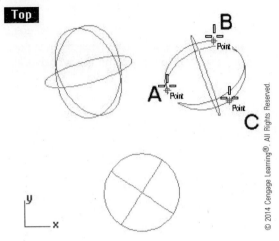

FIGURE 8–8 *Sphere being constructed by specifying three points on the surface*

Specifying Tangent Curves

Continue with the following steps by selecting three tangent curves.

9. Turn on Layer 03 and turn off Layer 02.
10. Select Solid > Sphere > Tangent to Curves.
11. Select the Tangent option at the command area.
12. Select curves A, B, and C, shown in Figure 8–9.

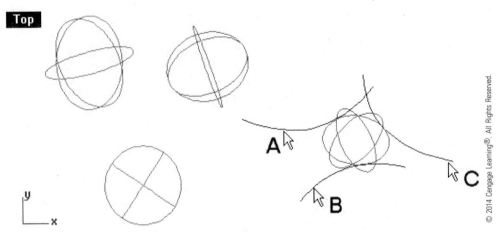

FIGURE 8–9 *Sphere being constructed by selecting three tangent curves*

Sphere around a Curve

Continue with the following steps to construct a sphere around a curve.

13. Turn on Layer 04 and turn off Layer 03.
14. Select Solid > Sphere > Around Curve.
15. Select curve A and points B and C, shown in Figure 8–10.

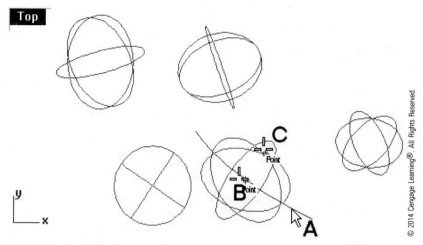

FIGURE 8–10 *Sphere being constructed around a curve*

Specifying Four Points

Continue with the following steps to construct a sphere by specifying four points. The first three specified points define a section of the sphere. The fourth point defines a point on the sphere.

16. Turn on Layer 05 and turn off Layer 04.
17. Select Solid > Sphere > 4 Points.
18. Select points A, B, C, and D, shown in Figure 8–11.

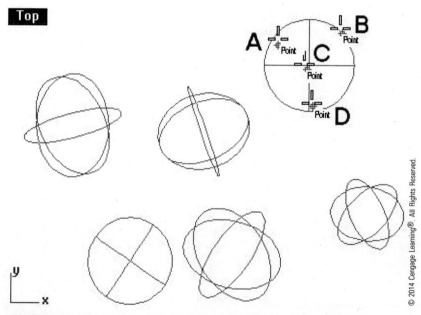

FIGURE 8–11 *Sphere being constructed by specifying four points on the surface*

Fitting to a Set of Points

Continue with the following steps to construct a sphere to best fit a set of selected points.

19. Turn on Layer 06 and turn off Layer 05.
20. Select Solid > Sphere > Fit Sphere to Points.
21. Click on location A (Figure 8–12) and drag to location B to select the point objects.
22. Do not save your file.

    A sphere best fitted to the selected points is constructed.

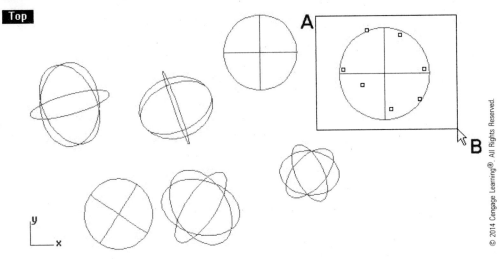

FIGURE 8–12 *Sphere being constructed to fit a set of points*

Cylinder

A cylinder is a polysurface consisting of three surfaces. The body is a cylindrical surface, and the top and bottom surfaces are circular planar surfaces. There are many ways to construct a cylinder.

Direction Constraints of Cylinder Axis

Direction constraint concerns the direction of the axis of the cylinder after its radius or diameter is specified. There are three direction constraints: none, vertical, and around curve. If direction constraint is set to "none," the other endpoint is freely determined. If direction constraint is set to "vertical," the other axis endpoint is set to be perpendicular to the base of the cylinder. If you want to construct a cylinder around a curve, you set direction constraint to "around curve." Perform the following steps.

1. Select File > Open and select the file Cylinder.3dm from the Chapter 8 folder on the Student Companion Website.
2. Select Solid > Cylinder.
3. Select the DirectionConstraint option at the command area.
4. Select the Vertical option at the command area.
5. Select points A, B, and C, shown in Figure 8–13.

    A is the center, AB is the radius, and C's Z coordinate defines the height of the cylinder.

6. Repeat the command.
7. Select the DirectionConstraint option at the command area.
8. Select the None option at the command area.
9. Select points D, E, and C, shown in Figure 8–13.

    D is the center, DE is the radius, and C is the location of the other axis's endpoint.

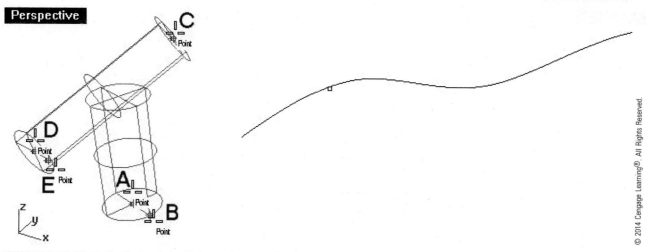

FIGURE 8–13 *Vertical cylinder and cylinder with user-defined axis endpoint being constructed*

10. Repeat the command.
11. Select the DirectionConstraint option at the command area.
12. Select the AroundCurve option at the command area.
13. Select curve A, shown in Figure 8–14.
14. Select point B to specify the center point.
15. Type 6 to specify the radius.
16. Type 30 to specify the height of the cylinder.

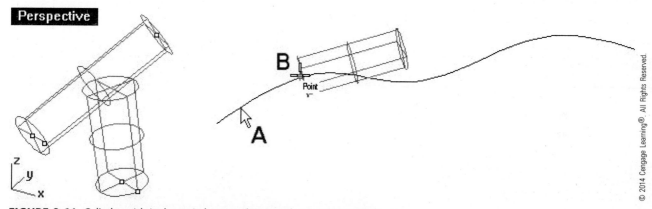

FIGURE 8–14 *Cylinder with its base circle around a curve being constructed*

Cylinder Base Tangent to Three Curves

Continue with the following steps to construct a cylinder with its base circle tangent to three curves.

17. Turn on Layer 02 and turn off Layer 00.
18. Select Solid > Cylinder.
19. Select the DirectionConstraint option at the command area.
20. Select the None option at the command area.
21. Select the Tangent option at the command area.
22. Select curves A, B, and C, shown in Figure 8–15.
23. Select point D.

A cylinder with its axis endpoint closest to the selected point is constructed.

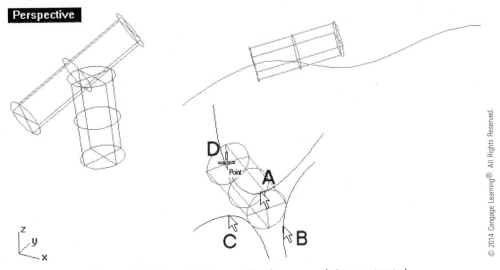

FIGURE 8–15 *Cylinder with its base circle tangent to three curves being constructed*

Cylinder Base Fitted to a Set of Points

Continue with the following steps to construct a cylinder with its base circle best fitted to a set of selected points.

24. Turn on Layer 03 and turn off Layer 02.
25. Select Solid > Cylinder.
26. Select the FitPoints option at the command area.
27. Click on A and drag to B (Figure 8–16) to select the point objects, and press the ENTER key.
28. Select point C.

    A cylinder with its axis endpoint closest to the selected point is constructed.

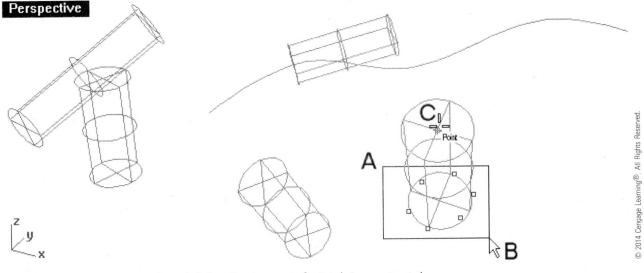

FIGURE 8–16 *Cylinder with its base circle best fitted to a set of points being constructed*

Cylinder Base Passing through Three Points

Continue with the following steps to construct a cylinder with its base circle defined by three selected points.

29. Turn on Layer 04 and turn off Layer 03.
30. Select Solid > Cylinder.

31. Select the 3Point option at the command area.
32. Select points A, B, and C, shown in Figure 8-17
33. Select point D.

    A cylinder with its axis endpoint closest to the selected point is constructed.
34. Do not save your file.

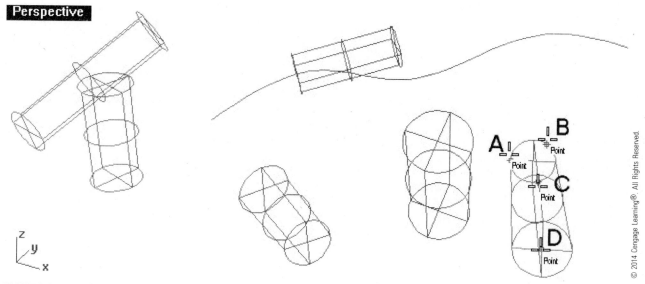

FIGURE 8-17 Cylinder with its base circle defined by three selected points being constructed

Radius and Diameter Option

By default, the base circle is defined by a center point and a radius. If you want to specify the diameter, you should select the Diameter option from the command line.

Cone

A cone is a polysurface consisting of two joined surfaces. The slant surface is a closed surface, and the base is a circular planar surface.

Direction Constraints of Cone Axis

The methods to construct a cone are similar to those used to construct a cylinder. You can define direction constraint with the apex vertical to the base, with the apex at a selected point, and with the axis of the base circle around a curve. You can also define a cone's base circle tangent to three curves, passing through three points, or best fitting to a set of points. Perform the following steps.

1. Select File > Open and select the file Cone.3dm from the Chapter 8 folder on the Student Companion Website. This template file is identical to the template file Cylinder.3dm.
2. Select Solid > Cone.
3. Select the DirectionConstraint option at the command area.
4. Select the Vertical option at the command area.
5. Select points A, B, and C, shown in Figure 8-18.

   A is the center, AB is the radius, and C's Z coordinate defines the height of the cone.
6. Repeat the command.
7. Select the DirectionConstraint option at the command area.
8. Select the None option at the command area.
9. Select points D, E, and C, shown in Figure 8-18.

   D is the center, DE is the radius, and C is the location of the apex.

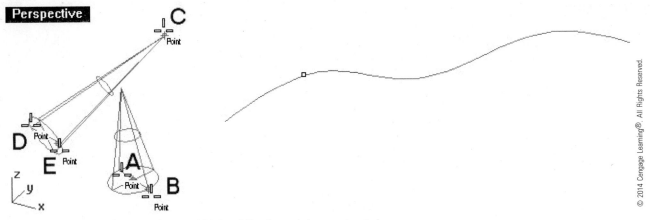

FIGURE 8–18 *Vertical cone and cone with user-defined apex being constructed*

10. Repeat the command.
11. Select the DirectionConstraint option at the command area.
12. Select the AroundCurve option at the command area.
13. Select curve A, shown in Figure 8–19.
14. Select point B to specify the center point.
15. Type 6 to specify the radius.
16. Type 30 to specify the height of the cone.

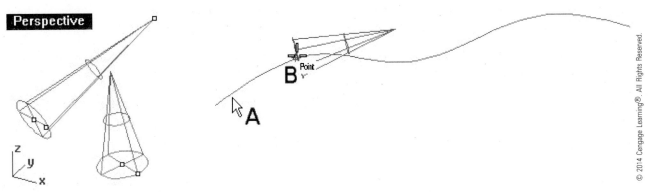

FIGURE 8–19 *Cone with its base circle around a curve being constructed*

Cone Base Tangent to Three Curves

Continue with the following steps to construct a cone with its base circle tangent to three curves.

17. Turn on Layer 02 and turn off Layer 00.
18. Select Solid > Cone.
19. Select the DirectionConstraint option at the command area.
20. Select the None option at the command area.
21. Select the Tangent option at the command area.
22. Select curves A, B, and C, shown in Figure 8–20.
23. Select point D.

    A cone with its apex closest to the selected point is constructed.

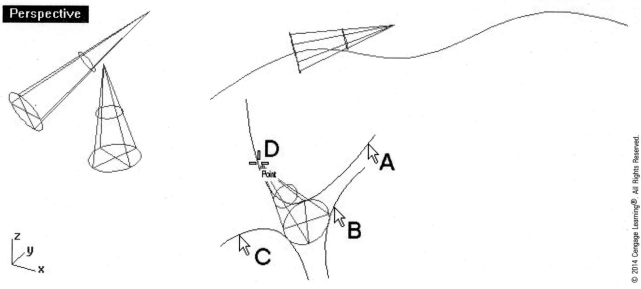

FIGURE 8–20 *Cone with its base circle tangent to three curves being constructed*

Cone Base Fitted to a Set of Points

Continue with the following steps to construct a cone with its base circle best fitted to a set of selected points.

24. Turn on Layer 03 and turn off Layer 02.
25. Select Solid > Cone.
26. Select the FitPoints option at the command area.
27. Click on A and drag to B (Figure 8–21) to select the point objects, and press the ENTER key.
28. Select point C.

    A cone with its apex closest to the selected point is constructed.

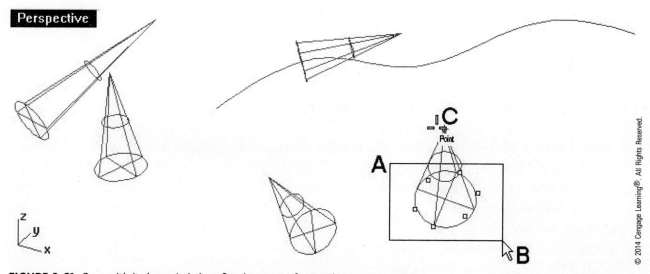

FIGURE 8–21 *Cone with its base circle best fitted to a set of points being constructed*

Cone Base Passing through Three Points

Continue with the following steps to construct a cone with its base circle defined by three selected points.

29. Turn on Layer 04 and turn off Layer 03.
30. Select Solid > Cone.

31. Select the 3Point option at the command area.
32. Select points A, B, and C, shown in Figure 8–22.
33. Select point D.

    A cone with its apex closest to the selected point is constructed.
34. Do not save your file.

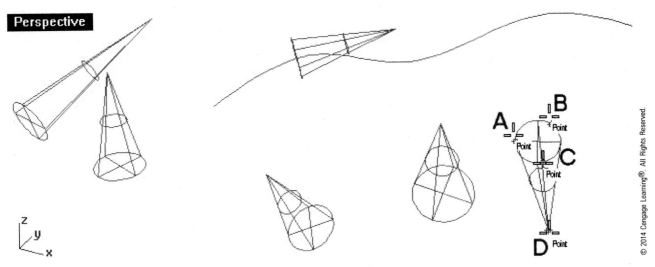

FIGURE 8–22 *Cone with its base circle defined by three selected points being constructed*

Truncated Cone

A truncated cone is a polysurface consisting of three joined surfaces. The slant surface is a closed surface, and the top and bottom surfaces are circular planar surfaces. The methods to construct a truncated cone are very similar to those used to construct a cone, with the exception that a second radius has to be specified.

Direction Constraints of Truncated Cone's Axis

There are three direction constraints. Perform the following steps.

1. Select File > Open and select the file TCone.3dm from the Chapter 8 folder on the Student Companion Website. This template file is identical to the template files Cylinder.3dm and Cone.3dm.
2. Select Solid > Truncated Cone.
3. Select the DirectionConstraint option at the command area.
4. Select the Vertical option at the command area.
5. Select points A and B, shown in Figure 8–23, to define the base of the truncated cone.
6. Select point C to define the height of the cone.
7. Type 4 at the command area to specify the other radius.

    A truncated cone is constructed.
8. Repeat the command.
9. Select the DirectionalConstraint option at the command area.
10. Select the None option at the command area.
11. Select points D and E, shown in Figure 8–23, to specify the center and radius of the truncated cone.
12. Select point C to define the location of the other axis's endpoint.
13. Type 4 at the command area to specify the other radius.

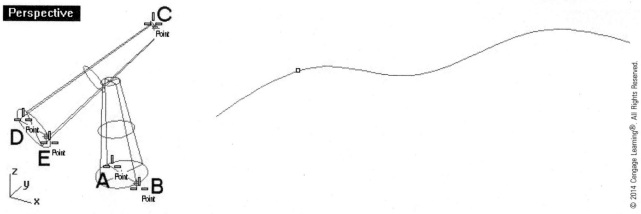

FIGURE 8–23 *Vertical truncated cone and truncated cone with user-defined second axis point being constructed*

14. Repeat the command.
15. Select the DirectionalConstraint option at the command area.
16. Select the AroundCurve option at the command area.
17. Select curve A, shown in Figure 8–24.
18. Select point B to specify the center point.
19. Type 6 to specify the radius of the base circle.
20. Type 30 to specify the height of the truncated cone.
21. Type 4 to specify the radius of the top circle.

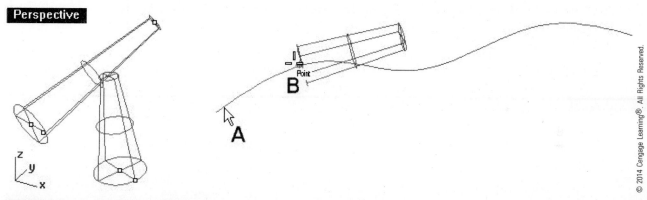

FIGURE 8–24 *Truncated cone with its base circle around a curve being constructed*

Truncated Cone Base Tangent to Three Curves

Continue with the following steps to construct a truncated cone with its base circle tangent to three curves.

22. Turn on Layer 02 and turn off Layer 00.
23. Select Solid > Truncated Cone.
24. Select the DirectionalConstraint option at the command area.
25. Select the None option at the command area.
26. Select the Tangent option at the command area.
27. Select curves A, B, and C, shown in Figure 8–25.
28. Select point D to specify the height of the truncated cone.
29. Type 4 at the command area to specify the other radius.

    A truncated cone with its second axis endpoint closest to the selected point is constructed.

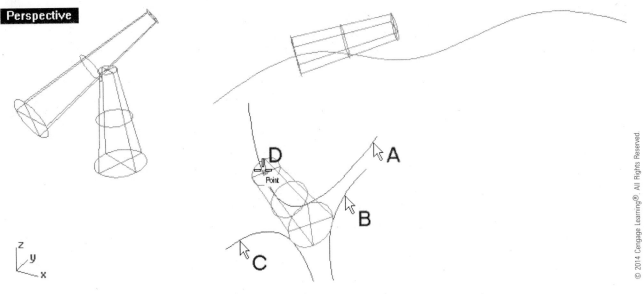

FIGURE 8–25 *Cone with its base circle tangent to three curves being constructed*

Fitting to a Set of Points

Continue with the following steps to construct a truncated cone with its base circle best fitted to a set of selected points.

30. Turn on Layer 03 and turn off Layer 02.
31. Select Solid > Truncated Cone.
32. Select the FitPoints option at the command area.
33. Click on A and drag to B (Figure 8–26) to select the point objects, and press the ENTER key.
34. Select point C to specify the height of the truncated cone.
35. Type 4 at the command area to specify the radius at the other end.

    A truncated cone with its second axis endpoint closest to the selected point is constructed.

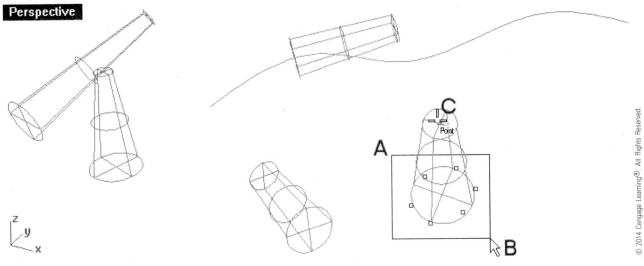

FIGURE 8–26 *Cone with its base circle best fitted to a set of points being constructed*

Truncated Cone Base Passing through Three Points

Continue with the following steps to construct a truncated cone with its base circle defined by three selected points.

36. Turn on Layer 04 and turn off Layer 03.
37. Select Solid > Truncated Cone.

38. Select the 3Point option at the command area.
39. Select points A, B, and C, shown in Figure 8-27.
40. Select point D to specify the height of the truncated cone.
41. Type 2 at the command area to specify the radius at the other end.

    A truncated cone with its second axis endpoint closest to the selected point is constructed.
42. Do not save your file.

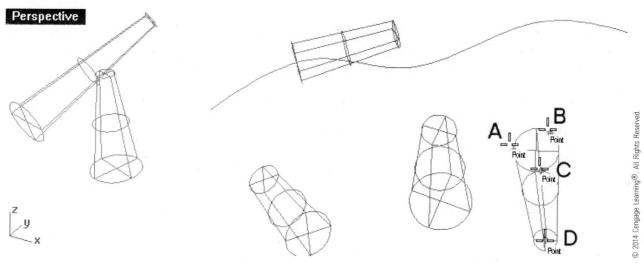

FIGURE 8–27 *Cone with its base circle defined by three selected points being constructed*

Pyramid

A pyramid is a polysurface with a polygon base and a number of slant triangular surfaces that encloses a volume without an opening or gap.

Direction Constraints of Pyramid Axis

Similar to constructing a cylinder, cone, and truncated cone, to construct a pyramid you need to select one of three ways to constrain the direction of the axis: none, vertical, and around curve. Perform the following steps.

1. Select File > Open and select the file Pyramid.3dm from the Chapter 8 folder on the Student Companion Website.
2. Select Solid > Pyramid.
3. Select the DirectionConstraint option at the command area.
4. Select the Vertical option at the command area.

Regular Polygon and Star Polygon of Pyramid Base

The base of a pyramid is a polygon, which can be a regular polygon or a star polygon. Continue with the following steps.

5. Select the NumberSides option at the command area.
6. Type 5 to specify the number of sides.
7. Select points A, B, and C, shown in Figure 8-28.

    Point A defines the center, point B is a corner of the base polygon, and point C defines the height.
8. Repeat the command.
9. Select the Star option at the command area.
10. Select points D, E, F, and G, shown in Figure 8-28.

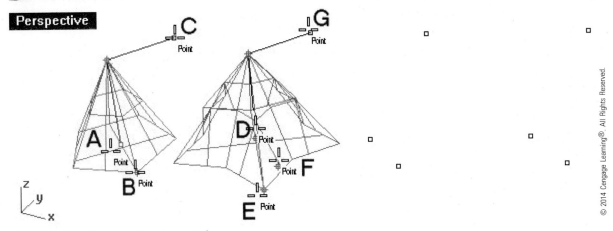

FIGURE 8–28 *Two pyramids constructed*

Edge, Inscribed, and Circumscribed Pyramid Base

There are three ways to define the base polygon of a pyramid: edge, inscribed circle, and circumscribed circle. The default way is to select the center point and the radius of the inscribed circle.

11. Repeat the command.
12. Select the Edge option at the command area.
13. Select points A, B, and C, shown in Figure 8-29.
14. Repeat the command.
15. Select the Circumscribed option at the command area.
16. Select points D, E, and F, shown in Figure 8-29.
17. Do not save your file.

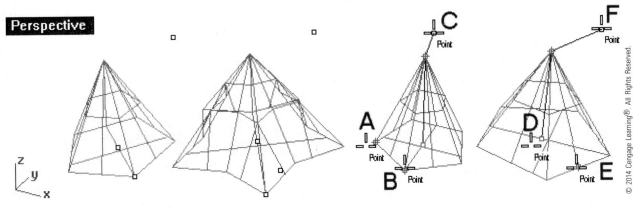

FIGURE 8–29 *Two more pyramids constructed*

Ellipsoid

An ellipsoid, like a sphere, is also a single closed-loop surface. You can construct an ellipsoid via five methods, as follows.

Ellipsoid's Center and Axis Endpoints

The first method creates an ellipsoid by specifying the ellipsoid's center and axis endpoints, as follows.

1. Select File > Open and select the file Ellipsoid.3dm from the Chapter 8 folder on the Student Companion Website.
2. Select Solid > Ellipsoid > From Center.
3. Select points A and B, shown in Figure 8-30, to specify the center and an axis endpoint.

4. Type 5 at the command area, press the ENTER key, and click on location C to specify the second axis radius and direction.
5. Move the cursor to location D and type 3 at the command area to set the third axis location.

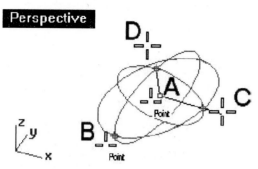

FIGURE 8–30 *Ellipsoid specified by its center and axis endpoints*
© 2014 Cengage Learning®. All Rights Reserved.

Ellipsoid's Diameter and Axis Endpoints

A second method creates an ellipsoid by specifying its diameter and two axis endpoints. Continue with the following steps.

6. Turn off Layer 00 and turn on Layer 02.
7. Repeat the command and select the Diameter option at the command area.
8. Select points A and B, shown in Figure 8–31, to specify the center and an axis endpoint.
9. Type 10 at the command area, press the ENTER key, and click on location C to specify the second axis diameter and direction.
10. Move the cursor to location D and type 3 at the command area to set the third axis location.

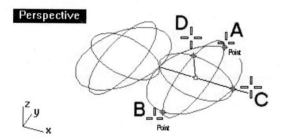

FIGURE 8–31 *Ellipsoid specified by its diameter being constructed*
© 2014 Cengage Learning®. All Rights Reserved.

Ellipsoid's Foci and a Point

A third method creates an ellipsoid by specifying its foci and a point on the ellipsoid. Continue with the following steps.

11. Turn on Layer 03 and turn off Layer 02
12. Select Solid > Ellipsoid > From Foci.
13. Select points A and B (shown in Figure 8–32) to specify the foci, and then select point C to specify a point on the ellipsoid.

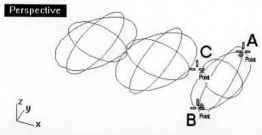

FIGURE 8–32 *Ellipsoid specified by its foci being constructed*
© 2014 Cengage Learning®. All Rights Reserved.

Ellipsoid's Bounding Box

A bounding box of an object is the smallest rectangular box that entirely encloses the object. A fourth method creates an ellipsoid by specifying a bounding box. Continue with the following steps.

14. Turn on Layer 04 and turn off Layer 03.
15. Select Solid > Ellipsoid > From Center and select the Corner option at the command area.
16. Select points A and B (shown in Figure 8–33).
17. Move the cursor to location C and type 2.5 at the command area to set the third axis location.

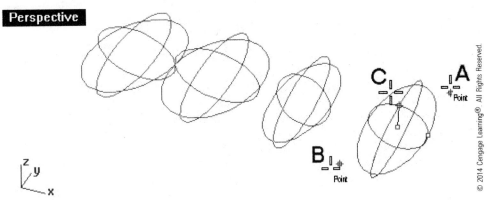

FIGURE 8–33 *Ellipsoid specified by its bounding box being constructed*

Ellipsoid on a Curve

A fifth method creates an ellipsoid on a curve. Continue with the following steps.

18. Turn on Layer 05 and turn off Layer 04.
19. Select Solid > Ellipsoid > From Center and select the AroundCurve option at the command area.
20. Select curve A (Figure 8–34).
21. Select point B along the curve.
22. Type 5 at the command area, press the ENTER key, and click on location C to specify the first radius.
23. Type 4 at the command area, press the ENTER key, and click on location D to specify the second radius.
24. Type 3 at the command area to specify the third radius.
25. Do not save your file.

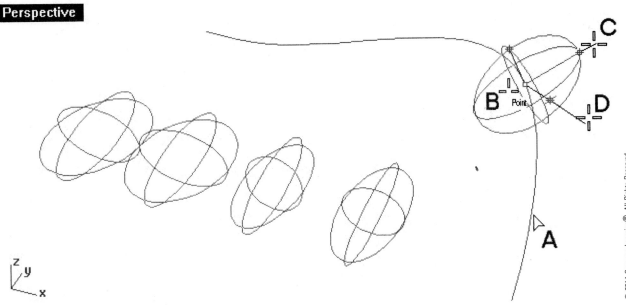

FIGURE 8–34 *Ellipsoid on a curve being constructed*

Paraboloid

A paraboloid is a polysurface consisting of a revolved parabola and a circular surface capping its end. To construct a paraboloid, perform the following steps.

1. Select File > Open and select the file Paraboloid.3dm from the Chapter 8 folder on the Student Companion Website.
2. Select Solid > Paraboloid > Focus, Direction.
3. If the prompt indicates Cap = No, select the Cap option at the command area to change it to Yes. (*Note:* If Cap = No, an open surface will be constructed instead.)
4. Select locations A, B, and C (shown in Figure 8–35) to indicate the focus, direction, and an endpoint of the paraboloid.
5. Select Solid > Paraboloid > Vertex, Focus.
6. Select locations D, E, and F (shown in Figure 8–35) to specify the vertex, focus, and an endpoint of the paraboloid.
   Two paraboloids are constructed.
7. Do not save your file.

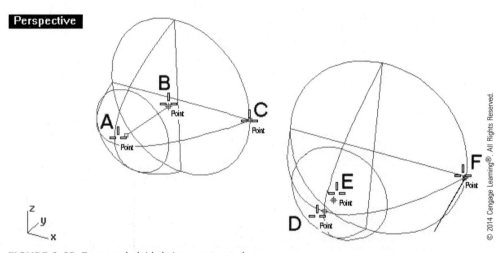

FIGURE 8–35 *Two paraboloids being constructed*

Tube

A tube is a polysurface consisting of two concentric cylindrical surfaces and two planar surfaces capping the ends. The options for constructing a tube are similar to those for constructing a cylinder, cone, and truncated cone.

Direction Constraints of Tube Axis

There are three direction constraints. Perform the following steps.

1. Select File > Open and select the file Tube.3dm from the Chapter 8 folder on the Student Companion Website.
2. Select Solid > Tube.
3. Select the DirectionConstraint option at the command area.
4. Select the Vertical option at the command area.
5. Select point A, shown in Figure 8–36.
6. Type 10 at the command area to specify the first radius.
7. Type 5 at the command area to specify the second radius.
8. Select point B to define the height of the cone.
9. Repeat the command.
10. Select the DirectionConstraint option at the command area.

11. Select the None option at the command area.
12. Select point C, shown in Figure 8-36.
13. Type 10 at the command area to specify the first radius.
14. Type 5 at the command area to specify the second radius.
15. Select point B.
16. Repeat the command.
17. Select the DirectionConstraint option at the command area.
18. Select the AroundCurve option at the command area.
19. Select curve D, shown in Figure 8-36.
20. Select point E to specify the center point.
21. Type 10 at the command area to specify the first radius.
22. Type 5 at the command area to specify the second radius.
23. Type 30 to specify the height of the tube.

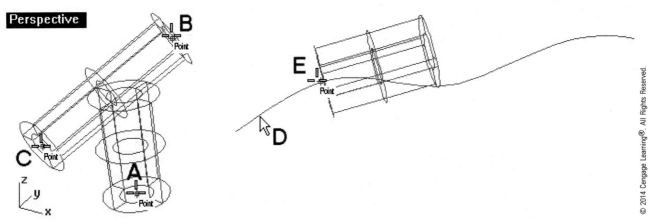

FIGURE 8-36 *Tube in three directions being constructed*

Axis of Tube's Base Tangent to Three Curves

Continue with the following steps to construct a tube with its base circle tangent to three curves.

24. Turn on Layer 02 and turn off Layer 00.
25. Select Solid > Tube.
26. Select the DirectionConstraint option at the command area.
27. Select the None option at the command area.
28. Select the Tangent option at the command area.
29. Select curves A, B, and C, shown in Figure 8-37.
30. Type 3 at the command area to specify the second radius.
31. Select point D.

    A tube cone is constructed.

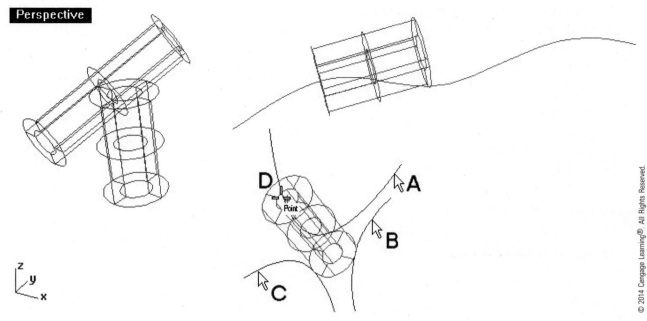

FIGURE 8–37 *Tube cone with its base circle tangent to three curves being constructed*

Tube Base's Outside Diameter Fitted to a Set of Points

Continue with the following steps to construct a cone with its base circle best fitted to a set of selected points.

32. Turn on Layer 03 and turn off Layer 02.
33. Select Solid > Tube.
34. Select the FitPoints option at the command area.
35. Click on A and drag to B (Figure 8–38) to select the point objects, and press the ENTER key.
36. Type 3 at the command area to specify the radius at the other end.
37. Select point C.

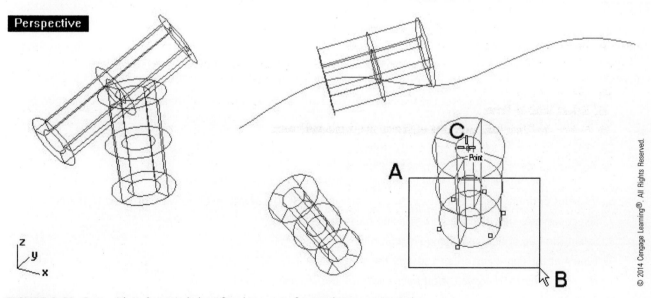

FIGURE 8–38 *Cone with its base circle best fitted to a set of points being constructed*

Tube Base's Outside Diameter Passing through Three Points

Continue with the following steps to construct a cone with its base circle defined by three selected points.

38. Turn on Layer 04 and turn off Layer 03.
39. Select Solid > Tube.
40. Select the 3Point option at the command area.
41. Select points A, B, and C, shown in Figure 8–39.
42. Type 3 at the command area to specify the radius at the other end.
43. Select point D.
44. Do not save your file.

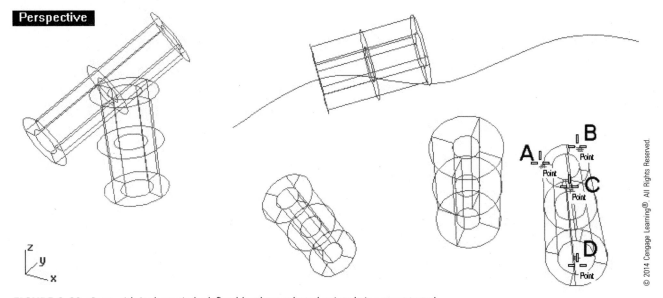

FIGURE 8–39 *Cone with its base circle defined by three selected points being constructed*

Pipe

A pipe is a polysurface with a variable-radius, circular cross-section that runs along a curve or a chain of curve/edges. The end caps of the pipe can be flat or spherical. To speed curve chain selection, you can select the AutoChain option and use the ChainContinuity option to help control how smoothly the curve segments need to be connected to get selected. The end cap can be flat or rounded. Perform the following steps to construct a pipe with spherical caps.

1. Select File > Open and select the file Pipe.3dm from the Chapter 8 folder on the Student Companion Website.
2. Set object snap mode to End.
3. Select Solid > Pipe.
4. Select the ChainEdge option at the command area.
5. If AutoChain=No, select it to change it to Yes. Otherwise, proceed to the next step.
6. If ChainContinuity is not Curvature, select it and change it to Curvature. Otherwise, proceed to the next step.
7. Select curve A, shown in Figure 8–40, and press the ENTER key.
8. Select the Cap option and change it to Round.
9. Type 4 at the command area to specify the start radius.
   The end nearer to where you select the curve is the starting point.
10. Type 3 at the command area to specify the end radius.
11. Select endpoint B.
12. Type 2 at the command area to specify the radius at point B.

13. Select endpoint C.
14. Type 2 at the command area.
15. Press the ENTER key.
16. Do not save your file.

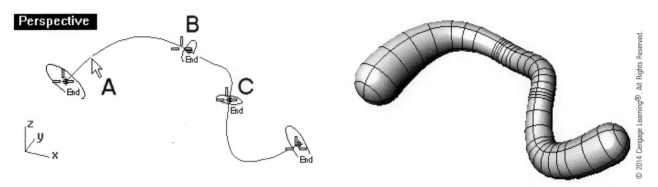

FIGURE 8–40 *Pipe with variable radii and round caps being constructed (left) and pipe constructed (right)*

Torus

A torus is single closed-loop surface in the shape of a doughnut, which resembles a cylinder bent in a ring. To construct a torus, you specify the pitch circle and the cylinder's radius or diameter. You can construct a torus in many ways.

Radius and Diameter of Torus's Pitch Circle and Vertical Torus

The implicit circle on which the cylinder is bent is called the pitch circle. By default, the pitch circle is specified by its radius, and the pitch circle is placed horizontally on the active construction plane. However, you can also specify pitch diameter and have the pitch circle placed vertically to the construction plane. Perform the following steps.

1. Select File > Open and select the file Torus.3dm from the Chapter 8 folder on the Student Companion Website.
2. Select Solid > Torus.
3. Select points A, B, and C, shown in Figure 8–41, where A is the center, AB is the radius of the pitch circle, and BC is the radius of the torus tube.
4. Repeat the command.
5. Select the Vertical option at the command area.
6. Select points D, E, and F, shown in Figure 8–41.
   A torus with its pitch circle vertical to the construction plane is constructed.
7. Repeat the command.
8. Select the 2Point option at the command area.
9. Select points G, H, and J, shown in Figure 8–41, where GH is the diameter of the pitch circle of the torus.

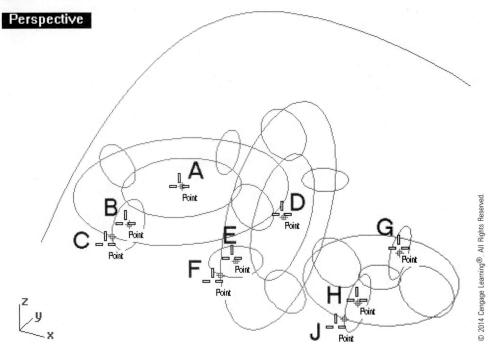

FIGURE 8–41 *From left to right: torus's pitch circle specified by radius, vertical torus, and torus's pitch circle specified by diameter*

Torus around a Curve

Continue with the following steps to construct a torus with its axis around a curve.

10. Turn on Layer 02 and turn off Layer 00.
11. Select Solid > Torus.
12. Select the AroundCurve option at the command area.
13. Select curve A, shown in Figure 8–42.
14. Select point B on curve A.
15. Type 9 at the command area to specify the pitch radius.
16. Type 3 at the command area to specify the radius of the torus tube.

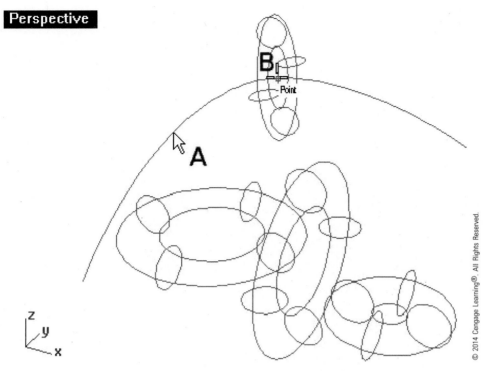

FIGURE 8–42 Torus with its pitch circle around a curve being constructed

Torus's Pitch Circle Tangent to Three Curves

Continue with the following steps to construct a torus with its base circle tangent to three curves.

17. Turn on Layer 03 and turn off Layer 02.
18. Select Solid > Torus.
19. Select the Tangent option at the command area.
20. Select curves A, B, and C, shown in Figure 8–43.
21. Type 3 at the command area to specify the second radius.

384 Inside Rhinoceros® 5

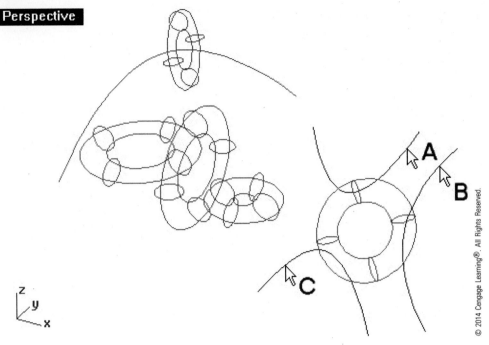

FIGURE 8–43 *Torus with its pitch circle tangent to three curves being constructed*

Torus's Pitch Circle Fitted to a Set of Points

Continue with the following steps to construct a torus with its base circle best fitted to a set of selected points.

    22. Turn on Layer 04 and turn off Layer 03.
    23. Select Solid > Torus.
    24. Select the FitPoints option at the command area.
    25. Click on A and drag to B (Figure 8–44) to select the point objects, and press the ENTER key.
    26. Type 3 at the command area to specify the radius at the other end.

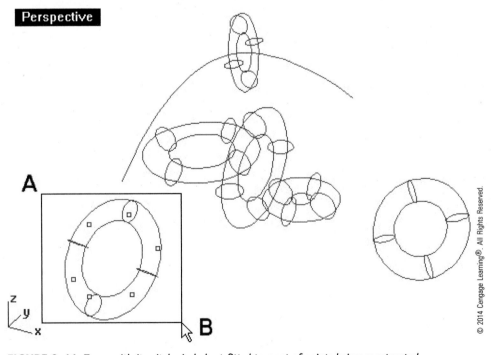

FIGURE 8–44 *Torus with its pitch circle best fitted to a set of points being constructed*

Torus's Pitch Circle Passing through Three Points

Continue with the following steps to construct a torus with its base circle defined by three selected points.

27. Turn on Layer 05 and turn off Layer 04.
28. Select Solid > Torus.
29. Select the 3Point option at the command area.
30. Select points A, B, and C, shown in Figure 8–45.
31. Type 3 at the command area.
32. Do not save your file.

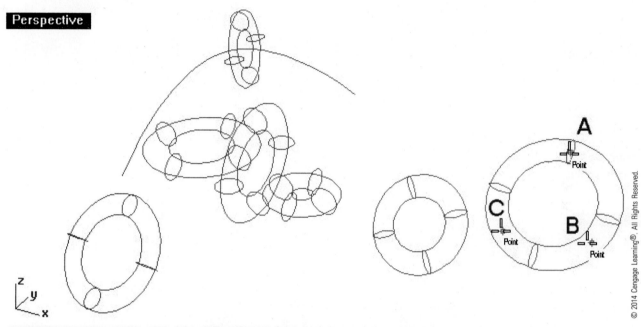

FIGURE 8–45 *Torus with its pitch circle defined by three selected points being constructed*

Solid Text Object

A text object is a polysurface consisting of three elements: top and bottom planar surfaces that resemble the text, and surfaces joining the top and bottom surfaces. To construct a text object, perform the following steps.

1. Start a new file. Use the Small Objects, Millimeters template.
2. Maximize the Perspective viewport.
3. Select Solid > Text.
4. In the Text Object dialog box, shown in Figure 8–46, type a text string, select a font and font style, specify text height and thickness, check the Solids box, and click on the OK button.
5. Click on a location in the Perspective viewport.

    A text object is constructed. (See Figure 8–47.)
6. Do not save your file.

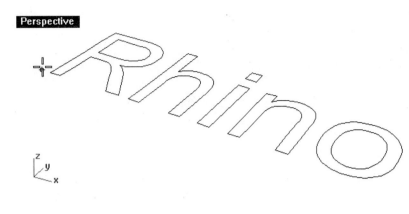

FIGURE 8-46 *Text Object dialog box*
Source: Robert McNeel and Associates Rhinoceros® 5

FIGURE 8-47 *Text object constructed*

CONSTRUCTING FREE-FORM SOLID OBJECTS

In terms of aesthetic design, free-form objects are more commonly used than objects of regular shapes. Naturally, it is important to know how to construct free-form polysurfaces and solids. The simplest way to construct a free-form solid is to extrude a closed planar curve. Another way to extrude a surface, in some other applications, is referred to as thickening. The third way to offset a polyline is to cap the ends and then extrude the curves. If you have a polysurface with planar openings, you can cap these planar holes. After capping, if the polysurface becomes a closed-loop polysurface, a solid is formed. If the solid object consists of free-form surfaces, you can construct the individual surfaces using methods explained in previous chapters, and then compose a solid from the surfaces.

Extrude Planar Curve

As explained in Chapter 3, there are a number of ways to extrude a curve. If the curve to be extruded is a closed planar curve, the same command can be used to obtain a solid, which is a closed polysurface consisting of a surface extruded from the planar curve and two planar surfaces enclosing the extruded surface. Perform the following steps to construct an extruded solid object.

1. Select File > Open and select the file ExtrudePlanarCurve.3dm from the Chapter 8 folder on the Student Companion Website.

2. Select Solid > Extrude Planar Curve > Straight.
3. Select curve A, shown in Figure 8–48, and press the ENTER key.
4. If Solid=No, select the Solid option at the command area to change it to Yes.
5. Type 10 at the command area to set the extrusion distance.

   An extruded solid is constructed.

6. Do not save your file.

   Note that this command is the same command used for constructing an extruded surface. Therefore, you can also indicate two points to specify an extrude direction, apply a taper angle in the extrusion process, extrude the curve in two directions, extrude the curve along a curve, extrude the curve along a subcurve, and extrude the curve to a point. (See Figure 8–49.) Remember to set the "Solid" option to "Yes."

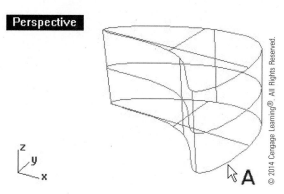

FIGURE 8–48 *Closed planar curve being extruded*

FIGURE 8–49 *From left to right: extrude in straight direction, extrude to a point, and extrude along a curve*

Extrude Free-Form Surface

By extruding a surface, you construct a solid, which is a polysurface consisting of two copies of the original surface a distance apart and a set of surfaces joining their edges. Perform the following steps.

1. Select File > Open and select the file ExtrudeSurface.3dm from the Chapter 8 folder on the Student Companion Website.
2. Select Solid > Extrude Surface > Straight.
3. Select surface A, shown in Figure 8–50, and press the ENTER key.
4. Type 20 at the command area to specify the extrusion distance.

   The surface is extruded to a solid.

5. Select Solid > Extrude Surface > To Point.
6. Select surface B, shown in Figure 8–50, and press the ENTER key.
7. Select point C (Figure 8–50).

   The surface is extruded to a point.

8. Select Solid > Extrude Surface > Along Curve.
9. Select surface D, shown in Figure 8–50, and press the ENTER key.
10. Select curve E (Figure 8–50).

    The surface is extruded along a curve.

11. Select Solid > Extrude Surface to Boundary > Straight.
12. Select surface F, shown in Figure 8–50, and press the ENTER key.
13. At the command area, select the ToBoundary option.
14. Select surface G (Figure 8–50).
15. Turn off the surface layer to appreciate the effect. (See Figure 8–51.)
16. Do not save your file.

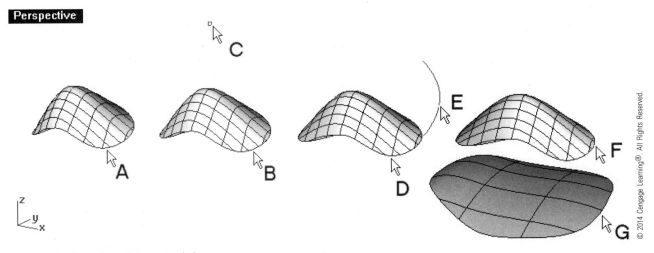

FIGURE 8–50 *Surfaces being extruded*

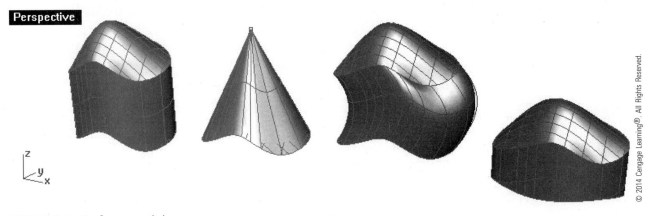

FIGURE 8–51 *Surfaces extruded*

Slab

The process of making a slab involves offsetting a polyline to produce a planar surface and then extruding the planar surface a specified distance. Perform the following steps.

1. Select File > Open and select the file Slab.3dm from the Chapter 8 folder on the Student Companion Website.
2. Select Solid > Slab.
3. Select polyline A, shown in Figure 8–52, and press the ENTER key.

4. Select the Distance option at the command area.
5. Type 5 to specify the offset distance.
6. Click on location B to indicate the offset direction.
7. Type 10 to specify the height of the slab.
8. Do not save your file.

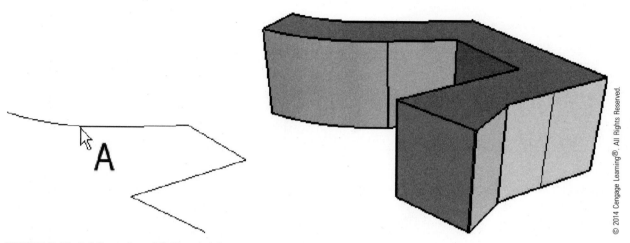

FIGURE 8–52 *Polyline selected (left) and slab constructed (right)*

Capping Planar Holes of a Polysurface

A planar hole is a hole with its boundary lying on a flat plane. Planar holes on a polysurface can be closed by capping, a process in which the planar openings are replaced by planar surfaces. If, after capping, the polysurface becomes a closed-loop object without openings or self-intersections, a solid is formed. Perform the following steps to cap a polysurface.

1. Select File > Open and select the file CapPlanarHole.3dm from the Chapter 8 folder on the Student Companion Website.
2. Select Solid > Cap Planar Holes.
3. Select surface A (shown in Figure 8–53) and press the ENTER key.
4. Do not save your file.

    The surface is capped. Note that capping works only on planar holes. For holes that are not planar, you can use the Patch command to construct patch surfaces instead.

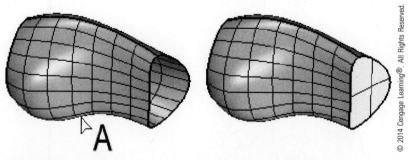

FIGURE 8–53 *Surface being capped (left) and capped surface (right)*

Solid from Closed-Loop Polysurface

You can construct a solid from a set of contiguous surfaces with common edges. The surfaces must not have any openings, gaps, or intersections except at their common edges and vertices. Perform the following steps.

1. Select File > Open and select the file CreateSolid.3dm from the Chapter 8 folder on the Student Companion Website.
2. Select Solid > Create Solid.

3. Click on A and drag to B, shown in Figure 8–54, and press the ENTER key.

   The surfaces are converted to a solid.

4. Do not save your file.

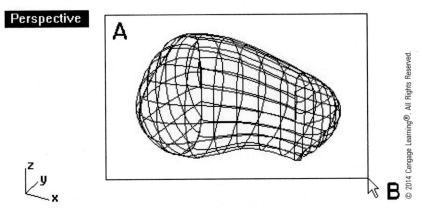

FIGURE 8–54 *Solid being constructed from a set of surfaces*

Joining and Exploding

Another way to form a solid is to join a set of contiguous surfaces that enclose a volume without any openings. You can examine an object's properties via the Detail button of the Object tab of the Properties dialog box. Naturally, you have to use the Properties command. To separate a polysurface or a solid into a set of individual surfaces, you explode it. Because a Rhino solid is a polysurface, you can use the same Explode command to separate it into a set of surfaces.

COMBINING RHINO SOLIDS

To compose a complex solid object, you can first construct two or more solid objects and then combine them using one of the three types of Boolean operations. Booleans generally work better if two objects overlap each other slightly.

Union

A union of a set of solids produces a solid that has the volume of all solids in the set. To construct a solid by uniting two solids, perform the following steps.

1. Select File > Open and select the file Union.3dm from the Chapter 8 folder on the Student Companion Website.
2. Select Solid > Union.
3. Select solids A and B (shown in Figure 8–55) and press the ENTER key.

   The solids are united.

4. Do not save your file.

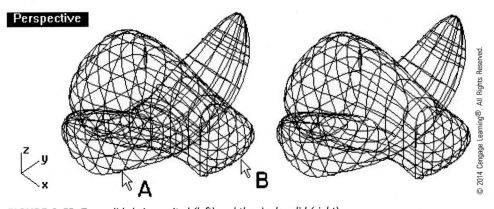

FIGURE 8–55 *Two solids being united (left) and the single solid (right)*

Difference

A difference of two sets of solids produces a solid that has the volume contained in the first set of solids but not in the second set of solids. To construct a solid by subtracting one solid from another solid, continue with the following steps.

1. Select File > Open and select the file Subtract.3dm from the Chapter 8 folder on the Student Companion Website.
2. Select Solid > Difference.
3. Select solid A (shown in Figure 8–56) and press the ENTER key.
   This is the first set of solids.
4. Click on B and drag to C (shown in Figure 8–56) and press the ENTER key.
   This is the second set of solids, and the second set of solids is subtracted from the first set of solids.
5. Do not save your file.

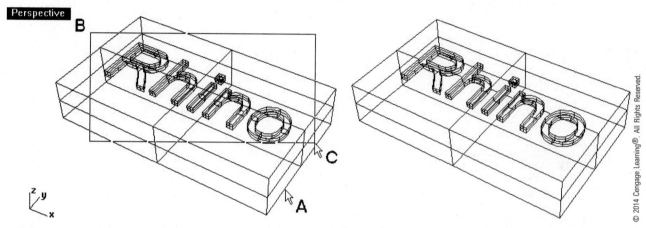

FIGURE 8–56 *One solid being subtracted from another (left) and the single solid (right)*

Intersection

An intersection of two sets of solids produces a solid that has the volume contained in both the first and second sets of solids. To construct a solid that has the volume contained in two solids, continue with the following steps.

1. Select File > Open and select the file Intersect.3dm from the Chapter 8 folder on the Student Companion Website.
2. Select Solid > Intersection.
3. Select solid A (shown in Figure 8–57) and press the ENTER key.
   This is the first set of solids.
4. Select solid B (shown in Figure 8–57) and press the ENTER key.
   This is the second set of solids. A solid of intersection is constructed.
5. Do not save your file.

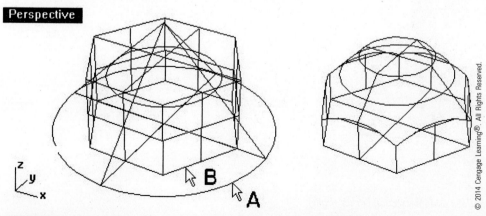

FIGURE 8–57 *Two solids being intersected (left) and the single solid (right)*

Boolean2Objects

If you are not quite sure which Boolean operation should be performed (union, difference, or intersection) to obtain a specific outcome, you can use the Boolean2Objects command, which enables you to cycle through the various possibilities in combining two solids. Perform the following steps.

1. Select File > Open and select the file Boolean2Objects.3dm from the Chapter 8 folder on the Student Companion Website.
2. Select Solid > Boolean Two Objects.
3. Select objects A and B, shown in Figure 8–58.
4. Click to cycle through various possibilities.
5. Press the ENTER key to accept.
   The objects are combined, using one of the Boolean operations chosen.
6. Do not save your file.

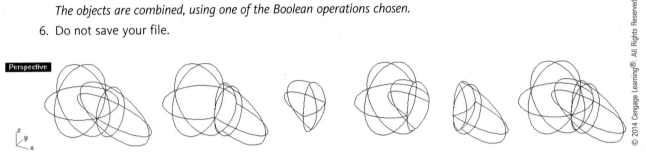

FIGURE 8–58 *From left to right: two objects being booleaned and five possibilities (join, intersect, subtract in one way, subtract in another way, and split)*

Intersect Two Sets

If you simply want to find the intersection of one set of objects with another set of objects instead of making a Boolean intersection, perform the following steps.

1. Select File > Open and select the file IntersectTwoSets.3dm from the Chapter 8 folder on the Student Companion Website.
2. Select Curve > Curve From Objects > Intersection of Two Sets.
3. Select A and B. indicated in Figure 8–59 and press the ENTER key.
4. Select C and D. indicated in Figure 8–59 and press the ENTER key.
5. Turn off the layer Solids to discover the result.
6. Do not save your file.

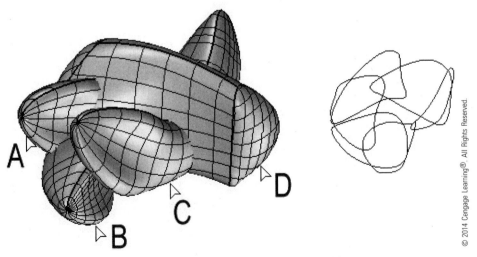

FIGURE 8–59 *Solids being selected (left) and intersection curves (right)*

Separating into Regions

If you have a non-manifold object and want to separate it into regions, perform the following steps.

1. Select File > Open and select the file IntersectTwoSets.3dm from the Chapter 8 folder on the Student Companion Website.
2. Type CreateRegions at the command area.
3. Select A, indicated in Figure 8–60.

   The non-manifold object is separated into two.
4. To discover the separation, click on A and drag it to a new location.
5. Do not save your file.

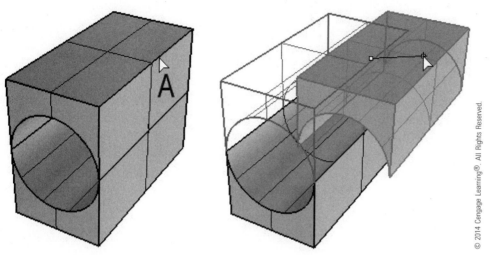

FIGURE 8–60 *Non-manifold object being separated into regions (left) and one of the separated regions being dragged apart (right)*

DETAILING A SOLID

To detail a Rhino solid, you can make it hollow; you can treat the edges by constructing constant/variable chamfers, fillets, and blends; you can construct bosses or ribs on selected faces; and you can make holes in various ways.

Constructing a Hollow Object

Hollowing a solid is a process of constructing a set of offset surfaces of each individual elements of the polysurface depicting a solid. Because whether offsetting is successful or not depends on the direction of offsetting and the amount of offset in such a way that the radius of curvature of the resulting surfaces must not be zero or negative, hollowing may not be always successful. Perform the following steps.

1. Select File > Open and select the file Hollow.3dm from the Chapter 8 folder on the Student Companion Website.
2. Type Shell at the command area or click on the Shell Polysurface button on the New in V5 toolbar.
3. Click on the Thickness option at the command area and set thickness 4.
4. Select face A, indicated in Figure 8–61, and press the ENTER key.

   The solid is made hollow, with the selected face removed.
5. Do not save your file.

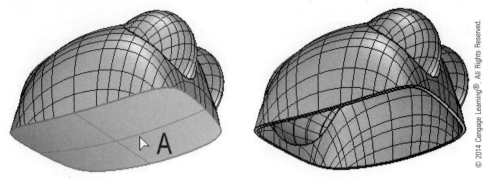

FIGURE 8-61 *Solid and one of its faces selected (left) and solid made hollow (right)*

Edge Treatment

There are three ways to treat the edges of a polysurface or a solid. You can construct a variable distance chamfer, a constant/variable radius fillet, or a constant/variable radius blend.

Variable Chamfer Edge

You can construct a variable distance chamfer surface on the edges of a polysurface or solid, as follows.

1. Select File > Open and select the file VariableChamfer.3dm from the Chapter 8 folder on the Student Companion Website.
2. Select Solid > Fillet Edge > Chamfer Edge.
3. Type 6 to set the next chamfer distance.
   The value affects edges selected subsequently.
4. Select edges A, B, C, D, and E, shown in Figure 8–62, and press the ENTER key.

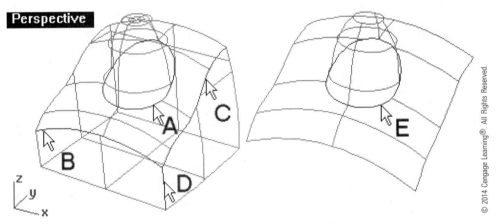

FIGURE 8-62 *Edges of a solid (left) and edge of a polysurface (right) selected*

5. Select the AddHandle option at the command area.
6. Click on location A (Figure 8–63) and press the ENTER key.
   Exact location is unimportant. A new handle is added.
7. Click on the added handle B (Figure 8–63).
8. Type 10 at the command area to set the chamfer distance at the added handle.
9. Select the RailType at the command area.
10. From the three options, select DistFromEdge.
11. Select the Preview option to display a preview.
12. Press the ENTER key.
    Variable chamfer edges are constructed. (See Figure 8–64.)
13. Do not save your file.

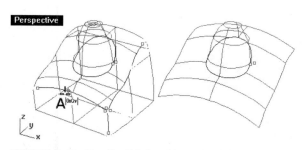
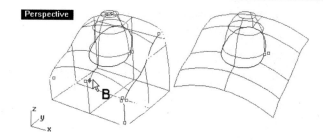

FIGURE 8–63 *Handle added and being modified*
© 2014 Cengage Learning®. All Rights Reserved.

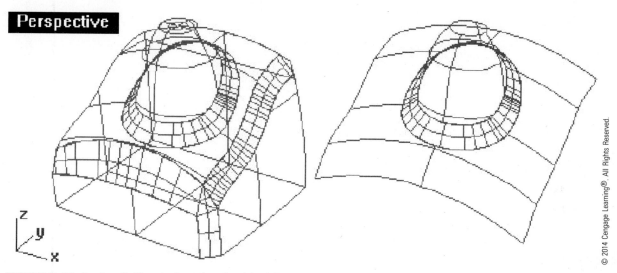

FIGURE 8–64 *Preview (left) and edges chamfered (right)*

Variable Fillet Edge

You can construct a variable radius fillet surface on the edges of a polysurface or solid, as follows.

1. Select File > Open and select the file VariableFillet.3dm from the Chapter 8 folder on the Student Companion Website.
2. Select Solid > Fillet Edge > Fillet Edge.
3. Select the Current Radius option at the command area.
4. Type 10 at the command area to set default value.
5. Select edges A, B, C, D, and E, shown in Figure 8–65, and press the ENTER key.

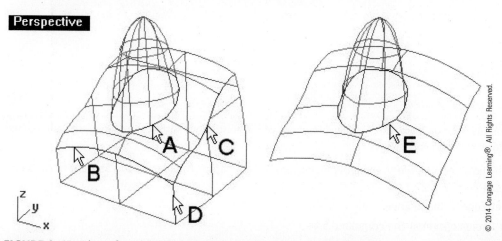

FIGURE 8–65 *Edges of a solid (left) and edge of a polysurface (right) selected*

6. Select the AddHandle option at the command area.
7. Click on edge A, shown in Figure 8–66, and press the ENTER key to add a handle.
8. Click on the added handle B, shown in Figure 8–66.
9. Type 6 at the command area to set the handle's value.
10. Select the RailType option at the command area.
11. Select the DistBetweenRails option.
12. Select the Preview option.

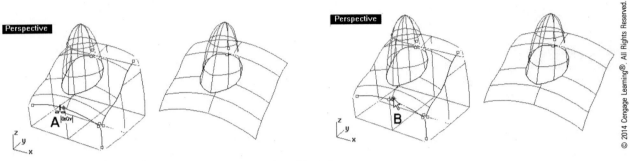

FIGURE 8–66 *A handle being added (left) and handle selected (right)*

13. If you are satisfied with the preview's edges, press the ENTER key.
    Edges are filleted. (See Figure 8–67.)
14. Do not save your file.

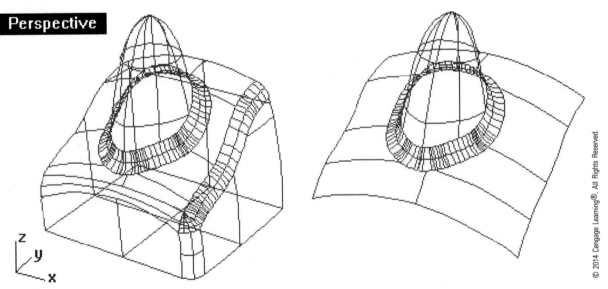

FIGURE 8–67 *Preview (left) and final filleted edges (right)*

Variable Blend Edge

A blended edge has G2 continuity. Not only are the tangency directions of the contiguous surfaces the same, the radii of curvature are also identical. Perform the following steps.

1. Select File > Open and select the file VariableBlend.3dm from the Chapter 8 folder on the Student Companion Website.
2. Select Solid > Fillet Edge > Blend Edge.
3. Select the Current Radius option at the command area.

4. Type 6 at the command area.
5. Select edges A, B, C, D, and E, shown in Figure 8–68, and press the ENTER key.

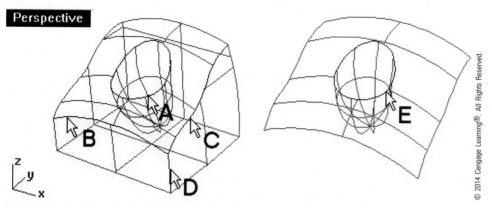

FIGURE 8–68 *Edges of a solid (left) and edge of a polysurface (right) selected*

6. Select the AddHandle option at the command area.
7. Click on edge A, shown in Figure 8–69, and press the ENTER key to add a handle.
8. Click on the added handle B, shown in Figure 8–69.
9. Type 10 at the command area to set the handle's value.
10. Select the RailType option at the command area.
11. Select the Rolling Ball option.
12. Select the Preview option.
13. If you are satisfied with the preview's edges, press the ENTER key.
    Edges are filleted. (See Figure 8–70.)
14. Do not save your file.

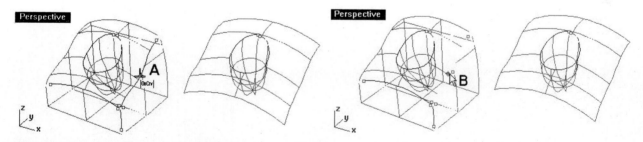

FIGURE 8–69 *A handle being added (left) and handle selected (right)*
© 2014 Cengage Learning®. All Rights Reserved.

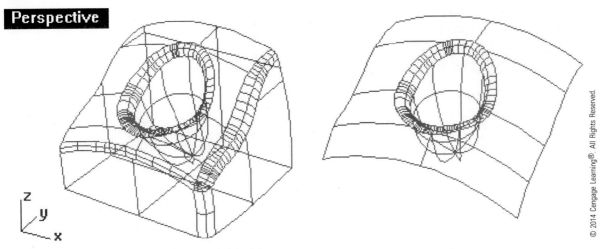

FIGURE 8–70 *Preview (left) and edges blended (right)*

Boss

A boss is constructed by extruding a planar closed curve in 3D space to a boundary solid object. Optionally, you may include a draft angle. The boss constructed will be united with the boundary solid. Perform the following steps.

1. Select File > Open and select the file Boss.3dm from the Chapter 8 folder on the Student Companion Website.
2. Select Solid > Boss.
3. Select the Mode option at the command area.
4. Select the Draft Angle option at the command area.
5. Type 3 at the command area to specify the draft angle.
6. Select closed planar curve A, shown in Figure 8–71, and press the ENTER key.
7. Select solid B, shown in Figure 8–71.

   A boss is constructed.

8. Do not save your file.

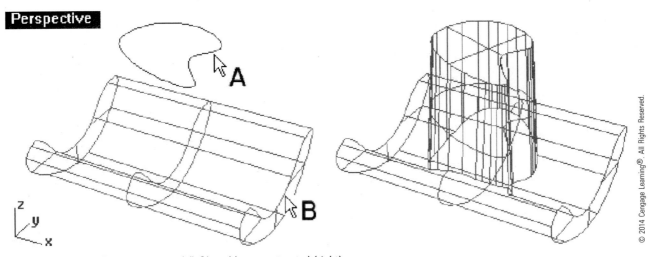

FIGURE 8–71 *Boss being constructed (left) and boss constructed (right)*

Rib

A rib is constructed by offsetting a planar curve (closed or open) and then extruding it toward a boundary solid. The shape of a rib is quite similar to a slab solid. While constructing the rib, you may include a draft angle. Perform the following steps.

1. Select File > Open and select the file Rib.3dm from the Chapter 8 folder on the Student Companion Website.
2. Select Solid > Rib.
3. Select the Distance option at the command area.
4. Type 3 at the command area to specify the offset distance.
5. Select the Mode option at the command area.
6. Select the DraftAngle option at the command area.
7. Type 1 at the command area to specify the draft angle.
8. Select curve A, shown in Figure 8–72, and press the ENTER key.
9. Select solid B, shown in Figure 8–72.

   A rib with a draft angle is constructed.
10. Do not save your file.

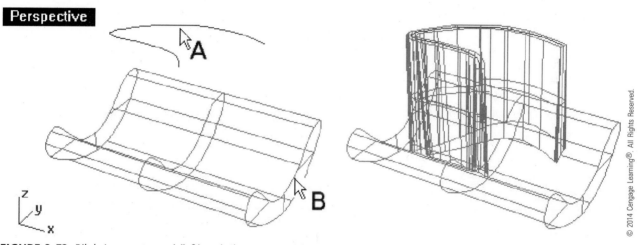

FIGURE 8–72 *Rib being constructed (left) and rib constructed (right)*

Hole

There are two general kinds of holes: round holes and holes of irregular shape. To construct a hole of irregular shape, you construct a planar closed curve to define the cross-section and then extrude it toward a solid or polysurface. Perform the following steps to construct a hole of irregular shape.

1. Select File > Open and select the file Hole.3dm from the Chapter 8 folder on the Student Companion Website.
2. Select Solid > Solid Edit Tools > Holes > Make Hole.
3. Select curve A, shown in Figure 8–73, and press the ENTER key.
4. Select closed polysurface B, shown in Figure 8–73.
5. Type 12 at the command area to indicate the depth of cut.
6. Click on location C, shown in Figure 8–73.

   A hole is cut.
7. Do not save your file.

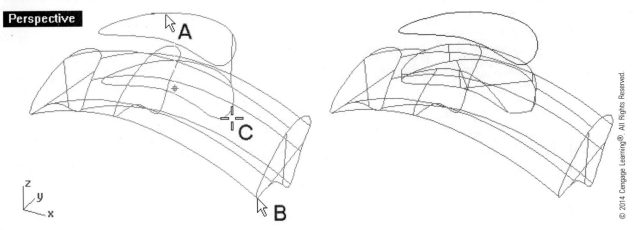

FIGURE 8–73 *Closed planar curve being used to cut a solid (left) and hole cut (right)*

Placing Multiple Holes

If a number of holes are required to be made on a solid, you can first construct a planar curve to depict the shape of the hole, and then construct multiple copies of the hole on a selected surface of a solid or polysurface. Perform the following steps.

1. Select File > Open and select the file PlaceHole.3dm from the Chapter 8 folder on the Student Companion Website.
2. Select Solid > Solid Edit Tools > Holes > Place Hole.
3. Select planar curve A, shown in Figure 8–74.
4. Click on a point along the curve to specify the base point.
   Exact location is unimportant for this tutorial.
5. Select target surface B.
6. Select target point C.
7. Type 4 at the command area to specify the depth.
8. Press the ENTER key to accept the default rotation angle.
9. Select points D, E, and F, and press the ENTER key.
   Four holes are placed.
10. Do not save your file.

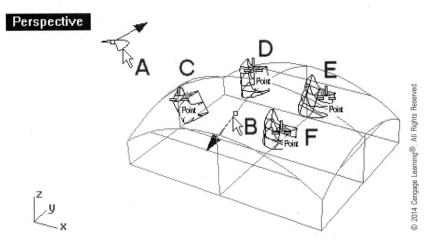

FIGURE 8–74 *Multiple copies of a hole being placed on a surface of a solid*

Constructing Round Holes

If the hole to be constructed is a round hole, you do not have to construct a circle to depict the hole's cross-section. You simply specify a radius and location on a surface of a solid or polysurface. Perform the following steps.

1. Select File > Open and select the file RoundHole.3dm from the Chapter 8 folder on the Student Companion Website.
2. Select Solid > Solid Edit Tools > Holes > Round Hole.
3. Select surface A, shown in Figure 8–75.
4. Click on the Direction option at the command area.
5. Select the SrfNormal at the command area.
6. Select the Depth option at the command area.
7. Type 5 to specify the depth of the hole.
8. Select the Radius option at the command area.
9. Type 3 to specify the radius.
10. Select points B, C, and D, and press the ENTER key.
    Three holes are placed.
11. Do not save your file.

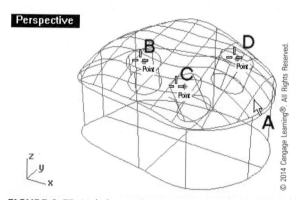

FIGURE 8–75 *Hole being placed on a surface of a solid*

Constructing Revolved Hole

Apart from constructing round holes and holes of irregular shape, you can construct a hole by revolving a curve. The axis of revolution of the revolved hole is determined by an imaginary line joining two endpoints of the curve. Perform the following steps.

1. Select File > Open and select the file RevolveHole.3dm from the Chapter 8 folder on the Student Companion Website.
2. Select Solid > Solid Edit Tools > Holes > Revolved Hole.
3. Select curve A, shown in Figure 8–76.
4. Select endpoint B as the base point.
5. Select target surface C.
6. Select point D and press the ENTER key.
   A revolved hole is constructed. Note that you can place multiple copies.
7. Repeat the command, using curve E as the profile curve, endpoint F as the base point, and point G of surface C as the target.
8. Do not save your file.

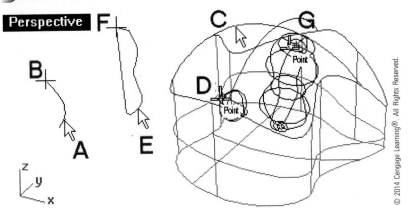

FIGURE 8–76 Revolved holes being placed

Arraying Holes on a Planar Face

Holes already constructed on a planar face of a solid or polysurface can be repeated by arraying in a rectangular pattern or a polar pattern. Perform the following steps.

1. Select File > Open and select the file ArrayHole.3dm from the Chapter 8 folder on the Student Companion Website.
2. Select Solid > Solid Edit Tools > Holes > Array Hole.
3. Select edge A, shown in Figure 8–77.
4. Type 4 at the command area to specify the number of holes in direction A. "A" is the first direction.
5. Type 3 at the command area to specify the number of holes in direction B. "B" is the second direction.
6. If the Rectangular option shown at the command area is Yes, click on it to change it to No.
7. Click on endpoints B and C to specify direction A.
8. Click on endpoints B and D to specify direction B.
9. Select the ASpacing option at the command area.
10. Type 5 to specify the spacing along direction A.
11. Select the BSpacing option at the command area.
12. Type 4 to specify the spacing along direction B.
13. Press the ENTER key.

    A rectangular pattern of holes is constructed.

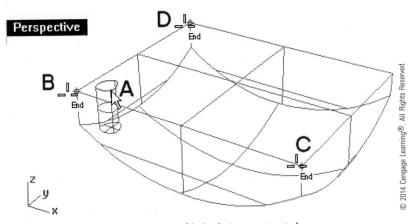

FIGURE 8–77 Rectangular pattern of holes being constructed

14. Select Solid > Solid Edit Tools > Holes > Array Hole Polar.
15. Select edge A, shown in Figure 8–78.

16. Select endpoint B as the center of array.
17. Type 4 to specify the number of holes.
18. Type -35 to specify the included angle.
19. Press the ENTER key.

    A polar pattern of holes is constructed.

20. Do not save your file.

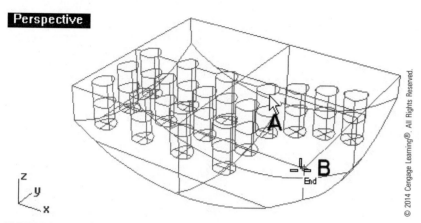

FIGURE 8–78 *Polar pattern of holes constructed*

When arraying, the following points have to be noted: Because this command simply repeats a hole that is already constructed, a rectangular or polar pattern of holes in a solid with curved surface on the opposite side of the hole may not result with a set of through holes. To avoid an unexpected outcome, do not have the repeated copies of holes overlap each other.

Mirror Holes on a Planar Surface

To construct mirror image of one or more holes lying on a planar face of a polysurface, perform the following steps.

1. Select File > Open and select the file MirrorHole.3dm from the Chapter 8 folder on the Student Companion Website.
2. Select Solid > Solid Edit Tools > Holes > Mirror Hole.
3. Select edge A, shown in Figure 8–79, and press the ENTER key.
4. Select Midpoint B and Midpoint C.

    A rectangular pattern of holes is constructed.

5. Do not save your file.

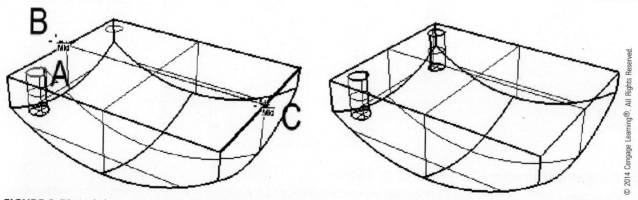

FIGURE 8–79 *Hole being mirrored (left) and hole mirrored (right)*

EDITING SOLIDS

There are several ways to edit solids. You can merge coplanar faces resulting from Boolean operations. You can split a solid by using a surface or a curve. To change the shape, you can manipulate its faces or its edges. If a surface is needed from a solid or polysurface, you can extract it. However, extracting a surface from a solid separates the surface from the solid or polysurface, leaving an opening.

Solid Cleaning Up: Merging Coplanar Faces

Solids resulting from the application of Boolean operations may have overlapping coplanar faces. You can clean up the solid by merging coplanar faces, as follows.

1. Select File > Open and select the file MergeFaces.3dm from the Chapter 8 folder on the Student Companion Website.
2. Select Solid > Union.
3. Select solids A and B (shown in Figure 8–80) and press the ENTER key. The solids are united. Note that overlapped faces resulted from the Boolean operation.
4. Select Solid > Solid Edit Tools > Faces > Merge face.
5. Select faces C and D, shown in Figure 8–80.
6. Repeat the command.
7. Select faces C and E, shown in Figure 8–80.
8. Right-click on the Merge two coplanar faces/Merge all coplanar faces button on the Solid Tools toolbar.
9. Select solid F, shown in Figure 8–80.

    All coplanar faces are merged. (See Figure 8–81.)

10. Do not save your file.

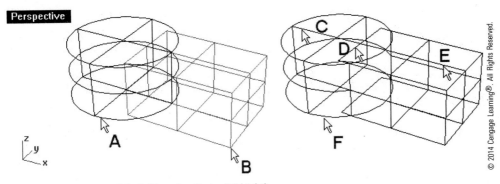

FIGURE 8–80 *Two solids (left) and united solid (right)*

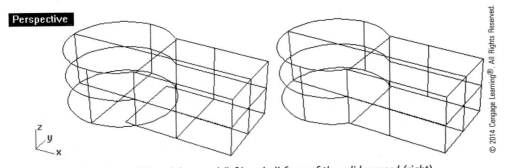

FIGURE 8–81 *Top faces of the solid merged (left) and all faces of the solid merged (right)*

Splitting and Trimming a Polysurface or a Solid

To split or trim a polysurface or a solid, you can use a curve, a surface, or a polysurface as cutting objects. If the cutting object is a curve and is residing on the surface or polysurface, the curve will be pulled back to cut. If the cutting curve is not on the surface or polysurface, the curve will be extruded through the surface or polysurface in a direction perpendicular to the active construction plane. Like splitting or trimming a surface, the cutting object, if it is an open-looped surface or polysurface, has to pass through the entire polysurface for the trimming action to carry out. However, splitting or trimming a closed-loop surface or polysurface produces an open-loop surface or polysurface, which is no longer a solid.

Boolean Splitting and Splitting

To split a solid into two solids, you use the Boolean split operation. To compare splitting a solid by using the BooleanSplit command on the Solid Tools toolbar with the Split command on the Main2 toolbar, perform the following steps.

1. Select File > Open and select the file Split.3dm from the Chapter 8 folder on the Student Companion Website.
2. Select Solid > Boolean Split.
3. Select solid A, shown in Figure 8–82, and press the ENTER key.
4. Select surface B, shown in Figure 8–82, and press the ENTER key.
5. Select Edit > Split.
6. Select solid C, shown in Figure 8–82, and press the ENTER key.
7. Select surface D, shown in Figure 8–82, and press the ENTER key.
8. Select Edit > Visibility > Hide.
9. Select A, B, C, and D, shown in Figure 8–82, and press the ENTER key.
10. Shade the display. (See Figure 8–83.)
11. Do not save your file.

The split command leaves the split polysurface open, but the Boolean Split command uses the splitting object to patch the opening.

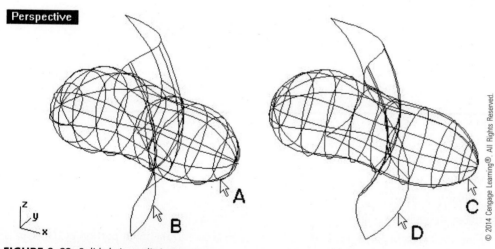

FIGURE 8–82 *Solids being split in two ways*

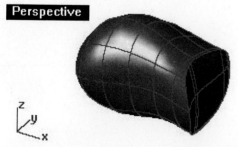

FIGURE 8–83 *Objects removed and viewport shaded*
© 2014 Cengage Learning®. All Rights Reserved.

Trimming and Wire-Cutting a Solid

A curve can be used to cut or trim a solid in two ways. Using the trim command on the Edit menu, the solid will become an open-loop polysurface after trimming. By wire-cutting, the resulting solid remains a closed-loop polysurface. Perform the following steps.

1. Select File > Open and select the file WireCut.3dm from the Chapter 8 folder on the Student Companion Website.
2. Select Edit > Trim.
3. Select curves A and B, shown in Figure 8–84, and press the ENTER key.
4. Select C and D, shown in Figure 8–84, and press the ENTER key.

   The solids are trimmed, and they become open-loop surfaces/polysurfaces.

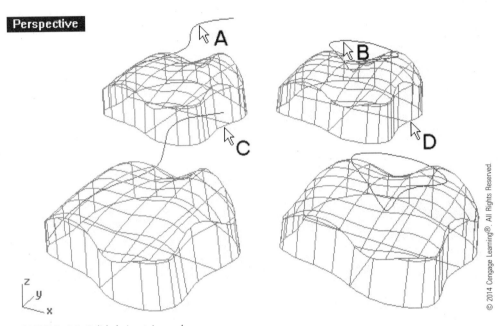

FIGURE 8–84 *Solids being trimmed*

5. Select Solid > Solid Edit Tools > Wire Cut.
6. Select curve A, shown in Figure 8–85.
7. Select solid B, shown in Figure 8–85.
8. Press the ENTER key to cut through the solid.
9. If the outer portion of the solid is highlighted, press the ENTER key to accept. If the inner portion of the solid is highlighted, select the Invert option at the command area and then press the ENTER key. The outer portion is wire-cut away.
10. Repeat the command.
11. Select curve C, shown in Figure 8–85.
12. Select solid D, shown in Figure 8–85, and press the ENTER key.
13. Press the ENTER key to cut through the solid.
14. Select the Direction option at the command area.
15. Select the X option at the command area.
16. Type 30 at the command area to specify the cut width in X direction.
17. Press the ENTER key to accept.

18. If the right-side portion of the solid is highlighted, press the ENTER key to accept. If the left-side portion of the solid is highlighted, select the Invert option at the command area and then press the ENTER key. The solid is wire-cut. (See Figure 8–86.)
19. Do not save your file.

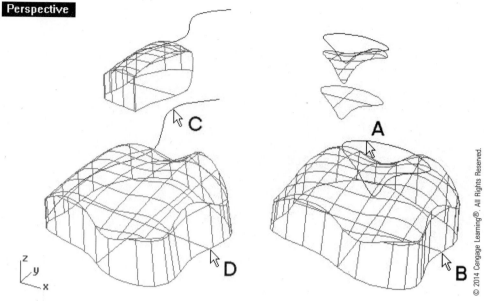

FIGURE 8–85 *Solids being wire-cut*

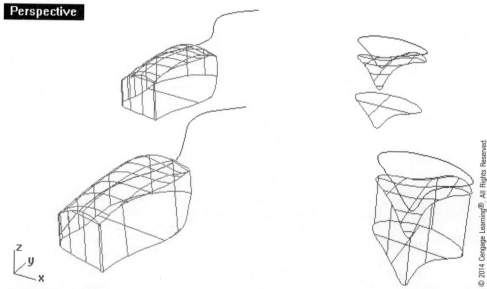

FIGURE 8–86 *Solids wire-cut*

Unjoining Edges

A way to separate a polysurface into two or more polysurfaces without exploding it is to unjoin its selected edges. Perform the following steps.

1. Select File > Open and select the file Unjoinedge.3dm from the Chapter 8 folder on the Student Companion Website.
2. Type Unjoinedge at the command area or click on Unjoin Edge on the New in V5 toolbar.

3. Select edges A and B, shown in Figure 8–87, and press the ENTER key.

   The polysurface is now separated into two along the selected edges.

4. Select C and drag it to a new position.
5. Do not save your file.

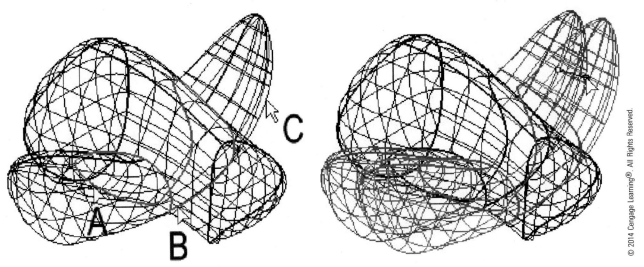

FIGURE 8–87 *Edges on a polysurface being selected (left) and separated polysurface dragged apart*

Manipulating Faces of a Solid

A solid's shape can be modified by moving, rotating, shearing, or extruding its faces. Because faces adjacent to the selected faces have to change in shape to respond, these operations may not always be successful if one of the adjacent faces cannot be changed.

Face Moving

A way to modify the shape of a solid is to move its face(s) to a new location. When a face or a number of faces are moved, its adjacent faces have to be stretched. If one of the adjacent faces cannot be stretched, the moving operation will be unsuccessful. Perform the following steps.

1. Select File > Open and select the file MoveFace.3dm from the Chapter 8 folder on the Student Companion Website.
2. Select Solid > Solid Edit Tools > Faces > Move Face.
3. Select face A, shown in Figure 8–88, and press the ENTER key.
4. Click on the Ortho button on the Status bar to turn on ortho mode.
5. Click on locations B and C. For the purpose of this tutorial, exact location is unimportant.

   The selected face is moved.

6. Do not save your file.

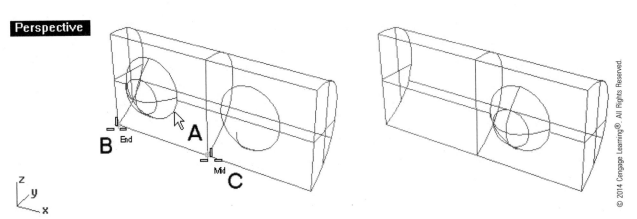

FIGURE 8–88 *Face selected (left) and face moved (right)*

Face Moving to a Boundary

A face can be moved to a boundary, thus changing the shape of the solid. Perform the following steps to understand this process.

1. Select File > Open and select the file MovetoBoundary.3dm from the Chapter 8 folder on the Student Companion Website.
2. Select Solid > Solid Edit Tools > Faces > Move Face to Boundary.
3. Select face A (Figure 8–89) and press the ENTER key.
4. If DeleteBoundary=No, click on it to change it to Yes.
5. Select surface B.
6. Turn off the Boundary layer to see the effect.
7. Do not save your file.

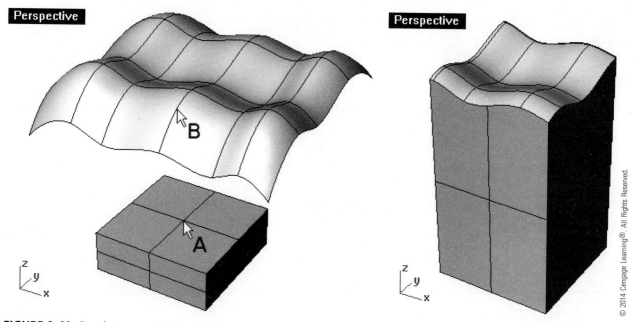

FIGURE 8–89 *Face being moved (left) and face moved to boundary surface (right)*

Face Rotation

Rotating selected faces of a solid around an axis is a way to modify a solid, as follows.

1. Select File > Open and select the file RotateFace.3dm from the Chapter 8 folder on the Student Companion Website.
2. Type RotateFace at the command area.
3. Select face A, shown in Figure 8–90, and press the ENTER key.
4. Select the 3DAxis option at the command area.
5. Select endpoints B and C.
6. Type 15 at the command area.
   The selected face is rotated.
7. Do not save your file.

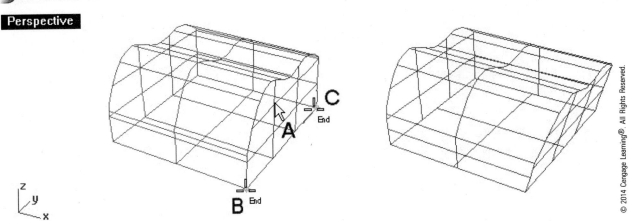

FIGURE 8–90 *Face being selected (left) and face rotated (right)*

Face Shearing

Some surfaces can be deformed by shifting at a specific angle through shearing. Because shearing applies only locally to selected surfaces, the process may not always be successful if any edge is moved out of face tolerance. Perform the following steps.

1. Select File > Open and select the file ShearFace.3dm from the Chapter 8 folder on the Student Companion Website.
2. Type ShearFace at the command area.
3. Select face A, shown in Figure 8–91, and press the ENTER key.
4. Select the 3DAxis option at the command area.
5. Select endpoints B, C, and D.
6. Type –15 at the command area.
   The selected face is sheared for an angle of 15 degrees.
7. Do not save your file.

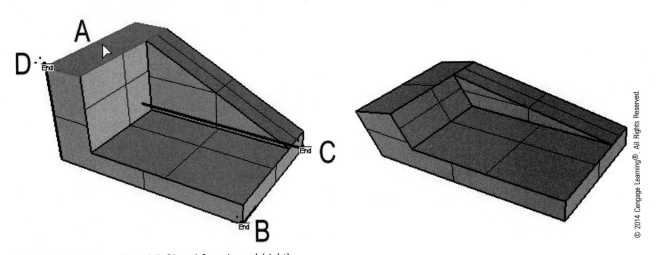

FIGURE 8–91 *Face selected (left) and face sheared (right)*

Face Extrusion

By extruding a face of a solid, the solid's shape is modified. Perform the following steps.

1. Select File > Open and select the file ExtrudeSide.3dm from the Chapter 8 folder on the Student Companion Website.

2. Select Solid > Extrude Surface > Straight.
3. Select face A, shown in Figure 8–92, and press the ENTER key.
4. Click on the Direction option at the command area.
5. Click on points A and B to specify a direction.
6. Click on point B to indicate the extrusion distance.
7. Do not save your file.

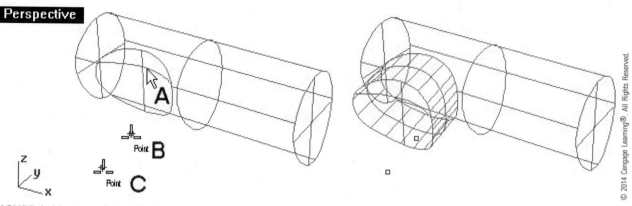

FIGURE 8–92 *Face selected (left) and face extruded (right)*

Face Extrusion along a Curve

This process is similar to extruding a curve or a surface along a curve. Here, selected surfaces are extruded along a curve. Perform the following steps.

1. Select File > Open and select the file ExtrudeFaceAlongCurve.3dm from the Chapter 8 folder on the Student Companion Website.
2. Select Solid > Extrude Surface > Along Curve.
3. Select face A, shown in Figure 8–93, and press the ENTER key.
4. Select curve B.
   The selected face is extruded along a path, and adjacent surfaces are modified.
5. Do not save your file.

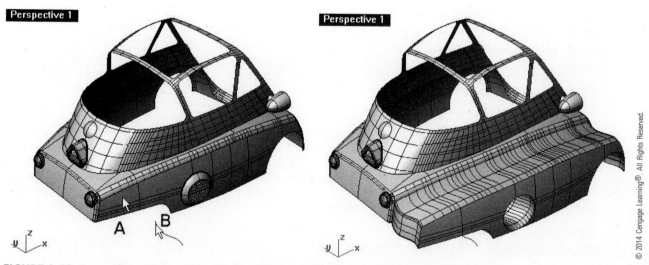

FIGURE 8–93 *Face and curve being selected (left) and face extruded along a curve*

Splitting a Face for Subsequent Operations

If you want to manipulate (move, rotate, or extrude) a portion of a surface instead of an entire face, you first have to split the face into two, as follows.

1. Select File > Open and select the file SplitFace.3dm from the Chapter 8 folder on the Student Companion Website.
2. Select Solid > Solid Edit Tools > Faces > Split Face.
3. Select face A, shown in Figure 8-94, and press the ENTER key.
4. Select the Curve option at the command area.
5. Select curve B and press the ENTER key.
   The curve is pulled to the face to split the selected face of the polysurface into two.
6. Select Solid > Extrude Surface > Straight.
7. Select face C, shown in Figure 8-94, and press the ENTER key.
8. Click on the Direction option at the command area and select points D and E to specify the direction.
9. Type 5 at the command area to specify the distance.
10. Do not save your file.

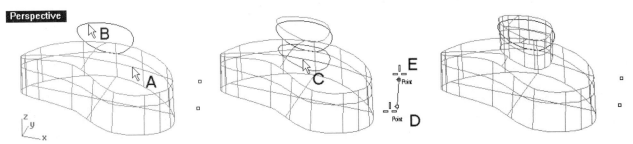

FIGURE 8-94 *From left to right: face being split, face split and being extruded, and split face extruded*
© 2014 Cengage Learning®. All Rights Reserved.

Folding Faces

An individual planar surface or planar face of a polysurface can be folded along a folding axis. If a planar surface is folded, the surface is simply split into two and rotated around the folding axis. If a face of a polysurface is folded, adjacent faces will be modified.

1. Select File > Open and select the file Fold.3dm from the Chapter 8 folder on the Student Companion Website.
2. Select Solid > Solid Edit Tools > Faces > Fold Face.
3. Select face A, shown in Figure 8-95.
4. Select points B and C.
5. Select the Symmetrical option at the command area.
6. Type 25 at the command area.
   The surface is folded.
7. Repeat the command.
8. Select face D, shown in Figure 8-95.
9. Select points E and F.
10. Select the Symmetrical option at the command area.
11. Type 25 at the command area.
    The surface is folded.
12. Do not save your file.

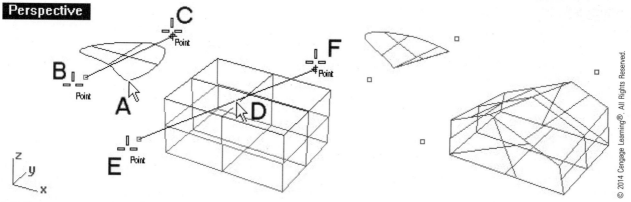

FIGURE 8–95 *Faces being folded (left) and faces folded (right)*

Rotating, Moving, and Scaling Edges of a Solid

A solid's shape can be modified by manipulating its edges. Naturally, faces adjacent to the manipulated edges have to change in shape. Therefore, edge manipulation may not always be successful if one of the affected faces cannot be changed.

Edge Rotation

Selected edges of a solid can be rotated around an axis to change the shape of a solid. Perform the following steps.

1. Select File > Open and select the file Edge.3dm from the Chapter 8 folder on the Student Companion Website.
2. Type RotateEdge at the command area.
3. Select edge A, shown in Figure 8–96, and press the ENTER key.
4. Select 3DAxis at the command area.
5. Select point B and then point C to specify the axis of rotation.
   The sequence of selection affects the direction of rotation.
6. Type 15 at the command area.
   The edge is rotated, and adjacent faces are modified.

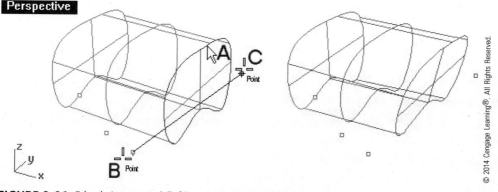

FIGURE 8–96 *Edge being rotated (left) and edge rotated (right)*

Edge Moving

If one or more edges are moved, the shape of the solid will be changed. Continue with the following steps.

7. Select Solid > Solid Edit Tools > Edges > Move Edge.
8. Select edge A, shown in Figure 8–97, and press the ENTER key.
9. Select point B and then point C.
   The edge is moved.

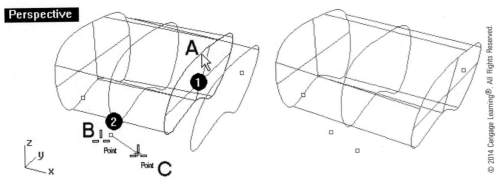

FIGURE 8-97 *Edge being moved (left) and edge moved (right)*

Edge Scaling

Scaling an edge changes the edge's length. As a result, the shape of the solid is changed. Continue with the following steps.

10. Type ScaleEdge at the command area.
11. Select edge A, shown in Figure 8-98, and press the ENTER key.
12. Select point B as the origin point.
13. Type 0.5 at the command area to specify the scale factor.
    The edge is scaled.
14. Do not save your file.

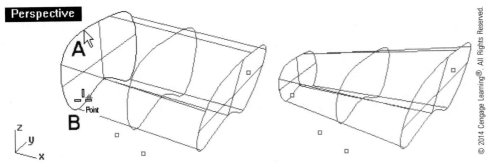

FIGURE 8-98 *Edge being scaled (left) and edge scaled (right)*

Moving Untrimmed Edges

An untrimmed edge of a surface or a polysurface can be moved, as follows.

1. Select File > Open and select the file MoveUntrimmedEdge.3dm from the Chapter 8 folder on the Student Companion Website.
2. Select Solid > Solid Edit Tools > Edges > Move Edge.
3. Select edge A, shown in Figure 8-99, and press the ENTER key.
   This is an untrimmed edge of a surface.
4. Select points B and C to define the distance and direction of move, respectively.
   The edge is moved.
5. Repeat the command.
6. Select edge D. This is the untrimmed edge of a polysurface.
7. Select points B and C.
   The edge is moved.
8. Do not save your file.

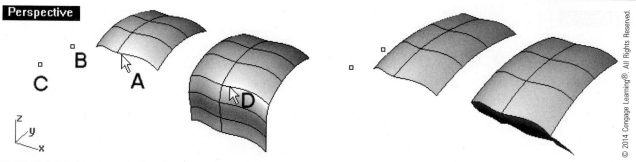

FIGURE 8–99 *Untrimmed edges being selected one by one (left) and edges moved (right)*

Rotating and Moving Holes

Holes residing on planar faces can be relocated by moving or rotation.

1. Select File > Open and select the file TranslateHole.3dm from the Chapter 8 folder on the Student Companion Website.
2. Select Solid > Solid Edit Tools > Holes > Rotate Hole.
3. Select edge A, shown in Figure 8–100.
4. Select point B.
5. Type –90 at the command area.

    The hole is rotated 90 degrees counter-clockwise.

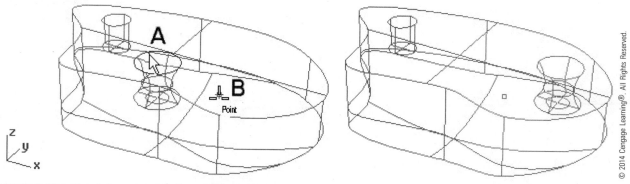

FIGURE 8–100 *Hole being rotated (left) and hole rotated (right)*

Continue with the following steps to move a hole.

6. Select Solid > Solid Edit Tools > Holes > Move Hole.
7. Select edge A, shown in Figure 8–101, and press the ENTER key.
8. Select point B.
9. Type r10<0 at the command area.

    The hole is moved a distance of 10 units in 0 degree direction.

10. Do not save your file.

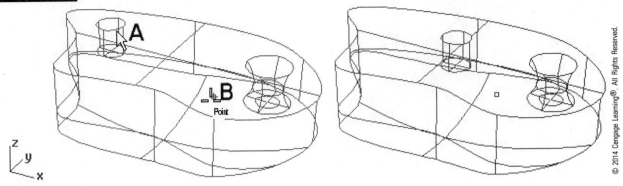

FIGURE 8–101 *Hole being moved (left) and hole moved (right)*

Copying Holes
Right-clicking the Move Hole/Copy Hole button on the Holes toolbar will copy a hole instead of moving it.

Deleting Holes
Unwanted holes in a solid can be deleted, as follows.

1. Select File > Open and select the file DeleteHole.3dm from the Chapter 8 folder on the Student Companion Website.
2. Select Solid > Solid Edit Tools > Holes > Delete Hole.
3. Select hole edges A, B, and C, shown in Figure 8–102.
   The holes are deleted.
4. Press the ENTER key to exit the command.
5. Do not save your file.

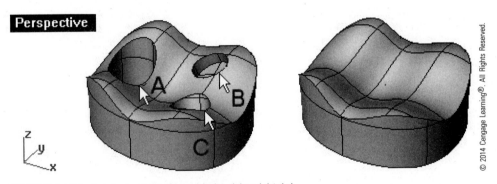

FIGURE 8–102 *Original solid (left) and holes deleted (right)*

Extracting a Surface from a Solid or Polysurface
Unlike exploding, which breaks down a joined polysurface into individual surfaces, extracting separates the selected surfaces from the polysurface, leaving the remaining surfaces joined. To extract a surface from a solid, perform the following steps.

1. Select File > Open and select the file ExtractSurface.3dm from the Chapter 8 folder on the Student Companion Website.
2. Select Solid > Extract Surface.
3. Select surface A (shown in Figure 8–103) and press the ENTER key.
   A surface is extracted.
4. Select Edit > Visibility > Hide.

5. Select the extracted surface, if it is not already selected.
   The extracted surface is hidden.
6. Do not save your file.

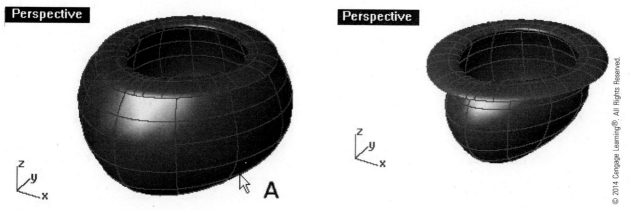

FIGURE 8–103 *Surface being extracted from a polysurface (left) and extracted surface hidden (right)*

LIGHTWEIGHT EXTRUSION OBJECTS

As mentioned in Chapter 3, Rhino 5 provides the option to use extrusion objects instead of ordinary surfaces in order to reduce computer resource requirement, and you can set the default extrusion as extrusion objects or polysurfaces.

The UseExtrusions command explained in Chapter 3 also affects the construction of box, cylinder, pipe, and slab.

CONSOLIDATION

A Rhino solid is a closed-loop NURBS surface or polysurface that encloses a volume with no gap or opening. In essence, you can use all the NURBS curves and surface tools delineated in the previous chapters to construct free-form NURBS surfaces and then use these surfaces to compose a solid. Because of its flexibility and ease of use, Rhino is particularly useful in making free-form solids.

The simplest way to construct a solid is to construct solids of regular pattern, including box, sphere, cylinder, cone, truncated cone, pyramid, ellipsoid, paraboloid, tube, pipe, torus, and text shape.

In addition to constructing solids of regular shapes, you can extrude a planar curve or a free-form surface to obtain an extruded solid, construct a slab solid from a curve, cap planar holes of a polysurface to convert it to a solid, and construct a solid from a set of contiguous surfaces that encloses a volume without any gap or opening.

To combine two or more solids to form a more complex solid, you use Boolean operations. To detail a solid, you can add variable distance chamfer, variable radius fillet, variable blend, boss, rib, and hole.

To clean up overlapped coplanar surfaces in a solid resulting from any Boolean operations, you merge them into a single face. If required, you can split a solid into two by using a surface or a curve. To modify a solid, you can manipulate its faces and edges, and move the location of the holes. Because a Rhino solid is a polysurface, you can extract a surface from a solid or a polysurface. However, extracting a surface reduces a solid to an open-loop polysurface or surface.

REVIEW QUESTIONS

1. Explain, in terms of Rhino definition, the difference between a polysurface and a solid.
2. List the kinds of regular shape solids available in Rhino.
3. How can solids of free-form shape be constructed in Rhino?
4. What are the ways to combine two or more solid objects into a single solid object?
5. In what ways can a solid be detailed?
6. Outline the ways to edit a Rhino solid.

CHAPTER 9

Polygon Meshes

INTRODUCTION

This chapter explains the ways to construct and manipulate polygon meshes that are used to approximate free-form objects.

OBJECTIVES

After studying this chapter, you should be able to

- Construct primitive meshed objects from scratch
- Derive polygon meshes from existing objects
- Combine and separate polygon meshes
- Manipulate mesh faces, edges, and vertices

OVERVIEW

As explained in Chapter 1, there are two ways to represent a surface in the computer: using NURBS surfaces to exactly represent surfaces or using polygon meshes to approximate surfaces.

Naturally, deployment of NURBS surfaces is the preferred method for designing and constructing 3D objects in the computer because it is the most accurate way of representing free-form objects. However, being an approximation of a smooth surface, polygon mesh is still used in many computer applications. In particular, rapid prototyping models in STL (Stereolithography) file format are a typical application of polygon meshes as a means of surface representation. In addition, when you import surfaces from other computerized applications, you may often import a set of polygon meshes. Therefore, you need to know how to deal with such meshes.

In this chapter, you will first learn how to construct mesh geometry of regular shapes. Then you will learn how to derive meshed objects from existing objects and how to combine meshed objects. Finally, and most important, you will learn how to edit meshed objects. Knowing how to edit meshed objects is particularly useful in repairing polygon meshes for rapid prototyping.

CONSTRUCTING POLYGON MESH PRIMITIVES

Mesh primitives are meshed objects of basic geometric shapes. Although it is not advisable to model in polygon mesh, learning how to construct these primitives lets you gain a basic understanding of meshed object construction from scratch.

In regard to basic geometric shapes, the simplest meshed objects are rectangular mesh planes and 3D mesh faces. In addition, there are seven types of mesh primitives: box, sphere, cylinder, cone, truncated cone, ellipsoid, and torus. The following sections take you through the construction process for each of these primitives.

Mesh Density Setting

Because a polygon mesh is a set of planar polygons that approximates a surface, the number of polygons used in the mesh has a direct impact on the accuracy of representation and the file size. A higher number of constituent polygons provide a more accurate model. However, file size increases quickly with a decrease in polygon size. (Smaller size means a greater number of polygons in the mesh.) Hence, you need to consider using the most appropriate mesh density to provide an optimal balance between the representational accuracy of the model and the size of the file.

Before constructing polygon meshes from scratch, you have to decide how dense you want the polygon mesh to be. After a polygon mesh is constructed, increasing the mesh density, if possible, does not increase the accuracy of the model. Conversely, you can reduce the mesh density after the model is constructed. However, reduction of mesh density simplifies the shape of the mesh.

Mesh 3D Face

A mesh 3D face has four corner points, which can be a planar quadrilateral or a 3D mesh face consisting of two triangular planes. The use of mesh 3D face is best applied to filling openings of existing polygon mesh, converting it to a closed-loop object for rapid prototyping. Because a 3D face is a polygon mesh of simplest structure, there is no mesh density to manipulate. Perform the following steps.

1. Select File > Open and select the file 3DFace.3dm from the Chapter 9 folder that you downloaded from the Student Companion Website.
2. Select Mesh > Polygon Mesh Primitives > 3-D Face.
3. Select points A, B, C, and D, shown in Figure 9–1, to define four vertices.

   A planar 3D face mesh is constructed.
4. Repeat the command.
5. Select points E, F, G, and H, shown in Figure 9–1.

   A 3D face mesh is constructed.
6. Do not save your file.

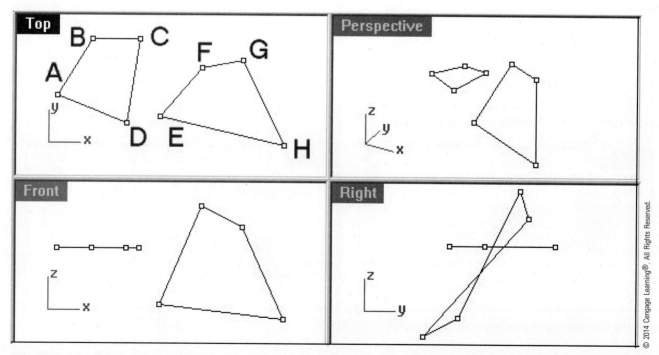

FIGURE 9–1 *3D faces being constructed*

Mesh Rectangular Plane

A mesh rectangular plane is a planar mesh in the shape of a rectangle. There are four ways to construct a mesh rectangular plane.

- Specifying the diagonal points
- Specifying two points to depict the base of the rectangle and a third point to depict the height of the rectangle
- Specifying two points to depict the base of the rectangle and a third point in the adjacent construction plane to construct a plane vertical to the active viewport
- Specifying the center and a corner of the rectangle

Mesh Density

Mesh density of a rectangular planar mesh is defined by two options: XFaces and YFaces. They determine the number of polygon mesh in the X and Y directions. To manipulate these options, select them from the command area while running the command. Perform the following steps to construct rectangular mesh planes in various ways.

1. Select File > Open and select the file PlaneMesh.3dm from the Chapter 9 folder on the Student Companion Website.
2. Select Mesh > Polygon Mesh Primitives > Plane.
3. Select points A and B, shown in Figure 9–2.
4. Repeat the command and select the 3Point option.
5. Select points C, D, and E, shown in Figure 9–2.
6. Repeat the command and use the Vertical option.
7. Select points F, G, and H, shown in Figure 9–2.
8. Repeat the command and use the Center option.
9. Select points J and K, shown in Figure 9–2.
10. Do not save your file.

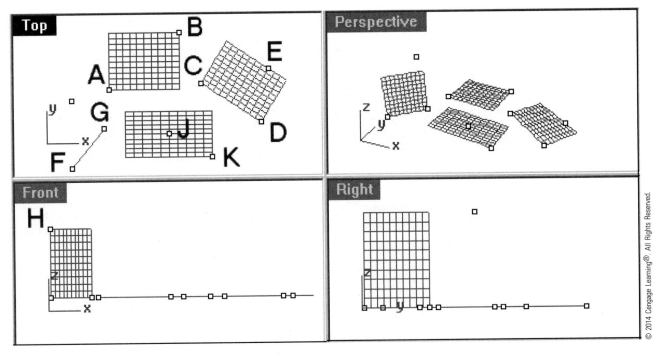

FIGURE 9–2 *Mesh rectangular planes being constructed*

Mesh Box

A mesh box consists of six rectangular planar meshes joined together. The methods to construct a mesh box are similar to those used to construct a NURBS surface box, explained in Chapter 8.

- Specifying the corners of the base rectangle and the height
- Specifying the diagonal of the box
- Using three points to specify the base rectangle and specifying the height
- Using two points on a viewport and a point in an adjacent viewport to specify the base rectangle and the height
- Specifying the center of the base rectangle, a corner of the base rectangle, and the height

Mesh Density

Mesh density of a rectangular planar mesh is defined by three options: XFaces, YFaces, and ZFaces. Perform the following steps to construct mesh boxes in various ways.

1. Select File > Open and select the file BoxMesh.3dm from the Chapter 9 folder on the Student Companion Website.
2. Select Mesh > Polygon Mesh Primitives > Box.
3. Select points A and B, shown in Figure 9–3, and type 15 at the command area to specify the height.
   A mesh box is constructed.
4. Repeat the command.
5. Select the 3Point option at the command area.
6. Select points C, D, and E, shown in Figure 9–3, and type 10 at the command area to specify the height.
   The second mesh box is constructed.
7. Repeat the command.
8. Select the Diagonal option.
9. Select points F and G, shown in Figure 9–3.
   The third mesh box is constructed.
10. Repeat the command.
11. Select the Vertical option.
12. Select points H and J, shown in Figure 9–3.
13. Type 15 to specify the vertical height.
14. Click on location K to indicate the direction.
15. Type 10 at the command area.
    A vertical mesh box is constructed.
16. Repeat the command.
17. Select the Center option at the command area.
18. Select points L and M.
19. Type 7 at the command area.
20. Do not save your file.

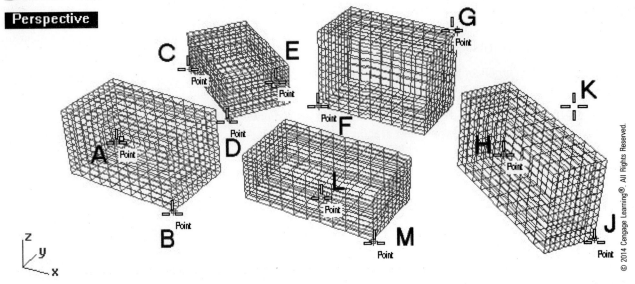

FIGURE 9-3 *Mesh boxes constructed*

Mesh Sphere

A mesh sphere is a single closed-loop meshed object. The ways to construct a mesh sphere are the same as those described in Chapter 8 to construct a NURBS surface sphere.

- Specifying the center and a point on the sphere
- Specifying the diameter of the sphere
- Specifying three points to define a section passing through the center of the sphere
- Specifying three tangent curves
- Specifying a point on a curve and the radius/diameter of the sphere
- Specifying four points on the surface of the sphere
- Fitting the surface of the sphere to a set of points

Mesh Density

A sphere's mesh density is defined by two options: VerticalFaces and AroundFaces. Perform the following steps to construct mesh spheres in various ways.

1. Select File > Open and select the file SphereMesh.3dm from the Chapter 9 folder on the Student Companion Website.
2. Select Mesh > Polygon Mesh Primitives > Sphere.
3. Referencing Figure 9-4 and using the default option, construct a mesh sphere with its center at A and radius AB; using the 2Point option, construct a mesh sphere with its diameter CD; and using the 3Point option, construct a mesh sphere with a section across the center passing through E, F, and G.

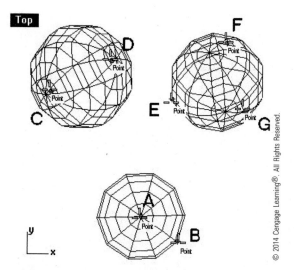

FIGURE 9–4 *From left to right: mesh sphere by specifying the diameter, mesh sphere by specifying the center and radius, and mesh sphere by specifying three points on the surface constructed*

4. Turn on Layer 02 and turn off Layer 00.
5. Referencing Figure 9–5 and using the Tangent option, construct a mesh sphere tangent to curves A, B, and C, and, using the AroundCurve option, construct a mesh sphere around curve D, center at E, and radius EF.

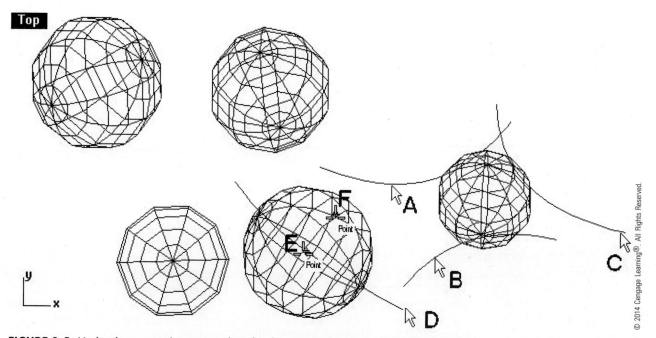

FIGURE 9–5 *Mesh sphere around a curve and mesh sphere tangent to three curves constructed*

6. Turn on Layer 03 and turn off Layer 02.
7. Referencing Figure 9–6 and using the 4Point option, construct a mesh sphere passing through points A, B, C, and D, and, using the FitPoints option, construct a mesh sphere fitted to a set of points enclosed by the rectangle EF.
8. Do not save your file.

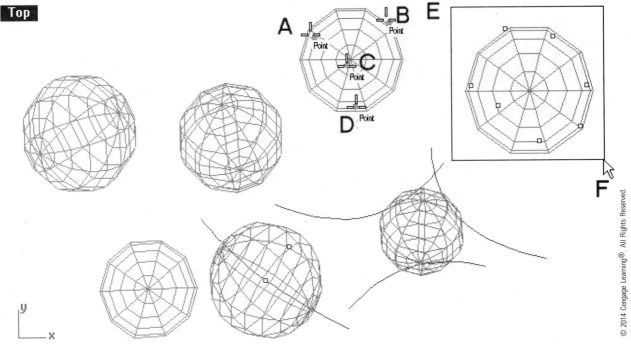

FIGURE 9–6 *Mesh sphere passing through four points and mesh sphere fitted to a set of points constructed*

Mesh Cylinder

A mesh cylinder consists of three polygon meshes, a cylindrical mesh and two circular planar meshes, joined together. The methods to construct a mesh cylinder are the same as those used to construct a NURBS surface cylinder.

Mesh Density

Mesh density can be set while constructing a meshed object. The mesh density in two directions of the polygon mesh is defined by two options: VerticalFaces and AroundFaces. Perform the following steps to construct mesh cylinders in various ways.

1. Select File > Open and select the file CylinderMesh.3dm from the Chapter 9 folder on the Student Companion Website.
2. Select Mesh > Polygon Mesh Primitives > Cylinder.
3. Select VerticalFaces option at the command area.
4. Type 8 at the command area to specify the number of meshes in the vertical direction.
5. Select AroundFaces option at the command area.
6. Type 8 at the command area to specify the number of meshes in the second direction.

Direction Constraints

Axis direction of a mesh cylinder can be defined in three ways. Continue with the following steps.

7. Select the DirectionConstraint option at the command area.
8. Select the Vertical option at the command area.
9. Set object snap mode to Point.
10. Select point A, shown in Figure 9–7, to specify the cylinder's center.
11. Type 10 at the command area to specify the radius.
    You can specify diameter instead of radius by selecting the Diameter option at the command area.
12. Select point B to use the point's Z coordinate as the height of the cylinder.
13. Repeat the command.

14. Select the DirectionConstraint option at the command area.
15. Select the None option.
16. Select point C, shown in Figure 9–7, to specify the cylinder's center.
17. Type 10 at the command area to specify the radius.
18. Select point C to define the other endpoint of the cylinder.

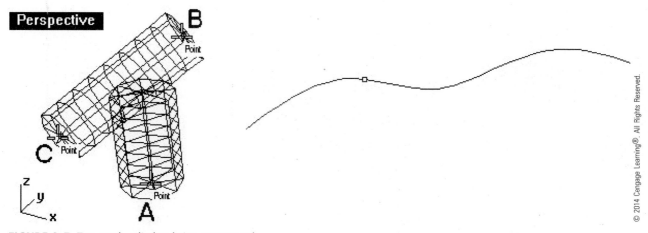

FIGURE 9–7 *Two mesh cylinders being constructed*

19. Repeat the command.
20. Select the DirectionConstraint option at the command area.
21. Select the AroundCurve option at the command area.
22. Select curve A and then point B, shown in Figure 9–8.
23. Type 10 at the command area to specify the radius.
24. Type 25 at the command area to specify the height.

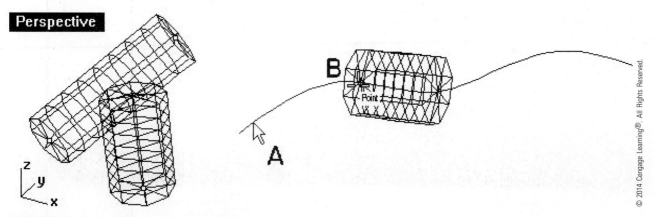

FIGURE 9–8 *Mesh cylinder around a curve being constructed*

Tangent Cylinder

The base circle of a mesh cylinder can be made tangent to three curves. Continue with the following steps.

25. Turn on Layer 02 and turn off Layer 00.
26. Select Mesh > Polygon Mesh Primitives > Cylinder
27. Select the Tangent option at the command area.
28. Select curves A, B, and C, shown in Figure 9–9.
29. Type 25 at the command area to specify the height.

Note that the sequence of selection of the tangent curves has an effect on the direction of the cylinder.

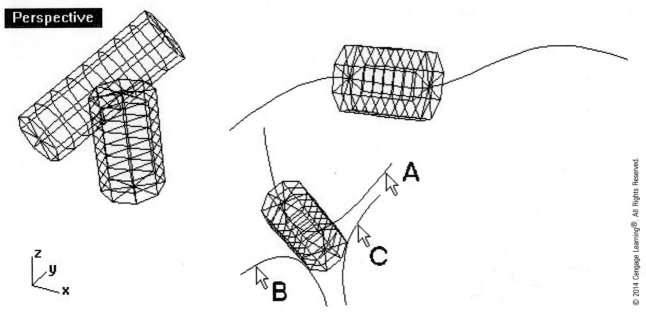

FIGURE 9–9 *Tangent mesh cylinder being constructed*

Mesh Cylinder with Base Circle Fitted to a Set of Points

The base circle of a mesh cylinder can be made to fit closest to a set of points. Continue with the following steps.

30. Turn on Layer 03 and turn off Layer 02.
31. Select Mesh > Polygon Mesh Primitives > Cylinder.
32. Select the FitPoints option at the command area.
33. Click on A and drag to B (Figure 9–10) to select the point objects, and press the ENTER key.
34. Select point C.

   A cylinder with its axis endpoint closest to the selected point is constructed.

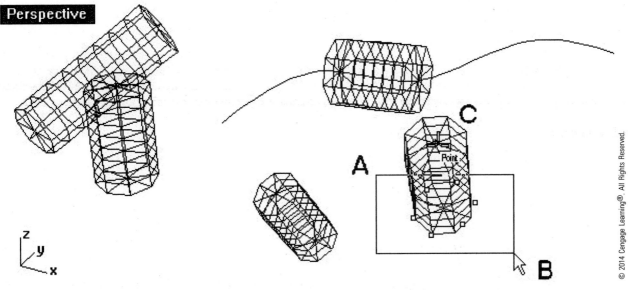

FIGURE 9–10 *Mesh cylinder with its base circle fitted to a set of points being constructed*

Mesh Cylinder with Base Circle Passing through Three Points

The base circle of a mesh cylinder can be made to pass through three points. Continue with the following steps.

35. Turn on Layer 04 and turn off Layer 03.
36. Select Mesh > Polygon Mesh Primitives > Cylinder.
37. Select the 3Point option at the command area.
38. Select points A, B, and C, shown in Figure 9–11.
39. Select point D.

    A cylinder with its axis endpoint closest to the selected point is constructed.

40. Do not save your file.

    If you compare the above delineations with the explanations for making NURBS cylinders, you will find that they are more or less the same, with the exception that you have to specify the mesh density and the outcome is a meshed object instead of a smooth polysurface.

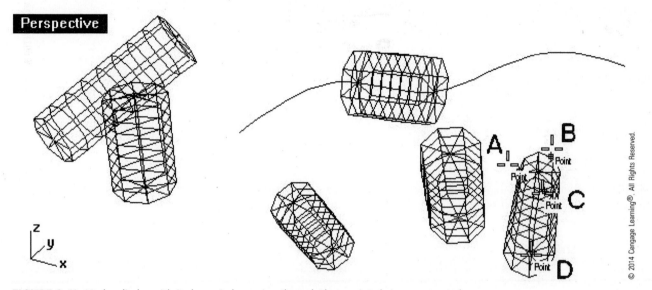

FIGURE 9–11 *Mesh cylinder with its base circle passing through three points being constructed*

Mesh Cone

A mesh cone consists of a slant conical mesh and a circular planar mesh joined together. Again, the methods to construct a mesh cone are the same as those used to construct a NURBS surface cone. Apart from specifying the mesh density, you can construct a cone without direction constraint, with vertical constraint, with its base circle around a curve, with its base circle tangent to three curves, with its base circle fitted to a set of points, and with its base circle passing through three points.

Mesh Density

A mesh cone's mesh density is defined by two options: VerticalFaces and AroundFaces. Perform the following steps to construct mesh cones in various ways.

1. Select File > Open and select the file CylinderMesh.3dm from the Chapter 9 folder on the Student Companion Website.
2. Select Mesh > Polygon Mesh Primitives > Cone.
3. Use the Vertical direction constraint option to construct a mesh cone (cone A in Figure 9–12) with a radius of 10 mm and a height of 25 mm.
4. Repeat the command and use the Around curve direction constraint option to construct a cone (cone B in Figure 9–12) with a base radius of 10 mm and a height of 25 mm.
5. Repeat the command and use the None direction constraint option to construct another mesh cone (cone C in Figure 9–12) with a radius of 10 mm.

6. Turn on Layer 02 and turn off Layer 00.
7. Construct a mesh cone (cone D in Figure 9–12) tangent to three curves. The height of the cone is 25 mm.
8. Turn on Layer 03 and turn off Layer 02.
9. Construct a mesh cone (cone E in Figure 9–12) with its base circle fitted to a set of points. The height of the cone is also 25 mm.
10. Turn on Layer 04 and turn off Layer 03.
11. Construct a mesh cone (cone F in Figure 9–12) with its base passing through three points.
12. Do not save your file.

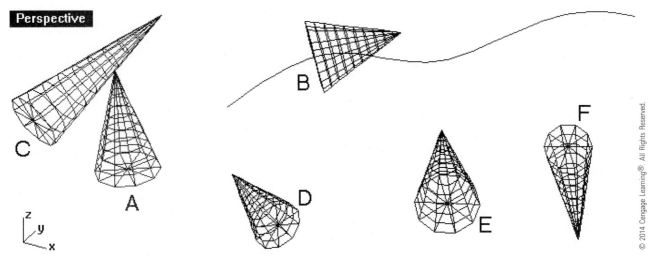

FIGURE 9–12 *From left to right: cone without direction constraint, cone with vertical constraint, cone with its base circle tangent to three curves, cone with its base circle around a curve, cone with its base circle fitted to a set of points, and cone with its base circle passing through three points*

Mesh Truncated Cone

A mesh truncated cone consists of a slant conical mesh and two circular planar meshes joined together. The methods to construct a mesh truncated cone are the same as those used for constructing truncated NURBS surface cone.

Mesh Density

A truncated cone's mesh density is defined by two options: VerticalFaces and AroundFaces. Perform the following steps to construct truncated mesh cones in various ways.

1. Select File > Open and select the file TConeMesh.3dm from the Chapter 9 folder on the Student Companion Website.
2. Select Mesh > Polygon Mesh Primitives > Truncated Cone.
3. Construct a truncated cone (cone A in Figure 9–13) with vertical direction constraint, having a base radius of 10 mm, a height of 32 mm, and a radius of 6 mm at the end.
4. Repeat the command to construct a truncated cone (cone B in Figure 9–13) without direction constraint, having a base radius of 10 mm, vertex at C, and a radius of 6 mm at the end.
5. Repeat the command to construct a truncated cone (cone D in Figure 9–13) around a curve. The base radius is 10 mm, the height is 20 mm, and the end radius is 6 mm.
6. Turn on Layer 02 and turn off Layer 00.
7. Construct a mesh truncated cone (cone E in Figure 9–13) tangent to three curves. The height is at F, and the end radius is 4 mm.
8. Turn on Layer 03 and turn off Layer 02.
9. Construct a truncated mesh cone (cone G in Figure 9–13) with its base circle fitted to a set of points. The height is at H, and the end radius is 4 mm.
10. Turn on Layer 04 and turn off Layer 03.

11. Construct a mesh cone (F in Figure 9–13) with its base passing through three points. The end radius is 2 mm, and the height is at G.
12. Do not save your file.

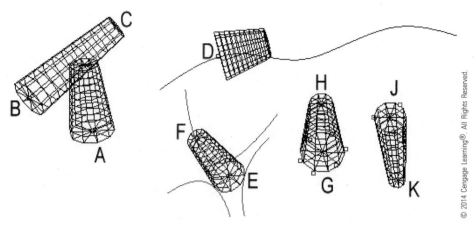

FIGURE 9–13 *From left to right: truncated cone without direction constraint, with vertical constraint, with its base circle tangent to three curves, with its base circle around a curve, with its base circle fitted to a set of points, and with its base circle passing through three points*

Mesh Ellipsoid

A mesh ellipsoid is a single meshed object. The methods to construct a mesh ellipsoid are the same as those for constructing a NURBS surface ellipsoid.

Mesh Density

A mesh ellipsoid's density is defined by two options: 1stDirFaces and 2ndDirFaces. Perform the following steps to construct mesh ellipsoids in various ways.

1. Select File > Open and select the file EllipsoidMesh.3dm from the Chapter 9 folder on the Student Companion Website.
2. Select Mesh > Polygon Mesh Primitives > Ellipsoid.
3. Select points A and B, shown in Figure 9–14, to specify the center and an axis endpoint, respectively.
4. Type 5 at the command area, press the ENTER key, and click any location on the current construction plane to specify the second axis radius and direction.
5. Type 3 at the command area and press the ENTER key.
6. Repeat the command.
7. Select the Diameter option at the command area.
8. Select points C and D, shown in Figure 9–14, to specify two axis endpoints.
9. Type 5 at the command area, press the ENTER key, and click anywhere on the active construction plane to specify the second axis diameter and direction.
10. Type 3 at the command area and press the ENTER key.

    An ellipsoid polygon mesh is constructed.

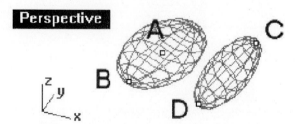

FIGURE 9–14 *Mesh ellipsoid specified by radii (left) and mesh ellipsoid specified by diameters*

© 2014 Cengage Learning®. All Rights Reserved.

11. Turn on Layer 02 and turn off Layer 00.
12. Select Mesh > Polygon Mesh Primitives > Ellipsoid.
13. Select the From Foci option at the command area.
14. Select points A and B (shown in Figure 9–15) to specify the foci, and then select point C to specify a point on the ellipsoid.

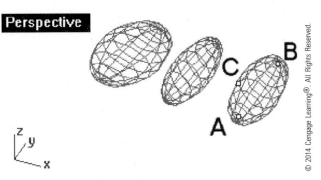

FIGURE 9–15 *Mesh ellipsoid specified by foci and corners constructed*

15. Turn on Layer 03 and turn off Layer 02.
16. Select Mesh > Polygon Mesh Primitives > Ellipsoid.
17. Select the AroundCurve option at the command area.
18. Select point A (Figure 9–16) along the curve to specify the center of the ellipsoid.
19. Turn on Ortho mode.
20. Type 5 at the command area, press the ENTER key, and click on location B to specify the first radius.
21. Type 4 at the command area, press the ENTER key, and click on location C to specify the second radius.
22. Type 3 at the command area and press the ENTER key.
23. Turn off Ortho mode.
24. Do not save your file.

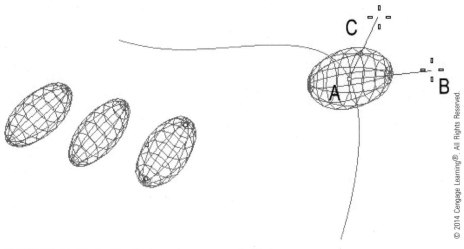

FIGURE 9–16 *Mesh ellipsoid around a curve constructed*

Mesh Torus

A mesh torus is a single closed-loop meshed object, and the ways to construct a mesh torus are the same as those for making a NURBS surface torus.

Mesh Density

The mesh density of a mesh torus is defined by two options: VerticalFaces and AroundFaces. Perform the following steps to construct mesh tori in various ways.

1. Select File > Open and select the file TorusMesh.3dm from the Chapter 9 folder on the Student Companion Website.
2. Select Mesh > Polygon Mesh Primitives > Torus.
3. Select points A and B, shown in Figure 9–17. Here A is the center and AB is the radius of the pitch circle.
4. Type 4 at the command area to specify the radius of the torus tube.
5. Repeat the command.
6. Select the Vertical option at the command area.
7. Select points C and D, shown in Figure 9–17.
8. Type 4 at the command area.
9. Repeat the command.
10. Select the 2Point option at the command area.
11. Select points E and F, shown in Figure 9–17.
12. Type 4 at the command area.

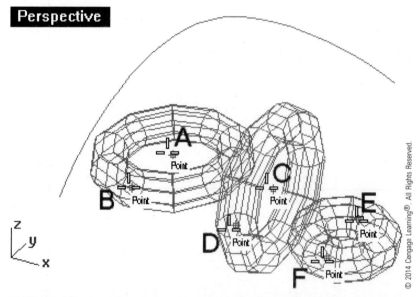

FIGURE 9–17 *From left to right: torus pitch circle specified by radius, vertical torus, and torus pitch circle specified by diameter*

13. Turn on Layer 02 and turn off Layer 00.
14. Select Mesh > Polygon Mesh Primitives > Torus.
15. Select the AroundCurve option at the command area.
16. Select curve A, shown in Figure 9–18.
17. Select point B on curve A.
18. Type 9 at the command area to specify the pitch radius.
19. Type 3 at the command area to specify the radius.

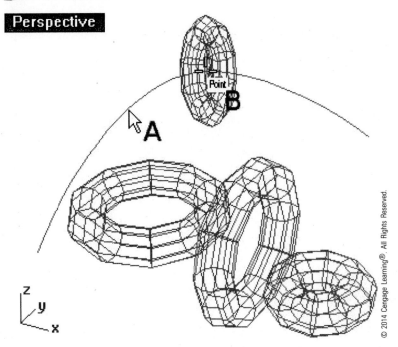

FIGURE 9–18 *Mesh torus with its pitch circle around a curve being constructed*

20. Turn on Layer 03 and turn off Layer 02.
21. Select Mesh > Polygon Mesh Primitives > Torus.
22. Select the Tangent option at the command area.
23. Select curves A, B, and C, shown in Figure 9–19.
24. Type 3 at the command area to specify the second radius.

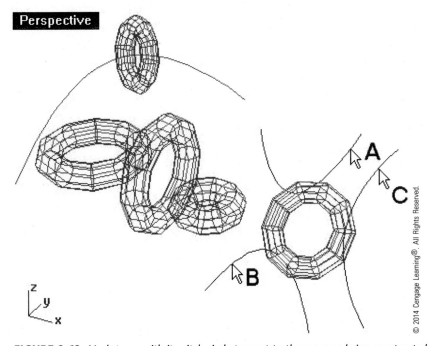

FIGURE 9–19 *Mesh torus with its pitch circle tangent to three curves being constructed*

25. Turn on Layer 04 and turn off Layer 03.
26. Select Mesh > Polygon Mesh Primitives > Torus.

27. Select the FitPoints option at the command area.
28. Click on A and drag to B (Figure 9–20) to select the point objects, and press the ENTER key.
29. Type 3 at the command area to specify the radius at the other end.

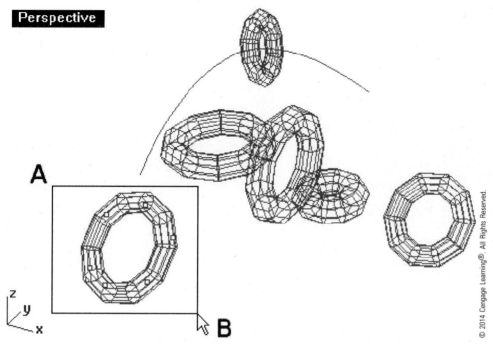

FIGURE 9–20 Mesh torus with its pitch circle best fitted to a set of points being constructed

30. Turn on Layer 05 and turn off Layer 04.
31. Select Mesh > Polygon Mesh Primitives > Torus.
32. Select the 3Point option at the command area.
33. Select points A, B, and C, shown in Figure 9–21.
34. Type 3 at the command area.
35. Do not save your file.

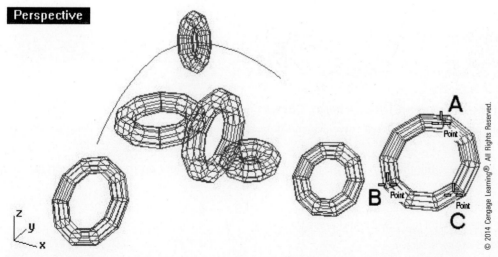

FIGURE 9–21 Mesh torus with its pitch circle defined by three selected points being constructed

CONSTRUCTING POLYGON MESHES FROM EXISTING OBJECTS

After learning ways to construct polygon mesh from scratch, you will now learn how to construct polygon mesh from existing objects, which can be done in six ways.

- Using a closed planar curve as boundary
- Using a closed curve as boundary and, optionally, curves and points to define mesh details
- Using the profile of a NURBS surface
- Using the control points of a NURBS surface
- Using the heightfield of a bitmap as vertices
- Offsetting an existing polygon mesh

Constructing a Polygon Mesh from a Set of Points

You can construct a polygon mesh to best fit to a set of point objects. This method is useful when you already have a cloud of points obtained by digitizing an object. Perform the following steps to construct a polygon mesh from a set of points.

1. Select File > Open and select the file PointMesh.3dm from the Chapter 9 folder on the Student Companion Website.
2. Select Mesh > Mesh from Points.
   Note that this command is a plug-in and you have to install this plug-in before you can use it.
3. Click on location A and drag to location B, shown in Figure 9–22, and press the ENTER key.
4. Press the ENTER key to accept the default options.
5. A polygon mesh is constructed. Do not save your file.

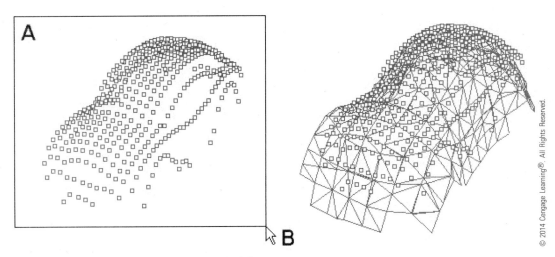

FIGURE 9–22 *Point objects (left) and polygon mesh constructed (right)*

Constructing a Polygon Mesh from a Closed Planar Curve

You can construct a planar polygon mesh from a closed curve, which can be a polyline or a NURBS curve. Perform the following steps.

1. Select File > Open and select the file PlanarMesh.3dm from the Chapter 9 folder on the Student Companion Website.
2. Type PlanarMesh at the command area.
3. Select curves A and B, shown in Figure 9–23, and press the ENTER key.
   Two planar mesh objects are constructed.
4. Do not save your file.

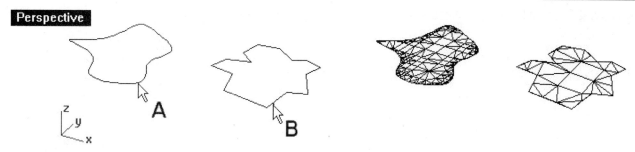

FIGURE 9–23 *NURBS curve and polyline (left) and planar meshes constructed (right)*
© 2014 Cengage Learning®. All Rights Reserved.

Constructing a Polygon Mesh from a Closed Polyline

You can construct a polygon mesh from a closed polyline or a curve, which can be planar or 3D. If a curve is used, a polyline will first be approximated to a polyline prior to outputting a mesh. The resulting polygon mesh will use the endpoints of the segments of the polyline as vertices of the meshes. Perform the following steps from a closed polyline.

1. Select File > Open and select the file PolylineMesh.3dm from the Chapter 9 folder on the Student Companion Website.
2. Select Mesh > Polygon Mesh > From Closed Polyline.
3. Select polyline A (shown in Figure 9–24).

    A polygon mesh is constructed.

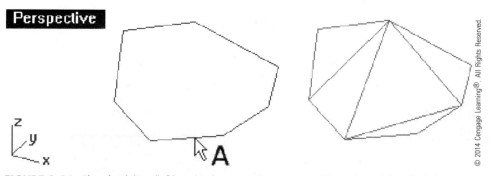

FIGURE 9–24 *Closed polyline (left) and polygon mesh constructed from the polyline (right)*

Constructing a Mesh Patch

Constructing a mesh patch is similar to making a NURBS surface patch. You select curves and points as defining objects. Perform the following steps.

1. Select File > Open and select the file MeshPatch1.3dm from the Chapter 9 folder on the Student Companion Website.
2. Click on the Mesh Patch button on the Mesh Creation toolbar.
3. Select curves A, B, C, and D and point E (Figure 9–25) and press the ENTER key.
4. Select curve B and press the ENTER key.

    This is the inner curve.

5. Select curve A.

    A mesh patch is constructed.

6. Do not save your file.

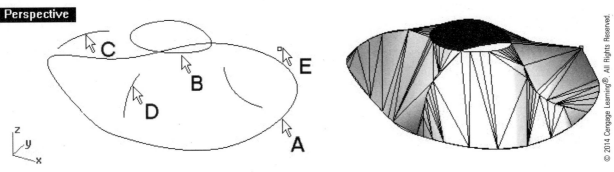

FIGURE 9–25 *Curves and points (left) and mesh patch constructed (right)*

To appreciate how point objects can be used in making a mesh object, perform the following steps.

1. Select File > Open and select the file MeshPatch2.3dm from the Chapter 9 folder on the Student Companion Website.
2. Click on the Mesh Patch button on the Mesh Creation toolbar.
3. Click on location B and drag to location C (Figure 9–26) to select all the point objects, and press the ENTER key.
4. Because there is no inner curve, press the ENTER key.
5. Select curve A.

   A mesh patch is constructed.
6. Do not save your file.

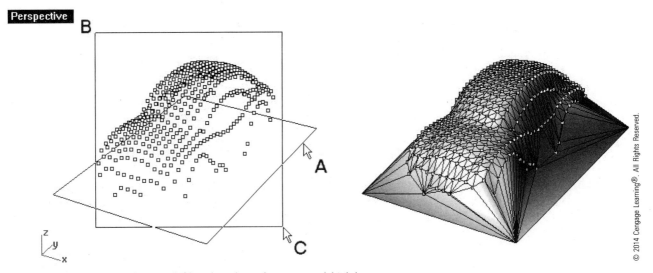

FIGURE 9–26 *Curves and points (left) and mesh patch constructed (right)*

Constructing a Polygon Mesh from a NURBS Surface

NURBS surfaces and polygon meshes are two different types of objects. A NURBS surface is an exact representation, but the polygon mesh is an approximated representation of the 3D object.

In Chapters 7 and 8, you already learned various ways to construct a NURBS surface model of complex shapes. Now, if you want to construct a polygon mesh of very complex shape, you may think about first building the model in NURBS surface or polysurface and then construct a polygon mesh from it. Perform the following steps to construct a polygon mesh from a NURBS surface.

1. Select File > Open and select the file MeshfromSurface.3dm from the Chapter 9 folder on the Student Companion Website.

2. Select Mesh > From NURBS Objects.
3. Select surface A (shown in Figure 9–27) and press the ENTER key.
4. In the Polygon Mesh Options dialog box, set the polygon density by moving the slider bar and then click on the OK button.

   Alternatively, you can click on the Detailed Controls button and manipulate various parameters.
5. A polygon mesh is constructed. Do not save your file.

   If you now set the display to shaded, you will not find any difference between the original NURBS surface and the derived polygon mesh, because shading and rendering are both making use of meshes.

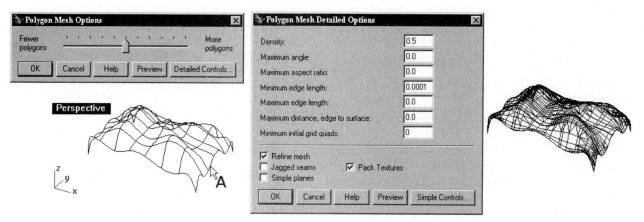

FIGURE 9–27 *NURBS surface (left), Polygon Mesh Options dialog box, and polygon mesh constructed from the surface (right)*
Source: Robert McNeel and Associates Rhinoceros® 5

Constructing a Polygon Mesh through the Control Points of a NURBS Surface

Apart from approximating the profile of a NURBS surface, you can use the NURBS surface control points as vertices to construct a polygon mesh. Perform the following steps to construct a polygon mesh from the control points of a NURBS surface.

1. Select File > Open and select the file MeshfromControlPolygon.3dm from the Chapter 9 folder on the Student Companion Website.
2. Select Mesh > From NURBS Control Polygon.
3. Select surface A (shown in Figure 9–28) and press the ENTER key.

   A polygon mesh is constructed.
4. Do not save your file.

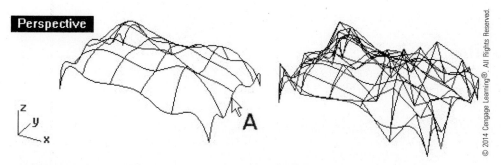

FIGURE 9–28 *Surface (left) and polygon mesh constructed from the control vertices (right)*

Constructing a Polygon Mesh from an Image's Heightfield

Similar to constructing a surface from a raster image, you can use the heightfield of an image to construct a polygon mesh. If the purpose of using the heightfield of an image is to construct a rapid prototype, you can directly construct a polygon mesh from the image, rather than first constructing a NURBS surface and then exporting/converting it to polygon mesh. Perform the following steps to construct a polygon mesh from the heightfield of an image.

1. Select File > Open and select the file HeightfieldMesh.3dm from the Chapter 9 folder on the Student Companion Website.
2. Select Mesh > Mesh Heightfield.
3. Select the file car.tga from the Chapter 9 folder on the Student Companion Website.
4. Select points A and B, shown in Figure 9–29.
5. Select the GridSize option at the command area.
6. Select the Width option at the command area.
7. Type 200 at the command area.
8. Select the Height option at the command area.
9. Type 200 at the command area.
10. Press the ENTER key.
11. Select the Elevation option at the command area.
12. Select the Value factor at the command area.

    This option determines the height values of the polygon mesh.

13. Type 0.01 at the command area.
14. Press the ENTER key twice.
15. Set the display to rendered to appreciate the visual effect.

    A polygon mesh is constructed from the heightfield of an image.

16. Do not save your file.

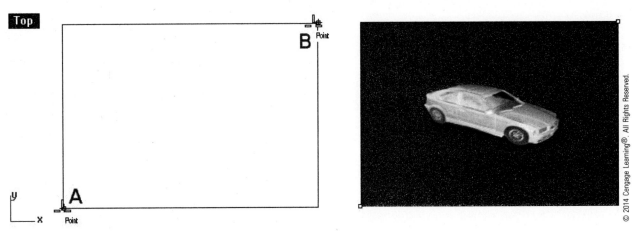

FIGURE 9–29 *Two points selected (left) and polygon mesh constructed (right)*

Constructing a Mesh by Offsetting a Mesh

By offsetting an existing polygon mesh, you obtain another polygon mesh. Using the cap option, you can construct a polygon mesh that encloses a volume from a single polygon mesh. Perform the following steps to construct an offset polygon mesh from an existing polygon mesh.

1. Select File > Open and select the file OffsetMesh.3dm from the Chapter 9 folder on the Student Companion Website.
2. Select Mesh > Offset Mesh.
3. Select mesh A, shown in Figure 9–30, and press the ENTER key.
4. Select the PickOffsetDistance option at the command area.
5. In the Offset Mesh dialog box, set offset distance to 2.
6. Click on the FlipAll button as necessary to change the offset direction.

7. Click on the OK button.

   An offset mesh is constructed.

8. Do not save your file.

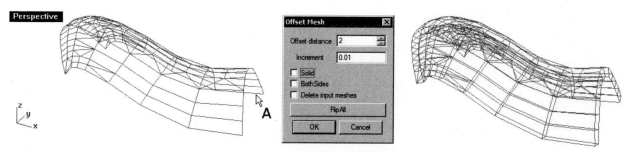

FIGURE 9–30 *From left to right: original polygon mesh, Offset Mesh dialog box, and offset with end cap polygon mesh*
Source: Robert McNeel and Associates Rhinoceros® 5

MAPPING POLYGON MESHES

You can map a polygon mesh object to surface in two ways: in accordance with vertex count and mesh structure and in accordance with the surface's UVN.

Vertex Count and Mesh Structure

In many animation programs that use polygon meshes to represent 3D objects, morphing of two different shapes requires the polygon meshes to have an identical vertex count and mesh structure. To construct two different shapes having identical vertex count and mesh structure, you construct two shapes as NURBS surfaces, construct a polygon mesh from one surface, and map the polygon mesh to the other surface. Perform the following steps to map a polygon mesh to a surface.

1. Select File > Open and select the file ApplyMesh.3dm from the Chapter 9 folder on the Student Companion Website.

   This file contains a polygon mesh constructed from a NURBS surface. To obtain another polygon mesh of identical vertex count and mesh structure, you will map the polygon mesh to another NURBS surface of different shape.

2. Select Mesh > Apply to Surface.
3. Select mesh A (shown in Figure 9–31).
4. Select surface B (shown in Figure 9–31).

   The polygon mesh constructed from a NURBS surface is applied to another surface, as shown in Figure 9–32.

5. Do not save your file.

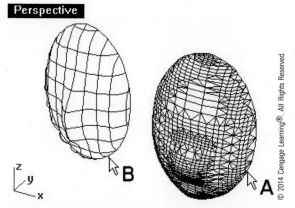

FIGURE 9–31 *Polygon mesh being applied to a surface*

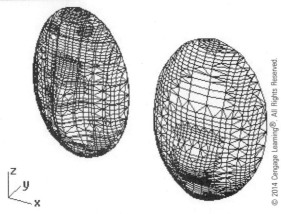

FIGURE 9–32 *Polygon mesh applied to a surface*

Surface's UVN

To map meshed objects and point objects accurately onto a surface in accordance with the surface's U and V coordinates, you use the ApplyMeshUVN command, as follows.

1. Select File > Open and select the file ApplyMeshUVN.3dm from the Chapter 9 folder on the Student Companion Website.
2. Select Mesh > Apply Mesh UVN.
3. Click on A and drag to B, shown in Figure 9–33, and press the ENTER key.
4. Select the VerticalScale option at the command area.
5. Type 0.5 at the command area to specify a vertical scale factor.
6. Select surface C (Figure 9–33).

    The selected meshed objects and point objects are mapped onto the surface. (See Figure 9–34.)

7. Repeat the command.
8. Click on A and drag to B, shown in Figure 9–33, and press the ENTER key.
9. Select the VerticalScale option at the command area.
10. Type 2 at the command area to specify a vertical scale factor.
11. Select surface D (Figure 9–33).

    The selected meshed objects and point objects are mapped onto the surface. (See Figure 9–34.)

12. Do not save your file.

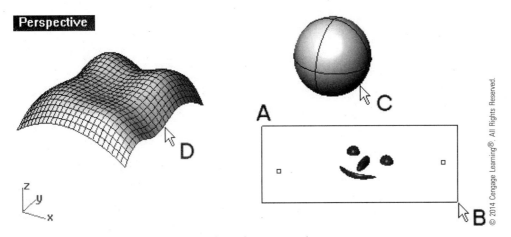

FIGURE 9–33 *Meshed objects and point objects being mapped*

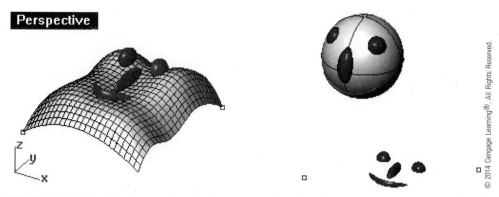

FIGURE 9–34 *Meshed objects and point objects mapped and scaled vertically*

COMBINING AND SEPARATING

More complex meshed objects can be composed by combining two or more individual meshed objects using Boolean operations (union, difference, and intersect). Apart from using Boolean operations, you can join contiguous meshed objects. Conversely, you can explode joined objects to individual meshed objects. To split or trim meshed objects, you use a curve, surface, polysurface, or polygon mesh.

Boolean Operations

There are three basic Boolean operations: union, difference, and intersection. To carry out these operations, the polygon meshes have to overlap each other in 3D space.

Mesh Union

A union of a set of polygon meshes produces a polygon mesh that has the volume of all polygon meshes in the set. Any overlapped portions of the polygon meshes will be trimmed away, and the final model will be a single polygon mesh. To construct a polygon mesh by uniting two polygon meshes, perform the following steps.

1. Select File > Open and select the file UnionMesh.3dm from the Chapter 9 folder on the Student Companion Website.
2. Select Mesh > Mesh Boolean > Union.
3. Select polygon meshes A and B, shown in Figure 9–35, and press the ENTER key.
   Two polygon meshes are united into one.
4. Do not save your file.

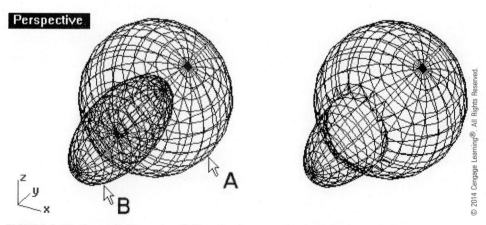

FIGURE 9–35 *Two polygon meshes (left) and polygon meshes united into one (right)*

Mesh Difference

A difference of two sets of polygon meshes produces a polygon mesh that has the volume contained in the first set of polygon meshes but not the second set of polygon meshes. In the operation, the second set of polygon meshes is subtracted from the first set of polygon meshes, and the result is a single polygon mesh. Perform the following steps to subtract a polygon mesh from another polygon mesh.

1. Select File > Open and select the file DifferenceMesh.3dm from the Chapter 9 folder on the Student Companion Website.
2. Select Mesh > Mesh Boolean > Difference.
3. Select mesh A, shown in Figure 9–36, and press the ENTER key.
   This is the first set of polygon mesh.
4. Select mesh B, shown in Figure 9–36, and press the ENTER key.
   This is the second set of polygon mesh, and the second set of polygon mesh is subtracted from the first set of polygon mesh.
5. Do not save your file.

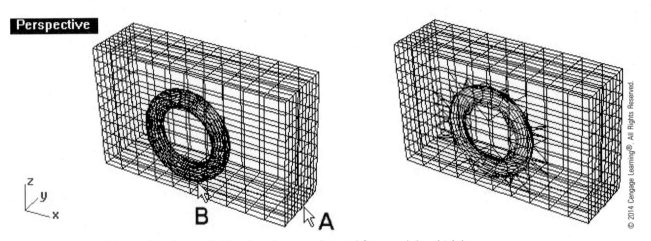

FIGURE 9–36 *Mesh box and mesh torus (left) and mesh torus subtracted from mesh box (right)*

Mesh Intersection

An intersection of two sets of polygon meshes produces a polygon mesh that has the volume contained in the first and second sets of polygon meshes. In other words, the volume common to two sets of polygon meshes will be retained, and the result is a single polygon mesh. Perform the following steps to construct an intersection polygon mesh from a pair of polygon meshes.

1. Select File > Open and select the file IntersectMesh.3dm from the Chapter 9 folder on the Student Companion Website.
2. Select Mesh > Mesh Boolean > Intersection.
3. Select mesh A, shown in Figure 9–37, and press the ENTER key.
   This is the first set of polygon mesh.
4. Select mesh B, shown in Figure 9–37, and press the ENTER key.
   This is the second set of polygon mesh, and an intersection of two polygon meshes is constructed.
5. Do not save your file.

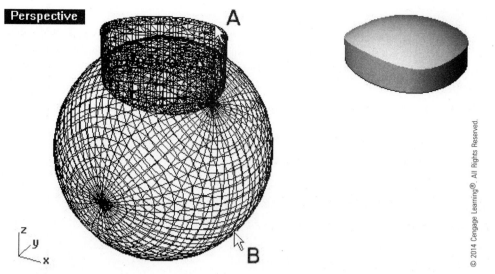

FIGURE 9–37 Two polygon meshes (left) and shaded display of an intersection of the polygon meshes (right)

Joining and Exploding Meshes

Quite often, you may have to join several meshed objects sharing common boundaries into a single meshed object. The opposite of joining is exploding joined meshes to obtain individual polygon meshed objects.

Perform the following steps to join two contiguous meshed objects.

1. Select File > Open and select the file JoinExplode.3dm from the Chapter 9 folder on the Student Companion Website.
2. Select Edit > Join.
3. Select meshed objects A, B, and C (Figure 9–38) and press the ENTER key.
   The selected meshes are joined.

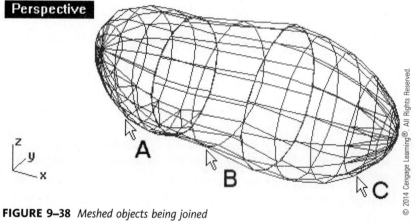

FIGURE 9–38 Meshed objects being joined

Continue with the following steps to explode joined meshed objects.

4. Select Edit > Explode, or click on the Explode button on the Main2 toolbar.
5. Select the joined polygon meshes and press the ENTER key.
   The meshes are exploded into individual polygon meshes.
6. Do not save your file.

Mesh Splitting

A polygon mesh can be split into two in two basic ways: Boolean split and splitting by a curve, surface, or polygon mesh.

Boolean Split

With the Boolean split method, you use a polygon mesh, surface, or polysurface to split a polygon mesh into two; the cutting object has to be large enough to cut through the polygon mesh. If the polygon mesh to be split is a closed-loop polygon mesh, the openings after splitting will be covered by a polygon mesh face resembling the shape of the cutting object. As a result, you obtain two closed-loop polygon meshes. Perform the following steps.

1. Select File > Open and select the file BooleanSplit.3dm from the Chapter 9 folder on the Student Companion Website.
2. Select Mesh > Mesh Boolean > Boolean Split.
3. Select mesh A, shown in Figure 9–39, and press the ENTER key.
   This is the mesh to be split.
4. Select mesh B, shown in Figure 9–39, and press the ENTER key.
   This is the cutting mesh.
5. Mesh A is split into two meshes. Hide A and B.
6. Shade the viewport to appreciate the result. Do not save your file.

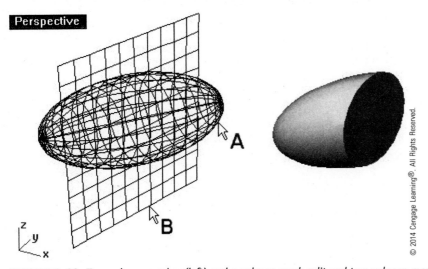

FIGURE 9–39 *Two polygon meshes (left) and a polygon mesh split and two polygon meshes hidden (right)*

Splitting by a Curve, Surface, or Polygon Mesh

The second method is to use a curve, surface, or polygon mesh as the cutting object. If a curve is used as the splitting tool, the curve is projected vertically onto the active construction plane, splitting the polygon mesh into two. Unlike Boolean split, the openings caused by splitting will not be filled up. Therefore, a closed-loop polygon mesh will become an open-loop polygon mesh after splitting this way. Perform the following steps.

1. Select File > Open and select the file MeshSplitTrim.3dm from the Chapter 9 folder on the Student Companion Website.
2. Select Mesh > Mesh Edit Tools > Mesh Split.
3. Select polygon meshes A, B, and C, shown in Figure 9–40, and press the ENTER key.
   These are the polygon meshes to be split.
4. Select curve D, surface E, and polygon mesh F, shown in Figure 9–40, and press the ENTER key.
   These are the splitting tools. The polygon meshes are split into two.
5. To appreciate the effect of splitting, move polygon meshes A, B, and C a small distance apart from their original position.
6. Do not save your file.

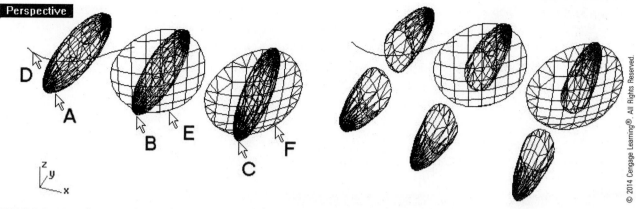

FIGURE 9–40 *Polygon meshes and splitting tools (left) and split polygon meshes moved apart (right)*

Mesh Trimming

You can split or trim a polygon mesh by using a curve, surface, or polygon mesh. Again, the opening left by trimming will not be filled up. Therefore, a closed-loop polygon mesh will become an open-loop polygon mesh after trimming.

1. Select File > Open and select the file MeshSplitTrim.3dm from the Chapter 9 folder on the Student Companion Website.
2. Select Mesh > Mesh Edit Tools > Mesh Trim.
3. Select curve A, surface B, and polygon mesh C, shown in Figure 9–41, and press the ENTER key.
   These are the trimming tools.
4. Select polygon meshes D, E, and F, and press the ENTER key.
   The selected portions of the polygon meshes are trimmed.
5. Do not save your file.

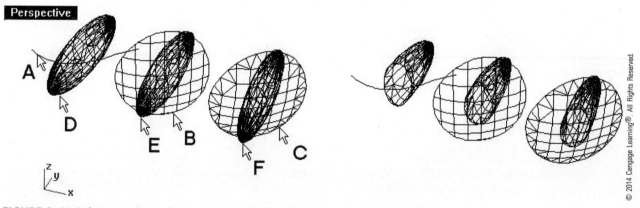

FIGURE 9–41 *Polygon meshes and trimming tools (left) and polygon meshes trimmed (right)*

MANIPULATING MESH FACES

Having learned to make primitive polygon meshes, construct polygon meshes from existing objects, and combine and separate polygon meshes, you will now work on the most important part of this chapter—editing polygon meshes. This information will help you when manipulating existing polygon mesh, in particular, and repairing polygon meshes for rapid prototyping.

Reducing Mesh Density

As explained earlier, a high polygon mesh density results in a large file size. If the polygon mesh is overly dense, you can reduce its density. Decreasing the polygon count means simplifying the mesh. To reduce a polygon density, perform the following steps. This process is irreversible, meaning that if you reduce the polygon count and later want to increase the count, you have to rebuild the mesh.

1. Select File > Open and select the file ReduceMesh.3dm from the Chapter 9 folder on the Student Companion Website.
2. Select Mesh > Mesh Edit Tools > Collapse > Reduce Vertex Count.
3. Select mesh A, shown in Figure 9–42, and press the ENTER key.
4. In the Reduce Mesh Options dialog box, set reduction percentage to 90, and click on the OK button.

   Note that you can first click on the Preview button to discover the outcome before confirming by clicking the OK button.
5. Do not save your file.

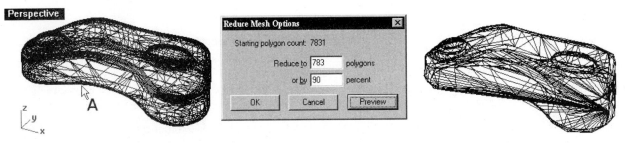

FIGURE 9–42 *From left to right: original meshed object, Reduce Mesh Options dialog box, and mesh count reduced*
Source: Robert McNeel and Associates Rhinoceros® 5

Collapsing Mesh Elements

Reduction of a meshed object's mesh count decreases the number of meshes evenly throughout the entire object. If you want to reduce mesh count in a more controlled way locally, you should use the mesh collapsing tools, which remove the selected quadrilateral mesh element's edge to turn it into a triangular mesh or remove selected mesh elements. When a mesh element is removed from the meshed object, unless the element lies on the boundary edge of the meshed object, those mesh elements adjacent to the removed element will be adjusted automatically to fill the gap left by the removed element. Therefore, do not confuse collapsing mesh elements with deleting mesh elements, which will be discussed later in this chapter.

Collapsing Mesh Vertex

You can select a vertex from a meshed object and collapse it to another selected vertex. By doing so, meshes sharing the selected vertex will be affected.

1. Select File > Open and select the file Collapse.3dm from the Chapter 9 folder on the Student Companion Website.
2. Select Mesh > Mesh Edit Tools > Collapse > Vertex.
3. Select vertices A and B, shown in Figure 9–43, and press the ENTER key.

   The selected vertices are collapsed.
4. Do not save your file.

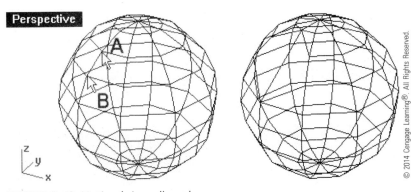

FIGURE 9–43 *Vertices being collapsed*

Collapsing Mesh Edge

Collapsing a mesh edge removes an edge of the meshed object. Any mesh element sharing that removed edge will be affected. If the affected edge is a quadrilateral edge, it will become a triangular mesh. If it is a triangular mesh, removing an edge means removing the mesh element.

1. Select File > Open and select the file Collapse.3dm from the Chapter 9 folder on the Student Companion Website.
2. Select Mesh > Mesh Edit Tools > Collapse > Edge.
3. Select edge A and then edge B, shown in Figure 9–44, and press the ENTER key.
4. Do not save your file.

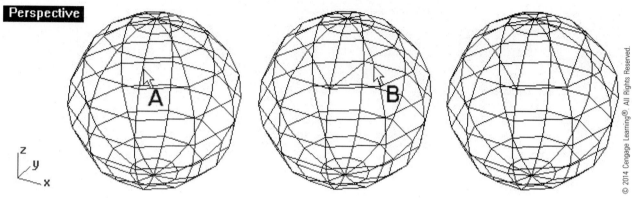

FIGURE 9–44 *From left to right: original meshed object, an edge collapsed, and two edges collapsed*

Collapsing Mesh Face

As the name implies, collapsing mesh face concerns the removal of the selected mesh element.

1. Select File > Open and select the file Collapse.3dm from the Chapter 9 folder on the Student Companion Website.
2. Select Mesh > Mesh Edit Tools > Collapse > Face.
3. Select face A, shown in Figure 9–45, and press the ENTER key.
4. Do not save your file.

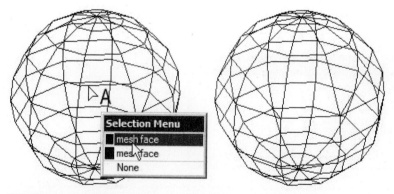

FIGURE 9–45 *Mesh face selected (left) and face collapsed (right)*
Source: Robert McNeel and Associates Rhinoceros® 5

Collapsing Mesh Element by Mesh Area

To have meshes collapsed by specifying a mesh element's area, perform the following steps.

1. Select File > Open and select the file Collapse.3dm from the Chapter 9 folder on the Student Companion Website.

2. Select Mesh > Mesh Edit Tools > Collapse > By Area.
3. Select meshed object A, shown in Figure 9–46.
4. In the Collapse mesh faces by area dialog box, click on the Select face button.
5. Click on the upper-right Select face button and then select face B, shown in Figure 9–46.
6. Click on the OK button.

   Faces greater than the selected face are collapsed.

7. Do not save your file.

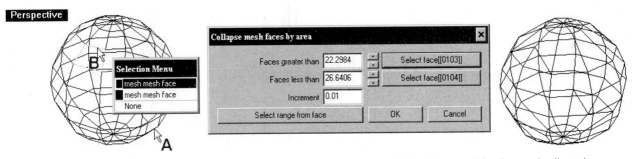

FIGURE 9–46 *From left to right: meshed object and a face selected, Collapse mesh faces by area dialog box, and collapsed meshed object*
Source: Robert McNeel and Associates Rhinoceros® 5

Collapsing Mesh Element by Mesh's Aspect Ratio

A method to remove mesh element is to state the aspect ratio of mesh elements to be collapsed.

1. Select File > Open and select the file Collapse.3dm from the Chapter 9 folder on the Student Companion Website.
2. Mesh > Mesh Edit Tools > Collapse > By Aspect Ratio.
3. Select meshed object A, shown in Figure 9–47.
4. Click on the Select Aspect Ratio From Face button and then select face B.
5. Click on the OK button.

   Faces equal to the selected face's aspect ratio are collapsed.

6. Do not save your file.

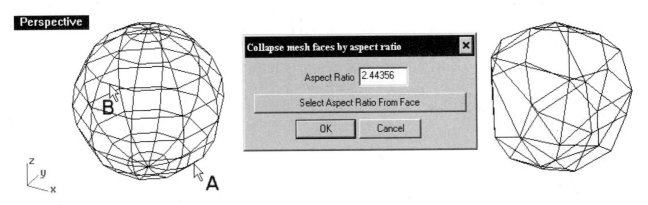

FIGURE 9–47 *From left to right: meshed object and mesh face selected, Collapse mesh face by aspect ratio dialog box, and collapsed meshed object*
Source: Robert McNeel and Associates Rhinoceros® 5

Collapsing Mesh Element by Edge Length

You can specify mesh elements' edge length to collapse, as follows.

1. Select File > Open and select the file Collapse.3dm from the Chapter 9 folder on the Student Companion Website.
2. Select Mesh > Mesh Edit Tools > Collapse > By Edge Length.
3. Select meshed object A, shown in Figure 9–48.
4. Click on the Select Edge button of the Collapse mesh faces by edge length dialog box.
5. Select edge B.
6. Click on the OK button.
    Edges longer than the selected edge are collapsed.
7. Do not save your file.

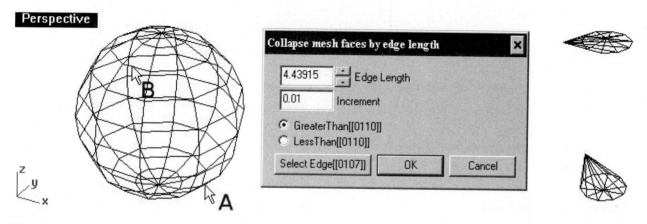

FIGURE 9–48 *From left to right: meshed object and edge selected, Collapse mesh faces by edge length dialog box, and collapsed meshed object*
Source: Robert McNeel and Associates Rhinoceros® 5

Increasing Mesh Faces by Edge Splitting

You can split the edge of a mesh to increase the number of meshes. By clicking along a selected edge, meshes shared by the edge will be split into more meshes. Perform the following steps.

1. Select File > Open and select the file SplitMeshEdge.3dm from the Chapter 9 folder on the Student Companion Website.
2. Select Mesh > Mesh Repair Tools > Split Edge.
3. Select edge A, shown in Figure 9–49.
4. Select a location near the midpoint of edge A, and press the ENTER key.
5. Select edge B, shown in Figure 9–49.
6. Select a location near the midpoint of edge A, and press the ENTER key.
7. Press the ENTER key.
8. Do not save your file.

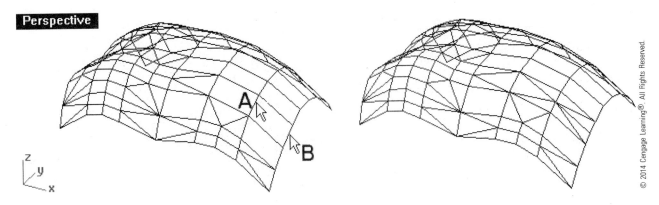

FIGURE 9–49 Original object (left) and edges split (right)

Deleting Mesh Faces

Unlike mesh face collapsing, deleting a mesh face removes a mesh element from the polygon mesh, leaving an opening. Perform the following steps to remove four mesh faces from a polygon mesh sphere.

1. Select File > Open and select the file DeleteMesh.3dm from the Chapter 9 folder on the Student Companion Website.
2. Select Mesh > Mesh Edit Tools > Delete Mesh Faces.
3. Select faces A, B, C, and D, shown in Figure 9–50, and press the ENTER key.
   The faces are deleted.
4. Do not save your file.

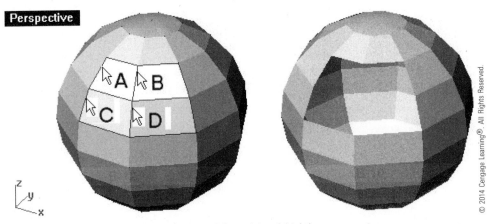

FIGURE 9–50 Mesh faces selected (left) and faces deleted (right)

Patching Single Mesh Face

Patching refers to filling up openings by adding individual triangular or quadrilateral mesh faces. You can construct triangular mesh face by selecting an edge and a vertex. If you select two edges, you construct a quadrilateral face. Perform the following steps.

1. Select File > Open and select the file PatchFace.3dm from the Chapter 9 folder on the Student Companion Website.
2. Select Mesh > Mesh Repair Tools > Patch Single Face.
3. If JoinMesh=No, select JoinMesh option at the command area to change it to Yes. Otherwise, proceed to the next step.

4. Select edge A and vertex B, shown in Figure 9–51.

   A triangular mesh face is patched.

5. Repeat the command.

6. Select edges C and D.

   A quadrilateral mesh face is patched.

7. Do not save your file.

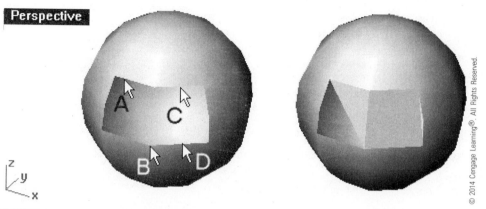

FIGURE 9–51 *Meshed object with a hole (left) and two mesh faces patched (right)*

Filling Holes

Instead of patching single faces on a meshed object, you can fill selected holes or all the holes.

Filling Individual Holes

To fill individual holes of a meshed object, perform the following steps.

1. Select File > Open and select the file FillHole.3dm from the Chapter 9 folder on the Student Companion Website.

2. Select Mesh > Mesh Repair Tools > Fill Hole.

3. Select edge A, shown in Figure 9–52.

   A hole is filled with triangular meshes.

Filling All Holes

To fill all the holes in a meshed object, continue with the following steps.

4. Select Mesh > Mesh Repair Tools > Fill Holes.

5. Select meshed object B, shown in Figure 9–52.

   All the holes are filled.

6. Do not save your file.

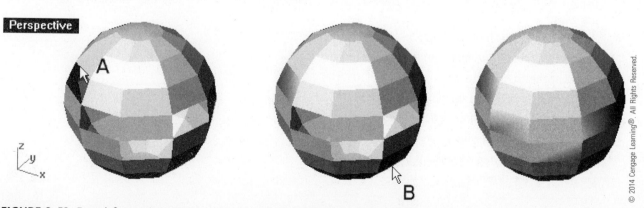

FIGURE 9–52 *From left to right: original meshed object, a hole filled, and all holes filled*

Matching Edges

Sometimes, a meshed object is not watertight because there are misaligned edges between contiguous mesh elements in the meshed object, leaving some openings along the edges. To repair such meshed objects, you can use the MatchMeshEdge command to match vertices within a specified tolerance and split the edges to make the edges match.

1. Select File > Open and select the file MatchMeshEdge.3dm from the Chapter 9 folder on the Student Companion Website.

   This meshed object consists of four polygon meshes joined together. Along the joints, there are gaps.

2. Select Mesh > Mesh Repair Tools > Match Mesh Edge.
3. Select the DistanceToAdjust option at the command area.
4. Type 4 at the command area.
5. Select the PickEdges option at the command area.
6. Select edge A, shown in Figure 9–53, and press the ENTER key.

   The edges of the mesh elements along the selected edge are adjusted and matched.

7. Repeat the command.
8. Select the meshed object and press the ENTER key.

   All edges are adjusted to match.

9. Do not save your file.

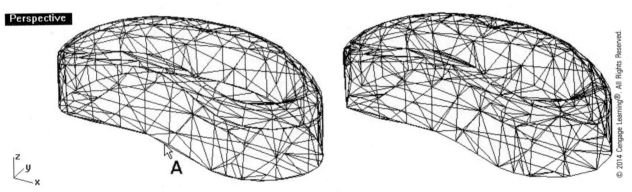

FIGURE 9–53 *Original meshed object (left) and matched object (right)*

Aligning Vertices

A way to close small gaps in a meshed object is to specify a tolerance value for vertices within such value to align into a single vertex. Care must be taken not to specify excessively large tolerance. Otherwise, too many mesh vertices will be collapsed, and the polygon mesh will be oversimplified. Perform the following steps.

1. Select File > Open and select the file AlignMeshVertices.3dm from the Chapter 9 folder on the Student Companion Website. This meshed object consists of four polygon meshes joined together. Along the joints, there are gaps.
2. Select Mesh > Mesh Repair Tools > Align Mesh Vertices.
3. Select the DistanceToAdjust option at the command area.
4. Type 1 at the command area.
5. Select meshed object A, shown in Figure 9-54.

   Vertices within the tolerance region are adjusted.

6. Repeat the command with a tolerance of 2 units.

   As can be seen, the meshed object is now oversimplified.

7. Do not save the drawing.

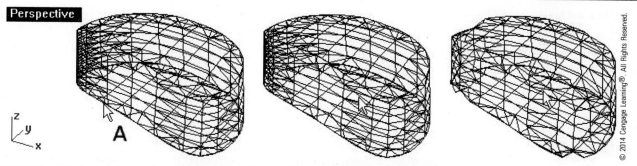

FIGURE 9–54 *From left to right: original meshed object, vertices within 1 unit tolerance collapsed, and vertices within 2 units tolerance collapsed*

Extraction of Mesh Elements

Extraction involves isolating selected mesh faces from a polygon mesh. The extracted elements are removed from the original mesh as separate polygon meshes. They are not deleted.

Extracting Mesh Face

You can extract selected faces of a meshed object. Extraction of mesh face isolates the mesh faces from the main body of the mesh object, leaving holes in the original object. Perform the following steps.

1. Select File > Open and select the file Extract01.3dm from the Chapter 9 folder on the Student Companion Website.
2. Select Mesh > Mesh Edit Tools > Extract > Faces.
3. Select mesh faces A and B, shown in Figure 9–55, and press the ENTER key.
   The selected mesh faces are extracted (detached and isolated) from the original mesh object.
4. To realize that the faces are extracted, press the DELETE key to delete the extracted faces and shade the viewport.
5. Do not save your file.

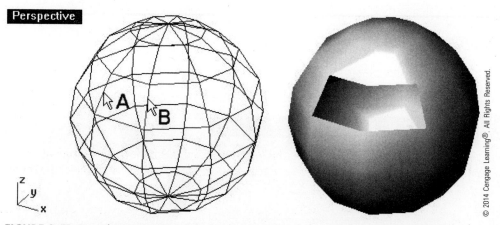

FIGURE 9–55 *Faces being extracted (left) and extracted faces deleted and viewport shaded (right).*

Extracting Mesh Faces by Specifying Mesh Face Area

Apart from extracting individual mesh faces, you can extract a set of mesh faces collectively by specifying the mesh area.

1. Select File > Open and select the file Extract01.3dm from the Chapter 9 folder on the Student Companion Website.
   This is the same file as the previous tutorial.
2. Select Mesh > Mesh Edit Tools > Extract > By Area.

3. Select meshed object A, shown in Figure 9–56.
4. In the Extract mesh faces by area dialog box, click on the Select range from face button.
5. Select face B, shown in Figure 9–56.
6. Click on the OK button. Mesh faces within area range are extracted.
7. Delete the extracted mesh faces and shade the viewport to appreciate the result.
8. Do not save your file.

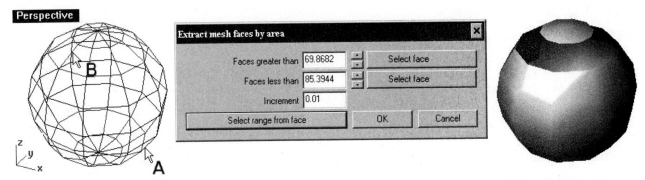

FIGURE 9–56 *From left to right: mesh faces selected, Extract mesh faces by area dialog box, and extracted faces deleted*
Source: Robert McNeel and Associates Rhinoceros® 5

Extracting Mesh Faces by Specifying Mesh Aspect Ratio

Another way of extracting a set of mesh faces is to specify mesh aspect ratio. Perform the following steps.

1. Select File > Open and select the file Extract01.3dm from the Chapter 9 folder on the Student Companion Website.
2. Select Mesh > Mesh Edit Tools > Extract > By Aspect Ratio.
3. Select meshed object A, shown in Figure 9–57.
4. Click on the Select Aspect Ratio From Face button of the Extract mesh faces by aspect ratio dialog box.
5. Select mesh face B.
6. Click on the OK button.
7. Delete the extracted faces and shade the viewport.
8. Do not save your file.

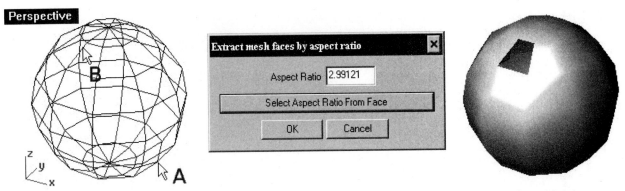

FIGURE 9–57 *From left to right: face selected, Extract mesh faces by aspect ratio dialog box, and extracted faces deleted*
Source: Robert McNeel and Associates Rhinoceros® 5

Extracting Mesh Faces by Specifying Mesh Draft Angle

In accordance with the mesh draft angle, you can extract a set of mesh faces. Perform the following steps.

1. Select File > Open and select the file Extract01.3dm from the Chapter 9 folder on the Student Companion Website.

2. Select Mesh > Mesh Edit Tools > Extract > By Draft Angle.
3. Select meshed object A, shown in Figure 9–58.
4. Set End angle from camera direction to 30 degree in the Extract Mesh Faces By Draft Angle dialog box.
5. Click on the OK button.
6. Delete the extracted faces and shade the viewport.
7. Do not save your file.

FIGURE 9–58 *From left to right: Meshed object, Extract Mesh Faces By Draft Angle dialog box, and extracted faces deleted*
Source: Robert McNeel and Associates Rhinoceros® 5

Extracting Mesh Faces by Specifying Mesh Face Edge Length

Mesh faces in a mesh object can be extracted by specifying mesh face aspect ratio. Perform the following steps.

1. Select File > Open and select the file Extract01.3dm from the Chapter 9 folder on the Student Companion Website.
2. Select Mesh > Mesh Edit Tools > Extract > By Edge Length.
3. Select mesh object A, shown in Figure 9–59.
4. In the Extract mesh faces by edge length dialog box, check the Greater Than button and click on the Select Edge button.
5. Select edge B, shown in Figure 9–59.
6. Click on the OK button.
7. Delete the extracted faces to see the result.
8. Do not save your file.

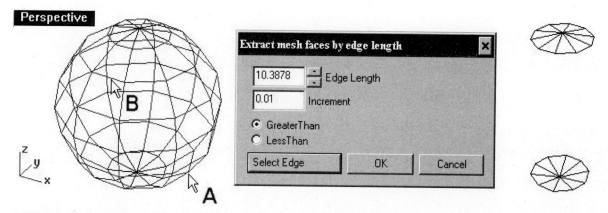

FIGURE 9–59 *From left to right: mesh face selected, Extract mesh faces by edge length dialog box, and extracted faces deleted*
Source: Robert McNeel and Associates Rhinoceros® 5

You can extract mesh faces that are connected to a selected face within a specified break angle. Perform the following steps.

1. Select File > Open and select the file Extract02.3dm from the Chapter 9 folder on the Student Companion Website.

2. Select Mesh > Mesh Edit Tools > Extract > Connected.
3. Select face A, shown in Figure 9–60.
4. In the Extract Connected Mesh Faces dialog box, set angle value to 35, check the Less than box, and click on the OK button.
5. Delete the extracted mesh faces to appreciate the result.
6. Do not save your file.

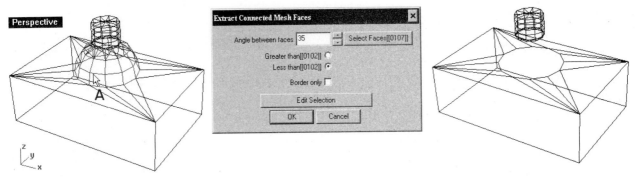

FIGURE 9–60 *From left to right: face selected, Extract Connected Mesh Faces dialog box, and extracted faces deleted*
Source: Robert McNeel and Associates Rhinoceros® 5

Extracting Mesh Faces Bounded by Unwelded Edges

You can extract faces of a meshed object that are not welded. Perform the following steps.

1. Select File > Open and select the file Extract03.3dm from the Chapter 9 folder on the Student Companion Website.
2. Select Mesh > Mesh Edit Tools > Extract > Part.
3. Select face A, shown in Figure 9–61.
4. Delete the extracted faces to see the result.
5. Do not save your file.

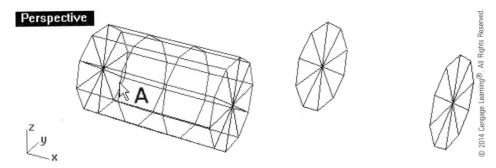

FIGURE 9–61 *Face selected (left) and extracted faces deleted (right)*

Converting Mesh Faces

Polygon mesh can be manifested as quadrilateral mesh elements or triangular mesh elements. You can convert all the meshes of a meshed object to triangular meshes or convert only nonplanar quadrilateral meshes to triangular meshes. Conversely, you can convert triangular mesh faces to quadrilateral mesh faces.

Converting Nonplanar Quadrilateral Mesh Faces to Triangular Mesh Faces

To convert only nonplanar quadrilateral mesh elements of a meshed object to triangular mesh elements, perform the following steps.

1. Select File > Open and select the file TriangulateMesh.3dm from the Chapter 9 folder on the Student Companion Website.

2. Select Mesh > Mesh Edit Tools > Triangulate Non-Planar Quads.
3. Select mesh A, shown in Figure 9–62.
4. Click on the OK button of the Triangulate Non-Planar Quads dialog box.

   Nonplanar quadrilateral meshes are converted to triangular meshes.

FIGURE 9–62 *From left to right: original meshed object, Triangulate Non-Planar Quads dialog box, and nonplanar quadrilateral meshes changed to triangular meshes*
Source: Robert McNeel and Associates Rhinoceros® 5

Converting All Mesh Faces to Triangular Mesh Faces

To convert all quadrilateral mesh elements of a meshed object to triangular mesh elements, continue with the following steps.

5. Left-click on the Triangulate Mesh/Triangulate non-planar quads button on the Mesh Tools toolbar.
6. Select B, shown in Figure 9–62, and press the ENTER key.

   All quadrilateral meshes are converted to triangular meshes. (See Figure 9–63.)

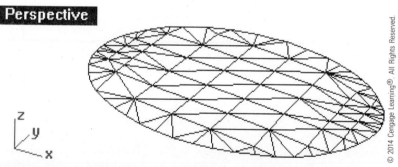

FIGURE 9–63 *All quadrilateral meshes changed to triangular meshes*

Converting Triangular Mesh Faces to Quadrilateral Mesh Faces

Contrary to converting quadrilateral mesh faces to triangular mesh faces, you can convert triangular mesh faces to quadrilateral mesh faces. However, not all triangular mesh faces can be converted. Continue with the following steps.

7. Click on the Quadrangulate Mesh button on the Mesh Tools toolbar.
8. Select the polygon mesh and press the ENTER key.

   Possible mesh faces are converted to quadrilateral faces, as shown in Figure 9–64.

9. Do not save your file.

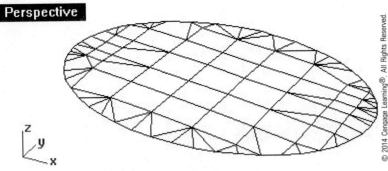

FIGURE 9-64 *Some faces converted to quadrilateral faces*

Rearranging Mesh Faces by Swapping Edges

You can improve mesh organization by swapping individual mesh edges. Corners of the selected triangle will swap.

1. Select File > Open and select the file SwapMeshEdge.3dm from the Chapter 9 folder on the Student Companion Website.
2. Select Mesh > Mesh Repair Tools > Swap Edge.
3. Select mesh edges A and B, shown in Figure 9–65, and press the ENTER key.

   The related meshes are rearranged. You can continue to swap other edges as appropriate.
4. Do not save your file.

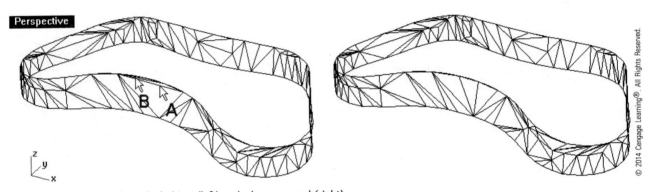

FIGURE 9-65 *Original meshed object (left) and edges swapped (right)*

Unifying Normal Direction

Quite often, a polygon mesh model is constructed from a set of polygon meshes joined together. Naturally, the normal directions of each individual polygon mesh may not be congruent. To unify the normal directions of the joined meshes, perform the following steps.

1. Select File > Open and select the file Normal.3dm from the Chapter 9 folder on the Student Companion Website.
2. Select Analyze > Direction.

   The normal direction of the meshed object is displayed.
3. Select Mesh > Mesh Repair Tools > Unify Normals.
4. Select polygon mesh A, shown in Figure 9–66.

   The normal directions of the meshes are unified.
5. Select Analyze > Direction.
6. Click on the Flip option at the command area as necessary.
7. Do not save your file.

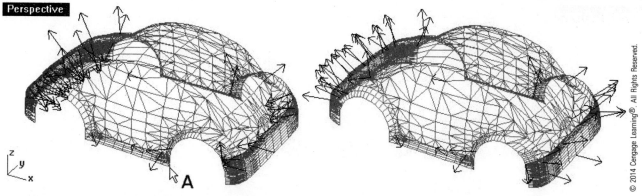

FIGURE 9–66 *Normal directions (left) and normal directions unified (right)*

Welding Polygon Meshes

At a glance, welding seems to be analogous to joining. However, these processes are different in that a joined set of polygon meshes still retains the common edges between contiguous meshes and a welded set of polygon meshes has its contiguous duplicated edges removed. To make rendering and shading look smoother, you can weld the entire meshed object, selected edges, or selected vertices.

Welding

To weld a meshed object, you have to specify angle tolerance that governs the angle between contiguous mesh elements. Perform the following steps.

1. Select File > Open and select the file Weld.3dm from the Chapter 9 folder on the Student Companion Website.
2. Select Mesh > Mesh Edit Tools > Weld.
3. Select meshed object A, shown in Figure 9–67, and press the ENTER key.
4. Type 35 at the command area to specify the angle tolerance.

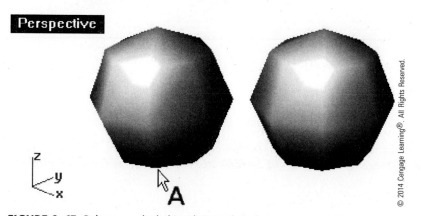

FIGURE 9–67 *Polygon meshed object being selected*

Welding Edges

To weld selected edges of a meshed object, continue with the following steps.

5. Click on the Weld Mesh Edge/Unweld Mesh Edge button on the Welding toolbar.
6. Select edges A, B, C, D, E, F, G, H, J, K, L, and M, shown in Figure 9–68.
   The selected edges are welded.

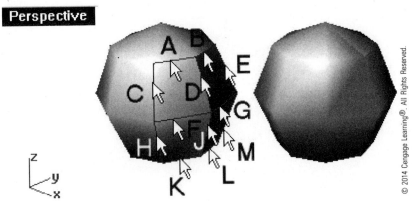

FIGURE 9-68 *Edges being welded*

Welding Vertices

To weld selected vertices, continue with the following steps.

7. Select Mesh > Mesh Edit Tools > Weld Selected Vertices.
8. Select vertices A, B, C, D, E, F, G, H, J, and K shown in Figure 9-69 and press the ENTER key. The selected vertices are welded.
9. Do not save your file.

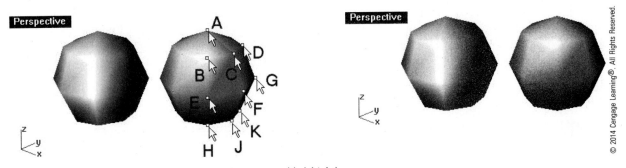

FIGURE 9-69 *Vertices being welded (left) and vertices welded (right)*

Unwelding

You can unweld vertices in accordance to a specified break angle. At the unwelded joints, no face will share a vertex with another. Perform the following steps.

1. Select File > Open and select the file Unweld.3dm from the Chapter 9 folder on the Student Companion Website.
2. Select Mesh > Mesh Edit Tools > Unweld.
3. Select meshed object A, shown in Figure 9-70, and press the ENTER key.
4. Type 10 to set the break angle.

   The meshed object is unwelded.

5. Do not save your file.

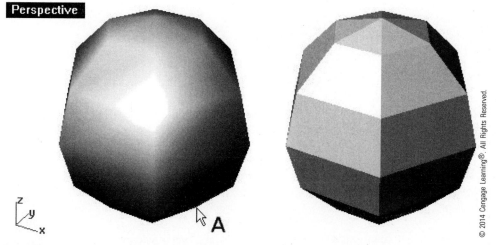

FIGURE 9–70 Original meshed object (left) and unwelded object (right)

General Repair Tools

Apart from the manipulation tools delineated so far, you can improve a mesh object by rebuilding the entire mesh object, rebuilding the mesh normals, extracting bad mesh faces or duplicated faces, culling bad faces, and splitting disjoint mesh.

Rebuilding Mesh

If a mesh is not acting properly, you may rebuild it. Rebuilding polygon mesh concerns stripping all information from a mesh and returning only the mesh geometry. You can rebuild a mesh object by

- Selecting Mesh > Mesh Repair Tools > Rebuild Mesh
- Clicking on the Rebuild Mesh button on the Mesh Tools toolbar

Rebuilding Mesh Normal

This is a way to improve the quality of a meshed object. The mesh normals are first removed, and then the face and vertex normals are reconstructed based on the orientation of the faces. Note that, after rebuilding, the direction of normal will not change. Normal can be rebuilt by

- Selecting Mesh > Mesh Repair Tools > Rebuild Mesh Normals
- Clicking on the Rebuild Mesh Normals button on the Mesh Tools toolbar

Extracting Duplicated Mesh Face

To remove duplicated mesh faces from a mesh object, click on the Duplicate Faces button on the Extract Mesh toolbar and select the mesh object.

Culling an Invalid Mesh

A meshed object may be invalid because one or more of its mesh elements have a zero area. To remove all such elements from the meshed object, you cull degenerated faces by

- Selecting Mesh > Mesh Edit Tools > Cull Degenerate Mesh Faces
- Clicking on the Cull degenerate mesh faces on the Mesh Tools toolbar

Splitting Disjoint Mesh

If a polygon mesh contains any disjoint mesh elements, one method to repair the polygon mesh is to first split the disjoint mesh element and then do subsequent repair work. To split disjoint mesh element, click on the Split disjoint mesh button on the Mesh Tools toolbar and select the mesh object.

Mesh Repair

To repair a mesh object imported from other applications, you can use the MeshRepair command.

Polygon Count

To find out the number of polygons in a mesh object, you use the PolygonCount command.

CONSOLIDATION

Although NURBS surfaces are the mainstream surface type in design and manufacturing, some systems still use polygon meshes to represent a free-form body, in particular, rapid prototyping systems. To cope with these systems, you may need to know how to construct and manipulate polygon meshes.

A polygon mesh approximates a smooth surface by employing a set of small planar polygonal faces. Accuracy of representation is inversely proportional to the size of the polygon—the smaller the polygon size, the more accurate the surface. One major disadvantage of the polygon mesh is that the file size increases tremendously with a decrease in the polygon size. Another disadvantage is that a surface can never be accurately represented with a mesh, no matter how small the polygon size.

You learned how to construct various kinds of polygon meshes from scratch. These polygon meshes are single mesh face, rectangular mesh plane, mesh box, mesh sphere, mesh cylinder, mesh cone, mesh truncated cone, mesh ellipsoid, and mesh torus.

You learned ways to derive polygon meshes from a set of points, from a closed polyline, from the profile of a surface/polysurface, from the control points of a surface, and from the heightfield of an image. You also learned how to offset a polygon mesh, map a polygon mesh to a surface, and apply polygon meshes and point objects to a surface's U, V, and N directions.

To combine two or more polygon meshes, you use Boolean union, difference, and intersection operations. Without changing the shape of the polygon meshes, you join contiguous polygon meshes and explode joined polygon meshes. In addition, you can split a polygon mesh into two and trim away unwanted portion of a polygon mesh. Finally, you learned various ways to manipulate individual mesh elements of a polygon mesh, which is particularly useful in repairing polygon meshes for rapid prototyping.

REVIEW QUESTIONS

1. Outline the methods of constructing various kinds of polygon meshes.
2. List the ways to construct polygon meshes from existing objects.
3. How can two or more polygon meshes be combined?
4. In what ways can a polygon be split and trimmed?
5. List the ways to manipulate elements of a polygon mesh.

CHAPTER 10

Advanced Modeling Methods— Transformation

INTRODUCTION

This chapter explores how curves, surfaces, and polygon meshes are transformed in terms of translation and deformation in order to achieve special modeling effects.

OBJECTIVES

After studying this chapter, you should be able to

- Translate curves, surfaces, and polygon meshes
- Deform curves, surfaces, and polygon meshes

OVERVIEW

An advanced way of modeling is to translate and deform curves, surfaces, and polygon meshes to achieve specific design outcomes. In essence, translation of objects concerns relocating selected objects or scaling selected objects without changing their basic shape. On the other hand, deformation changes an existing object's basic shapes.

TRANSLATION

Building on the basic translation methods (moving, copying, rotating, scaling, and mirroring) depicted in Chapter 2, this section will explain the following translation processes: soft moving, aligning, orienting, remapping, arraying, and making symmetric objects. In addition, history update will be reviewed.

Soft Moving

You can move a set of objects using a falloff value, so that objects farther away from the center will move less. Perform the following steps.

1. Select File > Open and select the file SoftMove.3dm from the Chapter 10 folder that you downloaded from the Student Companion Website.
2. Select Transform > Soft Move.
3. Select all the sphere objects and press the ENTER key.
4. Click on location A (shown in Figure 10–1) to specify the center of the move.

463

5. Click on location B (Figure 10–1) to specify the radius affected by the move.
6. Click on location C (Figure 10–2) to specify the offset point and press the ENTER key.
   The objects are moved.
7. Do not save your file.

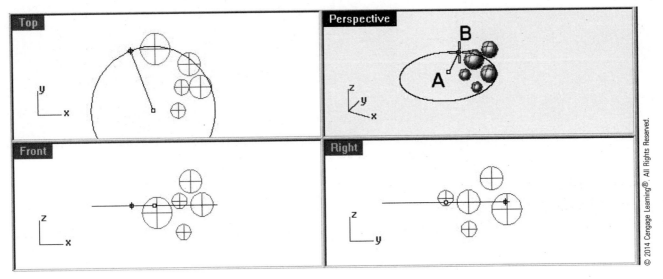

FIGURE 10–1 *Point to move from and radius point selected*

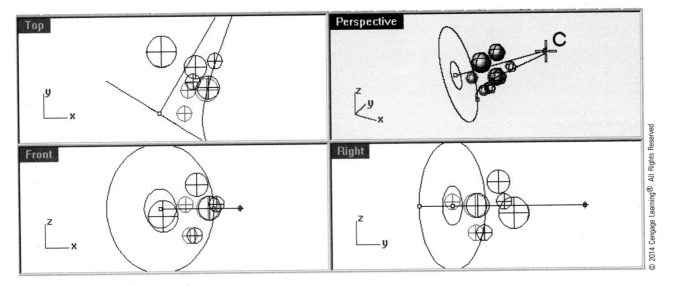

FIGURE 10–2 *Objects being moved*

Aligning Curves

Planar curves residing on one of the World construction planes (Top, Front, and Right) can be aligned, as follows.

1. Select File > Open and select the file AlignProfile.3dm from the Chapter 10 folder on the Student Companion Website.
2. Type AlignProfiles at the command area.
3. Select curve A, shown in Figure 10-3.
   This is the curve you will align to.

4. Select curve B.

   This is the curve to change, and the second curve is modified to align to the first curve.

5. Do not save your file.

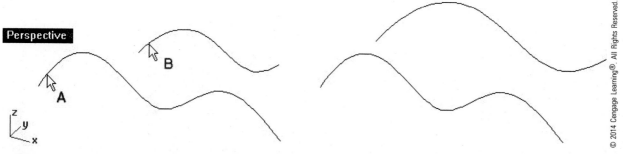

FIGURE 10–3 *Original curves (left) and aligned curve (right)*

Aligning Objects

Objects can be aligned by their bounding boxes, as follows.

1. Select File > Open and select the file Align.3dm from the Chapter 10 folder on the Student Companion Website.
2. Select Transform > Align.
3. Click on the Top viewport to make it the active viewport.
4. Select objects A, B, and C (Figure 10–4) and press the ENTER key.
5. Select the VertCenter option at the command area.

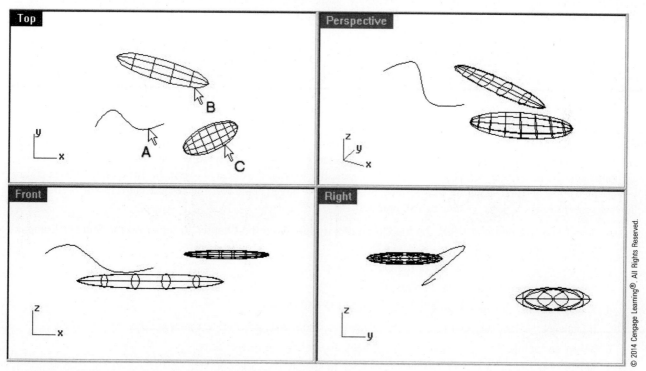

FIGURE 10–4 *Objects being aligned in the Top viewport*

6. Repeat the command.
7. Click on the Right viewport, select all three objects, and press the ENTER key.

8. Select the VertCenter option at the command area.
   Objects are aligned, as shown in Figure 10–5.
9. Do not save your file.

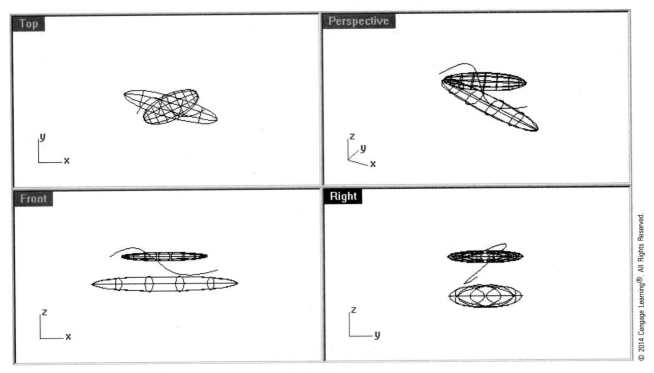

FIGURE 10–5 *Objects aligned in both the Top viewport and the Right viewport*

Orienting Objects

The Orient tool is used to change the orientation of selected objects and, where appropriate, change their scale.

Orient Two Points

A curve, surface, or polygon mesh can be repositioned and uniformly scaled by referencing two points and two target points. Basically, the object's first reference point will be relocated to the first target point, and the object's second reference point will be relocated to the second target point. As a result, the object's size is scaled by a ratio of the distances between the reference points and the target points. Optionally, you may choose not to scale the oriented object. In this case, the object is simply reoriented to a position along an axis formed from the first target point and the second target point. Perform the following steps to orient objects referencing two points.

1. Select File > Open and select the file Orient2Points.3dm from the Chapter 10 folder on the Student Companion Website.
2. Check the End and Point boxes of the Osnap dialog box.
3. Select Transform > Orient > 2 Points.
4. Select curve A, shown in Figure 10–6, and press the ENTER key.
5. If Scale=No, select the option to change Scale=Yes. Otherwise, proceed to the next step.
6. Select endpoints B and C, shown in Figure 10–6, to specify the reference points.
7. Select endpoints D and E, shown in Figure 10–6, to specify the target points.
   The selected curve is relocated and scaled.

   If Scale=No, the curve will simply reposition using BC as the reference axis and DE as the target axis.

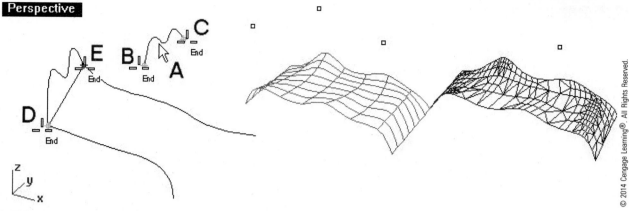

FIGURE 10–6 *Curve being oriented*

8. Select Transform > Orient > 2 Points.
9. Select surface A, shown in Figure 10-7, and press the ENTER key.
10. Select endpoints B and C (Figure 10-7) to specify the reference points.
11. Select endpoints D and E (Figure 10-7) to specify the target points.
    The selected surface is relocated and scaled.

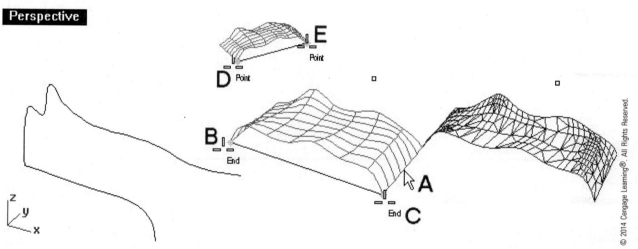

FIGURE 10–7 *Surface being oriented*

12. Select Transform > Orient > 2 Points.
13. Select polygon mesh A, shown in Figure 10-8, and press the ENTER key.
14. Select endpoints B and C (Figure 10-8) to specify the reference points.
15. Select endpoints D and E (Figure 10-8) to specify the target points.
    The selected polygon mesh is relocated and scaled.
16. Do not save your file.

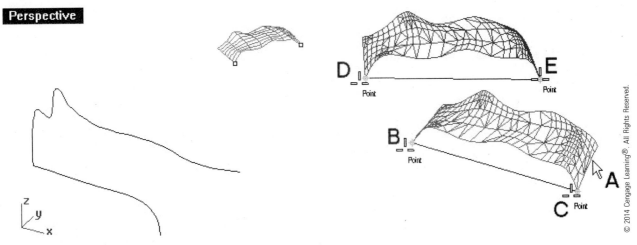

FIGURE 10–8 Polygon mesh being oriented

Orient Three Points

A curve, surface, or polygon mesh can be repositioned referencing an origin point, an axis, and a plane. The object is not scaled. (Note that you can also do this with the Orient 2 Points command using the No Scale option.) During orientation, you need to input three reference points and three target points. The object's first reference point will be the origin of translation; it will relocate to the first target point. The object's first and second reference points will specify an axis, which will align with an axis formed by the first and second target points. The object's first, second, and third reference points will specify a plane, which will align with a plane formed by the first, second, and third target points. Perform the following steps.

1. Select File > Open and select the file Orient3Points.3dm from the Chapter 10 folder on the Student Companion Website.
2. Check the End, Point, Cen, and Quad boxes, and clear all other boxes on the Osnap dialog box.
3. Select Transform > Orient > 3 Points.
4. Select circle A, shown in Figure 10–9, and press the ENTER key.
5. Select center point B, quadrant point C, and quadrant point D (Figure 10–9) to specify the reference points.
   Point C is the source origin, CD is the source axis, and BCD is the source plane.
6. Select endpoints E, F, and G (Figure 10–9) to specify the target points.
   The selected circle is relocated. Point E is the target origin, EF is the target axis, and EFG is the target plane.

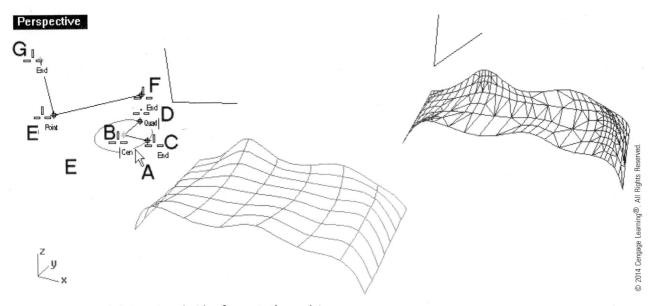

FIGURE 10–9 Circle being oriented with reference to three points

7. Select Transform > Orient > 3 Points.
8. Select surface A, shown in Figure 10-10, and press the ENTER key.
9. Select endpoints B, C, and D (Figure 10-10) to specify the reference points.
10. Select endpoints F, G, and H (Figure 10-10) to specify the target points.
    The selected surface is relocated.

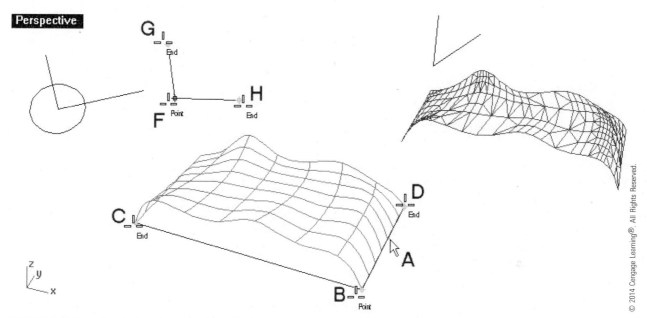

FIGURE 10-10 *Surface being oriented referencing three points*

11. Select Transform > Orient > 3 Points.
12. Select polygon mesh A, shown in Figure 10-11, and press the ENTER key.
13. Select endpoints B, C, and D (Figure 10-11) to specify the reference points.
14. Select endpoints E, F, and G (Figure 10-11) to specify the target points.
    The selected polygon mesh is relocated.
15. Do not save your file.

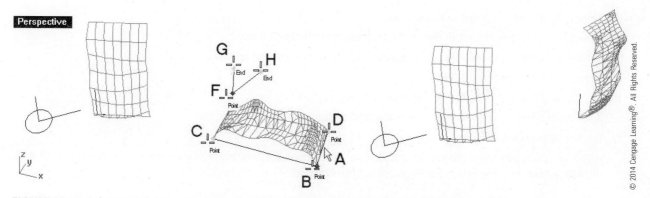

FIGURE 10-11 *Polygon mesh being oriented with reference to three points (left) and polygon mesh oriented (right)*

Orient on Surface

A simple way to construct an object (a curve, surface, or polygon mesh) on a surface is to first construct the object on any construction plane and then orient the object on the surface. After orientation, the reference plane (construction plane on which the objects are constructed) of the translated object will be tangent to the target surface. Perform the following steps.

1. Select File > Open and select the file OrientOnSurface.3dm from the Chapter 10 folder on the Student Companion Website.
2. Check the Point and Cen boxes, and clear all other boxes on the Osnap dialog box.
3. Select Transform > Orient > On Surface.
4. Select circle A, shown in Figure 10-12, and press the ENTER key.
5. Select center B (Figure 10-12) of the circle as the point to be oriented.
6. Select surface C (Figure 10-12).
7. Select points D and E (Figure 10-12) and press the ENTER key.

   Two copies of the selected circle are oriented on the surface.

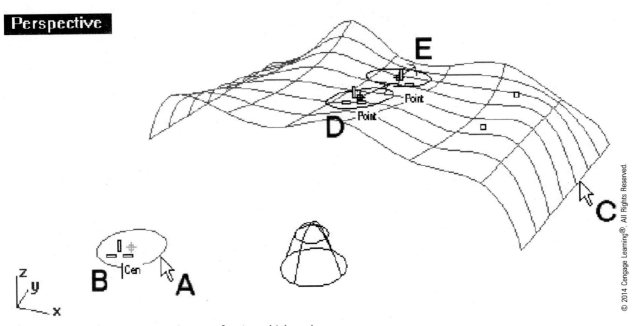

FIGURE 10-12 *Circle being oriented to a surface in multiple copies*

8. Select Transform > Orient > On Surface.
9. Select paraboloid A, shown in Figure 10-13, and press the ENTER key.
10. Select center B (Figure 10-13) of the paraboloid as the point to be oriented.
11. Select surface C (Figure 10-13).
12. Select points D and E (Figure 10-13) and press the ENTER key.

    Two copies of the selected paraboloid are oriented on the surface.
13. Do not save your file.

    Although a polygon mesh can also be oriented on a surface, it does not make much sense to mix polygon mesh and NURBS surface in a file, except for the purpose of illustration.

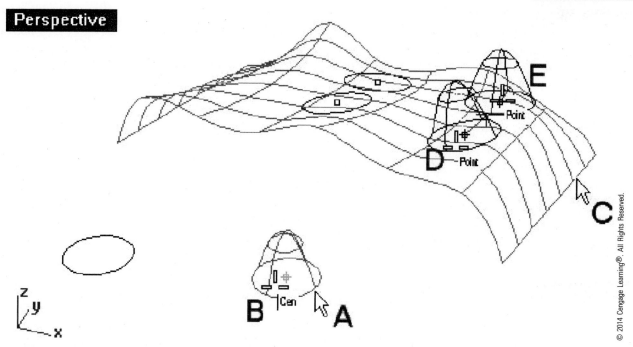

FIGURE 10–13 *Paraboloid being oriented to a surface in multiple copies*

Orient Perpendicular to Curve

Instead of setting up a construction plane perpendicular to a selected point on a curve for object construction, you may first construct objects on any construction plane and then orient them perpendicular to the target curve. This command is construction-plane dependent. The reference plane of the oriented object will become perpendicular to the target curve. Perform the following steps.

1. Select File > Open and select the file OrientPerpendicularToCurve.3dm from the Chapter 10 folder on the Student Companion Website.
2. Check the End, Point, and Cen boxes, and clear all other boxes on the Osnap dialog box.
3. Select Transform > Orient > Perpendicular to Curve.
4. Select circle A, shown in Figure 10–14, and press the ENTER key.
5. Select center B as the base point.
6. Select curve C.
7. Select point D of the curve.

   The circle is oriented perpendicular to the curve.

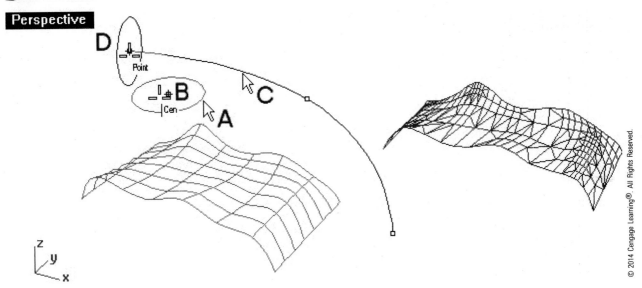

FIGURE 10–14 *Circle being oriented perpendicular to a curve*

8. Select Transform > Orient > Perpendicular to Curve.
9. Select surface A, shown in Figure 10–15, and press the ENTER key.
10. Select endpoint B, shown in Figure 10–15, as the base point.
11. Select curve C.
12. Select point D of the curve.
    The surface is oriented perpendicular to the curve.

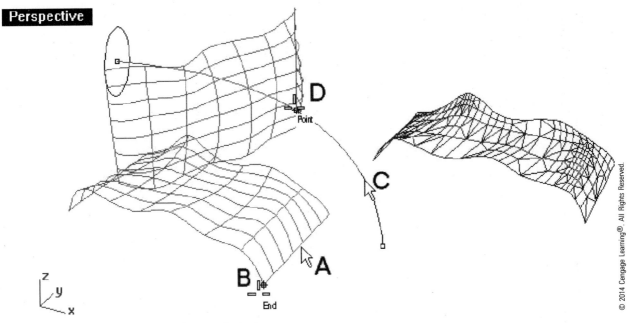

FIGURE 10–15 *Surface being oriented perpendicular to a curve*

13. Select Transform > Orient > Perpendicular to Curve.
14. Select polygon mesh A, shown in Figure 10–16, and press the ENTER key.
15. Select endpoint B, shown in Figure 10–16, as the base point.

16. Select curve C.
17. Select endpoint D of the curve.

    The polygon mesh is oriented perpendicular to the curve.
18. Do not save your file.

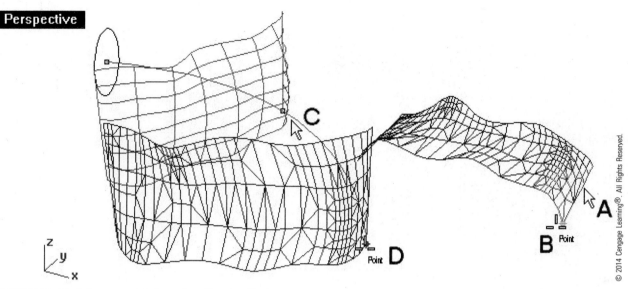

FIGURE 10–16 *Polygon mesh being oriented perpendicular to a curve*

Orient on Curve

You can orient selected curves, surfaces, and polygon meshes along curves, as follows.

1. Select File > Open and select the file OrientonCurve.3dm from the Chapter 10 folder on the Student Companion Website.
2. Set object snap mode to Cen and End.
3. Select Transform > Orient > On Curve.
4. Select circle A (Figure 10–17) and press the ENTER key.
5. Select center B.
6. Select curve C.
7. If the Copy option is No, click on it to change it to Yes.
8. Select point D and press the ENTER key.

    The circle is oriented on the curve.
9. Repeat the command.
10. Select surface E and press the ENTER key.
11. Select center F.
12. Select curve C and point G.

    The surface is oriented on the curve.
13. Repeat the command.
14. Select polygon mesh H and press the ENTER key.
15. Select endpoint J.
16. Select curve C and point K.

    The polygon mesh is oriented on the curve.
17. Do not save your file.

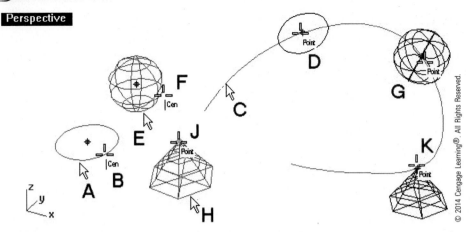

FIGURE 10–17 *Curve, surface, and polygon mesh being oriented on a curve*

Orient Curve to Edge

To construct curves tangent to an edge of a surface, you can first construct the curve and then orient it to the edge of the surface, as follows.

1. Select File > Open and select the file OrientCurveToEdge.3dm from the Chapter 10 folder on the Student Companion Website.
2. Check the Point box and clear all other boxes on the Osnap dialog box.
3. Select Transform > Orient > Curve to Edge.
4. Select curve A near its endpoint A (shown in Figure 10–18) and press the ENTER key.
   Note that orientation of the curve is based on which end of the curve you select.
5. Select surface edge B (Figure 10–18).
6. Select points C, D, E, and F (Figure 10–18) and press the ENTER key.
   Two copies of the curve are oriented to the edge of the surface.
7. Do not save your file.

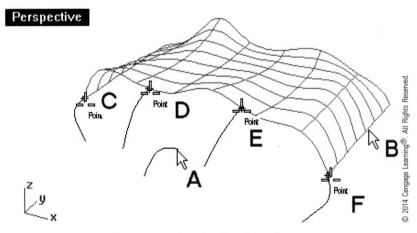

FIGURE 10–18 *Curve being oriented to the edge of a surface*

Remapping Objects from One CPlane to Another CPlane

Objects residing on one construction plane can be repositioned to another construction plane. This process is especially useful for orienting 2D drawings to 3D construction planes. You can map the front view to the Front viewport, the right view to the Right viewport, and so on. This is how most people use this command. Perform the following steps.

1. Select File > Open and select the file OrientRemapToCPlane.3dm from the Chapter 10 folder on the Student Companion Website.
2. Click on the Front viewport to set it as the current viewport.
3. Select Transform > Orient > Remap to CPlane.
4. Select curve A (shown in Figure 10-19) and press the ENTER key.
5. Select the CPlane option at the command area.
6. Type Right at the command area.

    The selected curve is remapped onto the Right viewport. (See Figure 10-20.)

7. Do not save your file.

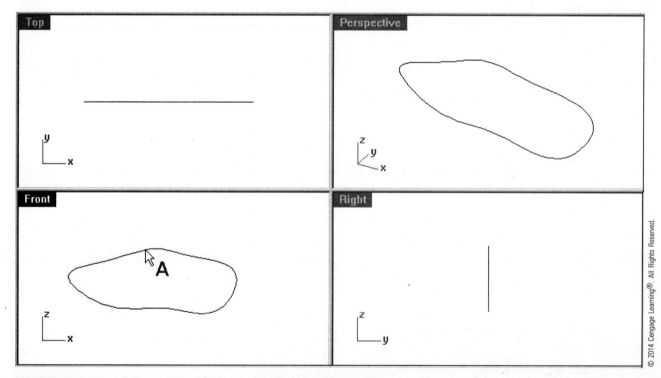

FIGURE 10-19 *Curve being remapped from one viewport to another viewport*

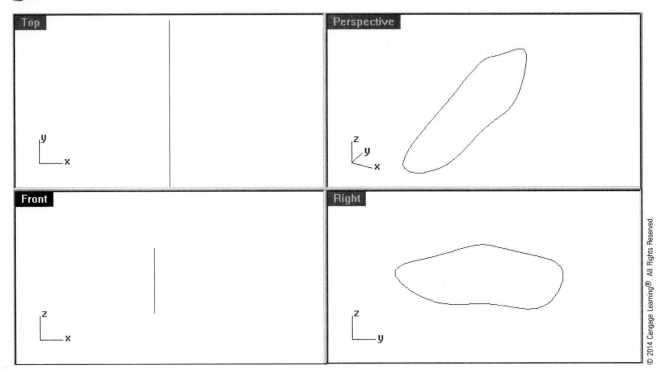

FIGURE 10–20 *Curve remapped from one viewport to another viewport*

DEFORMATION

Deformation processes that are going to be described in this section change the shape of selected curves, surfaces, polysurfaces, and polygon meshes. Processes are setting points, soft editing, projecting, shearing, twisting, bending, tapering, smoothing, making uniform, attaching to another object, stretching, cage editing, and maelstrom editing. (If a number of polygon meshes are to be deformed, welding them beforehand to eliminate duplicated edges will give a better result.)

Setting Points

Point-setting involves deforming the selected objects or their control points by setting selected control points of the objects to the X, Y, and Z coordinates of the World coordinate system or construction plane coordinate system. Perform the following steps.

1. Select File > Open and select the file SetPoints.3dm from the Chapter 10 folder on the Student Companion Website.
2. Select Edit > Control Points > Control Points On.
3. Select polygon mesh A, surface B, and curve C (shown in Figure 10–21) and press the ENTER key.
4. Select Transform > Set Points.
5. Click on location D and drag to location E, shown in Figure 10–21.

   The exact location is unimportant here because this is only an example serving to illustrate the concept.
6. Click on location F and drag to location G (Figure 10–21).
7. Select control points H and J, and press the ENTER key.

   A set of control points is selected.

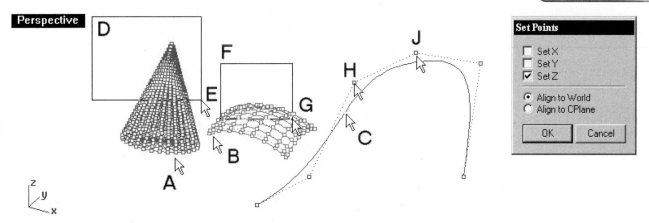

FIGURE 10–21 *Control points of polygon mesh, surface, and curve turned on and being selected (left) and Set Points dialog box (right)*
Source: Robert McNeel and Associates Rhinoceros® 5

8. In the Set Points dialog box, clear the Set X and Set Y boxes (if they are checked) and check the Set Z box.
9. Check the Align to World box and click on the OK button.
10. Set the display to four-viewport configuration by double-clicking the Perspective viewport's label.
11. Click on location A, shown in Figure 10–22.

The selected control points of the curve, surface, and polygon mesh are set to a specified Z value.

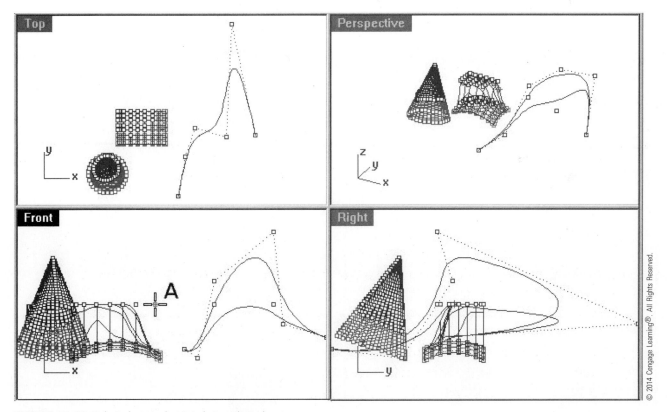

FIGURE 10–22 *Selected control points being aligned*

12. Maximize and shade the Perspective viewport to see the result. (See Figure 10–23.)
13. Do not save your file.

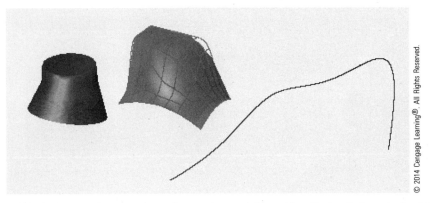

FIGURE 10–23 *Polygon mesh, surface, and curve deformed by aligning their control points*

Projecting to the Construction Plane

Using the set point operation described earlier, you set selected control points of a curve, surface, and polygon mesh. If you want to set the entire curve, surface, or polygon mesh to a construction plane, you project them. Perform the following steps.

1. Select File > Open and select the file ProjectToCplane.3dm from the Chapter 10 folder on the Student Companion Website.
2. Select Transform > Project to CPlane.
3. Click on the Front viewport to make it the active construction plane.
   Note that this command is view dependent.
4. In the Front viewport, select polygon mesh A, surface B, and curve C (shown in Figure 10–24) and press the ENTER key.
5. Select the Yes option at the command area.
   Two of the three selected objects are projected onto the construction plane, corresponding to the Front viewport, and the original curve, surface, and polygon mesh are deleted. (See Figure 10–25.) The objects that cannot be projected are reported at the command area.
6. Do not save your file.

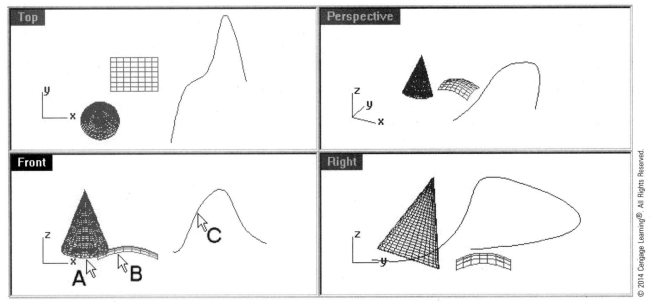

FIGURE 10–24 *Objects being projected*

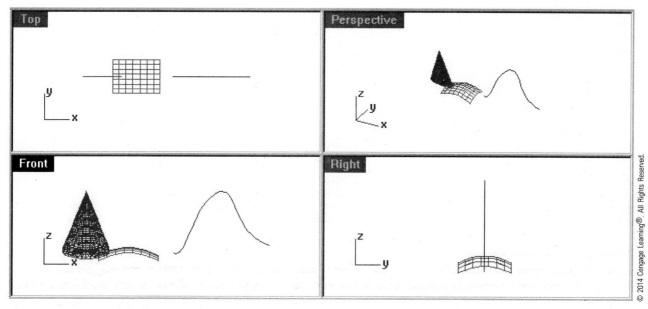

FIGURE 10–25 *Two objects projected*

Shearing along a Plane

Shearing is a deformation process that changes the shape of selected objects by defining a shearing axis and specifying a shearing angle. To define the shearing axis, you specify an origin point and a reference point. Perform the following steps to shear a curve, surface, and polygon mesh.

1. Select File > Open and select the file Shear.3dm from the Chapter 10 folder on the Student Companion Website.
2. Select Transform > Shear.
3. Select surface A, polygon mesh B, and curve C (shown in Figure 10–26), and press the ENTER key.
4. Select point D (Figure 10–26) as the origin point.
5. Select point E (Figure 10–26) as the reference point.
6. Type -35 to specify a shear angle of -35 degrees.

    A negative angle measures in the clockwise direction. The selected objects are sheared.

7. Do not save your file.

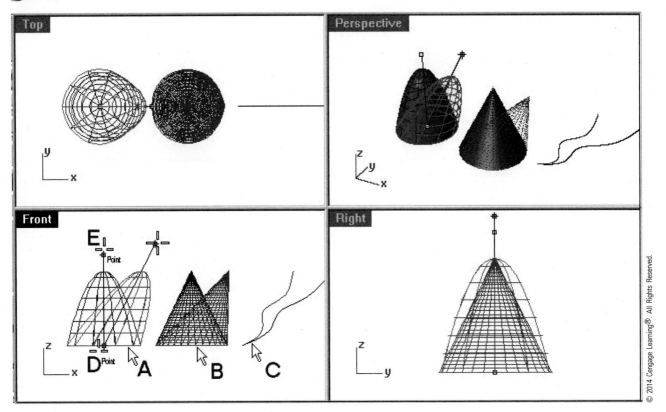

FIGURE 10-26 *Objects being sheared*

Twisting around an Axis

The twisting operation twists selected objects (curve, surface, polygon mesh, or control points). Naturally, you need to define a twisting axis and specify a twisting angle. A twisting axis is formed by picking two reference points. Perform the following steps to twist a curve, surface, and polygon mesh.

1. Select File > Open and select the file Twist.3dm from the Chapter 10 folder on the Student Companion Website.
2. Check the Point box on the Osnap dialog box and clear all other boxes.
3. Select Transform > Twist.
4. Select curve A (shown in Figure 10-27) and press the ENTER key.
5. Select points B and C (Figure 10-27) to indicate the twist axis.

NOTE Points B and C not only define a twist axis through which the object is twisted; they also specify the portion of the object between the points to be twisted. However, if the Infinite option of the command is set to Yes, the twist will happen throughout the object, even if the twist axis is shorter than the object.

6. Type 90 at the command area to specify the twist angle.
   The selected curve is twisted.
7. Select Transform > Twist.
8. Select polygon mesh D (Figure 10-27) and press the ENTER key.
9. Select points E and F (Figure 10-27) to indicate the twist axis.
10. Type 90 at the command area to specify the twist angle.
    The selected polygon mesh is twisted.

5. If Stretch=No, select Stretch option at the command area to change it to Yes.
6. Select curve E (Figure 10-44) to use it as the new backbone curve.

   The selected curves are transformed to flow along the new backbone curve. (See Figure 10-45.)

7. Repeat the command.
8. Select curve F (Figure 10-44) and press the ENTER key.
9. Select curve G (Figure 10-44) to use it as the original backbone curve.
10. If Stretch=No, select Stretch option at the command area to change it to Yes.
11. Select curve H (Figure 10-44) to use it as the new backbone curve.

    The selected curve is transformed to flow along the new backbone curve. (See Figure 10-45.)

12. Repeat the command.
13. Select polygon mesh J (Figure 10-44) and press the ENTER key.
14. Select curve K (Figure 10-44) to use it as the original backbone curve.
15. If Stretch=No, select Stretch option at the command area to change it to Yes.
16. Select curve L (Figure 10-44) to use it as the new backbone curve.

    The selected polygon mesh is transformed to flow along the new backbone curve. (See Figure 10-45.)

17. Repeat the command.
18. Select surface M (Figure 10-44) and press the ENTER key.
19. Select curve N (Figure 10-44) to use it as the original backbone curve.
20. If Stretch=No, select Stretch option at the command area to change it to Yes.
21. Select curve P (Figure 10-44) to use it as the new backbone curve.

    The selected surface is transformed to flow along the new backbone curve. (See Figure 10-45.)

22. Do not save your file.

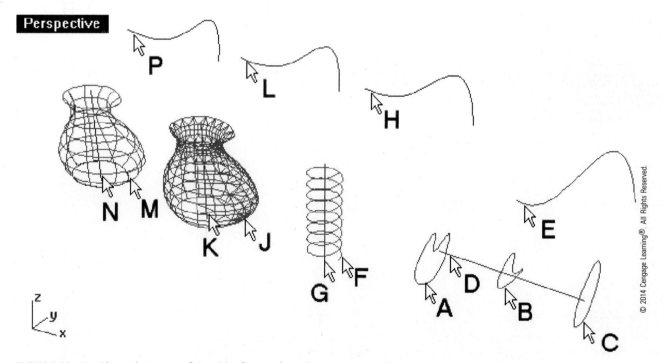

FIGURE 10-44 *Objects being transformed by flowing from their original backbone curves to new backbone curves*

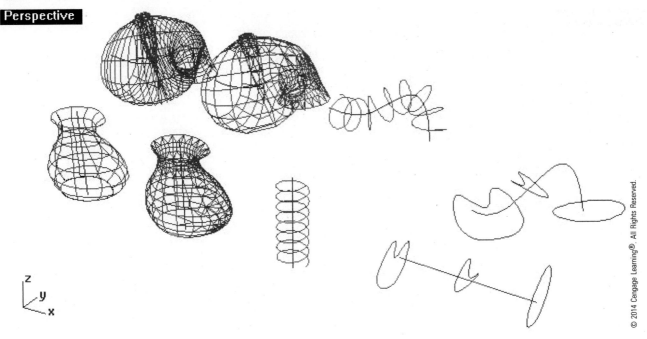

FIGURE 10–45 *Objects transformed from their original backbone curves to new backbone curves*

Flowing along a Surface

One way to transform a set of surfaces onto a free-form surface is to perform space morphing. The set of surfaces is first constructed on a planar backbone surface, which is used as reference. The surfaces are then space morphed onto a target surface. Perform the following steps.

1. Select File > Open and select the file FlowAlongSurface.3dm from the Chapter 10 folder on the Student Companion Website.
2. Select Transform > Flow along Surface.
3. Referencing Figure 10–45, click on location A and drag to location B to select the text object "Rhinoceros 4."
4. Select ellipsoid surface C, shown in Figure 10–46, and press the ENTER key.
5. Select surface D near corner D. This is the reference backbone surface, and the corner is the reference corner.
6. Select surface E near corner E.
   This is the target backbone surface, and the selected objects are flown onto the target surface.

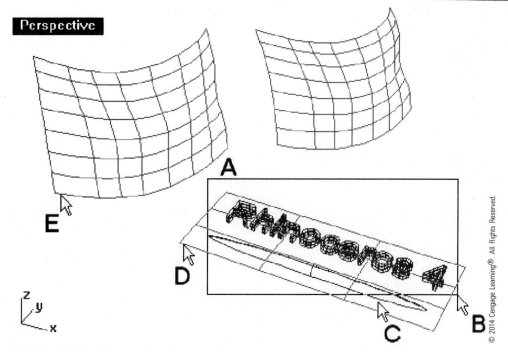

FIGURE 10–46 *Objects being morphed*

7. Repeat the command.
8. Click on location A, shown in Figure 10–47, and drag to location B.
9. Select C and press the ENTER key.
10. Select Rigid option at the command area.
11. Select surface D near corner D.
12. Select surface E near corner E.

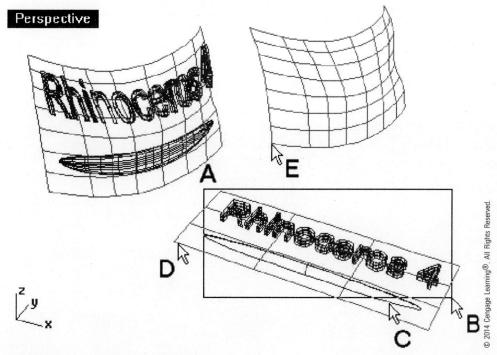

FIGURE 10–47 *Objects for rigid morphing*

The selected objects are space morphed. (See Figure 10-48.)

13. Do not save your file.

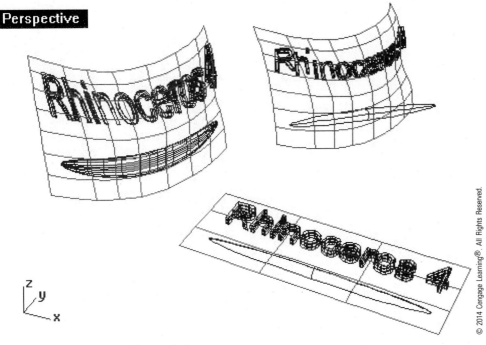

FIGURE 10–48 *Objects rigidly morphed*

Dropping onto a Surface

Perform the following steps to try another way of deforming objects and dropping onto a surface.

1. Select File > Open and select the file Splop.3dm from the Chapter 10 folder on the Student Companion Website.
2. Click on the Splop button on the Deformation Tools toolbar.
3. Select surfaces A and B, shown in Figure 10-49, and press the ENTER key.
4. Click on points C and D in the Front viewport to specify the reference sphere.
   Note that it is important to click on the construction plane parallel to the source objects.
5. Select surface E.
6. Click on F and G to describe a target sphere.
   A set of objects is dropped.
7. Click on H and J to describe another target sphere.
   Another set of objects is dropped.
8. Press the ENTER key.

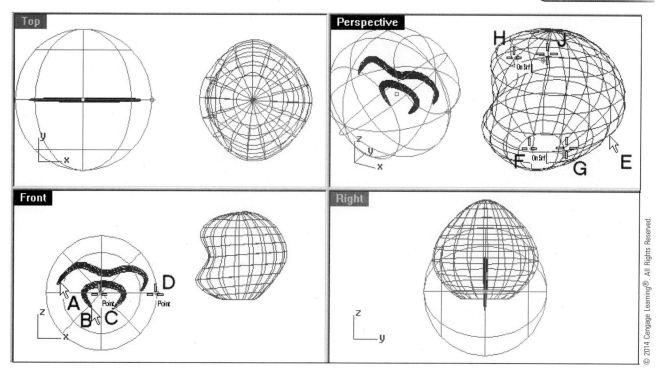

FIGURE 10–49 *Objects selected*

Two copies of surfaces are dropped onto a surface. (See Figure 10–50.)

9. Do not save your file.

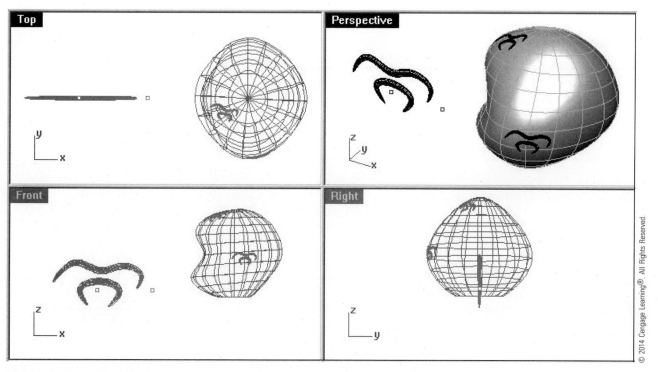

FIGURE 10–50 *Objects dropped*

CONSOLIDATION

To develop and modify your design, you can transform curves, surfaces, and polygon meshes and their control points as necessary. In terms of translation, in addition to the basic translation tools described in Chapter 2, you can soft move a set of objects, align planar curves, and align objects with reference to their bounding boxes. You can orient objects referencing two or three points onto a surface, perpendicular to a curve, tangent to the edge of a surface, and on a selected construction plane.

You can soft edit a curve or a surface. Control points of objects can be aligned by setting points or flattened to a plane by projection. Shapes of objects can be manipulated by shearing, twisting, bending, and tapering. To manipulate individual surface elements of a polysurface or a portion of a surface, it is advisable to shear, twist, bend, or taper selected control points. To simplify a curve or a surface, you apply a smoothing operation or make the control points uniform. To handle a set of polysurfaces collectively, you attach them to a backbone surface, stretch them to extend or shorten a portion of them, use cage for editing, and deform it as if the objects are being acted upon by a maelstrom. To achieve special effects, you can use combined translation and deformation operations—flowing along a curve, space morphing, applying to a surface's U, V, and N, and dropping onto a surface.

REVIEW QUESTIONS

1. List the methods used to orient and array objects.
2. List the ways objects can be deformed.
3. Describe the operations that translate as well as deform an object.

CHAPTER 11

Rhinoceros Data Analysis

INTRODUCTION

This chapter explains various analysis tools to examine curves, surfaces, and polygon meshes.

OBJECTIVES

After studying this chapter, you should be able to
- Use Rhinoceros tools to analyze curves, surfaces, and polygon meshes

OVERVIEW

To further improvise your design, you may need to examine the curves, surfaces, and polygon meshes that you already constructed. Tools for analysis can be divided into five major categories: general tools, dimensional analysis tools, surface profile analysis tools, mass properties tools, and diagnostics tools.

GENERAL TOOLS

The set of tools described next focuses on the form and shape of selected curves and surfaces.

What

To learn about the properties of selected objects and display them in a separate window, you can use the What command, as follows.

1. Select File > Open and select the file What.3dm from the Chapter 11 folder that you downloaded from the Student Companion Website.
2. Right-click on the List Object Database/What?! button on the Diagnostics toolbar or type What at the command area.
3. Select A and B, shown in Figure 11–1, and press the ENTER key.
   The Object Description dialog box is displayed.
4. If you want to save the information to a text file, click on the Save As button and then specify a file name.
5. Click on the Close button.
   Information about selected objects is retrieved.
6. Do not save your file.

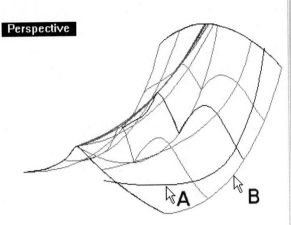 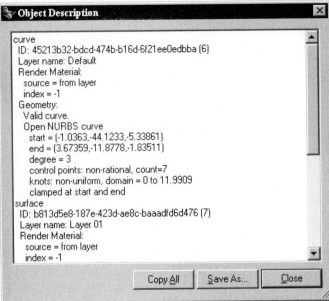

FIGURE 11–1 *What dialog box*
Source: Robert McNeel and Associates Rhinoceros® 5

Curvature Graph

One way to determine the smoothness of a curve or a surface and discern any irregularities is to display the curvature graph, which tells the radius of curvature of a curve along the curve or radius of curvature of a surface along its U and V isocurves. Perform the following steps.

1. Select File > Open and select the file general01.3dm from the Chapter 11 folder on the Student Companion Website.
2. Select Analyze > Curve > Curvature Graph On.
3. Select curve A and surface B (shown in Figure 11–2).
   The curvature graphs are displayed.
4. In the Curvature Graph dialog box, adjust the display scale and density as necessary.

To extract the curvature graph of a curve, continue with the following steps.

5. Type ExtractCurvatureGraph at the command area.
6. Select curve A, indicated in Figure 11–2, and press the ENTER key.
   The curve's curvature graph is extracted.
7. Select Analyze > Curve > Curvature Graph Off.
   The curvature graph is turned off.
8. Delete the curvature graph of curve A, indicated in Figure 11–3.

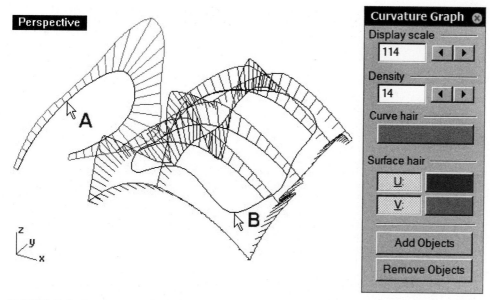

FIGURE 11–2 *Curvature graph displayed*
Source: Robert McNeel and Associates Rhinoceros® 5

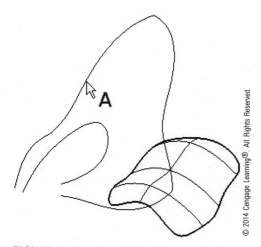

FIGURE 11–3 *Curve's curvature graph extracted and curvature graph displayed turned off*

Curvature Circle

Another way to evaluate the profile of a curve or a surface is to display or construct a curvature circle at a selected location along the curve or two curvature arcs at a selected location on the surface. Continue with the following steps.

9. Select Analyze > Curvature Circle.
10. Select surface A, shown in Figure 11–4.
11. If the MarkCurvature option is No, select the option at the command area to change it to Yes.
12. Select a point on the surface.
    The curvature arcs' radii are constructed.
13. Press the ENTER key to terminate the command.
14. Repeat the command.
15. Select curve B, shown in Figure 11–4.
16. Select a point on the curve.
    The curvature circle is constructed.
17. Press the ENTER key to terminate the command.
18. Select Edit > Undo twice to undo the last two commands.

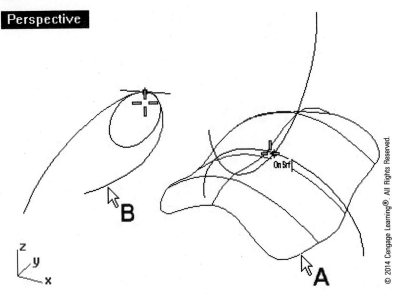

FIGURE 11–4 *Curvature arcs and curvature circle constructed*

Curvature Radius

You can find out the radius of curvature at a selected location along a curve or along the edge of a surface. Continue with the following steps.

19. Check the Point option of the Osnap dialog box.
20. Select Analyze > Radius.
21. Select point A on the curve, shown in Figure 11-5.
    The radius and diameter of curvature of the selected location are displayed.
22. Repeat the command.
23. Select point B on the edge of the surface, shown in Figure 11-5.
    The radius of curvature of the point on the surface's edge is displayed.

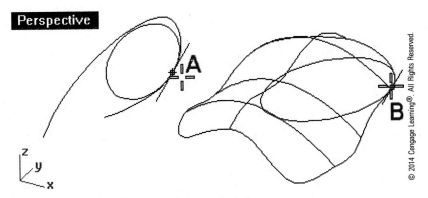

FIGURE 11–5 *Radius of curvature being evaluated*

Direction

Curves have a direction, from one endpoint to the other endpoint or vice versa. Direction of curves may have an effect on the final outcome of the constructed surfaces.

Surfaces have three directions: U, V, and N. The U and V directions are two orthogonal directions along the surface, and the N direction is the normal direction. You can change the U, V, and N directions and swap the U and V directions. Although polygon mesh does not have U and V directions, it has the N direction. Continue with the following steps.

24. Turn on Layer02.
25. Select Analyze > Direction.
26. Select curve A (shown in Figure 11-6).
    The direction of the curve is displayed.
27. Click to change the direction.
28. Press the ENTER key to terminate the command.
29. Select Analyze > Direction.
30. Select surface B. The surface's U, V, and N direction arrows are displayed, shown in Figure 11-6.
31. Click to change the normal direction.
32. Select the UReverse option to reverse the U direction, select the VReverse option to reverse the V direction, or select SwapUV option to swap U and V.
33. Press the ENTER key to terminate the command.
34. Select Analyze > Direction.
35. Select polygon mesh C.
    The normal direction of the polygon mesh is displayed.
36. Click to change the normal direction.
37. Press the ENTER key to terminate the command.
38. Do not save your file.

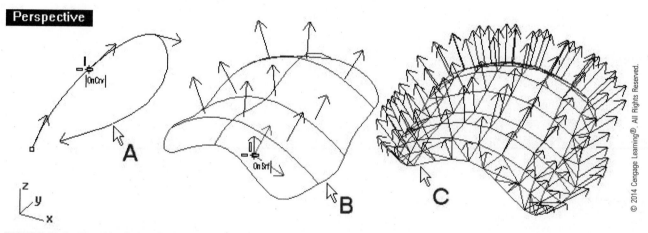

FIGURE 11-6 *Directions being displayed and manipulated*

Curves' Direction Matching

You can change the direction of a curve to match that of another curve. Perform the following steps.

1. Select File > Open and select the file MatchDirection.3dm from the Chapter 11 folder on the Student Companion Website.
2. Select Analyze > Direction.
3. Select curves A and B and press the ENTER key.
   The directions of the curves are displayed.
4. Select Curve > Curve Edit Tools > Match Direction.
5. Select A and then B, indicated in Figure 11-7, and press the ENTER key.
   The direction of curve B is changed.
6. Display the curves' direction to see the change.
7. Do not save your file.

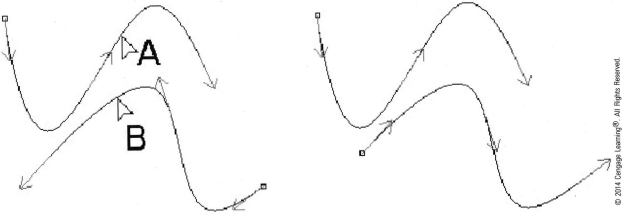

FIGURE 11–7 *Curves' original directions (left) and curves' direction matched (right)*

Geometric Continuity

To discern the continuity of a joint in a curve, perform the following steps.

1. Select File > Open and select the file general02.3dm from the Chapter 11 folder on the Student Companion Website.
2. Select Analyze > Curve > Geometric Continuity.
3. Select curves A and B, shown in Figure 11–8.
   The geometric continuity is displayed.

Deviation between Curves

If you have two curves that closely resemble each other, you can find out their deviation by performing the following steps.

4. Select Analyze > Curve > Deviation.
5. Select curves A and C (shown in Figure 11–8).
   A report is displayed.
6. Do not save your file.

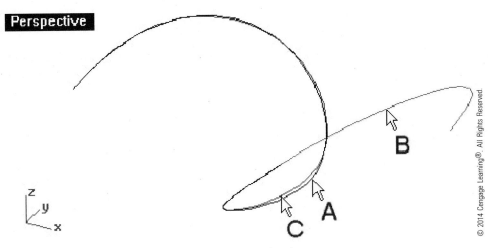

FIGURE 11–8 *Geometric continuity of curves being analyzed*

DIMENSIONAL ANALYSIS TOOLS

You can find out the coordinates of a selected point, the bounding box of an object, the length of a curve or an edge of a surface, the distance between two selected points, and the angle between two selected lines.

Evaluate Point

It is sometimes necessary to find out the coordinates of a selected point. Perform the following steps.

1. Select File > Open and select the file Dimensional.3dm from the Chapter 11 folder on the Student Companion Website.
2. Check the End box in the Osnap dialog box.
3. Select Analyze > Point.
4. Select endpoint A, shown in Figure 11-9.

   The coordinates of the selected location are displayed in the command line area.

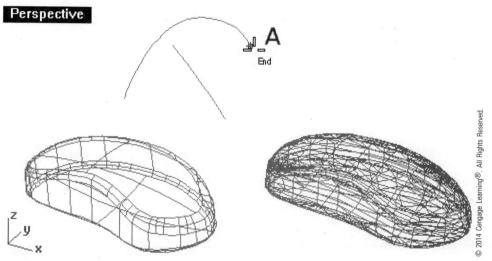

FIGURE 11-9 *Coordinates of the selected location being evaluated*

Construction of a Bounding Box

A bounding box of a curve, surface, or polygon mesh is the minimum size of a rectangular box that encloses the entire curve, surface, or polygon mesh. Continue with the following steps.

5. Select Analyze > Bounding Box.
6. Select curves A and B, shown in Figure 11-10, and press the ENTER key twice.

   A bounding box enclosing the selected objects is constructed, and the box's dimensions are displayed at the command area.

7. Repeat the command.
8. Select surface C (Figure 11-10) and press the ENTER key twice.
9. Repeat the command.
10. Select polygon mesh D (Figure 11-10) and press the ENTER key twice.

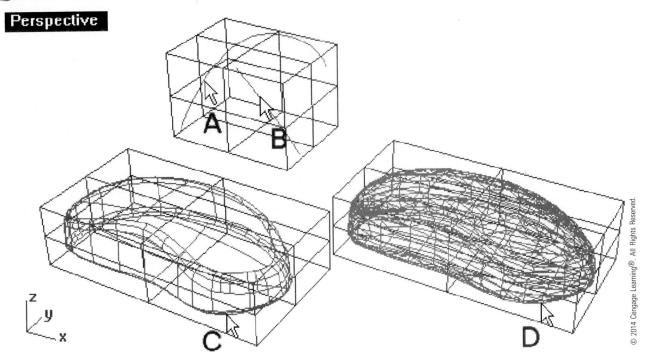

FIGURE 11-10 *Bonding boxes constructed*

Length Measurement

You can measure the length of a curve or the length of an edge of a surface. Continue with the following steps.

11. Select Analyze > Length.
12. Select curve A (shown in Figure 11-11).
    The length of the selected curve is displayed.
13. Select Analyze > Length.
14. Select surface edge B (Figure 11-11).
    The length of the selected surface edge is displayed.

Distance Measurement

You can measure the distance between two selected points. Continue with the following steps.

15. Select Analyze > Distance or click on the Distance button on the Analyze toolbar.
16. Select points C and D (Figure 11-11).
    The distance between the selected points is displayed.

Angle Measurement

You can measure the angle between two lines defined by four selected points. Continue with the following steps.

17. Select Analyze > Angle.
18. Select points E and F and then points C and D (Figure 11-11) to specify the first and second lines.
    The angle between the first and second lines is displayed.
19. Do not save your file.

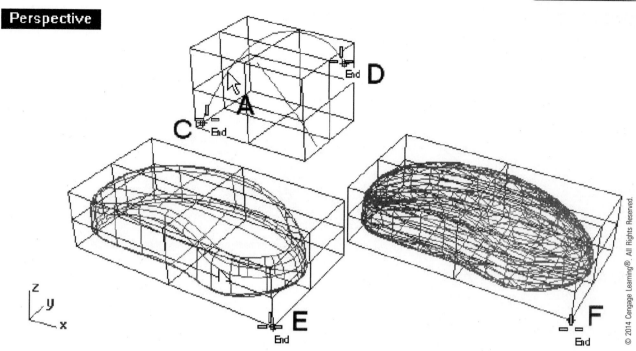

FIGURE 11–11 Length, distance, and angle measurement being taken

Miscellaneous Measurement

You can measure the area within a closed planar curve or surface, length of a curve, and angle between two planar surfaces, as follows.

1. Select File > Open and select the file Measurement.3dm from the Chapter 11 folder on the Student Companion Website.
2. Select Dimension > Area Dimension.
3. Select curve A, indicated in Figure 11–12.
4. Set Osnap to near and click on B and then C (Figure 11–12), and press the ENTER key.
   Area within the closed curve is dimensioned.
5. Select Dimension > Curve Length Dimension.
6. Select curve A, indicated in Figure 11–12.
7. Click on D and then E (Figure 11–12), and press the ENTER key.
   Length of the curve is dimensioned.
8. Maximize the Perspective viewport.
9. Select Dimension > Crease Angle Dimension.
10. Select surface F and G and click on H (Figure 11–12).
    Angle between the two planar surfaces is dimensioned.
11. Do not save your file.

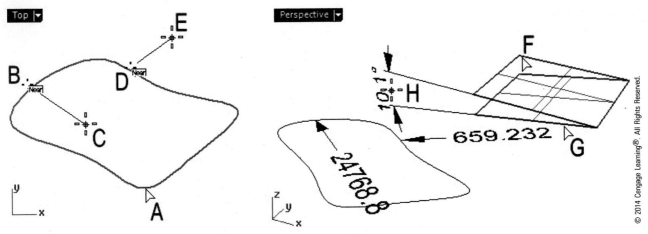

FIGURE 11–12 *Area and length being dimensioned (left) and angle between two planar surfaces being dimensioned (right)*

SURFACE ANALYSIS TOOLS

Because a NURBS surface is a smooth surface with no facets, the profile and silhouette of a NURBS surface are displayed using U and V isocurves. To gain a visual picture of the surface, you use curvature rendering, draft angle rendering, environmental map rendering, and zebra stripes rendering, and then construct curvature arcs.

In regard to individual points on a surface, you construct a point on a surface by specifying the U and V coordinates of the point. Given a point on the surface, you evaluate its U and V coordinates. On the other hand, you evaluate the deviation of any point from a surface. You can also evaluate a solid's thickness.

Curvature Rendering

You can display a color rendering to indicate the curvature values of a surface. In the rendering, different curvature values are represented by different colors. Perform the following steps.

1. Select File > Open and select the file Profile.3dm from the Chapter 11 folder on the Student Companion Website.
2. Select Analyze > Surface > Curvature Analysis.
3. Select the surface and press the ENTER key.
4. In the Curvature dialog box, select Mean in the Style pull-down list box.

    A rendering showing the curvature values in various colors is displayed, shown in Figure 11–13.

5. Close the Curvature dialog box.

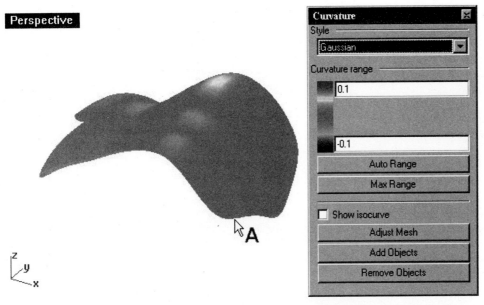

FIGURE 11–13 *Curvature values displayed in various colors (left) and Curvature dialog box (right)*
Source: Robert McNeel and Associates Rhinoceros® 5

Draft Angle Rendering

Draft angles are an essential consideration in mold and die design. You can display a color rendering to illustrate the draft angle value of a surface or a polygon mesh with respect to the construction plane. Continue with the following steps.

6. Select Analyze > Surface > Draft Angle Analysis.
7. Select the surface and press the ENTER key. Draft angle values on the surface are displayed in various colors, shown in Figure 11–14.

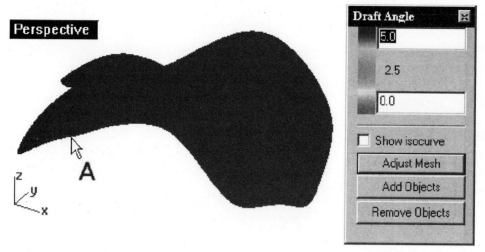

FIGURE 11–14 *Draft angle values displayed in various colors (left) and Draft Angle dialog box*
Source: Robert McNeel and Associates Rhinoceros® 5

Environment Map Rendering

To easily inspect the smoothness of a surface or a polygon mesh, you place a bitmap image on the surface. Continue with the following steps.

8. Select Analyze > Surface > Environment Map.
9. Select the surface and press the ENTER key.
10. In the Environment Map Options dialog box, select an image.
    The selected image is mapped onto the surface, shown in Figure 11–15.

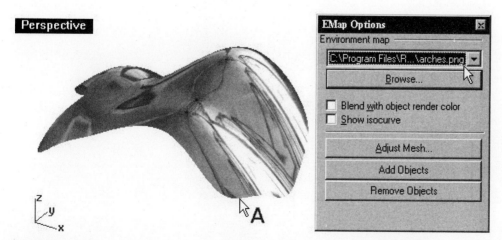

FIGURE 11–15 *Bitmap mapped onto a surface (left) and EMap Options dialog box (right)*
Source: Robert McNeel and Associates Rhinoceros® 5

Zebra Stripes Rendering

To effectively view the smoothness of a surface or a polygon mesh, you use a set of zebra stripes instead of a bitmap. Continue with the following steps.

11. Select Analyze > Surface > Zebra.
12. Select the surface and press the ENTER key.
13. In the Zebra Options dialog box, set the stripe direction and stripe size.

    Zebra stripes are placed on the surface, shown in Figure 11–16. This method is most valuable for checking the continuity between the surfaces of polysurfaces.

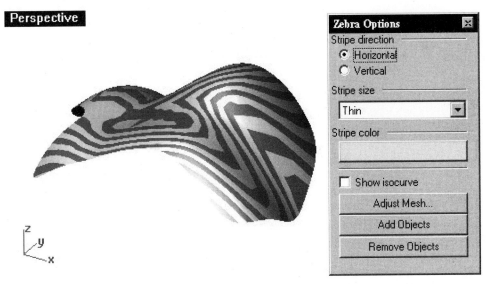

FIGURE 11–16 *Zebra stripes (left) and Zebra Options dialog box (right)*
Source: Robert McNeel and Associates Rhinoceros® 5

Extract Analysis Mesh

Rendering involves the construction of meshes from the NURBS surface by the system. To extract these meshes, perform the following steps.

14. Type ExtractAnalysisMesh at the command area.
15. Select surface A, indicated in Figure 11–17, and press the ENTER key.
    Analysis mesh of the surface is constructed.
16. Undo the last command.

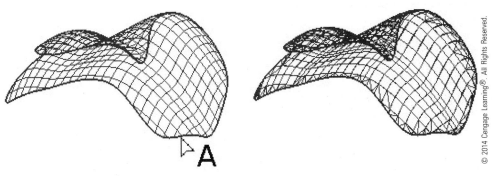

FIGURE 11–17 *Thickness analysis*

Points on a Surface

A surface has two sets of isocurves in two directions: U and V. Correspondingly, there are two axis directions: U and V. You can construct a point on a surface by specifying the point's U and V coordinate values. Continue with the following steps.

17. Select Analyze > Surface > Point from UV Coordinates.
18. Select surface A and press the ENTER key.
19. Type 20 to specify the U value.
20. Type 10 to specify the V value.

    A point is constructed on the surface, shown in Figure 11–18.

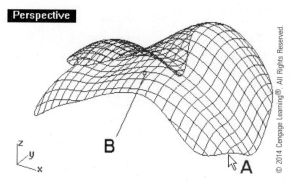

FIGURE 11–18 *Point constructed on the surface*

Evaluating the U and V Coordinates of a Point

Contrary to constructing a point on a surface by specifying the U and V coordinates, you can evaluate the U and V coordinates of a point on a surface. Continue with the following steps.

21. Select Analyze > Surface > UV Coordinates of a Point.
22. Select surface A.
23. Select point B (shown in Figure 11–18) and press the ENTER key.

    The U and V coordinates of the selected point on the surface are displayed at the command area.

Checking the Deviation of a Point from a Surface or a Curve

You can find out the deviation of a point or set of points from a surface. Perform the following steps.

1. Open the file AnalyzeSurface.3dm from the Chapter 11 folder on the Student Companion Website.
2. Select Analyze > Surface > Point Set Deviation.
3. Select point A (shown in Figure 11–19) and press the ENTER key.
4. Select surface B (shown in Figure 11–19) and press the ENTER key.
5. Do not save your file.

    A Point/Surface Deviation dialog box is displayed. The information is useful in identifying the deviation of a point or a set of point clouds from the surface constructed from the point objects.

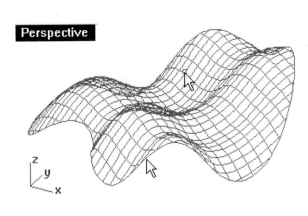
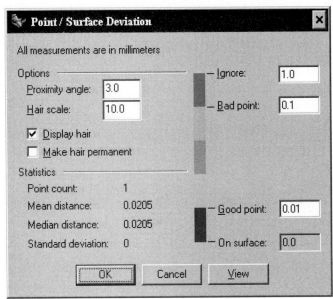

FIGURE 11-19 *Deviation of a point from a surface being evaluated (left) and Point/Surface Deviation dialog box (right)*
Source: Robert McNeel and Associates Rhinoceros® 5

Thickness Evaluation

If you have a solid with thickness, you can use the thickness evaluation tool to find out the variation of thickness, as follows.

1. Open the file Thickeness.3dm from the Chapter 11 folder on the Student Companion Website.
2. Right-click on the Thickness Analysis/Thickness Analysis Off button on the Surface Analysis toolbar.
3. Select object A, shown in Figure 11-20, and press the ENTER key.
   Variation in thickness is displayed, shown in Figure 11-20.
4. Do not save your file.

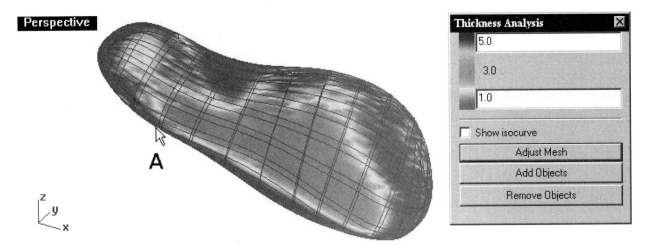

FIGURE 11-20 *Thickness analysis (left) and Thickness Analysis dialog box (right)*
Source: Robert McNeel and Associates Rhinoceros® 5

MASS PROPERTIES TOOLS

You can measure the area of a surface or a polygon mesh and the area centroid and area moments of a surface. Volume measure concerns objects that are closed, such as a solid or a polygon mesh that closes a volume without any openings. For such objects, you can measure their volumes. For solids, you can measure their volume centroid, volume moments, and hydrostatics.

Area Inquiry

You can determine the area of a surface or a polygon mesh. Perform the following steps.

1. Select File > Open and select the file Area.3dm from the Chapter 11 folder on the Student Companion Website.
2. Select Analyze > Mass Properties > Area
3. Select surface A (shown in Figure 11–21) and press the ENTER key.
4. Select Analyze > Mass Properties > Area.
5. Select polygon mesh B (Figure 11–21) and press the ENTER key.

Area Centroid Inquiry

You can determine the area centroid of a selected surface. Continue with the following steps.

6. Select Analyze > Mass Properties > Area Centroid.
7. Select surface A (Figure 11–21) and press the ENTER key.

   The area centroid is displayed, and a point (C) is constructed to indicate the location of the centroid.

Area Moments Inquiry

You can determine the area moments of a selected surface. Continue with the following steps.

8. Select Analyze > Mass Properties > Area Moments.
9. Select surface A (Figure 11–21) and press the ENTER key.

   Area moment information is displayed in the Area Moments dialog box, shown in Figure 11–21. You can save the information to a file and then click on the Close button to close the dialog box.

10. Do not save your file.

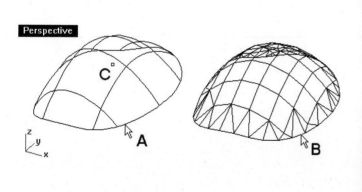
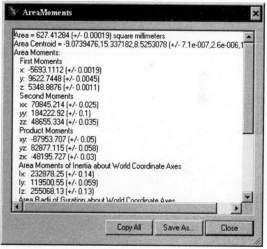

FIGURE 11–21 *Area properties inquiry*
Source: Robert McNeel and Associates Rhinoceros® 5

Volume Inquiry

You can determine the volume of a solid (a closed polysurface) or a closed polygon mesh. Perform the following steps.

1. Select File > Open and select the file Volume.3dm from the Chapter 11 folder on the Student Companion Website.
2. Select Analyze > Mass Properties > Volume.
3. Select solid A (shown in Figure 11–22) and press the ENTER key. Volume information of the solid is displayed in the command area.

4. Select Analyze > Mass Properties > Volume.
5. Select polygon mesh B (Figure 11–22) and press the ENTER key.
   Volume information of the polygon mesh is displayed in the command area.

Volume Centroid Inquiry

You can determine the volume centroid of a solid. Continue with the following steps.

6. Select Analyze > Mass Properties > Volume Centroid.
7. Select solid A (Figure 11–22) and press the ENTER key. Volume centroid (C) information is displayed at the command area.

Volume Moments Inquiry

You can determine the volume moments of a solid. Continue with the following steps.

8. Select Analyze > Mass Properties > Volume Moments.
9. Select solid A (Figure 11–22) and press the ENTER key. Volume moment information is displayed in the dialog box shown in Figure 11–22.

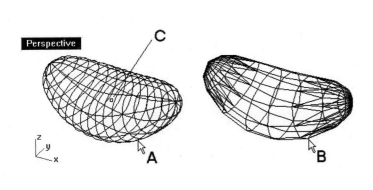
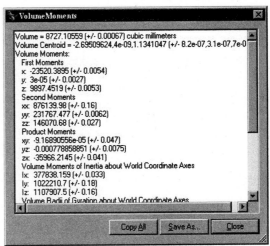

FIGURE 11–22 *Volume analyze being taken (left) and VolumeMoments dialog box (right)*
Source: Robert McNeel and Associates Rhinoceros® 5

Cut Volume

You can evaluate the volume of intersection of a closed object and an existing box, as follows.

1. Select File > Open and select the file CutVolume.3dm from the Chapter 11 folder on the Student Companion Website.
2. Type CutVolume at the command area.
3. Select object A (shown in Figure 11–23) and press the ENTER key.
4. Select box B. The cut volume is displayed at the command area.
5. Do not save your file.

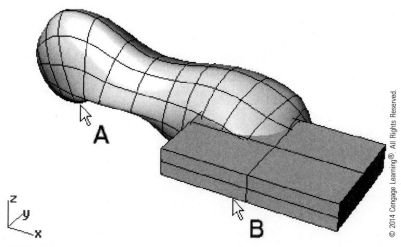

FIGURE 11–23 *Closed object and a box*

Hydrostatics

Assuming the waterline is set at the World coordinate plane, you can change the setting by typing W at the command area and specifying a new waterline. You can evaluate the hydrostatic value of a solid, as follows.

6. Select Analyze > Mass Properties > Hydrostatic.
7. Select the solid and press the ENTER key.
   Hydrostatics information is displayed in the Hydrostatics dialog box, shown in Figure 11–24.
8. Do not save your file.

> For professional marine application, it is advised to use RhinoMarine as a tool.   **NOTE**

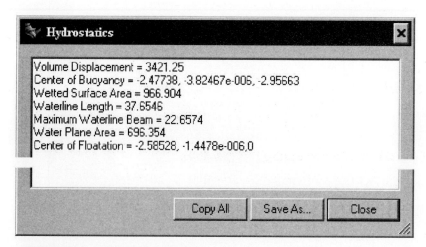

FIGURE 11–24 *Hydrostatics dialog box*
Source: Robert McNeel and Associates Rhinoceros® 5

DIAGNOSTICS TOOLS

Diagnostics tools concern database listing, finding bad objects, and auditing.

Listing a Database

You can display the database of a selected object in list form. Perform the following steps.

1. Select File > Open and select the file Diagnostics.3dm from the Chapter 11 folder on the Student Companion Website.
2. Select Analyze > Diagnostics > List, or click on the List Object Database/What?! button on the Diagnostics toolbar.
3. Select surface A (shown in Figure 11–25) and press the ENTER key.
   The database of the selected objected is displayed as a list, shown in Figure 11–25.
4. Click on the Close button.

Bad Object Report

You can check an object for errors and highlight erroneous objects. Continue with the following steps.

5. Select Analyze > Diagnostics > Check.
6. Select curve A (shown in Figure 11–25).
   The selected curve is checked.
7. Click on the Close button.

To check the entire file for bad objects, continue with the following.

8. Select Analyze > Diagnostics > Select Bad Objects.
   A report is displayed.

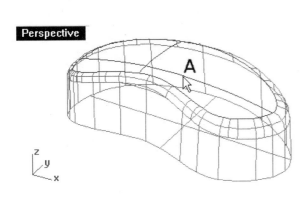

FIGURE 11–25 *Curve data listed*
Source: Robert McNeel and Associates Rhinoceros® 5

Auditing

You can audit the current Rhino file or other Rhino files.

- To audit the current file, select Analyze > Diagnostics > Audit, or click on the Audit button on the Diagnostics toolbar.
- To audit a different Rhino file, select Analyze > Diagnostics > Audit 3DM Files, or click on the Audit 3DM File button on the Diagnostics toolbar.

Checking New Objects

Before inserting objects or files into a Rhino file, you can set CheckNewObjects to On in order to check the errors of any new objects.

- Right-click on the Select Bad Objects/Check All New Objects button on the Analyze toolbar, or type CheckNewObjects at the command area and set it to Yes.

CONSOLIDATION

You can evaluate the profile of a curve or surface by displaying curvature graph, curvature circle, and curvature radius. You can manipulate the directions of a curve or a surface. You can also determine the continuity of contiguous curves or surfaces and the deviation between two curves. To perform dimensional analysis, you can evaluate a point's coordinates, an object's bounding box, the length of a curve or an edge of a surface, the distance between two points, and the angle between two lines.

To analyze a surface's profile, you can display a curvature rendering, draft angle rendering, environmental map rendering, and zebra stripe rendering. You can construct a point on a surface by specifying its U and V coordinates and determine the coordinates of a point on a surface in terms of the U and V coordinates. A point's deviation from a surface can also be determined.

In terms of mass properties, you can evaluate a surface's area, volume, and hydrostatics value. In addition, you can display an object's database list, report bad objects in a file, perform file auditing, and check the errors of any new objects.

REVIEW QUESTIONS

1. What is the difference between displaying a database list and reporting bad objects in a file?
2. List the methods you might use to analyze a NURBS surface.

CHAPTER 12

Group, Block, and Work Session

INTRODUCTION

This chapter covers group, block, and work session and suggests how these tools can help enhance object management, enable assembly simulation, and facilitate design collaboration.

OBJECTIVES

After studying this chapter, you should be able to

- Organize Rhino objects into data groups
- Define and manipulate data blocks and use block definitions to help construct an assembly of objects
- Carry out design collaboration via the work session manager

OVERVIEW

In Rhino, a group is a collection of geometric objects put together so that selection on any part of the objects causes the entire set of objects to be selected. A block is a data definition for a set of objects residing within a file's database and optionally referencing to an external file. To simulate an assembly of components in the computer, you may consider using block definitions. To handle a large project, you can divide it into a number of smaller files and organize them in a work session. Work sessions can help facilitate design collaboration.

GROUP

To help select a set of objects (curves, surfaces, and/or polygon meshes) repeatedly in a file, you can think about organizing them into groups. After putting objects in a group, selecting any of them will automatically cause the entire group of objects to be selected. However, you may still select objects individually by holding down the CONTROL and SHIFT keys while clicking on the objects.

Constructing Groups

Group construction is very simple. You use the group command and select objects that you want to group together. Perform the following steps.

1. Select File > Open and select the file Group.3dm from the Chapter 12 folder that you downloaded from the Student Companion Website.

2. Select Edit > Groups > Group.
3. Select objects A, B, and C, shown in Figure 12-1, and press the ENTER key.
   Selected objects are put in a group.

Adding and Removing Group Elements

After a group is constructed, you may add new group elements to the group as well as remove existing group elements from the group. Continue with the following steps.

4. Select Group > Add to Group.
5. Select curve D, shown in Figure 12-1, and press the ENTER key.
6. Select object C (a member of a group already constructed).
   Curve D is added to the group.
7. Click on the Remove from Group button on the Grouping toolbar.
8. Select curve A, shown in Figure 12-1, and press the ENTER key.
   The selected object is removed from the group.

Assigning Group Names

Groups can be named for easy identification in case there are too many groups in a file. Continue with the following steps.

9. Select Edit > Groups > Set Group Name.
10. Select object B, shown in Figure 12-1, and press the ENTER key.
11. Type Group_A at the command area.
    The selected group is assigned a group name. Note that space is not allowed in a group's name.

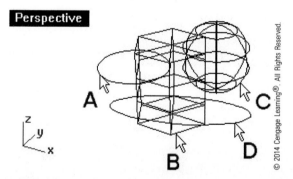

FIGURE 12-1 *Objects being put in a group, added to a group, and removed from a group, and group being assigned a group name*

To appreciate how group names can help identify different groups in a file, continue with the following steps.

12. Turn on Layer 01. Objects constructed in this layer are already put in a group.
13. Select Transform > Move.
14. Click on location A, shown in Figure 12-2.
    A pop-up menu is displayed. You may select one of the two groups.
15. Press the ESC key to cancel the command because we do not intend to move the objects.

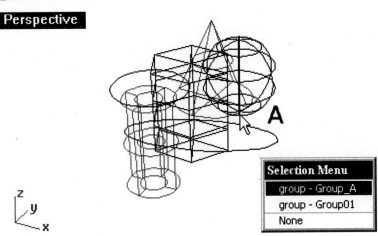

FIGURE 12-2 *Group names displayed*
Source: Robert McNeel and Associates Rhinoceros® 5

Combining Groups

By assigning the name of an existing group to a new group, two groups are combined into one.

16. Select Edit > Groups > Set Group Name.
17. Select A, shown in Figure 12-3, and press the ENTER key.
18. Type Group01 at the command area.

    Because "Group01" is already assigned to an existing group, the selected objects/groups are combined with Group01.

Ungrouping

Grouped objects can be ungrouped, thus reverting back to individual objects.

19. Select Edit > Groups > Ungroup, or click on the Ungroup button on the Grouping toolbar.
20. Select A, shown in Figure 12-3, and press the ENTER key.
21. Do not save your file.

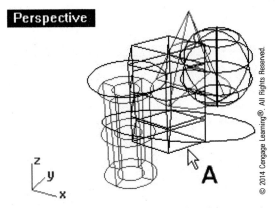

FIGURE 12-3 *Groups being combined and ungrouped*

BLOCK

Think about a Rhino file with many identical copies of geometric objects, including mirrored objects and scaled objects. Although each set of geometric objects may take up very little computer memory space, a hundred sets of such copies will increase the file size significantly. Now think about another scenario. You constructed a set of geometric objects in a Rhino file, and you have to repeat the same set of objects in another Rhino file. Both cases require proper management of the geometric objects.

Block and Block Instances

One way to save memory space in a Rhino file with many repeated objects is to define a block definition from the objects, give the block a name, and insert the block at designated locations. We call each insertion of the block an instance of the block definition. Even though there may be many instances of the blocks in your Rhino file, they take up very little memory space because each insertion instance stores only the information about the name of the block and the block's location, scale, and orientation. Using blocks, you can repeat a set of objects of a drawing in another drawing file as well.

Defining a Block from Existing Objects

A block is a set of objects residing in the Rhino file's database. In the block, you can include curves, surfaces, and/or polygon meshes. Blocks can be defined in two ways: by selecting a set of objects from within a Rhino file and by inserting a Rhino file. In the first method, you select a set of geometric objects from within a file, put them in a block, assign an insertion base point, and specify a block name.

1. Select File > Open and select the file Block.3dm from the Chapter 12 folder on the Student Companion Website.
   In the file, you will find two objects, a cup and a pyramid. You will now examine their properties.
2. Select Edit > Object Properties.
3. Select A, shown in Figure 12–4, and then click on the Details button of the Properties dialog box. This will display the Object Description dialog box. As can be seen, this is a polysurface.
4. Click on the Close button, select B, and click on the Details button of the Properties dialog box.
   As shown, this is a block definition. This block definition is predefined in this file.
5. Close the dialog boxes.

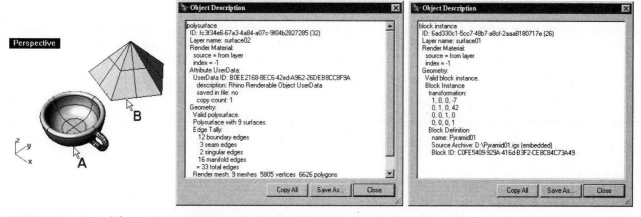

FIGURE 12–4 *From left to right: two objects in the file and Object Description of the objects*
Source: Robert McNeel and Associates Rhinoceros® 5

6. Set object snap mode to End.
7. Select Edit > Blocks > Create Block Definition.
8. Select object A, shown in Figure 12–5, and press the ENTER key.
9. Select endpoint B as the insertion base point.

| Insertion base point is the reference point when you insert the block. | NOTE |

10. In the Block Definition Properties dialog box, assign a block name "Cup."
11. Type www.rhino3d.com in the URL field of the dialog box.
12. Click on the OK button.

A block is constructed from the selected object. Although there is no visual change in the display, what you see now is an instance of the block. You can verify that this is a block definition by repeating steps 2 and 3 above.

As a hyperlink is constructed, you can open the hyperlink by selecting Tools > Hyperlink > Open Hyperlink and selecting the block object. If your computer is already connected to the Internet, your default internet browser will open and go to www.rhino3d.com.

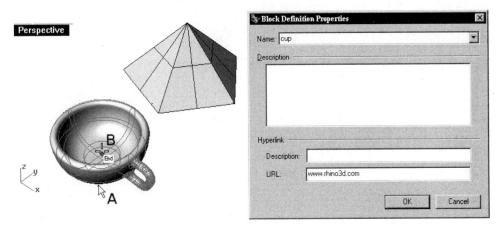

FIGURE 12–5 *Block being constructed*
Source: Robert McNeel and Associates Rhinoceros® 5

Block Insertion

After a block definition is defined in a file, you can insert it as an instance. Optionally, you may specify a scale factor, which can be uniform or nonuniform, and a rotation angle. Being an instance of the block, objects will be referenced to the block definition. Changes in the block definition will cause automatic update of the block instance. Continue with the following steps to insert a block instance.

13. Select Edit > Blocks > Insert Block Instance.
14. Select the block "Cup" from the Name pull-down list box.
15. In the Insertion Point area of the dialog box, check the Prompt box if it is not already checked.
16. In the Scale area, check the Uniform box and type 0.8 in the X box.
17. In the Rotation area, type 45 in the Angle box.
18. Select Block Instance in the Insert As area, if it is not already selected.

As shown in the dialog box, the block can be inserted as three kinds of objects: Block instance, Group, or Individual Objects. If you click on the Group button, the objects defined in the block definition will be inserted and then grouped. If you click on the Individual Objects button, the objects are simply copied and pasted at the inserted location.

19. Click on the OK button.
20. Select location A, shown in Figure 12–6 (exact location is unimportant for the purpose of this tutorial).

    The selected block is inserted as an instance, as shown in Figure 12–7.

Chapter 12 • Group, Block, and Work Session 521

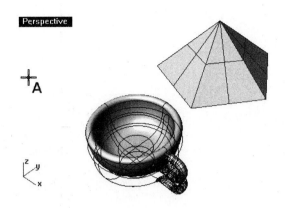
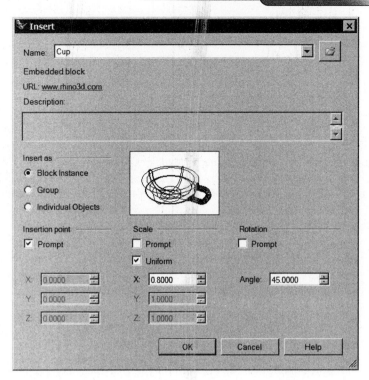

FIGURE 12–6 *Block definition being inserted*
Source: Robert McNeel and Associates Rhinoceros® 5

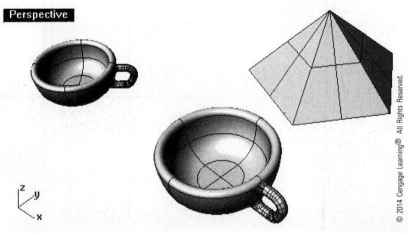

FIGURE 12–7 *Block instance inserted*

Inserting an External File

The second way to define a block definition is to insert an external file by using the same command for inserting a block instance from an internal block definition.

File Type

The type of file that you can insert is not restricted to Rhino files. You will learn more about file types in Chapter 13. Now continue with the following steps.

21. Select Edit > Blocks > Insert Block Instance.
22. In the Insert dialog box, click on the button A indicated in Figure 12–7.
23. In the Import dialog box that opens, select the file Plate01.3dm from the Chapter 12 folder of the Student Companion Website and click on the OK button.
24. On returning to the Insert dialog box, check on the Block instance box in the Inserted As area, if it is not already checked, and set scale to 1 and rotation angle to 0.

Reference to the External File

Similar to inserting an internal block, a block can be inserted as a block instance, a group, or individual objects. If the file is inserted as a block, there are three ways to treat the block definition:

- Embed
- Link and Embed
- Link

If the Embed option is selected, the file is first imported into the current file. The imported data is then used to construct a block definition, and an instance of the block is inserted. Although a block definition is already constructed in the active file from the external file, there is still a passive reference to the external file. You can always update the block via the Block Manager, described in the next section.

If the Link and Embed option is chosen, the inserted block data is always linked to the external file. Depending on the settings in the Block Manager, the inserted block can be set to update automatically if the external file changes.

Finally, if you choose the Link option, the data is linked only to the external file and is not saved in the current file. Naturally, the current file size is much smaller because only the link information is saved. However, the update also depends on the settings in the Block Manager. The disadvantage of linking is that in order for the external file to open with a linked block, the external file must always be in the same linked location..

25. Click on the OK button.
26. On the Insert File Options dialog box that opens, click on the Link button and then click the OK button. (See Figure 12-8.)
27. Click on location A, indicated in Figure 12-9; exact location is unimportant here.

    A block instance is inserted by linking to an external file. (See Figure 12-9.)

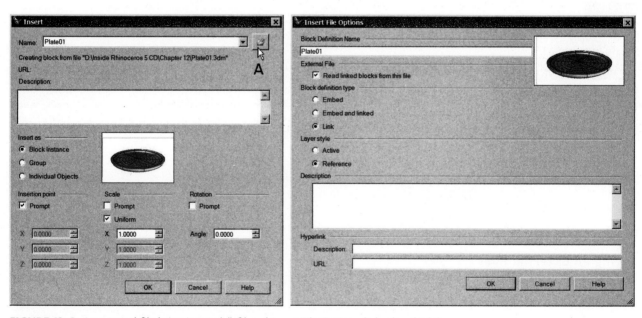

FIGURE 12-8 *An external file being inserted (left) and Insert File Options dialog box (right)*
Source: Robert McNeel and Associates Rhinoceros® 5

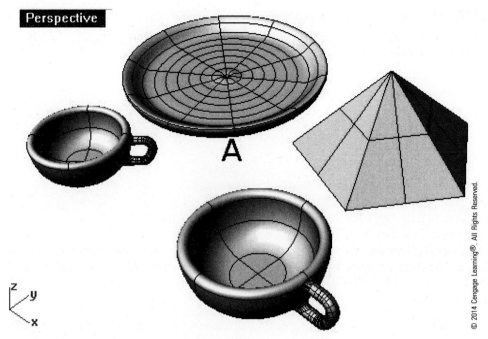

FIGURE 12–9 *File imported and instance inserted*

Block Manager

You can use the block manager to manage blocks in the following ways.

- Modify a block's properties
- Export a block to form an individual file
- Delete a block from the computer's memory

If the block refers to an external file, you can update the link information so that the latest version of the imported file is reflected. If a block is nested in another block, you can determine the nesting information. (When a block is used as a drawing object in constructing another block, the block is said to be nested.) You can find out how many instances of the block are in the file. Continue with the following steps.

28. Select Edit > Blocks > Block Manager.

    The Block Manager dialog box is displayed, shown in Figure 12–10.

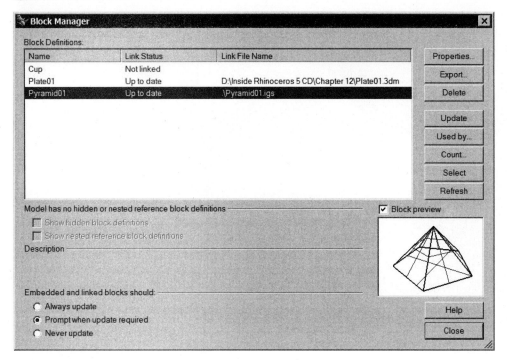

FIGURE 12-10 *Block Manager dialog box*
Source: Robert McNeel and Associates Rhinoceros® 5

The lower-left section of the Block Manager dialog box includes three buttons:

- Always update
- Prompt when update required
- Never update

If the Always button is checked, all linked external files will be updated automatically each time the file is opened. If the Prompt button is checked, you will be prompted to update when the external file is modified. If the Never button is checked, no action will be taken by the system. As shown in the Block Manager dialog box, there are three block definitions in the file. The block definition "Cup" is not linked, because it is a block defined and saved in the file. The block definition "Plate01" is up–to-date, meaning that the block is defined from an external file and that the content of the block is referenced to the most up-to-date external file. Finally, the block definition "Pyramid01" has a status of "File Not Found," meaning that it was defined from an external file but the file does not exist anymore.

29. In the Block Manager dialog box, select the block "Pyramid01" and click on the Properties button.
30. In the Block Definition Properties dialog box, shown in Figure 12-11, click on button A.
31. In the Open dialog box that follows, select the file Pyramid02.3dm from the Chapter 12 folder of the Student Companion Website, and click on the OK button.
    The referenced file is changed.
32. Click on the OK button of the Block Definition Properties dialog box.

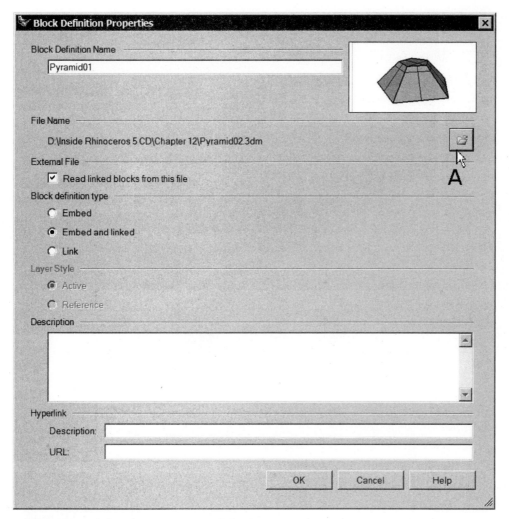

FIGURE 12-11 Block Definition Properties dialog box
Source: Robert McNeel and Associates Rhinoceros® 5

33. On returning to the Block Manager dialog box, click on the Update button.
34. Click on the OK button of the Rhinoceros 4 Update Block dialog box.
35. Click on the Close button of the Block Manager dialog box.

    The unreferenced block definition is now referenced to the file Pyramid02.igs. (See Figure 12–12.)

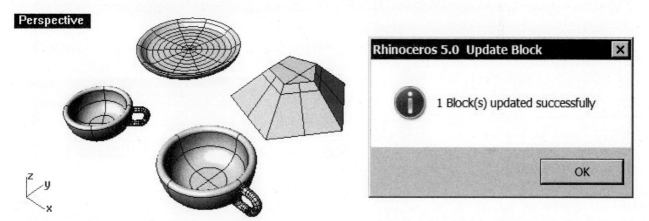

FIGURE 12-12 Unreferenced file changed and updated
Source: Robert McNeel and Associates Rhinoceros® 5

Nested Block

When an instance of a block definition is used in the construction of another block definition, the block definition is said to be nested. Continue with the following steps.

36. Referencing Figure 12–13, drag an instance of the block definition "Cup" to a new location.
37. Press the ESC key to clear any selection and select Edit > Blocks > Create Block Definition.
38. Select A and B, shown in Figure 12–13, and press the ENTER key.
39. Select endpoint C.
40. In the Block Definition Properties dialog box, type TeaSet in the name box and click on the OK button.

    A block definition is constructed from two block instances.

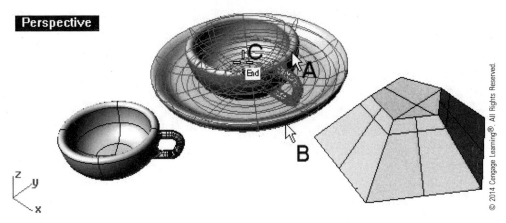

FIGURE 12–13 *An instance moved and nested block being constructed*

41. Select Edit > Blocks > Block Manager, or click on the Block Manager button on the Block toolbar.
42. Select the block definition "Cup" and click on the Used By button.

    The Nested Block Definition dialog box is displayed, showing that it is nested in the block definition "TeaSet."

43. Click on the Close button.
44. While the block definition "Cup" is still selected, click on the Count button.

    The Block Instance Count dialog box is displayed, indicating that there is one instance referenced to the block definition, shown in Figure 12–14.

45. Click on the Close button of the Count Blocks dialog box.

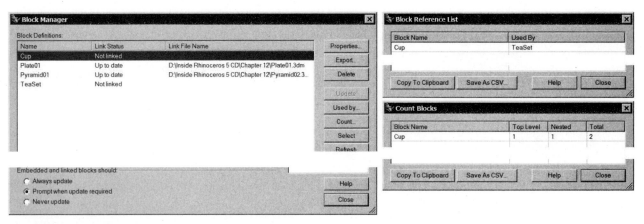

FIGURE 12–14 *Nested block*
Source: Robert McNeel and Associates Rhinoceros® 5

Deleting Block Definition

If you delete a block definition via the Block Manager, all the instances and the block will be removed. However, if a block is nested, it cannot be deleted this way. Continue with the following steps.

46. Open the Block Manager dialog box, if it is closed.
47. Select the block definition "Pyramid01" and click on the Delete button.
48. A warning dialog box is displayed. Click on the Yes button if you really want to delete the block definition and its instances. If you click on the No button, only the inserted instance of the block is deleted.
49. Close the Block Manager dialog box.

    The block definition is removed.

Exploding a Block Instance

As mentioned, an internal block definition can be inserted only as a block instance, and an external file can be inserted as an instance as well as individual objects or grouped objects. If you want to make a copy of the objects from an internal block definition so that the copied objects are independent of the block definition, you explode the block instance. After an instance is exploded, the instance becomes a set of objects without any reference to the block instance. Continue with the following steps.

50. Select Edit > Explode, or click on the Explode button on the Main2 toolbar.
51. Select A, shown in Figure 12–15, and press the ENTER key.

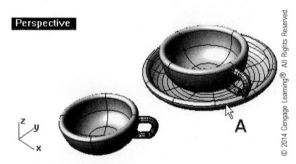

FIGURE 12–15 *An instance being exploded*

To appreciate what happens to the block definition, continue with the following steps.

52. Open the Block Manager dialog box.
53. Select the block definition "TeaSet" and click on the Count button.

    The Block Instance Count dialog box is displayed, indicating that there is 0 instance in the file, indicating that the block definition is still retained but there is no inserted instance.

54. Close the Block Instance Count dialog box and the Block Manager.

Purging Unreferenced Blocks

Because block definitions are residing in the memory space, even when the instances of the block are deleted or exploded, the block definitions are still in the file. If all the instances of a block definition are removed, either by deleting or exploding, the block is called an unreferenced block. You can remove unreferenced blocks by purging, as follows.

55. Select Tools > File Utilities > Purge Unused Information.
56. Check that all the options are Yes and press the ENTER key.

    The unreferenced block is purged. If you open the Block Manager, you will find that the block definition "TeaSet" is purged (removed).

Block Edit

You can modify a block definition by adding or removing elements from it as well as resetting the inserting base point by double-clicking it. Continue with the following steps.

57. With reference to Figure 12-16, double-click A.
    The Block Edit dialog box is displayed.
58. Click on the Point button.
59. Click on location B to reset the insertion point and click on the OK button.
60. Click on the Yes button to confirm the change.
    Note that the locations are changed because of the change in the insertion point.

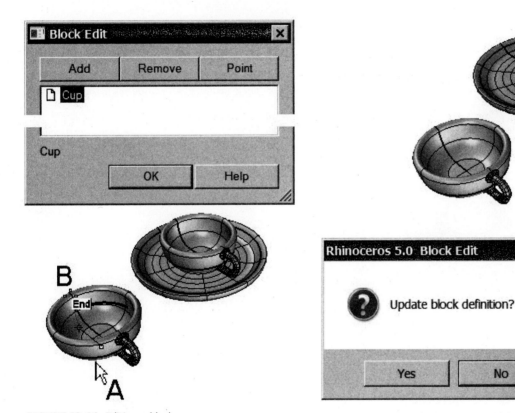

FIGURE 12-16 *Editing a block*
Source: Robert McNeel and Associates Rhinoceros® 5

Replace Block

You can replace a block instance with a different definition. Continue with the following steps.

61. With reference to Figure 12-17, click on instance A and drag it to location B; exact location is unimportant because the purpose is to separate it from the other instance.
62. Click on the Replace Block button on the Block toolbar.
63. Select C, indicated in Figure 12-17, and press the ENTER key.
64. Press the ENTER key again to accept the default of None.
65. Click on D.
    An selected Cup instance is replaced by the Plate01 block definition.
66. Do not save your file.

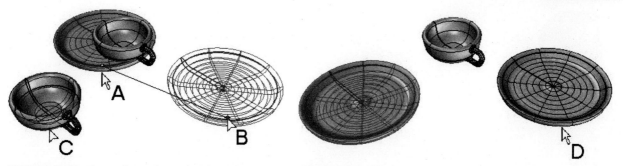

FIGURE 12-17 *Plate01 being dragged and Cup being replaced (left) and a cup instance replaced (right)*
© 2014 Cengage Learning®. All Rights Reserved.

Model Base Point

If a Rhino file is inserted into another file, the default base point is the origin. However, you can redefine the base point, as follows.

1. Open the Rhino file, which will be used as a block for insertion into another file.
2. Click on the Set Model Base Point button on the Block toolbar.
3. Specify a point and save the file.

Although the objects in the file are unchanged, the specified base point will be used next time you insert this file.

Advantages of Insertion as Instances

The advantages of inserting as instances of the block rather than copying as individual objects are that file size is smaller (because block insertion does not actually duplicate the entities) and that change to a block can cause global change to all instances of the blocks (because instances refer to block definitions). You save a considerable amount of memory space when a block with numerous drawing objects is inserted many times in a drawing. In addition, redefining the block definition causes all the referenced instances to change.

Assembly Simulation

With the exception of very simple objects, such as a ruler, most objects have more than one component put together to form a useful whole. The set of components put together is called an assembly. There are three approaches to designing a product or a system. In the first approach, you start from making the individual components and work upward to build the assembly. In the second approach, you start from the assembly as a whole and work downward to construct individual components. In the third approach, you construct some components and work upward as well as downward to construct individual components in the context of the assembly.

Components of an Assembly

For complex devices that have many parts, it is common practice to organize the parts into a number of smaller subassemblies such that each subassembly has fewer parts. Therefore, an assembly set may consist of a file depicting the assembly and a number of files depicting individual components, or an assembly file together with a number of subassembly files and component files. Collectively calling the individual parts or subassemblies components, you can define an assembly in the computer as a data set containing information about a collection of components linked to the assembly and how the components are assembled together. Figure 12-18 shows an assembly of a robot toy from a set of components residing on a number of individual files.

FIGURE 12–18 *An assembly of components*

The Bottom-Up Approach

The bottom-up approach is used when you already have a good idea on the size and shape of the components of an assembly or you are working as a team on an assembly. Using Rhino as a tool, you construct all the individual components in individual files. Then you start a new file to depict the assembly, insert the individual files into the assembly file as block references that linked to the original files, and orient the inserted instances as may be appropriate. Perform the following steps.

1. Select File > Open and select the file Robot.3dm from the Chapter 12 folder on the Student Companion Website.
2. Select Edit > Blocks > Insert Block Instance.
3. Click on the File button on the Insert dialog box (Figure 12–19).
4. Select the file Robot-Body.3dm from the Chapter 12 folder on the Student Companion Website.
5. Clear the Prompt box in the Insertion point area.
6. In the Scale area, click on the Uniform button and set X to 1.
7. In the Rotation area, set the rotation angle to 0.
8. Click on the OK button.
9. On the Insert File Options dialog box that opens, click on the Link button and then the OK button.

    A block definition is constructed from the external file, and an instance of the block definition is inserted.

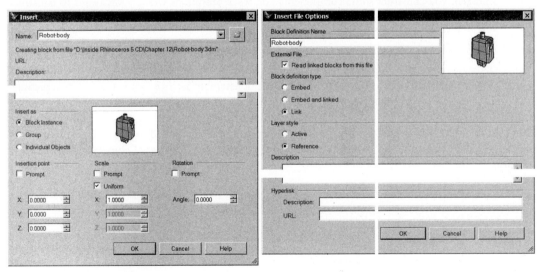

FIGURE 12–19 *Block definition being constructed from an external file and an instance being inserted*
Source: Robert McNeel and Associates Rhinoceros® 5

10. Referencing Figure 12-20, insert and link the files Robot-Arm.3dm and Robot-Leg.3dm from the Chapter 12 folder on the Student Companion Website, shown in Figure 12-20.
11. Repeat the Insert command to insert the block definition Robot-Arm. While inserting, clear the Uniform box and set the X scale to -1, Y scale to 1, and Z scale to 1 in the Insert dialog box.
12. Repeat the command to insert the block definition Robot-Leg, with X scale = −1, Y scale = 1, and Z scale = 1.

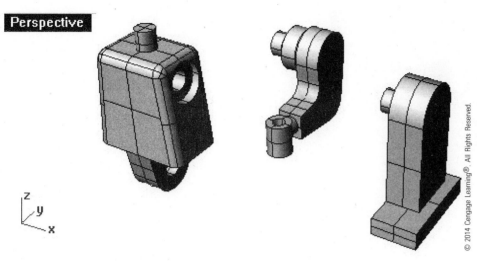

FIGURE 12-20 *Robot-Arm and Robot-Leg inserted*

Instances with X scale equal to −1 are inserted, shown in Figure 12-21.

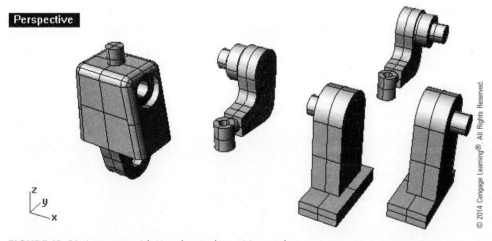

FIGURE 12-21 *Instances with X scale equal to −1 inserted*

Orient Objects

After constructing three block definitions and five instances, refer to what you have learned in Chapter 10 to orient the instances, as shown in Figure 12-22. Now you have constructed an assembly from three external files by inserting them into the current file as block definitions and block instances. Because the block definitions are constructed from external files, they are referenced to the external files in a passive way. Here "passive" means that the block definitions remember the source files, but an update has to be made explicitly and manually by clicking on the Update button of the Block Manager dialog box.

FIGURE 12–22 *Instances oriented*

The Top-Down Approach

Sometimes you have a concept in your mind, but you have no concrete ideas about the component parts. You use the top-down approach—starting a Rhino file and designing some component parts there. From the preliminary component parts, you improvise. Upon finalizing the design, you construct block definitions for each of the components and export the block instances as external files. The main advantages in using this approach are that you see other parts while working on an individual part, and you can continuously switch from designing one part to another. Continue with the following steps to construct a component.

13. Set the current layer to Robot-Head. A curve is already constructed on this layer.
14. Referencing Figure 12–23, revolve curve A around an axis defined by BC.

    A revolved surface is constructed. This will become the head of the robot toy.

FIGURE 12–23 *Robot-Head layer turned on and revolved surface being constructed*

Defining a Block Definition

Now a component is constructed in the assembly file. If all the components are constructed this way, you are using the top-down approach. To put the component that you construct in a block definition, continue with the following steps.

15. Select Edit > Blocks > Create Block Definition.
16. Select A (Figure 12–24) and press the ENTER key.
17. Select Center B.
18. Set the block definition name to Robot-Head.

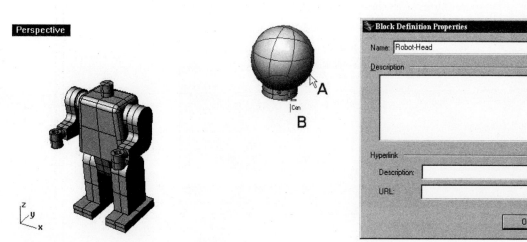

FIGURE 12–24 *Block definition being constructed*
Source: Robert McNeel and Associates Rhinoceros® 5

19. Referencing Figure 12–25, orient the robot's head.

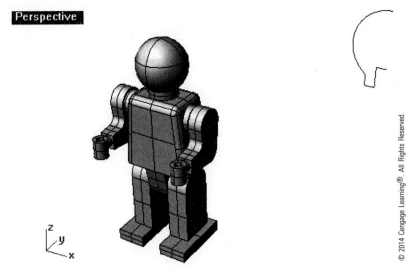

FIGURE 12–25 *Robot's head oriented*

Exporting and Referencing

To export the block definition to an external file and reference it to the exported file, continue with the following steps.

20. Open the Block Manager dialog box.
21. Select the block definition "Robot-Head" and click on the Export button. (See Figure 12–26.)
22. Save the file to a folder in your computer.
23. Select the block definition "Robot-Head" and click on the Properties button.
24. In the Block Definition Properties dialog box that follows, click on the Browse button and select the exported file to reference the block to the exported file.

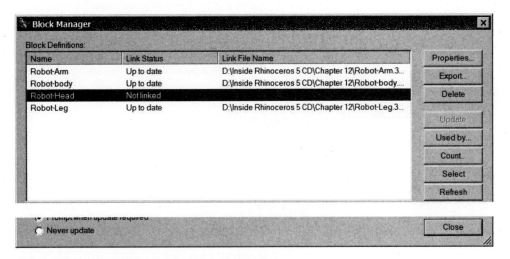

FIGURE 12–26 *Internal block definition being exported*
Source: Robert McNeel and Associates Rhinoceros® 5

25. Click on the OK button and then click the Close button.

    The assembly is complete. (Figure 12–27 shows the rendered image of the assembly.)
26. Do not save your file.

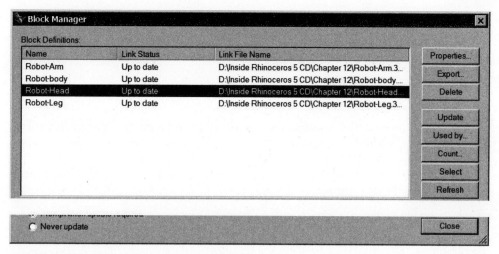

FIGURE 12–27 *Block referenced to an external file*
Source: Robert McNeel and Associates Rhinoceros® 5

Hybrid Approach

Because the design you are now working on concerns making a component in the assembly as well as inserting components from external files, you are using a hybrid approach.

Named Position

In a file consisting of many objects, you can save the positions of selected objects and assign names to those arrangements, enabling you to store several arrangements of objects and recall them at will. Perform the following steps.

1. Select File > Open and select the file NamedPosition.3dm from the Chapter 12 folder on the Student Companion Website.
2. Select Panels > Named Positions if the Named Positions panel is not already displayed.
3. In the Named Positions dialog box (Figure 12–28), two positions are already saved. Select Position 2, click on the Restore button, and then click the OK button. The second saved position is restored.
4. Do not save your file.

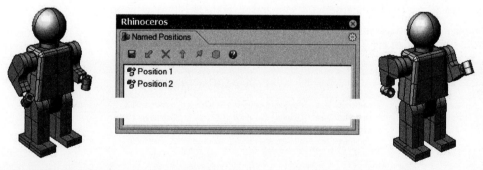

FIGURE 12–28 *Two saved positions of objects and Named Positions dialog panel*
Source: Robert McNeel and Associates Rhinoceros® 5

WORKSESSION MANAGER AND DESIGN COLLABORATION

In an industrial environment, it is a common practice to team up a group of designers to work collaboratively on a large project. In Rhino, a way to enable design work to be shared among the designers is to divide the project into a number of smaller projects. To work collaboratively, each designer works on and controls a file but can view other portions of the project. To facilitate such design collaboration activity, you can use the Worksession Manager to perform four tasks: file attachment, file detachment, attached file activation, and refreshing attached files.

Working Directory

If a project involves a number of files, it is necessary to set the working directory as follows.

1. Click on the Set Working Directory button on the File toolbar.
2. Select a directory from the Select Directory dialog box.

File Attachment

You can use the Worksession Manager to attach external files of various formats that Rhino supports. (Supported files will be covered in the next chapter.) The same file can be attached concurrently by a number of designers working on different computers. Geometry in the attached file cannot be modified but can be used as input for constructing geometry.

Attached files are listed in the Worksession Manager dialog box. The attached file list will not be saved in the current file; you have to save the list in a worksession file. Perform the following steps.

1. Select File > Open and select the file carbody02.3dm from the Chapter 12 folder on the Student Companion Website.
2. Select File > Worksession Manager.

   The Worksession Manager dialog box is displayed, shown in Figure 12–29. The current file is listed in the Worksession Manager dialog box. There is a tick mark in the first column of the list, indicating that this is the active file.

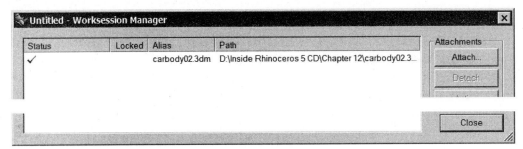

FIGURE 12–29 *File opened (left) and Worksession Manager dialog box displayed (right)*
Source: Robert McNeel and Associates Rhinoceros® 5

3. In the Worksession Manager dialog box, click on the Attach button.
4. Select the file carbody01.3dm from the Chapter 12 folder on the Student Companion Website.

   A file is attached.
5. Repeat steps 3 and 4 twice to attach the files carbody04.3dm and carbody05.3dm one by one.

   Now, the current working file has three external files attached. (See Figure 12–30.)

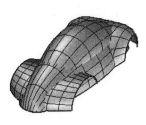
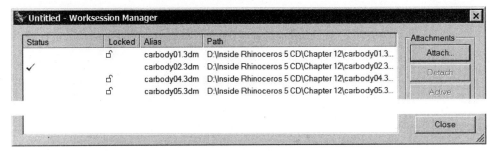

FIGURE 12–30 *Three external files attached (left) and one of them being detached (right)*
Source: Robert McNeel and Associates Rhinoceros® 5

File Detachment

Contrary to file attachment, you can detach attached files, thus removing them from the Worksession Manager dialog box. It is important to note that if you close the Worksession Manager without saving it, closing the current file will detach the attached files automatically. Continue with the following steps.

6. Select the file carbody05.3dm from the list in the Worksession Manager dialog box.
7. Click on the Detach button.

   The selected file is detached.

Attached File Activation

If you want to modify the geometry of an attached file, you can activate it via the Worksession Manager. Although a file can be attached by many persons for use as reference, only one person can activate a file to modify it. Once an attached file is activated, the current file will be closed. Naturally, you will be prompted to save changes. Continue with the following steps.

8. Select the file carbody01 from the list in the Worksession Manager dialog box.
9. Click on the Active button.

   The selected file becomes the active file. If any changes are made to the previous active file, you will be prompted to save the changes. Because there is no modification made to the file carbody02.3dm, the active file simply changes to carbody01.3dm without any prompt.

10. Click on the Close button of the Worksession dialog box to close it. (See Figure 12–31.)

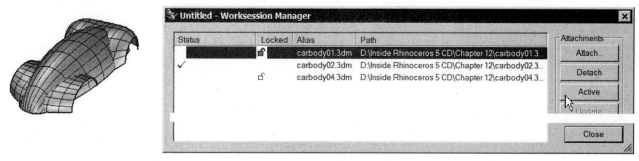

FIGURE 12–31 *Two external files attached (left) and one attached file made active (right)*
Source: Robert McNeel and Associates Rhinoceros® 5

To appreciate how the attached file can be used to work on the current active file, continue with the following steps.

11. Referencing Figure 12–32, rotate the display.

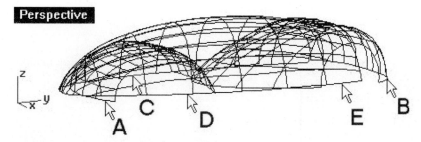

FIGURE 12–32 *Display rotated and surface being trimmed*
© 2014 Cengage Learning®. All Rights Reserved.

12. Select Edit > Trim, or click on the Trim/Untrim Surface button on the Main 1 toolbar.
13. Select A and B and press the ENTER key.
14. Select locations C, D, and E, and press the ENTER key.

    The selected portions of the surface in the active file are trimmed by surfaces from the attached file. (See Figure 12–33.)

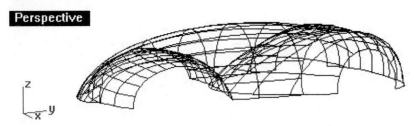

FIGURE 12–33 *Surface in the current file trimmed by surfaces from the attached files*
© 2014 Cengage Learning®. All Rights Reserved.

Refreshing Attached Files

Because a file can be attached by many designers working concurrently, it may have been activated and modified by one of the designers who attached the file. Therefore, you may need to refresh the attached files from time to time to display the most up-to-date version of the file. Continue with the following steps.

15. Select File > Worksession Manager or right-click on the Attach/Worksession button of the File toolbar.
16. Click on the Refresh All button.

    You will be prompted to save the file. If you click on the OK button, the attached files are refreshed. If these files are modified by another person, any change will be reflected.

17. Do not save your file.

Worksession File

The list of files, including the currently open file or activated file and all the attached files, together with their layer states, can be saved in a Worksession file.

Saving a Worksession File

To save a worksession file, open the Worksession Manager dialog box, click on the Save button, and specify a file name. If the worksession has already been saved and you want to save it to a new file, click on the Save As button and specify a new file name.

NOTE If you are using the evaluation version of Rhino 5, saving a worksession file will be counted as saving once.

The saved worksession file will have the extension rows. It remembers the list, layer states, and the window placement and size of the attached files. Now perform the following steps to open a worksession file.

1. Create a folder named Inside Rhinoceros 5 CD in your computer.
2. Copy the folder Chapter 12 on the Student Companion Website.
3. Select File > Open and select the file worksession.rws from the Chapter 12 folder on the Student Companion Website.

    You need to set the Files of Type to Rhino Worksession (.rws).*

When you open a Worksession file, the following operations will be performed.

- The current file will be closed. Naturally, you will be prompted to save changes.
- The current Worksession will be cleared.
- The files listed in the opened Worksession file will be opened, with the active file ready for editing and the other files attached for reference.
- The layer states of the listed files will be set.
- You can save the list of files (the currently open file and all attached files) and their layer states in a worksession file.

Layer States

Although you cannot edit the geometry of the attached files, you can modify their layer states, such as changing layer color, turning layers on and off, and locking layers. This way, you can change the color of the displayed geometry and control the display of some geometry to enhance visual effect.

If you want the computer to remember the attached file list in the Worksession Manager dialog box and the layer states of the attached file, you have to save the worksession. However, even if you save a worksession, layer states will not be saved in the attached files themselves. In other words, if a layer of an attached file is red and you change its color to blue in the worksession, the color of the layer will remain red if you open the file individually, and the color of the layer will become blue if you open the file in the context of a worksession. Continue with the following steps.

4. Open the Layer panel
5. Use the Worksession Manager to make active the attached files one by one and modify the color settings of the layers as necessary. (See Figure 12–34.)

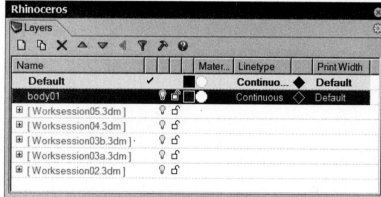

FIGURE 12–34 *Layer state being modified (left) and Edit Layer dialog box (right)*
Source: Robert McNeel and Associates Rhinoceros® 5

6. If you wish to save the layer states, open the Worksession Manager dialog box and click on the Save button. Otherwise, do not save your files.

Because the layer states of the attached files saved in a worksession file do not affect the layer state of the attached files themselves, by attaching the same file to different worksession files and setting different layer states in the worksession files, you can save numerous layer states for a single file.

Reference Geometry Limit

Understanding that substantial computer resources are required if there are many attached files, you may limit the amount of reference geometry by constructing a sphere, outside which objects will be purged. To set the limit, select File > Work Session > Limit and then construct a sphere.

File Attachment and File Insertion

The major difference between file attachment and file insertion is that an attached file is not copied to the current file. If the worksession is not saved, the attachment information will be lost after the current file is closed. On the other hand, an inserted file is copied to the current file to become a set of objects or a block definition. If the file is inserted as a block definition, the external file is referenced. Change in the external file can be reflected by updating.

Design Collaboration

To sum up, the Worksession Manager can help enhance design collaboration as follows.

- Many designers can attach the same file to display it as reference.
- Designers can change the layer states of the attached files to enhance visual effect.
- One of the designers can activate an attached file to edit it.
- Other designers can refresh the attached files to display the most up-to-date version of the attached files.
- The attached file list and their layer states in the context of the worksession can be saved, thus saving time to reattach files and reset the layer states.

DRAG AND DROP

You can drag any file supported by Rhino from the Windows explorer to Rhinoceros windows to perform one of the following tasks: open, insert, import, and attach. After dragging, the File Options dialog box, shown in Figure 12–35, will display.

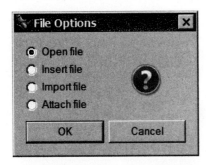

FIGURE 12–35 *File Options dialog box*
Source: Robert McNeel and Associates Rhinoceros® 5

If the dragged file is to be opened, you will be prompted to save the current file before closing it. If the dragged file is to be inserted, the Insert dialog box will be displayed. You can insert the file in one of three ways (Insert, Import, or Attach). If the dragged file is to be imported, the content of the file will be imported and merged with the current file. If the dragged file is to be attached, the file is attached and its file name will display in the list of the Worksession dialog box. You can reference to its geometry and change its layer state in the context of the current file.

TEXTUAL INFORMATION

You can attach textual information to either object geometry or the attributes of an object by using the SetUserText command, and retrieve the attached textual information by using the GetUserText command.

CONSOLIDATION

By putting a set of objects in a group, you can select the entire set of objects collectively by picking an element in the group. To select an individual object of the group, you hold down the CONTROL and SHIFT keys simultaneously while picking the object. After a group is formed, you can add elements to it or remove elements from it. Groups can be named for easy identification, grouped objects can be ungrouped, and groups can be merged.

One way to manage repeated sets of geometric objects is to define block definitions and insert instances referenced to the block definitions. A block definition can be defined by selecting objects from the current file or inserting an external file. When an external file is inserted, it can be inserted in one of three ways: as an instance, as a group of objects, or as individual objects.

Because inserted instances are referenced to the block definition, instances will change automatically if the block definition is modified. Furthermore, if the block definition is defined from an external file and the external file is modified, you can update the change via the Block Manager.

There are three approaches in designing an assembly of components: the bottom-up approach, the top-down approach, and the hybrid approach. The term "bottom" refers to the individual components of an assembly and "top" refers to the assembly as a whole. Using the bottom-up approach, you construct a set of Rhino files to depict individual components of the assembly, start a new file for the assembly, and insert the component files into the assembly file. Using the top-down approach, you construct components in a Rhino file, construct block definitions for each component, export the block definitions as individual external files, and link the blocks to the exported files.

The Worksession Manager is a useful design collaboration tool. It enables you to divide a large project into a number of small files. By opening one of the files and attaching the remaining files via the Worksession Manager, you can work on a small part of the project and reference to the other parts of the project.

In a collaborative design environment in which a group of people work on a project, each people can open a file and attach other files. The latest version of the files done by other people can be seen by refreshing the attached files. By activating different files, design work can pass from one designer to another.

REVIEW QUESTIONS

1. Differentiate between a group and a block.
2. What are the two ways to construct a block?
3. How can an external referenced block be updated?
4. What are the ways to remove block definitions in a file?
5. Briefly explain how bottom-up and top-down design approaches can be carried out using Rhino.
6. Outline how the Worksession Manager helps facilitate design collaboration.

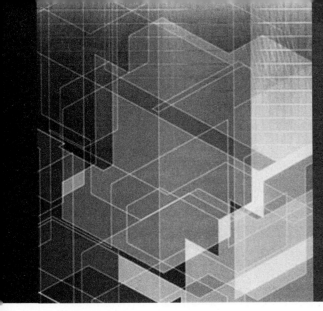

CHAPTER 13

2D Drawing Output and Data Exchange

INTRODUCTION

This chapter explores the process of constructing 2D engineering drawings and delineates data exchange methods.

OBJECTIVES

After studying this chapter, you should be able to

- Construct 2D engineering drawings
- Perform data exchange

OVERVIEW

2D orthographic engineering drawings are the conventional means of communication among engineering personnel. Although the advent of computer-aided design applications replaced some of the uses of 2D drawings, there remain many situations in which 2D drawings and/or 2D drawing output are useful or necessary. In a drawing, you have to add dimensions and annotations to help delineate details of objects depicted by the drawing views. To address downstream and upstream operations, you export Rhino files to various data formats and read various data formats into Rhino.

ENGINEERING DRAWING

Apart from delineating the geometric shape of objects in two dimensions, one important function of engineering drawings is to specify precisely the dimensions of the objects they represent, along with annotations that convey other information about the object or objects represented. Engineering drawing construction consists of two major tasks:

- Constructing 2D orthographic projection drawings
- Dimensioning and annotating the 2D drawings

Methods to Construct 2D Drawings

Using Rhino as a tool, you can construct 2D engineering drawings in three ways, as follows.

- You can let the computer generate 2D drawings from 3D objects automatically.
- You can set out a page layout and construct detail viewports to delineate various orthographic projection views.

- You can construct 2D drawings from scratch, using the curve tools described in Chapters 4 through 6. Construction of 2D drawings from scratch requires having a thorough understanding of orthographic projection principles and a good perception of how 3D objects would look when projected onto a planar face. Erroneous engineering drawings are quite often the result of human errors.

Generating 2D Engineering Drawings from 3D Objects

The process of outputting a 3D model as an engineering (2D) drawing is fairly simple using Rhino. Basically, you use an appropriate command and let the computer do all of the 2D drawing construction work. Because the 2D drawings are produced automatically, you can rely on Rhino and your computer in projecting 3D objects onto a plane, and sound knowledge in orthographic projection is not necessary. Perform the following steps.

1. Open the file 3Dto2D.3dm from the Chapter 13 folder on the companion CD.
2. Select Dimension > Make 2-D Drawings.
3. Referencing Figure 13–1, click on A and drag to B, and then press the ENTER key.

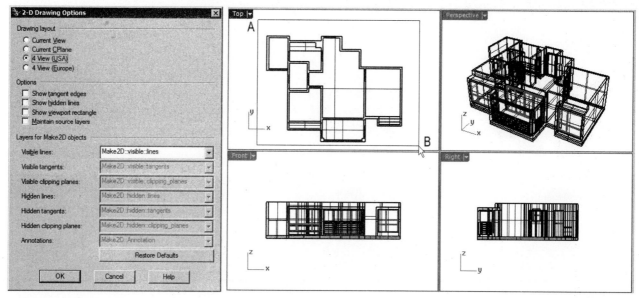

FIGURE 13–1 *2-D Drawing Options dialog box (left) and 3D object being selected (right)*
Source: Robert McNeel and Associates Rhinoceros® 5

4. In the Drawing layout area of the 2-D Drawing Options dialog box, select 4 View (USA) and then click on the OK button.
5. Set the current layer to Make2D::visible::lines and turn off all other polysurface layers.
6. Maximize the Top viewport. A 2D drawing is constructed, shown in Figure 13–2.
7. Do not save your file.

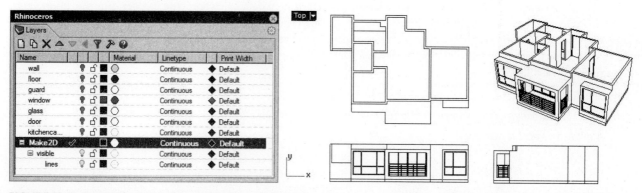

FIGURE 13–2 *Front, side, top, and isometric views constructed*
Source: Robert McNeel and Associates Rhinoceros® 5

As you can see, generation of 2D drawings from 3D objects is very simple. However, this method of producing a drawing has a drawback—the choice of viewing directions is restricted.

Constructing Page Layout and Detail Viewports

Another method to produce a 2D engineering drawing from a 3D model is to use page layout to depict a piece of drawing paper and detail viewports to depict various orthographic drawing views.

Constructing a Page Layout

A page layout is analogous to a sheet of paper on which you construct 2D engineering drawings. Perform the following steps to construct a page layout for a Rhino file.

1. Open the file Locomotive.3dm from the Chapter 13 folder on the companion CD.
2. Select View > Layout > New Layout.
3. In the New Layout dialog box, shown in Figure 13-3, select a printer (here the Microsoft XPS Document Writer is chosen), set the paper size to A4 Landscape, set the Initial Detail Count to 4, and click on the OK button.

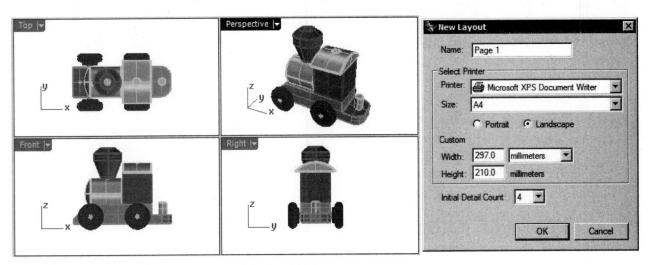

FIGURE 13-3 *3D model (left) and New Layout dialog box (right)*
Source: Robert McNeel and Associates Rhinoceros® 5

Initial Detail Count refers to the number of detail viewports that will be constructed automatically by the system. If you choose 0, you will have a blank page layout. A page layout with four detail viewports is constructed. (See Figure 13-4.)

Manipulation of Detail Viewports

Detail viewports are viewports constructed on a page layout. By activating a detail viewport, you can perform pan, zoom, and rotate commands in order to obtain an appropriate view. At the lower-left corner of a detail viewport is a lock/unlock icon. Clicking on the icon locks/unlocks the viewport in terms of pan, zoom, and rotate. In a page layout, you can have many detail viewports, each depicting a different direction of viewing. Detail viewports can be deleted and can be constructed anywhere on the page layout, as follows.

4. Select Edit > Delete.
5. Select detail viewport A, shown in Figure 13-4, and press the ENTER key.

   A detail viewport is deleted.

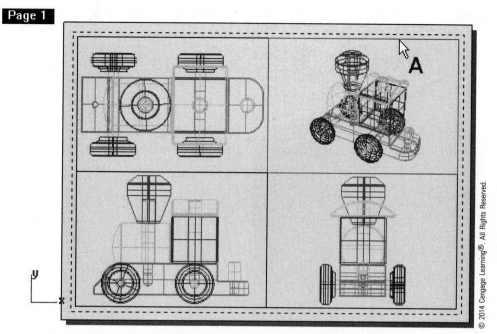

FIGURE 13–4 *Page layout with four detail viewports and a viewport being deleted*

6. Select View > Layout > Add Detail View.
7. Click on locations A and B, shown in Figure 13-5. Exact location is unimportant.
   A detail viewport is constructed.

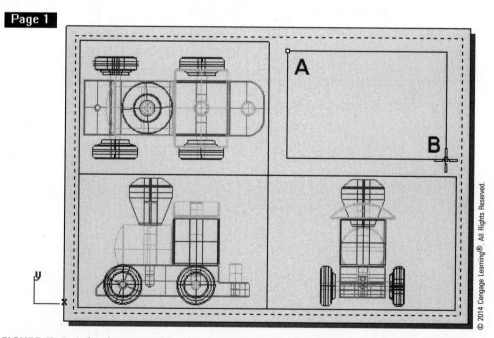

FIGURE 13–5 *A detail viewport deleted and another detail viewport being constructed*

To comply with engineering drawing standards, the display scale of the detail viewports has to be consistent and the detail viewports have to be aligned properly.

8. Select View > Layout > Scale Detail View.
9. Select detail viewport A, shown in Figure 13-6, and press the ENTER key.

10. Set the distance on page to 1 unit.
11. Type 2 at the command area.

    This will set the display scale to "half" because one unit displayed in the detail viewport is equivalent to 2 units displayed in the page layout.

12. Repeat the command to set the scale of detail for viewports B, C, and D.

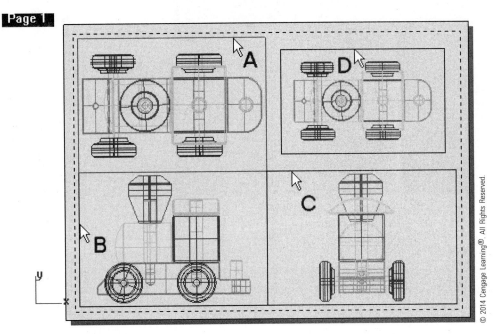

FIGURE 13–6 *Detail viewports being scaled*

To help align the detail viewports, a dummy box is constructed and put in a layer that is turned off. Continue with the following steps.

13. Turn on the Reference layer, keeping the current layer unchanged.
14. Select Transform > Move.
15. Select detail viewport A and press the ENTER key.
16. Select endpoint B (Figure 13–7) as the base point and endpoint C as the point to move.

    The viewport is moved.

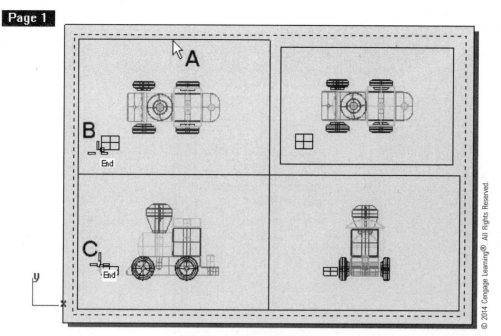

FIGURE 13–7 *Detail viewport being moved*

17. Turn on Ortho mode by checking the Ortho box at the Status bar.
18. Referencing Figure 13–8, select and drag the moved detail viewport from A to B. The exact location is unimportant as long as the detail viewports are orthogonal to each other.

    The Top detail viewport is aligned properly with the Front detail viewport.

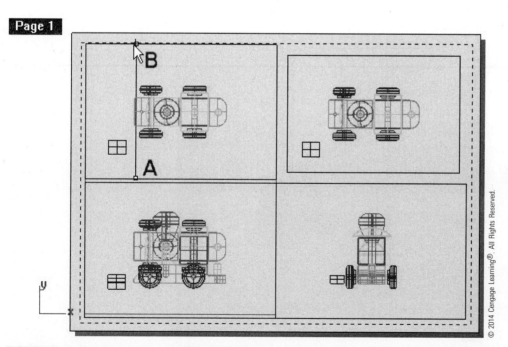

FIGURE 13–8 *Top viewport moved and being dragged*

19. Repeat steps 14 through 18 above to align the right-side detail viewport, shown in Figure 13–9.
20. Turn off the Reference layer.
21. Turn off Ortho mode.

To modify the display of the detail viewport, continue with the following steps.

22. Select View > Layout > Enable detail view and select detail viewport A, shown in Figure 13–9, or double-click on the interior of viewport A.

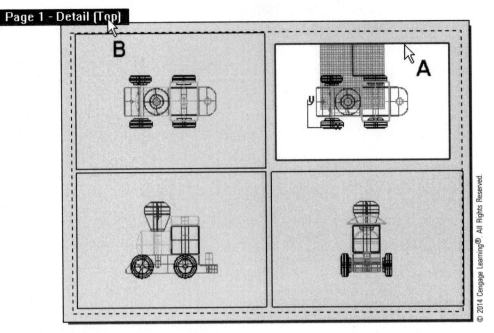

FIGURE 13–9 *Detail viewport being enabled*

23. Right-click on B, the detail viewport's label, and select Shaded.
    The detail viewport is changed to shaded mode.
24. Referencing Figure 13–10, rotate the viewport.
25. To exit the viewport, double-click anywhere outside the active detail viewport.
    The drawing is complete.
26. Do not save your file.

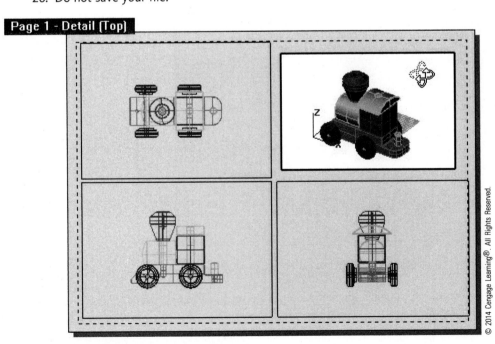

FIGURE 13–10 *Detail viewport shaded and rotated*

In order not to print the detail viewports, they are residing on a layer with print color set to white, the color of the paper on which the drawing is printed.

Copying Page Layout

In a Rhino file, you can have more than one page layout. Therefore, you can use a layout for a set of detail viewports showing a set of objects and use another layout for another set of drawing viewports.

- To facilitate construction of a page layout from an existing page layout, you can use the CopyLayout command.

Import Page Layout

To save time in drawing construction, you can import a layout from another file. Perform the following steps.

1. Open the file Import Layout.3dm from the Chapter 13 folder on the companion CD.
2. Select View > Layout > Import.
3. Select the file Locomotive 1.3dm from the Chapter 13 folder on the companion CD. (See Figure 13–11.)

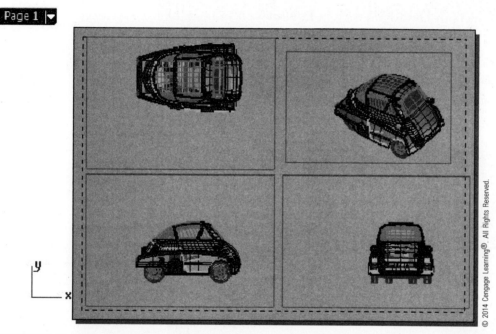

FIGURE 13–11 *Layout imported*

Manipulating Layers in Detail Views of a Layout

To turn on and off selected layers in a detail view, continue with the following steps.

4. Double-click on A to activate a detail view.
5. Type HideLayersInDetail at the command area or click on the Hide Layers in Detail button on the New in V5 toolbar.
6. Click on the Layers options at the command area.
7. In the Hide Layers in Detail dialog box (Figure 13–12), select layer Wheel and click on the OK button.
   Layer Wheel is hidden in the detail view.
8. Type ShowLayersInDetails at the command area or click on the Show Layers in Detail button on the New in V5 toolbar.
9. In the Show Layers in Detail dialog box (Figure 13–12), select layer Wheel and click on the OK button.
   Layer Wheel is shown in the detail view.

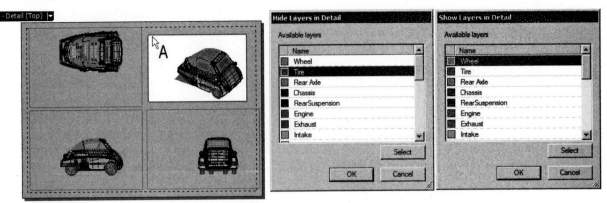

FIGURE 13-12 *From left to right: layers in detail view being manipulated, Hide Layers in Detail dialog box, and Show Layers in Detail dialog box*
Source: Robert McNeel and Associates Rhinoceros® 5

Hiding and Showing Objects in Detail Views of a Layout

Apart from layer manipulation, you can hide and show objects in a detail view, as follows.

10. Type HideInDetail at the command area.
11. Select A, B, and C, indicated in Figure 13-13, and press the ENTER key.
    Selected objects are hidden.

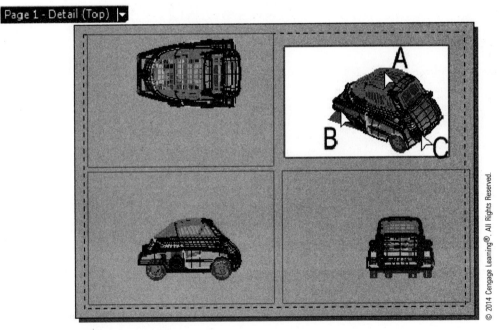

FIGURE 13-13 *Objects being hidden*

12. Type ShowSelectedInDetail at the command area.
13. Select A, indicated in Figure 13-14, and press the ENTER key.
    The selected hidden object is shown.
14. Type ShowInDetail at the command area.
    All the hidden objects are shown.
15. Do not save your file.

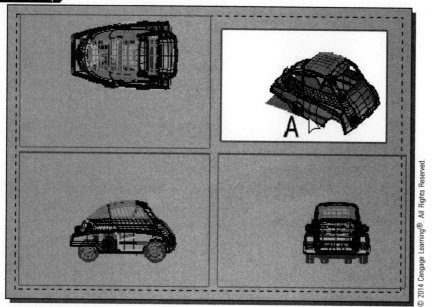

FIGURE 13–14 *Hidden objects being shown*

Constructing 2D Drawings from Scratch

To produce a 2D drawing from scratch, you are essentially using the computer as an electronic drawing board. As when you are drawing using pencil and paper, you need to have excellent knowledge of orthographic projection and good perception of 3D objects in terms of how they would look when they are projected onto a plane. To appreciate a 2D drawing constructed from scratch, perform the following steps.

1. Open the file 2Ddrawing.3dm from the Chapter 13 folder on the companion CD. (See Figure 13–15.)
2. Maximize the Perspective viewport.
3. Select View > Pan, Zoom, and Rotate > Rotate View, or click on the Rotate View button on the Standard toolbar.
4. Rotate the view to appreciate how the 2D drawing looks.

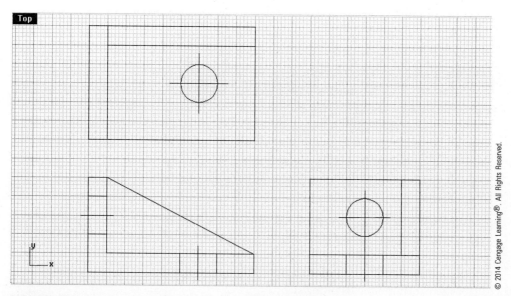

FIGURE 13–15 *2D drawing constructed from scratch using curve tools*

Hatching

Hatching is a way to highlight a section face of an object cut by an imaginary plane. No matter which way you construct a drawing, you can construct a hatch pattern within a closed planar boundary curve. Continue with the following steps.

5. Maximize the Top viewport.
6. Select Dimension > Hatch.
7. In the command area, if Boundary=No, click on it to change it to Yes. Otherwise, proceed to the next step.
8. Select curves A, B, C, D, E, F, G, H, J, and K (Figure 13–16) and press the ENTER key.
9. Click on locations L, M, and N (Figure 13–16) and press the ENTER key.
10. In the Hatch dialog box (Figure 13–16), select Hatch1, set Pattern rotation to 45° and Pattern scale to 20, and click on the OK button. The selected locations of the drawing are hatched. (See Figure 13–17.)

    Note: If solid hatch pattern is used and subsequently exploded, the solid hatch will become a surface.

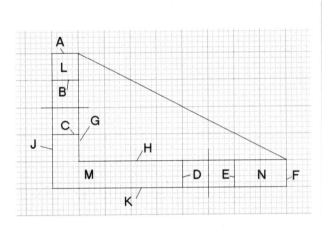
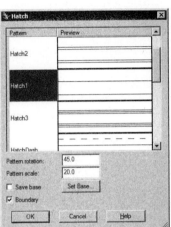

FIGURE 13–16 *Boundaries selected (left) and Hatch dialog box (right)*
Source: Robert McNeel and Associates Rhinoceros® 5

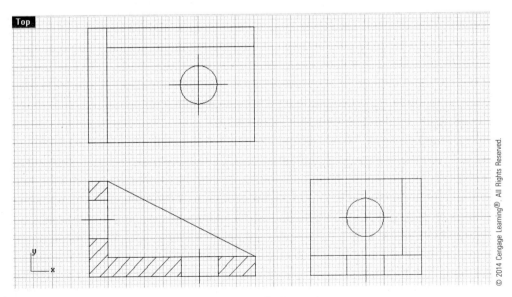

FIGURE 13–17 *Selected locations of the drawing hatched*

Modifying a Hatch

If you want to load or unload a hatch pattern, perform the following steps.

11. Select Tools > Options.
12. In the Options dialog box, click on Document Properties > Annotations > Hatch.

You can modify the base point and the scale of a hatch, as follows.

13. Select Edit > Object Properties.
14. Select the hatch pattern that you want to modify.
15. In the Properties dialog box, click on the Hatch button, shown in Figure 13-18.
16. Make any modification as may be necessary.

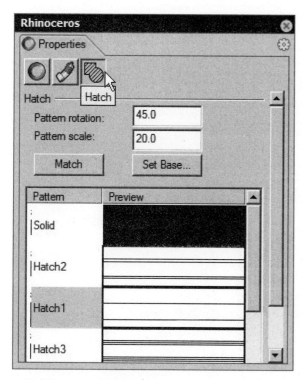

FIGURE 13-18 *Modifying a hatch*
Source: Robert McNeel and Associates Rhinoceros® 5

Line Types

Because 2D orthographic projection drawings are line drawings, it is necessary to use various line types to depict different kinds of line objects, as follows.

- Thick continuous lines for visible outlines and edges
- Thin continuous lines for dimension lines, leaders, hatchings, outlines of adjacent parts, and outlines of revolved sections
- Thin dash lines for hidden outlines and hidden edges
- Thin chain lines for center lines and extreme positions of moving parts

To use these lines in your drawing, you might need to load them into your Rhino file, if such line types are not available in the template file that you use to construct the drawing, as follows.

17. Select File > Properties.
18. Click on Document Properties > Annotations > Linetypes.
19. Click on the Load button. In the Load Linetype dialog box, select the line type that is needed and click on the Add button.
20. Click on the Close button to return to the Options dialog box.
21. Click on the OK button of the Options dialog box.

By default, line type is By Layer, meaning that objects will have a line type assigned to the layer. Therefore, you can simply change the layer's line type to set the line type of objects residing on a layer. To set the line type of an object individually, continue with the following steps.

22. Select Edit > Object Properties.
23. Hold down the SHIFT key and select lines A, B, C, D, E, and F (Figure 13–19).
24. In the Properties dialog box, select the Object tab.
25. Select Center from the Linetype pull-down list box.
    The selected curve's line type is changed.
26. Do not save your file.

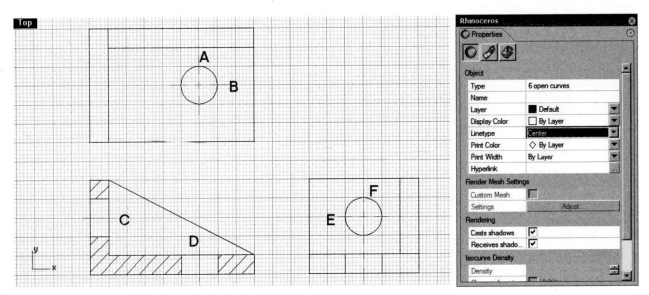

FIGURE 13–19 *Line type being changed*
Source: Robert McNeel and Associates Rhinoceros® 5

Dimensioning and Annotation

To complement 2D drawing views that depict the shape of an object projected orthogonally, you add dimensions and annotations. Dimensions indicate the size of the object; annotations incorporate additional information other than dimensions. There are seven kinds of dimensions: linear, aligned, rotated, radial, diameter, angle, and ordinate. There are four kinds of annotations: leader, text box, dots, and annotation at the curve's endpoints.

Dimension Styles

Before adding dimensions and annotations to a drawing, it is recommended to check the Dimensions tab of the Document Properties dialog box and change the settings as necessary. Perform the following steps.

1. Open the file Dimension1.3dm from the Chapter 13 folder on the companion CD.
2. Select Dimension > Dimension Styles.
3. In the Document Properties dialog box, shown in Figure 13–20, click on Annotation > Dimensions.
    As shown in Figure 13–20, there are three dimension styles in this file. If you wish to use some other dimension styles from another Rhino file, click on the Import button, select a file, and select the dimension styles from the Import Dimension Styles dialog box that opens.

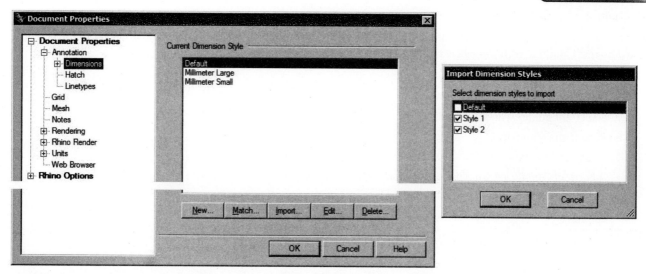

FIGURE 13–20 *Dimension style manipulation (left) and Import Dimension Styles dialog box (right)*
Source: Robert McNeel and Associates Rhinoceros® 5

4. Click on the Default dimension style, set Text height to 3 units, Arrow length to 3 units, Extension line offset to 1 unit, and Extension line extension to 1 unit, and then click on the OK button, shown in Figure 13–21.

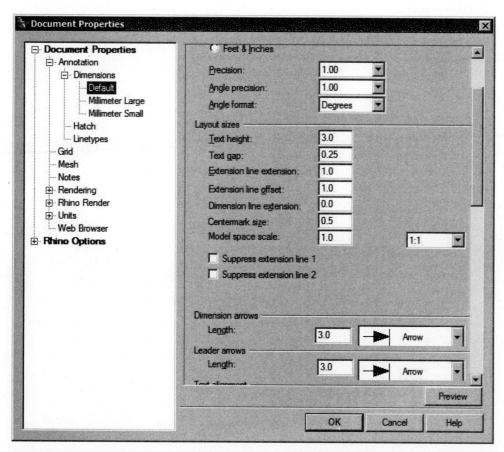

FIGURE 13–21 *Dimension options*
Source: Robert McNeel and Associates Rhinoceros® 5

Linear Dimension

The most common way to specify the dimensions of an object is to state the vertical or horizontal distance between two selected points. Continue with the following steps.

5. Set object snap mode to End.
6. Select Dimension > Linear Dimension.
7. Select endpoints A and B and click on location C (Figure 13-22).
8. Repeat the command.
9. Select endpoints A and D and click on location E (Figure 13-22).

   A horizontal and a vertical linear dimension are constructed.

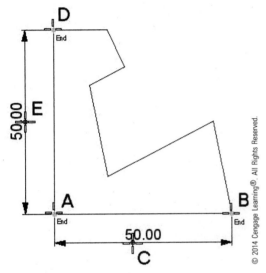

FIGURE 13-22 *Horizontal and vertical linear dimensions being constructed*

Aligned Dimension

To specify the distance between two selected points, you use the aligned dimension. The dimension line is parallel to an imaginary line connecting the two selected points. Continue with the following steps.

10. Select Dimension > Aligned Dimension.
11. Select endpoints A and B and click on location C (Figure 13-23).

    An aligned dimension is constructed.

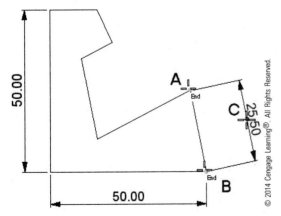

FIGURE 13-23 *Aligned dimension being constructed*

Rotated Dimension

To specify the distance between two selected points measured in a direction defined by a rotation angle, you use the rotated dimension. Continue with the following steps.

12. Select Dimension > Rotated Dimension.
13. Check the Perp box on the Osnap dialog box.
14. Referencing Figure 13–24, select endpoint A and point B perpendicular to A.

    The rotation angle is defined. Alternatively, you can type a value at the command area to specify the rotation angle.

15. Select endpoints A and C and click on location D (Figure 13–24).

    A rotated dimension is constructed.

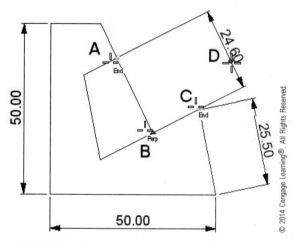

FIGURE 13–24 *Rotated dimension being constructed*

Angle Dimension

To specify the angle between two nonparallel lines, you use angle dimension. Continue with the following steps.

16. Select Dimension > Angle Dimension.
17. Select lines A and B and click on location C (Figure 13–25).

    An angle dimension is constructed.

18. Do not save your file.

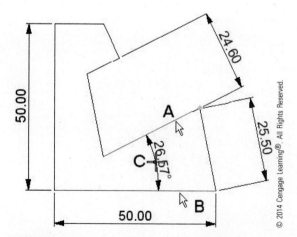

FIGURE 13–25 *Angle dimension being constructed*

Radial Dimension

To specify the radius of an arc or a circle, you use radial dimension. Perform the following steps.

1. Open the file Dimension2.3dm from the Chapter 13 folder on the companion CD.
2. Select Dimension > Radial Dimension.
3. Select arc A and click on location B (Figure 13–26).
4. Repeat the command.
5. Select arc C and click on location D (Figure 13–26).

   Two radial dimensions are constructed.

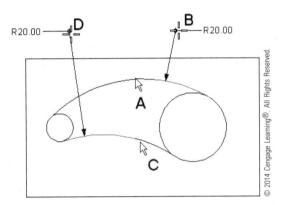

FIGURE 13–26 *Radial dimensions being constructed*

Diameter Dimension

To specify the diameter of an arc or circle, you use diameter dimension. Continue with the following steps.

6. Select Dimension > Diameter Dimension.
7. Select circle A and click on location B (Figure 13–27).
8. Repeat the command.
9. Select circle C and click on location D (Figure 13–27).

   Two diameter dimensions are constructed.

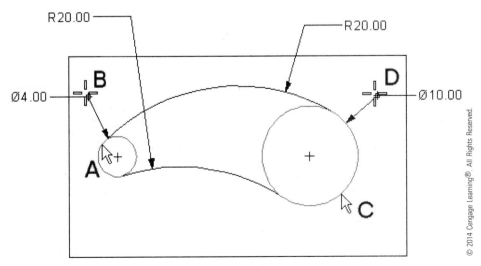

FIGURE 13–27 *Diameter dimensions being constructed*

Ordinate Dimension

You can construct ordinate dimensions, depicting the X coordinates and Y coordinates with reference to a specified base point. To facilitate the placement of ordinate dimensions, you can turn on Ortho mode. Continue with the following steps.

10. Set object snap mode to End and Cen and clear other boxes on the Osnap toolbar.
11. Check the Ortho box on the status line.
12. Select Dimension > Ordinate Dimension.
13. Select the Basepoint option at the command area.
14. Select endpoint A, shown in Figure 13–28.
15. Select center B and click on location C.
16. Repeat steps 12 through 15 two more times to construct a vertical ordinate dimension for center B at D and a horizontal ordinate dimension for center E at F.
17. Clear the check mark on the Ortho button of the Status bar.

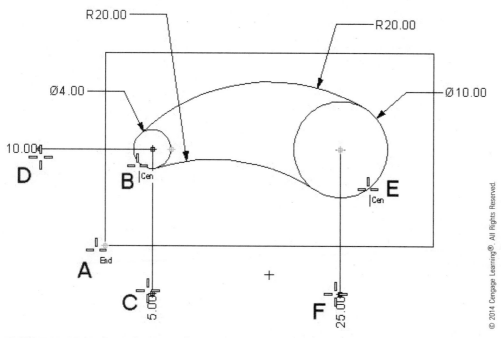

FIGURE 13–28 *Ordinate dimensions being constructed*

Leader

A leader is a set of lines with an arrowhead at one end of the connected lines and a text string at the other end. Continue with the following steps.

18. Select Dimension > Leader.
19. Set object snap mode to Mid and clear all other check boxes.
20. Referencing Figure 13–29, select midpoint A, click on locations B and C, and press the ENTER key.
21. Type the text string "RECTANGLE" in the Leader Text dialog box and click on the OK button.
    A leader is constructed.

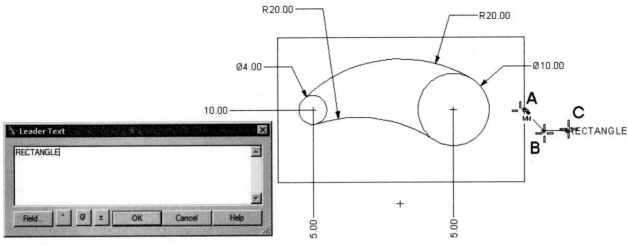

FIGURE 13–29 Leader being constructed
Source: Robert McNeel and Associates Rhinoceros® 5

Text Box

A text box is a set of text string and is 2D; you must not confuse a text box with a 3D text object. If you explode a text box, a set of curves is obtained. On the other hand, if you explode a 3D text object, you obtain a set of surfaces. Continue with the following steps.

22. Select Dimension > Text Block.
23. Click on location A, shown in Figure 13-30.
24. In the Text dialog box, type the text string "RHINO 5" and click on the OK button.

    A text string is constructed.

Instead of inputting text string in the Text dialog box, you can click on the Import File button of the Text dialog box to import a text file (with a file extension of .txt).

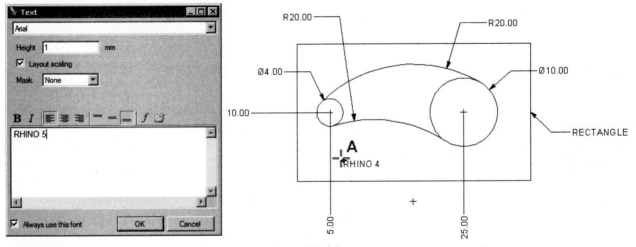

FIGURE 13–30 Text dialog box (left) and text being constructed (right)
Source: Robert McNeel and Associates Rhinoceros® 5

Editing Text

Text that is already constructed can be modified, as follows.

25. Select Edit > Object Properties, and select A, indicated in Figure 13-31.
26. In the Properties dialog box, click on the Text button B and change the text string as shown.
27. While the Properties dialog box is still open, select C.
28. In the Properties dialog box, click on the Leader button D and change the text string accordingly.

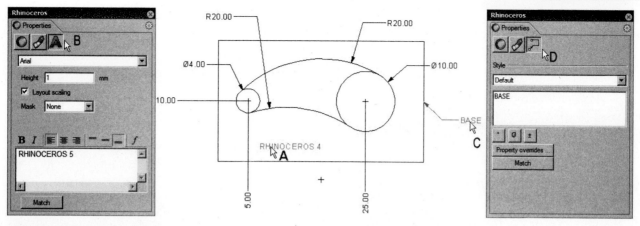

FIGURE 13–31 *Text and leader being modified*
Source: Robert McNeel and Associates Rhinoceros® 5

Another method of editing the text and leader text is to double-click on the text or leader text to display the on-the-spot editing box, shown in Figure 13–32.

29. Do not save your file.

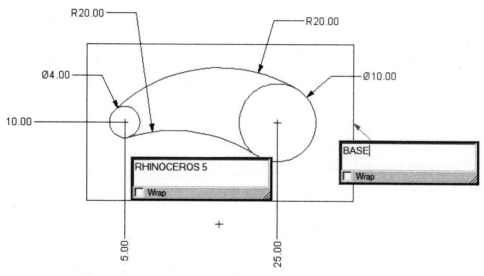

FIGURE 13–32 *On-the-spot editing box*
Source: Robert McNeel and Associates Rhinoceros® 5

Miscellaneous Annotation Commands

If you wish to search for a certain text string in a Rhino document, you can use the FINDTEXT command. If you want to scale the text size in relation to printing scale, you can use the TEXTSCALE command. To add a revision cloud to highlight a text string, you can use the REVCLOUD command.

Dots

It is possible to place a dot with annotation that stays parallel to the viewport and sizes appropriately with the view.

1. Open the file Dot.3dm from the Chapter 13 folder on the companion CD.
2. Select Dimension > Annotate Dot.
3. Type "CURVE" in the Dot dialog box and click on the OK button.
4. Select the midpoint of curve A, shown in Figure 13–33.

   A dot is constructed.

Arrow Head

You can add an arrow head at the endpoint of a curve.

5. Click on the Add/remove arrowhead on curve button on the Annotate toolbar.
6. Select B, shown in Figure 13-33.

   An arrowhead is constructed.

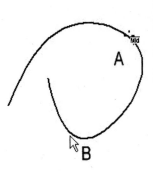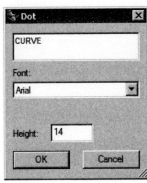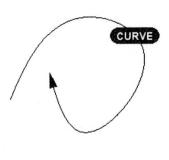

FIGURE 13-33 *From left to right: curve being manipulated, Dot dialog box, and dot and arrow constructed*
Source: Robert McNeel and Associates Rhinoceros® 5

To appreciate how the annotation dot, curve end annotation, and arrowhead are parallel to the viewport's viewing direction, continue with the following steps.

7. Select View > Pan, Zoom, and Rotate > Rotate View, or click on the Rotate View button on the Standard toolbar.
8. Rotate the view.

   You will find that no matter how you rotate the view, the arrow and the dot are always parallel to the viewing plane.

Conversion

To convert a dot to a text object, perform the following steps.

9. Type ConvertDots at the command area.
10. Select dot A, indicated in Figure 13-34, and press the ENTER key.
11. At the command area, set DeleteInput=Yes and TextHeight=3.
12. Press the ENTER key.

    The selected dot is converted to a text object.

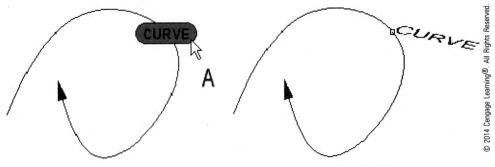

FIGURE 13-34 *Dot being converted (left) and dot converted to text object (right)*

13. Do not save your file.

Exploding Dimensions

The explode command can be applied to dimensions, text, and hatch patterns, changing the dimensions to curves and text, the text to curves, and the hatch patterns to lines. However, this process is irreversible; exploded dimensions cannot be reconstructed as a single object.

Printing

To facilitate communication, you may need to output printed copies of your drawing. Perform the following steps.

1. Open the file Locomotive 1.3dm from the Chapter 13 folder on the companion CD.
2. To appreciate how the drawing looks printed, select View > Print Preview.
3. Select File > Print. (See Figure 13–35.)

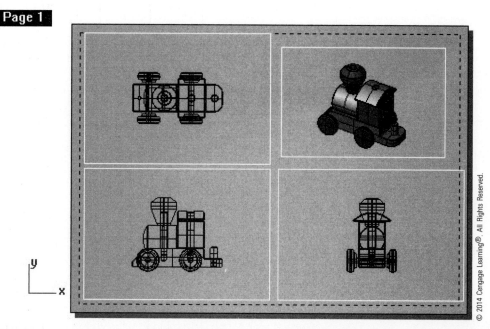

FIGURE 13–35 *Display set to print*

Figure 13–36 shows the Print Setup dialog box. At the left of the dialog box are six tabs: Destination, View and Output Scale, Margins and Position, Linetypes and Line Widths, Visibility, and Printer Details. Click on these tabs to make necessary changes.

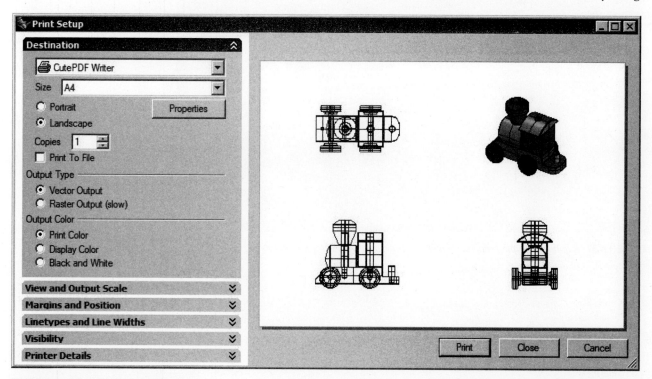

FIGURE 13–36 *Print Setup dialog box*
Source: Robert McNeel and Associates Rhinoceros® 5

4. Click on the Print button to print the document or click on the Close button to abort printing.
5. Do not save your file.

DATA EXCHANGE

The prime objective of exchanging data in various file formats is to facilitate downstream and upstream computerized operations.

Downstream Operations

Constructing 3D models and producing 2D drawings from 3D models are not necessarily the end of the digital modeling process. In a computerized manufacturing system, for example, the same digital model can be (and often should be) designed to be used in all downstream operations, such as finite element analysis, rapid prototyping, CNC (computer numeric control) machining, and computerized assembly. Figure 13–37 shows a rapid prototyping machine making a 3D object, and Figure 13–38 shows the CNC machining of a free-form 3D object.

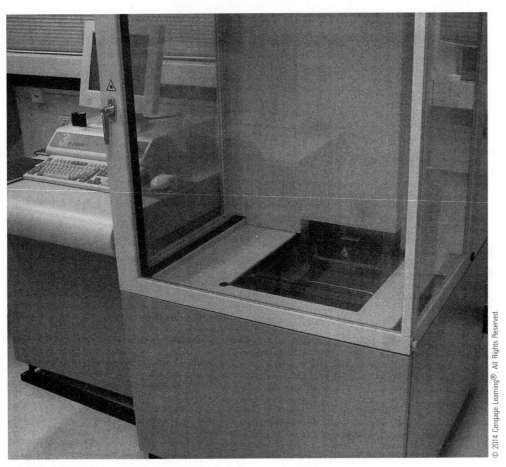

FIGURE 13–37 *Rapid prototyping machine making a rapid prototype from a 3D model*

FIGURE 13-38 *CNC machining of a free-form 3D object*

To enhance illustration of the 3D object, you construct renderings and animations. Because these operations may be done using different types of computer applications, and because each application may use a unique type of data format, the digital modeling system must enable the conversion of the 3D digital model into various file formats.

Rhino File Exporting

If your free-form models are constructed to facilitate downstream computerized operations, you can output them to various file formats for this purpose. You can save the entire Rhino file or export selected objects as various file formats, as discussed below.

- To output the entire Rhino file in another format, select File > Save as.
- To export selected objects, select File > Export Selected.

Save as Template and Export with Origin

You must not confuse the Save as command with the Save as Template command and the Export selected command with the Export with Origin command.

- By selecting File > Save As Template, the entire drawing is saved as a Rhino template.
- By selecting File > Export with Origin, you export selected objects with origin point to Rhino format.

Upstream Operations

To take advantage of other computerized operation's strength, the digital modeling system must enable the opening of various file formats so that digital models constructed in other systems can be used for further elaboration of the design.

Other Files Importing

To reuse digital data from other applications, you can:

- Open a file by selecting File > Open
- Import a file by selecting File > Import
- Insert a file by selecting File > Insert

Naturally, opening a file converts the file to Rhino format, whereupon you can continue with the design. If you want to incorporate data into an existing Rhino file, you import or insert the file. Importing a file converts the imported data to individual Rhino curves or surfaces. Inserting a file also converts the data to Rhino format, but the file can be inserted as a block definition, a group of objects, or a set of individual objects. File types that can be opened, imported, or inserted can be found by expanding the file type pull-down list when executing the open, insert, export, and save commands.

Objects Import and Export

In essence, objects that you construct in a Rhino file are points, curves, surfaces, polysurfaces, and polygon meshes. In addition, you can incorporate dimensions, annotations, and rendering objects, such as material properties, light objects, and environment information.

Objects to Be Imported and Exported

Before saving the entire Rhino file or exporting selected Rhino objects into a particular file format, you need to know what kind of Rhino objects can be exported and whether such object types will be changed after exporting.

For example, exporting to 3DS format will export rendering objects together with the 3D geometry, but this process converts all NURBS surfaces to polygon meshes. If you want to keep the NURBS surface data in the target file, you should export the file to IGES, SAT, or STP format. However, rendering objects may not be exported.

Another example is exporting to Adobe Illustrator format. The target file will become a 2D drawing projected in the active viewport. Importing the exported Adobe Illustrator file back into Rhino will convert the original 3D object into planar drawings of the 3D object.

When you work in the other direction, you have to understand what kinds of data are in the source file before importing. You cannot expect to have a set of NURBS surfaces by importing a file that contains only polygon meshes.

Conversion to Bézier Curves/Surfaces

By default, Rhino curves and surfaces are NURBS curves and surfaces. To convert such curves or surfaces to Bézier curves or surfaces, you use the ConvertToBeziers command.

File Formats

Because there are so many kinds of file formats that Rhino can export to and import from, you are advised to keep the source Rhino file before exporting and experiment with various kinds of exporting and importing before proceeding to real project work. The following delineation serves to provide some information about the commonly used file formats that you may export to or import from.

STEP (.stp, .step) Files

STEP stands for Standard for the Exchange of Product model data. It is an international standard (ISO 10303) to provide a complete definition of the physical and functional characteristics of an object.

Direct X (.x) Files

Direct X (.x) is a set of technologies designed for multimedia elements. Rhino NURBS surfaces exported to Direct X will become a set of polygon meshes. Naturally, any surfaces that are imported via Direct X are polygon meshes.

Slice (.slc) Files

Slice files are StereoLithography Contour (.slc) files. It slices a 3D object to produce 2D contours. The contour lines are polylines. NURBS surfaces have to be translated into either .slc format or .stl format (described later) before the data can be read into the solid imaging machine software. Rhino solids exported will become a set of contour lines.

Point (*.asc, *.csv, *.txt, *.xyz, *.cgo_ascii, *.cc) Files

Point files are coordinates of point objects. Only point objects in a Rhino file are exported. If you open a point file, you get only points.

General Hydrostatic System (.gf, .pm) Files

GHS stands for General Hydrostatic System. These GHS files concern hydrostatic properties, stability, and other marine-related properties. If the Rhino file has such properties, you can export them to GHS formats. On the other hand, GHS files can be imported and attached to Rhino surfaces.

XGL Files

XGL files represent 3D objects for the purpose of visualization. They include all the 3D information related to SGI's OpenGL rendering library. NURBS surfaces are exported to polygon meshes.

Verband Der Automobileindustrie (.vda) Files

A VDA file is a neutral file format defined by the German Association of Automobile Industries for the exchange of computer-aided design data across various systems.

RenderMan (.rib) Files

RIB stands for RenderMan Interface Bytestream. RenderMan is a technical specification for the interface between modeling application and rendering application. The file conveys model data, lights, camera, shaders, attributes, and options to rendering applications for it to produce a photorealistic image. Refer to *https://renderman.pixar.com/* for more information.

Object Properties (.csv) Files

CSV stands for Comma-Separated Value. CSV file is used to exchange data between applications. It has a tabular format, with fields separated by a comma and quoted by a double-quote character. Rhino exports object properties to this format.

Adobe Illustrator (.ai) and Window Metafile (.wmf) Files

These two formats are viewport dependent. Exporting produces either a 2D Adobe Illustrator file or a 2D Window Metafile from the selected viewport. Therefore, you have to position the 3D object properly in the active viewport prior to exporting. Before importing from Adobe Illustrator, convert the text object in Illustrator to curves because Rhino does not read Adobe Illustrator text.

STL (.stl) Files

STL stands for Stereolithography. NURBS surfaces exported to an STL file will become triangular meshes, and imported STL files are triangular meshes. Although open objects in Rhino can also be exported to STL format, you should try to ensure the objects are watertight by applying join, weld, and unifymeshnormals commands. To discover the location of open edges, use the SelNakedMeshEdgePt command.

VRML (.wrl, .vrml) Files

VRML stands for Virtual Reality Modeling Language. This is a standard for representing 3D scenes for delivery via the World Wide Web. Objects in VRML format are represented by polygon meshes. Naturally, NURBS surfaces will be exported as polygon meshes.

POV-Ray Mesh (.pov) Files

POV stands for Persistence of Vision. It is a ray tracer that reads a source file describing the scene to be rendered and outputs a rendered image. A Rhino file exported to POV format becomes a set of data describing the rendering objects and geometry in polygon meshes. The file is written in ASCII format. In other words, you can open a POV file with the notepad or any word processor. Refer to *http://www.povray.org/* for more details about POV files.

Cult 3D (.cd) Files

Cult 3D is an application that enables you to apply interactivity to the geometry that is constructed in another 3D modeler and exported to .cd format. A Rhino file exported to Cult3D format is a polygon mesh. Refer to *http://www.cult3d.com/* to learn more about Cult 3D.

Moray UDO (.udo) Files

UDO stands for User-Defined Objects. It is a description on how objects should look in Moray and tells Moray how to render the wireframe of the object. For more information about Moray, refer to *http://www.stmuc.com/moray/*.

Light Wave (.lwo) Files

LWO stands for Light Wave Object, a file format used by Light Wave. The file stores information about points, NURBS surfaces, polygon meshes, splines, and surface attributes.

Raw Triangles (.raw) Files

This file format is used by Persistent of Vision applications for inputting triangular facet geometry. Naturally, surfaces are exported or imported as meshes.

ACIS (.sat, .sab) Files

ACIS is a 3D modeling kernel developed by Spatial Technology (*http://www.spatial.com/*). It stores modeling information (wireframe, surface, and solid) in two kinds of file formats: Standard ACIS Text (.sat) and Standard ACIS Binary (.sab).

IGES (.igs, .iges) Files

IGES stands for Initial Graphics Exchange Specifications. It was first introduced in 1979 as a standard platform to exchange computer-aided design data among various applications.

Parasolid (.x_t) Files

Parasolid is a kernel modeler used by many solid modeling applications. More information about it can be obtained from *http://www.ugs.com/products/open/parasolid/*.

3D Studio Files

Exporting Rhino surfaces to 3D Studio file format will convert the NURBS surfaces into polygon meshes. Importing a file back to Rhino will give a set of meshes.

CONSOLIDATION

2D orthographic engineering drawings are a standard means of communication among engineers. In Rhino, you create 2D drawings in two ways: by selecting a 3D object and letting the computer generate the drawing views, and by using the curve tools to construct a drawing from scratch. A Rhino 2D drawing can incorporate information on the dimensions of the object or objects the drawing represents as well as other annotation (textual information).

Rhino is a 3D digital modeling tool. You use it to construct points, curves, surfaces, polysurfaces, solids, and polygon meshes. In addition, you output photorealistic renderings, 2D drawings, and file formats of various types for downstream computerized operations. You also reuse upstream digital models by opening various file formats.

To export Rhino objects, you can use Save as command to export the entire file or use the Export selected command to export selected objects. To import other objects into Rhino, you can open, import, or insert a file.

REVIEW QUESTIONS

1. Explain how 2D engineering drawings are produced.
2. In conjunction with the drawing views, what kinds of objects have to be incorporated in a drawing?
3. In what ways can Rhino objects be exported?
4. What are the ways to import other objects into Rhino?
5. Give a brief account of the file formats supported by Rhino.

CHAPTER 14

Rendering

INTRODUCTION

This chapter describes the use of Rhino as a tool in rendering and animation.

OBJECTIVES

After studying this chapter, you should be able to

- State the key concepts involved in rendering
- Use Rhino to produce rendered images and animations
- Manipulate the imaginary camera in a scene
- Add lighting effects to a scene
- Include environment elements in a scene
- Apply digital material properties to objects in a scene

OVERVIEW

Digital rendering is a method of producing a photorealistic image of an object in a 3D scene in the computer. Unlike shading, which simply applies a shaded color to the surface of an object, rendering takes into account the material properties (color and texture) assigned to the object, the effect of lighting in the scene, and any additional environment factors. To produce a photorealistic image, you need to

- Manipulate artificial camera's lens length and location as well as the target's location
- Use artificial lighting in the scene
- Include environment objects to enhance reality in the scene
- Assign material properties to objects in the scene

Rhino, together with Flamingo nXt, incorporates the capacity to produce output from two types of renderers: Rhino and Flamingo nXt. The Rhino renderer is the basic renderer. In producing a rendered image, Rhino takes into account the materials assigned to objects, lights added to a scene, and simple environment objects (ambient light and background color). Flamingo nXt is sold separately as plug-ins to the Rhino application. It provides additional rendering, materials, lightings, environments, and utilities that greatly enhance the rendering process, producing a very realistic image.

DIGITAL RENDERING AND ANIMATION

Using Rhinoceros, rendering and animation can be done in any viewport, using the default camera setting, lights, environment material, and material properties.

Rendering

Rendering can be done in two basic ways: preview rendering and full rendering.

Render Resolution

Prior to performing rendering, you may wish to set the resolution of the image to be rendered.

1. Select Render > Render Properties. (See Figure 14–1.)

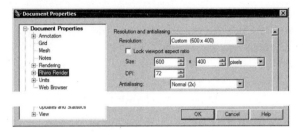

FIGURE 14–1 *Rhino Render tab of the Document Properties dialog box*
Source: Robert McNeel and Associates Rhinoceros® 5

Render Preview

Because a fully rendered image takes into account all the digital data of a file, it may take a very long time to produce a good image, if the file includes substantial geometry and information on how the geometry is rendered. For the sake of previewing, you may want to speed up rendering time by applying minimal rendering settings. Perform the following steps.

2. Select File > Open and select the file BubbleCarRendering.3dm from the Chapter 14 folder on the companion CD.
3. Select Render > Render Preview, or click on the Render Preview button on the Render toolbar.

   A render preview is produced, shown in Figure 14–2. The rendering resolution can be specified from the Rhino Render tab of the Document Properties dialog box accessed by selecting Render > Render Properties. Note that, because this is a preview rendering, the edges of the bubble car in the image are jagged.

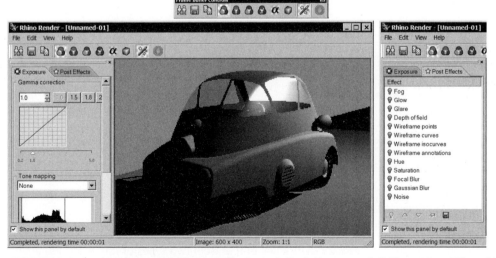

FIGURE 14–2 *Render preview with Frame Buffer Controls (top), Exposure tab (left), and Post Effects (right)*
Source: Robert McNeel and Associates Rhinoceros® 5

In the Rhino Render dialog box shown in Figure 14–2, apart from the rendered preview in the main area, you will find the Frame Buffer Controls toolbar at the top and an Effects panel at the right consisting of two tabs.

Frame Buffer Controls

The Frame Buffer Controls toolbar has eleven buttons; their functions are delineated in the following Table 14–1. Now continue with the following step.

4. Try out all of these buttons one-by-one to appreciate their effects.

TABLE 14–1 *Frame Buffer Controls toolbar buttons*

Button (from Left to Right)	Function
Clone Render Window	This button clones the existing window for the purpose of trying out various exposure and post effects.
Save Image As	This button saves the rendered image.
Copy Image to Clipboard	This button copies the rendered image to the clipboard for pasting it to other applications.
Show RGB Channel	Graphics systems contain four channels for pixel information; among them, three are for storing red, green, and blue, and there is an alpha channel, which is a mask, specifying how the pixel's colors should be merged with another when the two are overlaid, one object on top of another. This button shows the red, green, and blue channels.
Show Red Channel	This button shows only the red channel.
Show Green Channel	This button shows only the green channel.
Show Blue Channel	This button shows only the blue channel.
Show Alpha Channel	This button shows only the alpha channel.
Show Distance Channel	This button shows a gray scale image, delineating the distance of various objects in the scene from the camera.
Show/Hide Effects Panel	This button shows or hides the Effects panel.
Stop Rendering	This button terminates the rendering process.

Effects Panel

As shown in Figure 14–2, the Effects panel has two tabs: Exposure tab and Post Effects tab.

Exposure Tab

The Exposure tab has two areas: Gamma correction and Tone mapping. Gamma correction refers to the compensation required to cope with human perception under various conditions. Now continue with the following steps.

5. Click on the Clone Render Window button on the Frame Buffer Controls dialog box six times to clone six more rendered windows.
6. Click on one of the cloned window to activate it.
7. In the Exposure tab, click on the 2.2 button of the Gamma correction area. (See Figure 14–3.)

Tone mapping is a technique used to adjust an image's contrast locally in such a way that the entire range of maximum contrast is applied to each region of the image. How tone mapping is done depends on the tone mapping algorithm used. Continue with the following steps.

8. Click on the second cloned window.
9. In the Tone mapping pull-down list box, select Reinhard 2005. (See Figure 14–3.)

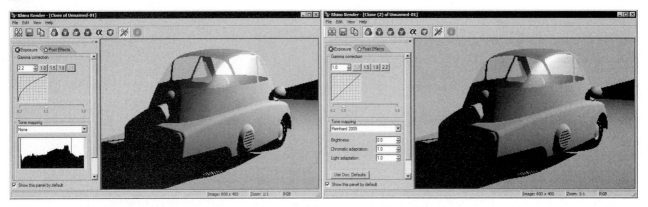

FIGURE 14-3 *Cloned windows with Gamma correction (left) and Tone mapping (right)*
Source: Robert McNeel and Associates Rhinoceros® 5

Post Effects Tab

The Post Effects tab has thirteen buttons, enabling fog, glow, glare, and depth of field effects to be applied, displaying points, curves, isocurves, and annotations in the rendered image, allowing the adjustment of hue and saturation, and applying focal blur, Gaussian blur and noise to the image. Continue with the following steps.

10. Click on the third cloned rendered window and click on the Post Effects tab.
11. Click on the Glow button to activate it, right-click on it, and select Properties.

    In the right-click menu, apart from selecting Properties, you can select Move Up or Move Down to change the sequence of application of effects if more than one effect is selected.

12. In the Glow Properties dialog box, right-click on the first color palette and select Pick From Image.
13. Click on the tail light A (color red), shown in Figure 14–4 (left), of the bubble car image and then click on the OK button.

    Glow effect is applied to the selected area of the image, shown in Figure 14–4 (right).

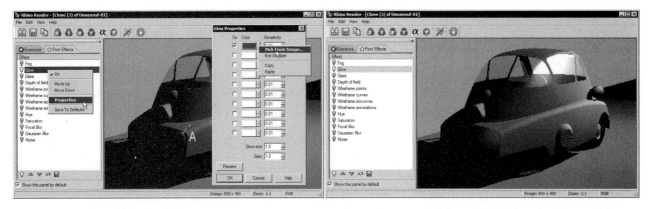

FIGURE 14-4 *Right-click on the Glow button and selecting area of glow (left) and glow effect applied to the selected area (right)*
Source: Robert McNeel and Associates Rhinoceros® 5

14. Click on the remaining cloned rendered windows one-by-one and apply Fog, Glare, Depth of field, and Noise effects to them individually. (See Figure 14–5.)

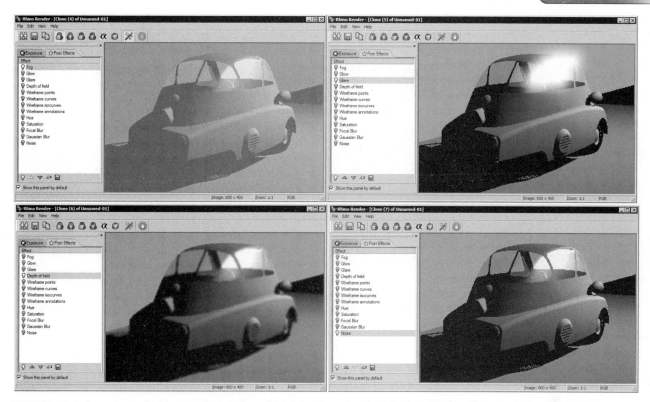

FIGURE 14-5 *From top to bottom and from left to right: Fog, Glare, Depth of field, and Noise effects applied to cloned rendered windows individually*
Source: Robert McNeel and Associates Rhinoceros® 5

Preview Render Window and Preview Render in Window

To further reduce rendering time, you can preview a part of the viewport by describing a rectangular area in the screen instead of rendering the entire viewport.

15. Type RenderPreviewWindow at the command area.
16. Click on locations A and B, shown in Figure 14–6.

    A preview window is constructed.

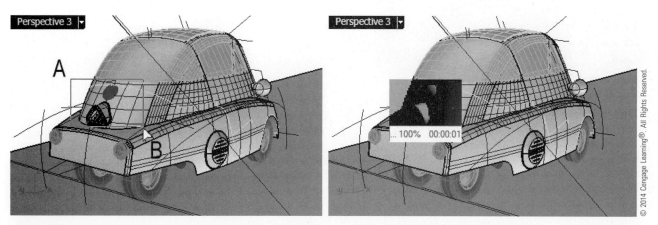

FIGURE 14-6 *Preview render window selected (left) and selected portion preview rendered in viewport (right)*

17. Type RenderPreviewInWindow at the command area.
18. Click on locations A and B, shown in Figure 14–7.

    Another preview is constructed.

Note the difference that one command simply shows a preview window while the other enables you to save the preview as a bitmap image as well as to fine-tune that rendered portion using the tools available in the Render dialog box.

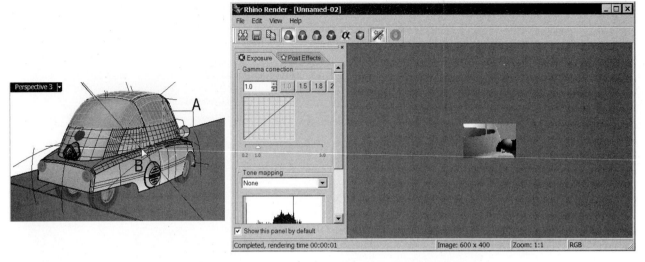

FIGURE 14–7 *Preview render window selected (left) and selected portion preview rendered (right)*
Source: Robert McNeel and Associates Rhinoceros® 5

Full Render

A full render takes into consideration all the rendering information incorporated in the Rhino file.

19. Select Render > Render or click on the Render button on the Render toolbar.

    A rendered image is produced, shown in Figure 14–8.

Similar to preview rendering, you can use the tools available in the Render dialog box to fine-tune your image.

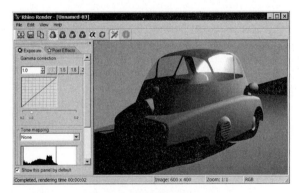

FIGURE 14–8 *Full rendered image*
Source: Robert McNeel and Associates Rhinoceros® 5

Render Window and Render in Window

You will notice that it takes much longer to produce the rendered image than the render preview you did previously. However, if you compare the rendered image with the render preview image, you will find that the latter has a better quality. In particular, anti-aliasing is applied to remove the jagged edges. (Anti-aliasing will be explained later in this chapter.)

In the process of fine-tuning your image, you can render only a part of the viewport, as follows.

20. Type RenderWindow at the command area.
21. Click on locations A and B, shown in Figure 14–9.

    The selected rectangular area is rendered.

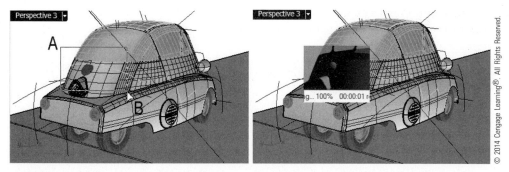

FIGURE 14–9 *Render window selected (left) and selected portion rendered in viewport (right)*

22. Type RenderInWindow at the command area.
23. Click on locations A and B, shown in Figure 14–10.
    The selected rectangular area is rendered.

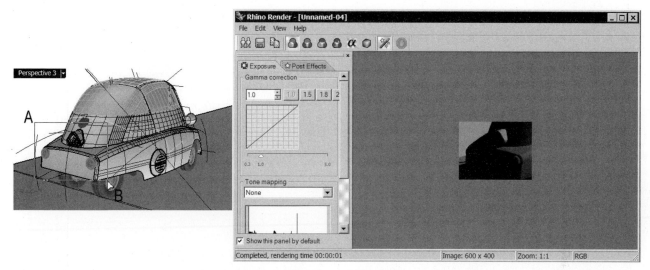

FIGURE 14–10 *Render window selected (left) and selected portion rendered (right)*
Source: Robert McNeel and Associates Rhinoceros® 5

Flamingo nXt Rendering

Flamingo nXt is a very advanced rendering tool, which serves as a plug-in in Rhinoceros. If you already installed this application in your computer, you can make use of it to produce photorealistic images. If you already installed Flamingo nXt in your computer, proceed with the following steps.

24. Select Render > Current Renderer > Flamingo nXt.

Flamingo nXt Rendering Resolution

25. Select Flamingo > Control Panel.
26. Select the Render tab of the Flamingo nXt control panel and set the resolution to 600 × 400 pixels, shown in Figure 14–11.

FIGURE 14–11 *Setting rendering resolution by using the Render tab of the Flamingo nXt control panel*
Source: Robert McNeel and Associates Rhinoceros® 5

Flamingo nXt Rendering

27. Select Render > Render or Flamingo nXt > Render.
28. In the Resume Rendering dialog box, expand the Render Constraints area and set the number of passes to 30. (This is an arbitrary number. You can set it to a higher number to obtain a much better image.)
29. After rendering passes are complete, you may expand the Adjust Image and Post Effects areas to make necessary adjustments and apply effects to your image, view information about the image by expanding the Information area, and view different channels of the image by expanding the Channels area.

    Figure 14–12 *shows the rendered image after 30 passes (left) and various post effects applied (right).*
30. Do not save the file.

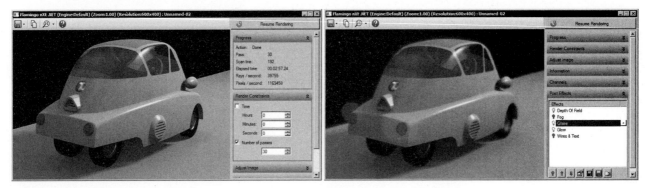

FIGURE 14–12 *Rendering passes being constrained (left) and Post Effects area expanded (right)*
Source: Robert McNeel and Associates Rhinoceros® 5

Digital Animation

Simply speaking, an animation simulates motion by displaying a set of images sequentially at such a speed that the observer perceives movement of the objects in the images. Naturally, construction of an animation involves producing a set of sequential rendered images. In Rhino, the sequential rendered images can be done by manipulating the camera in several ways (turntable, path, and fly-through animation) and by manipulating the sun's light source.

Turntable Animation

The turntable animation is the simplest way of animating the viewport. The imaginary camera simply rotates around the center of the viewport. Perform the following steps.

1. Select File > Open and select the file FoodGrinder.3dm from the Chapter 14 folder on the companion CD.
2. Click on the Set up 360 Degree Turntable Animation button on the Animation Setup toolbar or type SetTurnTableAnimation at the command area.

3. In the Set Turntable Animation dialog box, set the number of frames to 36, accept Clockwise direction, select jpg format in the File type pull-down list box, select RenderFull in the Capture method pull-down list box, accept the Perspective viewport to render, and type Turn Table in the Animation name box, shown in Figure 14–13 (left).
4. Click on the OK button.
5. Click on the Play Animation button on the Animation Preview toolbar or type PlayAnimation at the command area to observe the preview.
6. Click on the Record Animation button on the Animation toolbar or type RecordAnimation at the command area to save the animation to file.
7. At the command area, click on the TargetFolder option and then select a folder in your computer.
8. Press the ENTER key.

Rendering using the current renderer takes place frame by frame, Figure 14–13 (right) shows the Flamingo nXt rendering dialog box.

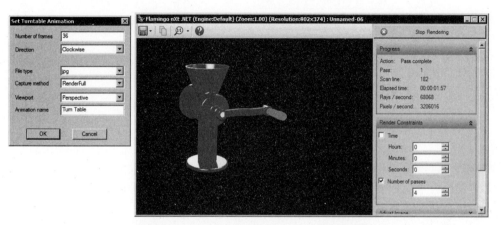

FIGURE 14–13 *Set Turntable Animation dialog box (left) and Flamingo nXt dialog box (right)*
Source: Robert McNeel and Associates Rhinoceros® 5

The time to render depends on the number of frames, the resolution (specified in the Rhino Render tab of the Document Properties dialog box of Rhino renderer or the Render tab of Flamingo nXt's Control Panel), and the kind of rendering. After rendering is completed, a set of rendered images, together with an html document, will be saved in the designated folder.

9. Open the Turn Table folders of Chapter 14 folder on the companion CD and click on the html file. (See Figure 14–14.)
10. Click on the Start button to run the animation, the > button to move to the next frame, the < button to go to the previous frame, the >> button to go to the last frame, and the << button to return to the first frame. To stop the animation while it is running, click on the Stop button.

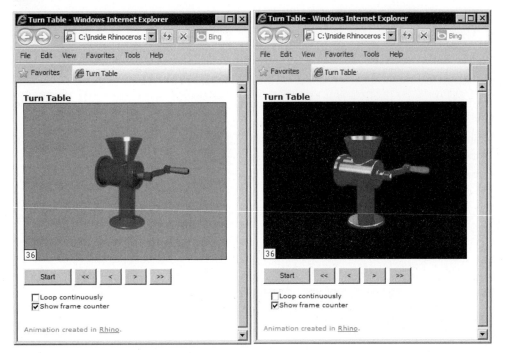

FIGURE 14–14 *Turntable Animation html file opened: Rhino renderer (left) and Flamingo nXt renderer (right)*
Source: Robert McNeel and Associates Rhinoceros® 5

Path Animation

A path animation requires two separate paths: one for guiding the motion of the camera and one for the camera's target point. You can use a curve path for the camera and a point for the target. The target will be fixed, and the camera will be moving. If you use a point for the camera and a curve for the target, the camera's location will be fixed and the target point will be moving. If you use curves for both the camera and the target, both will be moving. (If you use point objects for both the camera and the target, you will not get an animation effect.) Continue with the following steps.

11. Click on the Path Animation button on the Animation Setup toolbar or type SetPathAnimation at the command area.
12. Select curve A, shown in Figure 14–15, to specify the path for the camera.
13. Select point B, shown in Figure 14–15, to specify the target point.

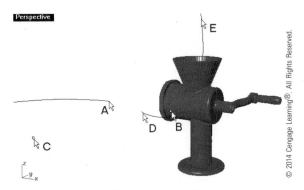

FIGURE 14–15 *Specifying camera and target paths*

Choosing a curve for the camera path and a point for the target path causes the camera to move along the path while targeting at a fixed point.

14. Specify frame numbers, rendered image file type, capture method, viewport to render, and an animation sequence name.

15. Click on the Record Animation button on the Animation toolbar to save the rendered files.

    Figure 14–16 *shows the Set Animation dialog box (left) and a frame of the animation (right).*

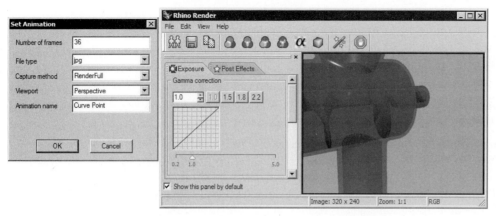

FIGURE 14–16 *Set Animation dialog box (left) and an image of the curve/point animation (right)*
Source: Robert McNeel and Associates Rhinoceros® 5

Continue with the following steps to construct a path animation with the camera located at a fixed point and the target point moving along a path, and a path animation with both the camera and the target point moving along path curves.

16. Repeat the procedure outlined in steps 11 through 15, but select point C, shown in Figure 14–15, as the camera path and curve D, shown in Figure 14–15, as the target path.

    Figure 14–17 *shows the Set Animation dialog box (left) and a frame of the animation (right).*

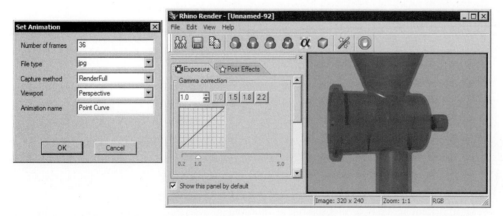

FIGURE 14–17 *Set Animation dialog box (left) and an image of the point/curve animation (right)*
Source: Robert McNeel and Associates Rhinoceros® 5

17. Repeat the procedure outlined in steps 11 through 15, but select curve A, shown in Figure 14–15, as the camera path and curve D, shown in Figure 14–15, as the target path.

    Figure 14–18 *shows the Set Animation dialog box (left) and a frame of the animation (right).*

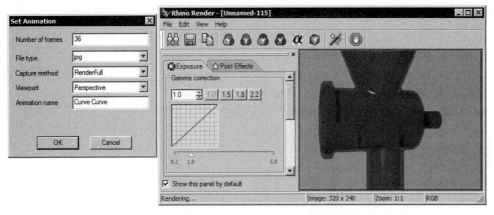

FIGURE 14–18 *Set Animation dialog box (left) and an image of the curve/curve animation (right)*
Source: Robert McNeel and Associates Rhinoceros® 5

Fly-Through Animation

A fly-through animation is a special kind of path animation. It requires only a path because the target is always tangent to the path. Continue with the following steps.

18. Click on the Fly-through Animation button on the Animation Setup toolbar.
19. Select curve ED, shown in Figure 14–15.
20. Specify the number of frames, file type, capture method, viewport to render, and an animation sequence name.
21. Click on the Record Animation button on the Animation toolbar to save the animated images.

    Figure 14–19 shows the Set Animation dialog box (left) and a frame of the Flamingo nXt animation (right). Rhino Render frame will be similar to that shown in Figure 14–18.

22. Do not save the file.

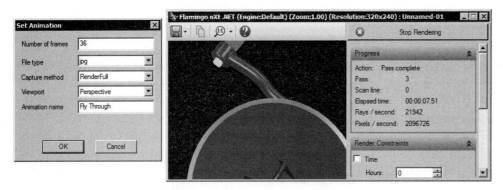

FIGURE 14–19 *Set Animation dialog box (left) and fly-through animation (right)*
Source: Robert McNeel and Associates Rhinoceros® 5

Sun Study Animation

You can animate the sun shadow cast in a scene in a day or in a season, as follows.

1. Select File > Open and select the file LivingRoomStudy.3dm from the Chapter 14 folder on the companion CD.
2. Click on the One Day Sun Study/Seasonal Sun Study button of the Animation Setup toolbar or type SetOne-DaySunAnimation at the command area.
3. In the One Day Sun Animation dialog box, click on the Set button. (See Figure 14–20.)
4. In the Set Animation Location dialog box that follows, select London, United Kingdom, or, if you prefer to set your current location, click on the Here button.
5. Click on the OK button.
6. Set your Date, Start time, End time, and File type; specify the file name; and click on the OK button.
7. Click on the Record Animation button on the Animation toolbar to save the animated images.

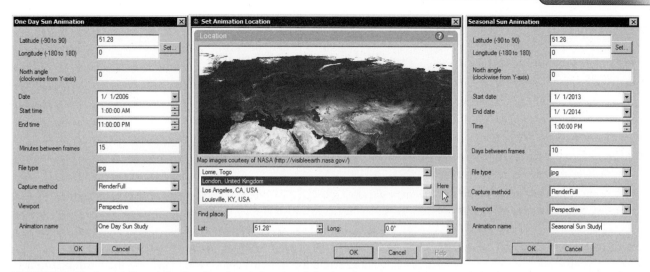

FIGURE 14–20 *From left to right: One Day Sun Animation dialog box, Set Animation Location dialog box, and Seasonal Sun Animation dialog box*
Source: Robert McNeel and Associates Rhinoceros® 5

8. Right-click on the One Day Sun Study/Seasonal Sun Study button on the Setup Animation toolbar to set up seasonal sun study animation.
9. Click on the Record button on the Animation toolbar. Season study animation is saved.

   Sets of rendered images are saved in the Chapter 14/Animation folder on the companion CD. You may open the folder to view the animations. Figure 14–21 shows a frame of the One Day Sun Study Animation (left) and Seasonal Sun Study Animation (right).

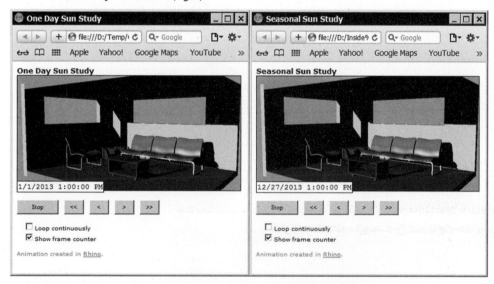

FIGURE 14–21 *One Day Sun Study (left) and Seasonal Sun Study (right)*
Source: Robert McNeel and Associates Rhinoceros® 5

Setting Safe Frame

On older, television screens and display unit's screens, which are not flat, the areas around the edges of a picture may not be clearly seen. "Safe frame" is a term used in television production to depict the areas of the television picture that can be seen on older television screens. Continue with the following steps to set and display safe frame information.

10. Select Render > Safe Frame Settings.
11. In the Safe Frame tab of the Document Properties dialog box, make appropriate settings, shown in Figure 14–22, and click on the OK button.
12. Select Render > Show Safe Frame.

    Safe frames are displayed.

13. Do not save the file.

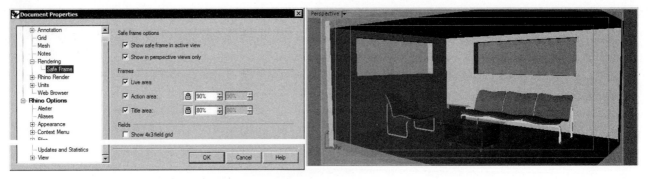

FIGURE 14-22 *Safe frame settings (left) and safe frame displayed (right)*
Source: Robert McNeel and Associates Rhinoceros® 5

Flamingo Rendering Farm

If you are constructing a photorealistic rendered animation, rendering time may be very long because there will be so many frames to render and each frame will require a lengthy period of time to produce. To save time, you can connect two or more computers together via the Flamingo rendering farm.

Using Flamingo rendering farm, you can configure a number of connected computers, and the farm will dispatch the rendering tasks to the computers, using the combined CPUs.

The basic farm allows the connection of two computers, and a full deployment Flamingo nXt allows the connection of an unlimited number of computers.

CAMERA SETTING

In each viewport, there is an imaginary camera, which is defined in the Viewport Properties dialog box, delineating the camera's lens size, location, and target point. You may open the Viewport Properties dialog box by right-clicking on a viewport's label and selecting Viewport Properties from the menu.

Lens Setting

Viewports can be classified as Parallel projection viewport or Perspective viewport. If you produce a photorealistic rendered image, it is natural that you should set the viewport to be perspective, because this viewport allows you to set the lens size to simulate real camera photo-taking. To set the lens size, you can input a value in the Lens length field in the Viewport Properties dialog box or use the Lens Length toolbar. Now perform the following steps.

1. Select File > Open and select the file SkateScooter.3dm from the Chapter 14 folder on the companion CD.
2. Click on the 17 mm Camera button on the Lens Length toolbar and then render the Perspective viewport.
3. Click on other buttons on the Lens Length toolbar to appreciate the effect of having different lens lengths. (See Figure 14-23.)

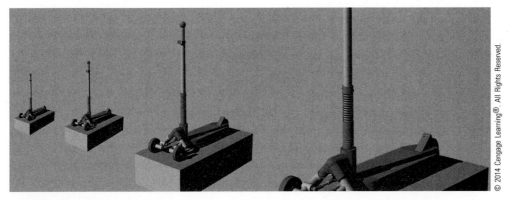

FIGURE 14-23 *From left to right: 17 mm lens, 25 mm lens, 50 mm lens, and 100 mm lens*

Camera Orientation—Walkabout

Imagining that you are holding a camera and watching a scene through the Perspective viewport, you click on the buttons on the Walkabout toolbar to simulate the effect of walking about in the scene. To control the step size, you can click on the buttons on the Step Size toolbar. Continue with the following steps.

4. Click on the 25 mm lens button on the Lens Length toolbar.
5. Click on the Medium-large Steps button on the Step Size toolbar.
6. Click on the Walk Forward button on the Walkabout toolbar.
7. Click on the Walk Back button on the Walkabout toolbar. (See Figure 14-24.)
8. Click on the Walk Right button on the Walkabout toolbar.
9. Click on the Walk Left button on the Walkabout toolbar. (See Figure 14-24.)

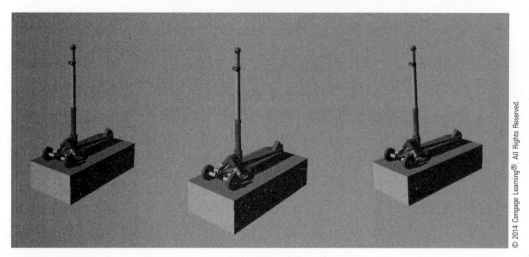

FIGURE 14-24 *From left to right: 25 mm lens, walk forward, and walk right*

You can also elevate your perspective up and down, as follows.

10. Click on the 50mm lens button on the Lens Length toolbar and render the Perspective viewport. (See Figure 14-12.)
11. Click on the Medium-large Steps button on the Step Size toolbar.
12. Click on the elevator up button on the Walkabout toolbar. (See Figure 14-25.)
13. Click on the elevator down button on the Walkabout toolbar twice. (See Figure 14-25.)

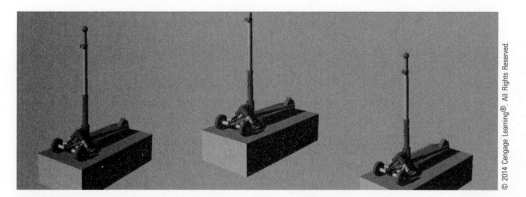

FIGURE 14-25 *From left to right: 50 mm lens, elevator up, and elevator down*

Camera Orientation—Look About

In addition to simulating the effect of walking about in the scene, you can simulate the effect of looking about. That is, you stand still but turn your head around. The rotation increment of looking about has to be set beforehand. Continue with the following steps.

14. Click on the Set Rotate Increment button on the LookAbout toolbar.
15. Type 30 at the command area to set the rotation increment to 30 degrees.
16. Click on the Look Right button on the LookAbout toolbar and render the viewport. (See Figure 14–26.)
17. Click on the Look Left button on the LookAbout toolbar.
18. Click on the Look Up button on the LookAbout toolbar. (See Figure 14–26.)
19. Click on the Look Down button on the LookAbout toolbar twice.
20. Click on the Look Up button on the LookAbout toolbar. (See Figure 14–26.)

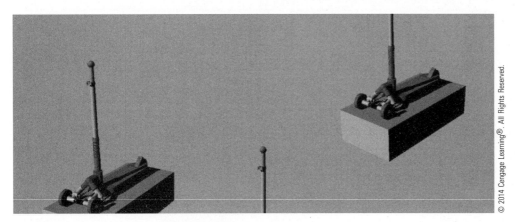

FIGURE 14–26 *From left to right: look right, look up, and look down*

Camera Orientation—Moving Target Point

To focus, you can move the target point of the camera toward the center of the bounding box of the target object, as follows.

21. Select View > Set Camera > Move Target to Objects, or type Movetargettoobjects at the command area.
22. Select polysurface A, shown in Figure 14–27, and press the ENTER key.

    You may choose more than one object. The target of the camera is set to the center of the bounding box of the selected object.

23. Click on Undo View Change on the Standard toolbar to reset the viewport.

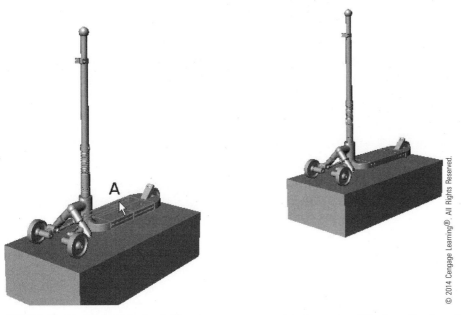

FIGURE 14–27 *Original display (left) and camera target set to the center of the bounding box of object A (right)*

Camera Orientation—To Surface

You can move a camera and align it with the normal direction of a selected surface, as follows.

24. Select View > Set Camera > Orient Camera to Surface, or type Orientcameratosrf at the command area.
25. Select surface A and location B on the surface (Figure 14–28).
    The camera and its target direction are aligned to the normal direction of the surface at a designated location.
26. Click on Undo View Change on the Standard toolbar to reset the viewport.

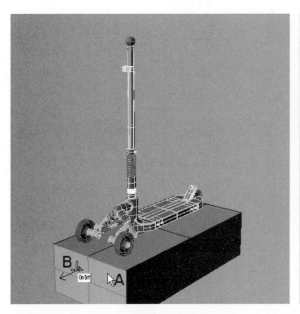
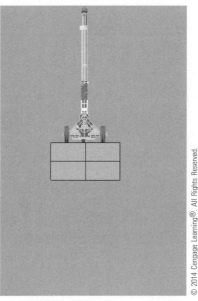

FIGURE 14–28 *Face and location selected (left) and camera and target aligned (right)*

Camera Manipulation—Camera Pyramid

A more interactive way of manipulating the camera of a selected viewport is to display its graphical representation in terms of a pyramid in other viewports. Continue with the following steps.

27. Double-click on the Perspective viewport's label to reset the display to 4 viewport display.
28. Check View > Set Camera > Show Camera, or right-click on any viewport's label and check Set Camera > Show Camera.
    The camera's graphical pyramid is displayed.

This command toggles between showing and hiding the camera of a selected viewport. The pyramid represents the camera's location, angle, and field of view. It has four control points. By dragging the control point at the tip of the pyramid, you adjust the camera's viewpoint; by dragging the control point at the opposite end of the straight line that runs through the axis of the pyramid, you adjust the target point location; by dragging the control point on the middle of one side of the pyramid base, you roll the camera; and by dragging the control point on one of the corners of the base of the pyramid, you adjust the field of view/lens angle. In addition, there are two rectangular planes in dotted lines that depict the far and near clipping planes.

29. Drag control point A, shown in Figure 14–29.
    The camera's viewpoint is changed.
30. Drag control point B.
    The camera's target point is changed.
31. Drag control point C.
    The camera's field of view is changed.
32. Drag control point D.
    The camera is rolled.
33. Do not save the file.

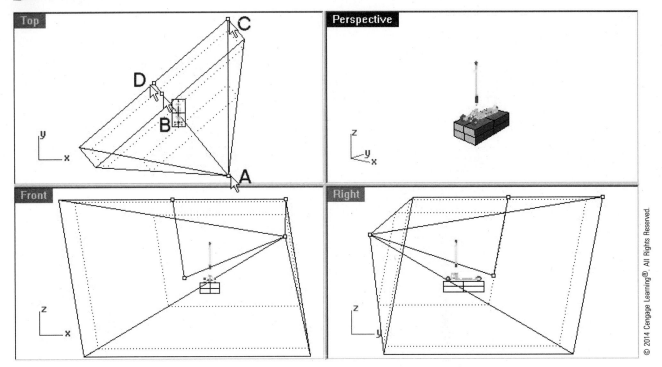

FIGURE 14-29 *Camera's pyramid representation being manipulated*

DIGITAL LIGHTING

The purpose of lighting is to illuminate the 3D scene and to achieve specific lighting effects. In a Rhino scene, you can add virtual lights and apply sunlight and environmental lighting effects. There are five categories of virtual lighting: spotlights, point light, directional light, rectangular light, and linear light.

Default Light

In a Rhino file, there is a default light. If you imagine the screen as a camera viewer through which you look into the 3D space in which you construct the model, the default light is located over your left shoulder. The intensity of this light is always good enough to illuminate the entire 3D space, no matter how large it is. This is why you always have a well-illuminated scene in any rendering process, even though you have not added any light. If you are satisfied with the default lighting effect, you do not need to add any light.

Note that once a light of any kind is added, the default light is turned off automatically. It will turn on again if you delete all added lights. Therefore, to simulate a dark environment, you need to add a very dim light in the scene.

Spotlights

A spotlight is characterized by a light source emitting a conical beam of light toward a target. It illuminates a conical volume in the scene. In the scene, a spotlight is represented by two cones with a common vertex. The cones represent the direction of the light, not the range of the light. The inner cone represents the area of full brightness of the light, and brightness will decrease from the inner cone toward the outer cone. To adjust the size and location of the spotlight, you can turn on the control points and drag them to new locations. A spotlight that has narrower cones produces more detail than a spotlight with wider cones. Perform the following steps.

1. Select File > Open and select the file Modelcar.3dm from the Chapter 14 folder on the companion CD.
2. Select Render > Create Spotlight.
3. Referencing Figure 14-30, click on location A to specify the base of the cone of the spotlight and location B to specify the radius of the cone of the spotlight.

    Exact locations are unimportant.

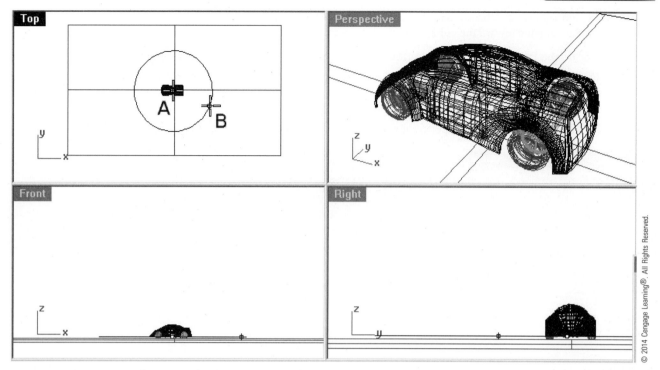

FIGURE 14–30 *Base cone location and radius being specified*

As explained earlier in this book, you can construct a 3D point by holding down the CONTROL key while clicking on the screen. Continue with the following steps to specify a 3D point to locate the spotlight.

    4. Hold down the CONTROL key and then click on location C and then D, shown in Figure 14–31, to specify the end of the cone of the spotlight.

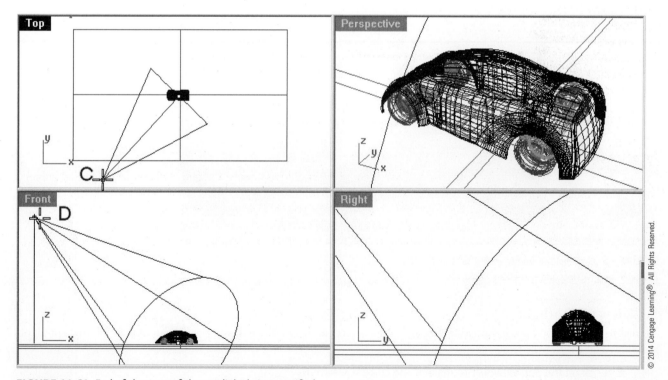

FIGURE 14–31 *End of the cone of the spotlight being specified*

As can be seen on your screen, the spotlight is displayed as two cones. The inner cone depicts the area of full intensity of the light. From the inner cone toward the outer cone, the intensity of the light weakens. Continue with the following steps.

5. Click on the Edit Light Properties button on the Properties toolbar or the Lights toolbar.

    As shown in Figure 14–32, the spotlight's properties include color, on/off, intensity, and hardness. Among them, the hardness slider bar determines the size of the inner cone of the spotlight. The larger the value, the harder is the light's edge.

6. Click on the X mark at the upper-right corner of the dialog box to close it.

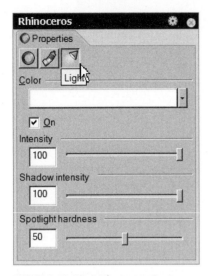

FIGURE 14–32 *Light properties*
Source: Robert McNeel and Associates Rhinoceros® 5

7. Render the Perspective viewport. (See Figure 14–33.)

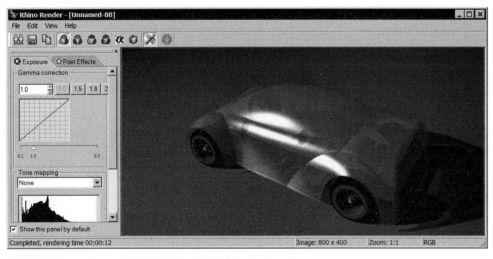

FIGURE 14–33 *Rendered image with a spotlight added*
Source: Robert McNeel and Associates Rhinoceros® 5

Set View to Spotlight

To realize how the spotlight is located in relation to the objects to be illuminated, you can set the view to selected spotlight, as follows.

8. Click on the Top viewport to make it the current viewport.
9. Select Render > Set View to Spotlight.

10. Select the spotlight. The viewport is set to the spotlight's direction, shown in Figure 14–34.
11. Right-click on the Top viewport's label and select Set View > Top to reset the Top viewport's orientation.
12. Zoom the viewport to its extent.

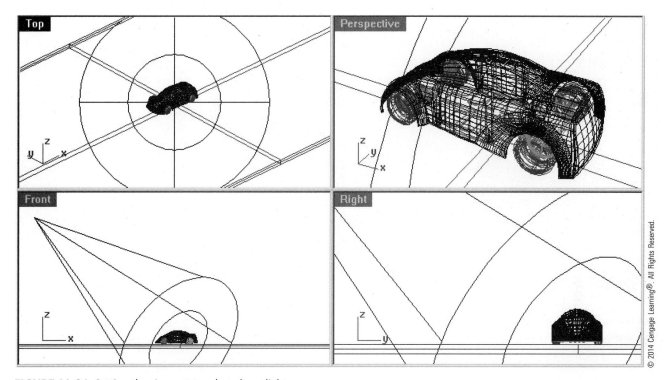

FIGURE 14–34 *Setting the viewport to selected spotlight*

Edit Light by Looking

Edit light by looking is a way to position a spotlight. You can adjust the spotlight's projection direction as if you are manipulating the spotlight at the spotlight location, as follows.

13. Select Render > Edit Light by Looking.
14. Select the spotlight in the Front viewport.
15. Click on the Rotate View button on the Standard toolbar.
16. Manipulate the Front viewport. (See Figure 14–35.)
17. Press the ENTER key.

    The spotlight direction is set, and the display is returned to its previous orientation.

18. Undo the last command.

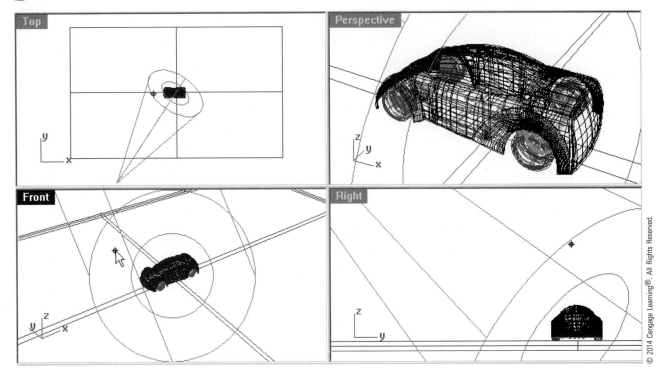

FIGURE 14–35 *Manipulating spotlight*

Set Spotlight to View

Another way to position a spotlight is to set the spotlight to a specified view, as follows.

19. Select Render > Set Spotlight to View, or click on the Set spotlight to view/Add a spotlight from current view button on the Light Tools toolbar.
20. Click on the Perspective viewport.
21. Select spotlight A, shown in Figure 14–36.
    The spotlight is oriented in accordance with the orientation of the selected viewport.
22. Undo the last command.

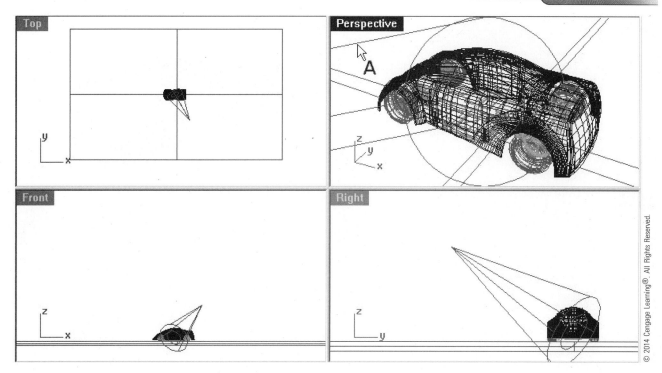

FIGURE 14–36 *Spotlight manipulated*

Contrary to setting a spotlight to the viewport's viewing direction, you can change the viewport's orientation to match the lighting direction by selecting Render > Set View to Spotlight, thus allowing you to see how the light is shining onto the scene.

Point Light

A point light is analogous to an ordinary electric light bulb that emits light in all directions. Point light illuminates an entire scene. To construct a point light, you specify a location of the light source, as follows.

23. Turn off the spotlight by clicking on the Edit Light Properties button of the Properties toolbar or the Lights toolbar, and clearing the On check box.
24. Select Render > Create Point Light.
25. Hold down the CONTROL key and click on location A and then B, shown in Figure 14–37.

    A point light is constructed.

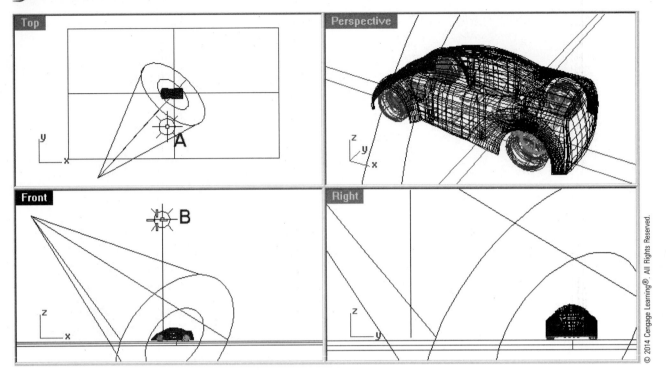

FIGURE 14–37 Point light being constructed

26. With reference to Figure 14–32, use the Properties dialog box to turn off the spotlight by unchecking the On box.
27. Render the Perspective viewport.

    The result is shown in Figure 14–38.

28. Do not save the file.

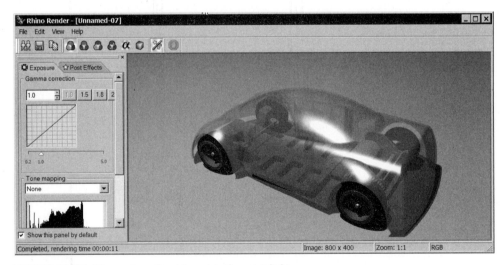

FIGURE 14–38 Viewport rendered by using the point light
Source: Robert McNeel and Associates Rhinoceros® 5

Directional Light

Directional lighting derives from a light source at a distance, such as the sun. This lighting, generally represented as a beam or parallel beams of light, illuminates an entire scene. You specify a direction vector for the directional light. A directional light in a scene is a symbol. It depicts the direction. It does not matter where it is placed.

1. Select File > Open and select the file LivingRoom.3dm from the Chapter 14 folder on the companion CD.
2. Select Render > Create Directional Light.
3. Select location A, shown in Figure 14–39, to specify the end-of-light direction vector. Exact location is unimportant.
4. Hold down the CONTROL key, and click on location B and then location C, shown in Figure 14–39, to specify the start-of-light direction vector.

   A direction light is constructed.
5. Render the Perspective viewport. (See Figure 14–40.)

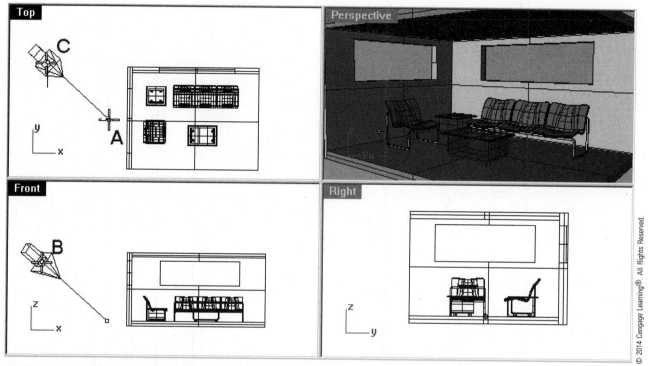

FIGURE 14–39 *Light direction vector being constructed*

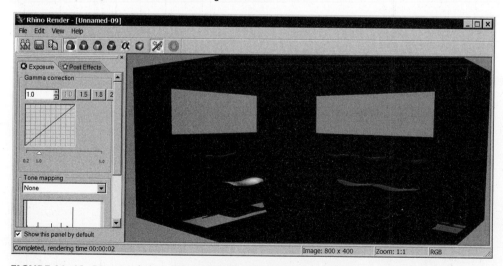

FIGURE 14–40 *Direction light added and Perspective viewport rendered*
Source: Robert McNeel and Associates Rhinoceros® 5

Rectangular Light

Rectangular lighting simulates the effect of a fluorescent light box. You specify the location, width, and length of the rectangular light. In a scene, a rectangular light is a rectangular symbol depicting a rectangular light box. Like other types of lights, a rectangular light is not shown in the rendered image. To add a rectangular light to a scene, continue with the following steps.

6. Click on the Edit Light Properties button on the Properties toolbar or the Lights toolbar.
7. Select the direction light and clear the On button of the Properties dialog box. The direction light is turned off.
8. Check the Ortho button on the Status bar.
9. Select Render > Create Rectangular Light.
10. Hold down the CONTROL key and select location A and then B, shown in Figure 14–41, to specify the location of the rectangular light.
11. Select locations C and D to specify the length and width of the light.

    A rectangular light is constructed.
12. Render the Perspective viewport, shown in Figure 14–42.

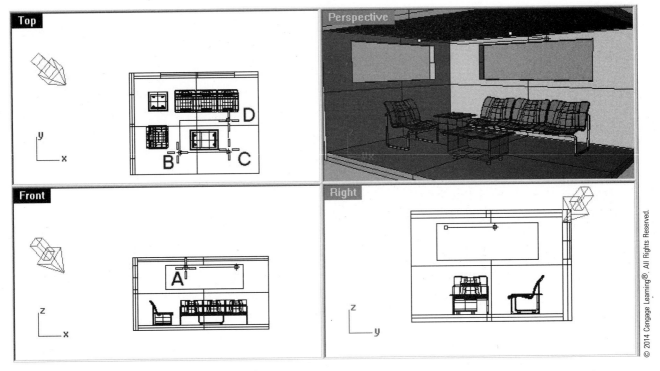

FIGURE 14–41 *Rectangular light being constructed*

FIGURE 14–42 *Rendered rectangular-light scene*
Source: Robert McNeel and Associates Rhinoceros® 5

Linear Light

Linear light simulates the effect of a fluorescent tube being added to a scene. You specify the start point and endpoint of the linear fluorescent light. In a scene, a linear light is a symbol in the shape of a cylinder. It is not displayed in the rendered image. To turn off the rectangular light and add a linear light to a scene, perform the following steps.

13. Click on the Edit Light Properties button on the Properties toolbar or the Lights toolbar.
14. Select the direction light and clear the On button of the Properties dialog box.
    The direction light is turned off.
15. Select Render > Create Linear Light.
16. Hold down the CONTROL key and select location A and then B, shown in Figure 14–43, to specify the origin of the linear light.
17. Select location C (shown in Figure 14–43) to specify the length and direction of the light.
    A linear light is constructed.
18. Render the scene, shown in Figure 14–44.
19. Do not save your file.

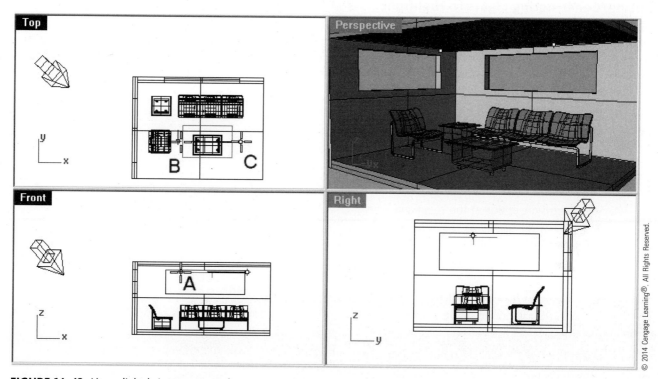

FIGURE 14–43 *Linear light being constructed*

FIGURE 14-44 *Rendered linear-light scene*
Source: Robert McNeel and Associates Rhinoceros® 5

Editing/Constructing Lights to Highlight

To enhance visual effect, you may want to edit an existing or construct a new directional light, spotlight, or point light such that it hits on a designated location of a surface and then reflects light toward the imaginary camera in the viewport, giving the effect of highlighting. Apart from editing or constructing light objects to give the effect of highlight on specific locations of a surface, you can also first construct a helper line and then place a new light along the line subsequently.

Now perform the following steps.

1. Select File > Open and select the file ModelCarHighlight.3dm from the Chapter 14 folder on the companion CD.
2. Click on the Edit Light by Highlight Location button on the Lights toolbar.
3. Press the ENTER key to construct a new light.
   If you have an existing light and you wish to position it so that it produces the effect of highlight, select the light.
4. Select the Type option at the command area.
   As shown in the command line, there are four options: directional light, point light, spotlight, and line. Naturally, directional light, point light, and spotlight options enable you to place such lights to produce the effect of highlight in the active viewport. The fourth option, line, constructs a helper light for you to place a light precisely.

Directional Light That Highlights

To construct a directional light to hit on a specific location of a surface to produce the effect of highlighting, continue with the following steps.

5. Select the DirectionalLight option at the command area.
6. Disable Osnap.
7. Select surface A, shown in Figure 14-45.

> **NOTE** You have to click on the viewport that you want to render.

8. Click on location B, where you want to have specular highlight.
   A directional light that produces the effect of highlight at a selected location is constructed.
9. Render the Perspective viewport. (See Figure 14-46.)
   Because the exact location where you place the light may not be the same, the rendered image shown in Figure 14-46 may be different from yours.

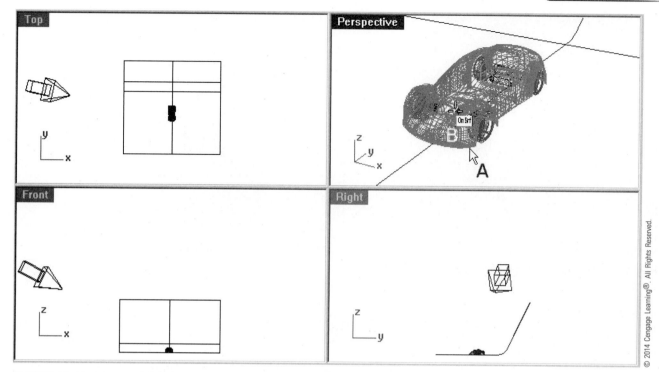

FIGURE 14–45 *Directional light being constructed to produce highlighting effect*

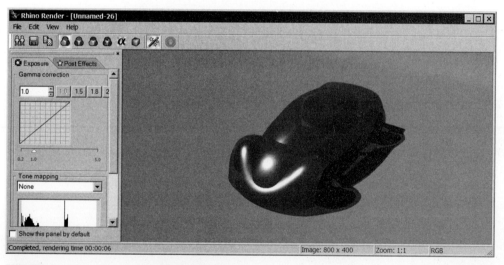

FIGURE 14–46 *Perspective viewport rendered*
Source: Robert McNeel and Associates Rhinoceros® 5

Spotlight That Highlights

To construct a spotlight that hits at a designated location in the Perspective viewport to produce highlight, continue with the following steps.

10. Set the current layer to Light02 and turn off the Light01 layer.

    The directional light is turned off. The default setting in the Render page of the Options dialog box does not use any light residing on layers that are turned off.

11. Click on the Edit Light by Highlight Location button on the Lights toolbar.
12. Press the ENTER key to construct a new light.
13. Select the Type option and then the Spotlight option at the command area.

14. Select surface A, shown in Figure 14–47.
15. Click on location B.

    A spotlight that produces highlighting effect at a selected location is constructed.
16. Render the Perspective viewport. (See Figure 14–48.)

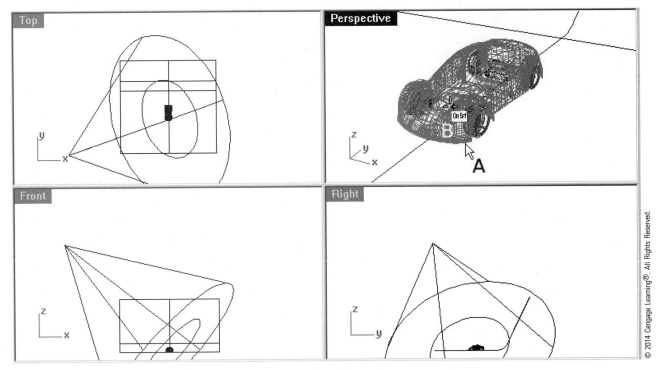

FIGURE 14–47 *Spotlight being constructed to produce highlighting effect*

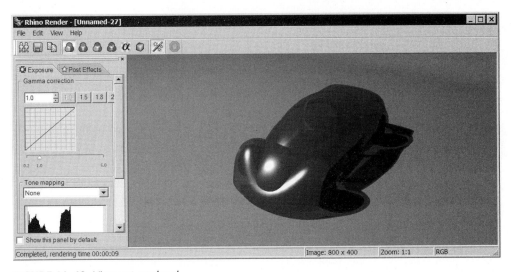

FIGURE 14–48 *Viewport rendered*
Source: Robert McNeel and Associates Rhinoceros® 5

Point Light That Highlights

To place a point light that highlights at a designated location, continue with the following steps.

17. Set the current layer to Light03 and turn off the Light02 layer to turn off the bouncing spotlight.
18. Click on the Edit Light by Highlight Location button on the Lights toolbar.

19. Press the ENTER key to construct a new light.
20. Select the Type option and then the PointLight option at the command area.
21. Select surface A, shown in Figure 14–49.
22. Click on location B.

    A spotlight that produces highlight at a selected location is constructed.

23. Render the Perspective viewport. (See Figure 14–50.)

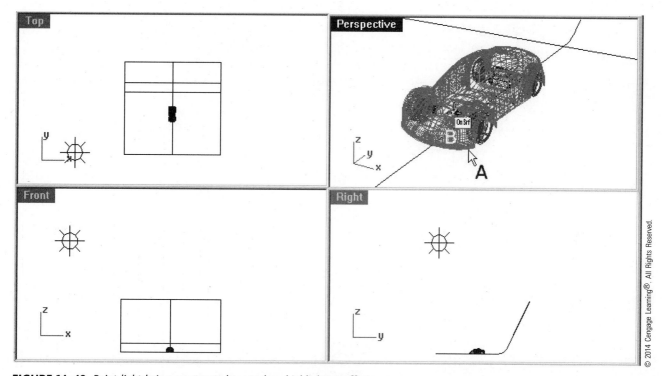

FIGURE 14–49 *Point light being constructed to produce highlighting effect*

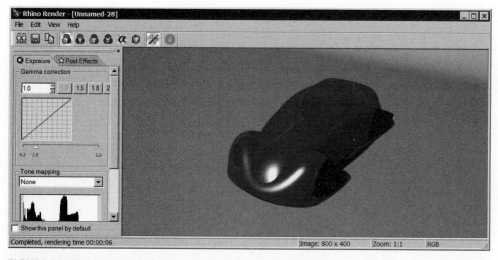

FIGURE 14–50 *Viewport rendered*
Source: Robert McNeel and Associates Rhinoceros® 5

Helper Line for Subsequent Construction of Lights to Highlight

The helper line is a line that depicts where the light source should be directed onto the surface to give highlight effect. With this line, you can place a light or place an object to reflect under the light. Now continue with the following steps.

24. Set the current layer to Light04 and turn off the Light03 layer.
25. Click on the Edit Light by Highlight Location button on the Lights toolbar.
26. Press the ENTER key to construct a new light.
27. Select the Line option at the command area.
28. Select surface A, shown in Figure 14–51.
29. Click on location B.

    A helper line for subsequent location of any light source is constructed.

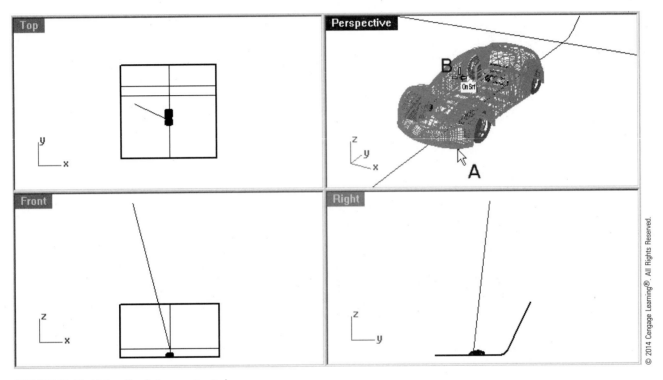

FIGURE 14–51 *Helper line being constructed*

Using the helper line, you can place a directional light, spotlight, or point light. Continue with the following steps to place a directional light along the helper line.

30. Select Render > Create Directional Light.
31. Set object snap mode to nearest.
32. Select location A and then location B, shown in Figure 14–52.

    A directional light is placed along the helper line.

33. Render the Perspective viewport. (See Figure 14–53.)

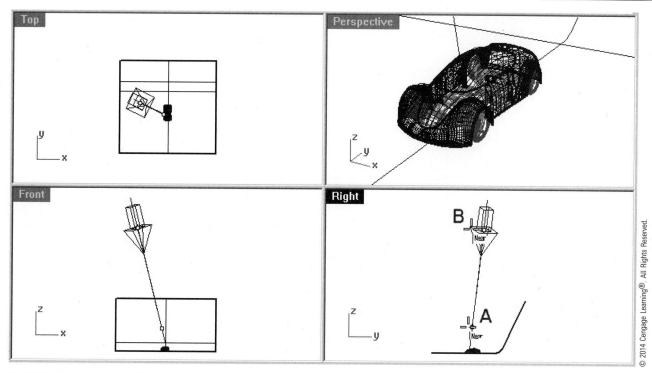

FIGURE 14–52 *Directional light being placed along the helper line*

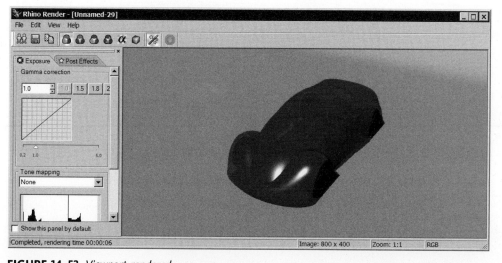

FIGURE 14–53 *Viewport rendered*
Source: Robert McNeel and Associates Rhinoceros® 5

Relocating a Light for It to Highlight

Because highlighting effect is associated with camera orientation, if you change the viewport orientation, it will be lost. To relocate an existing light so that it highlights light, continue with the following steps.

34. Set the current layer to Light05 and turn off the Light04 layer.
    A spotlight is already constructed on this layer.
35. Click on the Bounce Light button on the Lights toolbar.
36. Select A, shown in Figure 14–54, the spotlight.
37. Select location B (surface) and then location C (highlight), shown in Figure 14–54.
    The selected spotlight is relocated to bounce light in the Perspective viewport.

38. Render the Perspective viewport. (See Figure 14–55.)
39. Do not save the file.

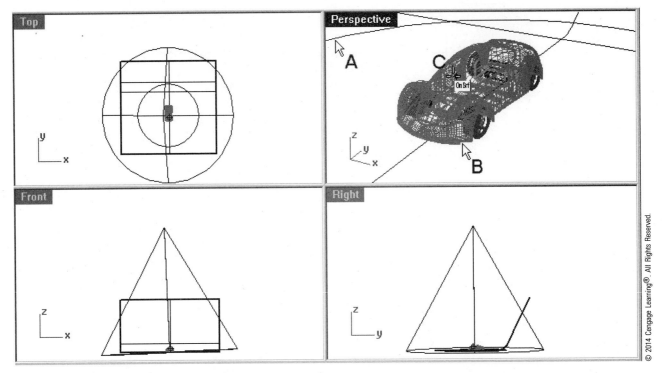

FIGURE 14–54 *Spotlight being relocated*

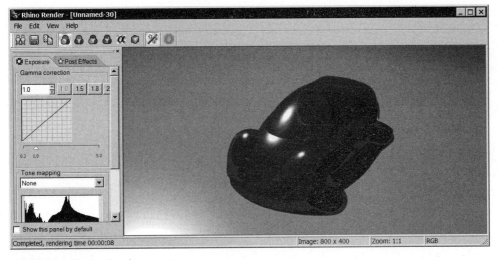

FIGURE 14–55 *Rendered viewport*
Source: Robert McNeel and Associates Rhinoceros® 5

Managing Lights

To sum up, there are five kinds of lights: spot, point, directional, rectangular, and linear. In addition, there are sunlight and skylight. All of them contribute to the overall light intensity in the Rhino scene. To manage lights, you can use the Light panes of the Properties dialog box by selecting the light and clicking on the Properties button on the Standard toolbar. To manage all the lights collectively, with the exception of not setting spotlight's hardness, you can use the Light panels.

1. Select File > Open and select the file LivingRoom_lights.3dm from the Chapter 14 folder on the companion CD.
   In this scene, together with skylight and sunlight, there are five kinds of lights.

2. Select spotlight A, indicated in Figure 14–56, and click on the Properties button on the Standard toolbar.
   The Light pane of the Properties dialog box is displayed.
3. Select Render > Panels > Lights.
   The Light panel is displayed, allowing you to turn on/off lights as well as setting their intensity and color.

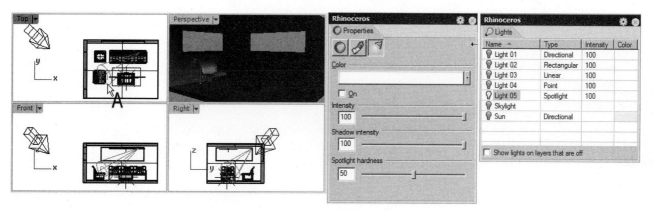

FIGURE 14–56 *From left to right: spotlight being selected, spotlight's properties, and Lights panel*
Source: Robert McNeel and Associates Rhinoceros® 5

Skylight and Sunlight

To manage sunlight and skylight, continue with the following steps.

4. In the Lights panel, turn on Skylight and Sun and turn off all other lights.
5. Select Render > Panels > Sun.
6. With reference to Figure 14–57, in the Sun panel, clear the Manual control box, if it is checked.
   As you can see, the Sun area is grayed out, depicting that you can manipulate its location only via the Date and Time and Location areas.
7. Expand the Date and Time area and specify a date and time.
8. Expand the Location area and specify a location.
9. Activate the Perspective viewport and render the scene. (See Figure 14–58.)
10. Do not save the file.

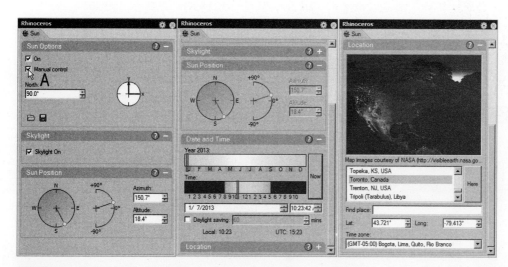

FIGURE 14–57 *From left to right: Manual control box being unchecked, Date and Time area, and Location area*
Source: Robert McNeel and Associates Rhinoceros® 5

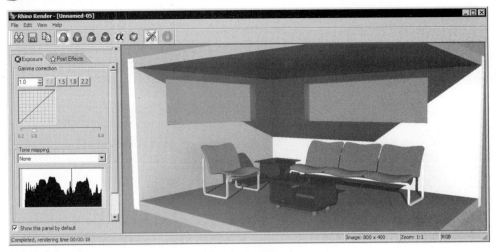

FIGURE 14–58 *Skylight and sun managed and rendered*
Source: Robert McNeel and Associates Rhinoceros® 5

Turning Off Lights

To reiterate, there are two ways to turn off lights in a scene.

- Click on the Edit Light Properties button on the Properties or Lights toolbar, select the light, and clear the On check box.
- Clear the Use lights on layers that are off button on the Rhino Render page of the Document Properties dialog box, put the lights on separate layers, and turn off those layers.

DIGITAL MATERIAL

Digital material assignment is a crucial element that contributes to the final digital rendered image. In Rhino, material properties can be assigned to an object in one of three ways, as follows.

- By default, material property is ByLayer, meaning that the material property of any object is in accordance with the material property assigned to the layer on which the object resides. Therefore, you can simply assign a material property to a layer and then have the object residing in that layer inherit the material properties of the layer.
- You can assign a material property to an object independent of the layer where it resides.
- You can have an object inherit material properties from its parent, which has its material properties assigned in one of the above two ways.

Assigning Material Properties

To assign digital material to a layer, you can click on the Material column of the layer in the Layer dialog box. To assign digital material to selected object(s), click on the Edit Material Properties button on the Properties toolbar and select an object. (See Figure 14–59.) If you click on the Material column of the Layer dialog box, the Material Editor dialog box is displayed. If you click on the Edit Material Properties button on the Properties toolbar and select an object, the Material tab of the Properties dialog box will display. (See Figure 14–60.)

FIGURE 14–59 *Clicking on the Material column of the Layer dialog box to assign material properties to a layer*
Source: Robert McNeel and Associates Rhinoceros® 5

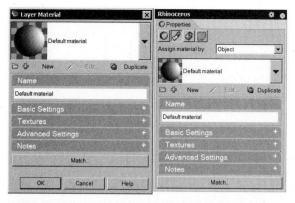

FIGURE 14–60 *Layer Material dialog box (left) and Material tab of the Properties panel (right)*
Source: Robert McNeel and Associates Rhinoceros® 5

As shown in Figure 14–60, the Material Editor dialog box is similar to the Material tab of the Properties panel. This is the case because basic material assignment method is the same for assigning material properties to a layer or to individual objects. In any case, material properties can be assigned in two ways.

- By using Rhino's material properties
- By using a plug-in, such as Flamingo nXt

Rhino Materials and Material Tab of the Properties Panel

A Rhino material consists of five elements: Name, Basic Settings, Textures, Advanced Settings, and Notes, shown in Figure 14–61. Among them, the Name area simply provides a designation to the material and the Notes area provides additional texture information about the material. You can access this tab by selecting an object and clicking on the Material tab of the Properties panel, or you can select Render > Material Editor to display a similar dialog box, shown in Figure 14–61.

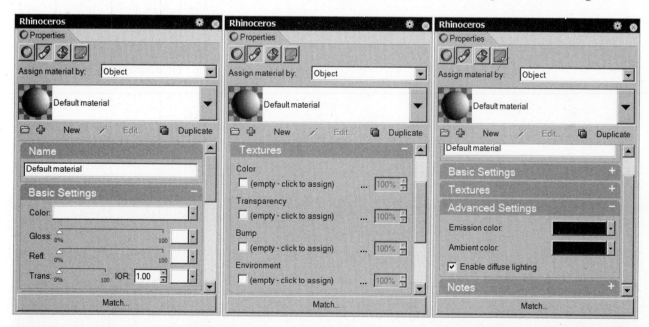

FIGURE 14–61 *Various areas of the Material tab of the Properties panel, from left to right: Name and Basic Settings, Textures, and Advanced Settings and Notes*
Source: Robert McNeel and Associates Rhinoceros® 5

Basic Settings

The basic settings area of the Material tab concerns the basic appearance of the material and specifies a color, a gloss color, gloss color finish level, reflectivity color, reflectivity color finish level, transparency color, transparency level, and index of refraction (IOR), as indicated in the Basic Settings area of the Materials tab of the Properties panel.

Color. The color of basic settings area of the Material tab is the basic color; sometimes it is also called the diffuse color, which is the color of an object that we perceive when viewing under indirect light. You set the basic color by clicking on the color swatch to bring out the Select Color dialog box and selecting a color via the HSB color system or the RGB color system, as discussed in Chapter 2.

Gloss Finish Level and Gloss Color. These two parameters establish the appearance of the object when it is viewed under direct light. The gloss finish level concerns the strength of glossiness, and the gloss color is the color of the glossy area. If you are using the HSB color system to assign the basic color, a way to set gloss color is to use the same hue (H) value color but with a less saturation (S) value. Obviously, you could use a different color to produce a special effect for materials that appear in a different color under direct light.

Reflection Level and Reflection Color. These two parameters determine how reflective the material is and the color of the material at the reflected areas. If a material is 100 percent reflective, it behaves more or less like a mirror and it will reflect objects. The reflection color is the coating color of the reflected objects.

Transparency Level, Index of Refraction, and Transparency Color. These three parameters set the overall opacity of the object, how light is refracted through the transparent material, and the transparency color.

To appreciate the effects of various basic settings, perform the following steps.

1. Select File > Open and select the file Material00.3dm from the Chapter 14 folder on the companion CD.
   The basic color of all the objects in this file is red. However, each of them has different gloss finish, reflectivity, and transparency.
2. Select Render > Render.
3. Select A, indicated in Figure 14–62, and then click on the Object Properties button on the Standard toolbar.
   As can be seen, this object's glossiness is 27 percent, reflectivity is 55 percent, and its reflective color is cyan. Therefore, it is glossy, it is reflective, and the reflective object's color is cyan.
4. Continue with selecting objects B, C, D, E, and F one-by-one to appreciate how various settings affect a material's appearance.
5. Do not save the file.

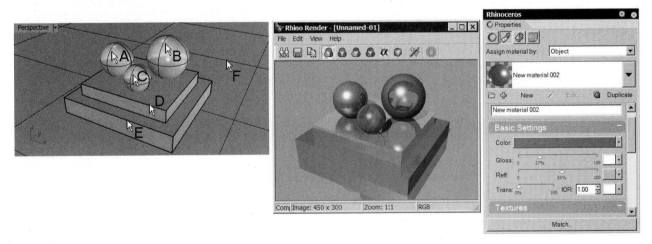

FIGURE 14–62 *From left to right: objects being selected, Rhino renderer, and Material pane*
Source: Robert McNeel and Associates Rhinoceros® 5

Texture

Apart from assigning overall color, gloss color, glossiness, and transparency, you can wrap bitmaps onto selected objects or objects in a layer to simulate four kinds of visual effects: surface texture, partial transparency, surface bumpiness, and environment reflection.

The way how a bitmap is wrapped around an object depends on the mapping methods applied, which is denoted by mapping channels and will be explained later. If there is no specific mapping method assigned to an object, the default mapping is along the U and V directions of the surface.

Reparameterizing. To improve the effect of bitmap mapping, it may be necessary to reparameterize the surface by typing Reparameterize at the command area and selecting the objects to be reparameterized. Now perform the following steps.

1. Select File > Open and select the file Material01.3dm from the Chapter 14 folder on the companion CD.
2. Type Reparameterize at the command area.
3. Select all the objects and press the ENTER key twice.

Color Texture Map and Intensity. Applying texture map to an object is analogous to putting wallpaper on a wall. The intensity of the bitmap ranges from 0 percent to 100 percent. In essence, a 0 percent intensity is equivalent to having no texture mapping at all. At the other extreme, a 100 percent intensity setting will cause the object to possess the color and texture of the texture bitmap, with the object's original color and transparency totally masked by the texture map. If the intensity percentage is anywhere between 0 and 100, the object's color and transparency setting will show off appropriately. Continue with the following steps.

4. Select Render > Render to appreciate the effect of rendering.
5. Select A, indicated in Figure 14–63, and then click on the Object Properties button on the Standard toolbar.
   The material for object A has a color texture map superimposed on the basic material.
6. Select B, indicated in Figure 14–63, and then click on the Object Properties button on the Standard toolbar.
   You will find that the basic settings for object B are the same except that they do not have a color texture map.

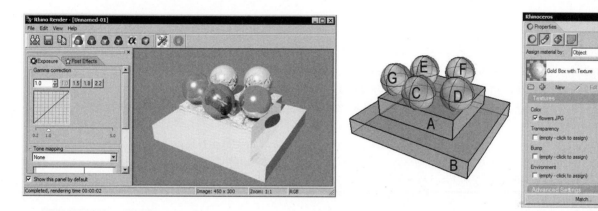

FIGURE 14–63 *From left to right: rendered image, objects, and Properties panel*
Source: Robert McNeel and Associates Rhinoceros® 5

Environment Texture Map and Intensity. To simulate reflective objects reflecting the scenes in the surrounding environment, you can use a color bitmap, delineating the imaginary environment in the scene and wrapping it around the objects.

To control how reflective the object is, you set the intensity of the environment map. A value of 100 percent intensity will make the object look like a perfect mirror, reflecting the environment. Any value between 100 percent and 0 percent will give the effect of varying degrees of reflectiveness. Continue with the following steps.

7. Select C, indicated in Figure 14–63, and then click on the Object Properties button on the Standard toolbar.
   This object is assigned with an environmental map, simulating the effect of reflecting its environment depicted by the map.
8. Select D, indicated in Figure 14–63, and then click on the Object Properties button on the Standard toolbar.
   This object simply reflects the environment around it.

Bump Texture Map and Intensity. If you want to have the objects look bumpy, you can wrap a bitmap onto the object, using the color value of the bitmap to adjust the apparent surface irregularity. The bitmap used is called bump map.

Because the bump map is only used to affect bumpiness, the original color assignment to the object is not affected. Again, a black-and-white or gray scale bitmap is preferred because you can better foresee and estimate the effect of bumpiness.

After applying a bump map, the object's surface will look bumpy, with the bumpiness degree dependent on the color value and the intensity of the bump map. It must be noted that a bump map only tells the computer to produce a bumpiness visual effect during rendering. In reality, the surface is still as it was. Therefore, if you really want to construct a bumpy surface, you should use appropriate surface manipulation tools. Continue with the following steps.

9. Select E, indicated in Figure 14–63, and then click on the Object Properties button on the Standard toolbar.
10. Select F, indicated in Figure 14–63, and then click on the Object Properties button on the Standard toolbar.

    These two objects use bitmaps to simulate the effect of bumpiness.

Transparency Texture Map and Intensity. If you want only a portion of the object to be transparent instead of the entire object, you should wrap a bitmap onto the object, using the bitmap's color value to tell the computer what portion of the object is transparent and the degree of transparency.

In essence, regardless of the color of the bitmap (red, green, or blue) only the color value is used to affect transparency. Therefore, it is more sensible to use a black-and-white or gray scale bitmap as the transparency map.

After wrapping the transparent map onto the object, the area of the object covered by the darker color will become transparent. The degree of transparency depends on the gray scale of the bitmap and the intensity of the transparency map. For example, if you use a black-and-white bitmap and set the transparency intensity to 100 percent, areas covered by the black spots of the bitmap will be totally transparent, and areas covered by the white spots of the bitmap will be totally opaque. Continue with the following steps.

11. Select G, indicated in Figure 14–63, and then click on the Object Properties button on the Standard toolbar.

    This object is a transparent object, with transparency being assigned with the basic settings.
12. Turn on Layer Curtain and render the scene.

    Note in Figure 14–64 that the curtain object is transparent.
13. Click on A, indicated in Figure 14–64, and then click on the Object Properties button on the Standard toolbar.
14. In the Texture tab of the Properties dialog box, check the Transparency texture map to turn it on.

    If the map is empty, you can click on it and select a bitmap. To temporarily disable the bitmap, uncheck the box. To remove the bitmap, right-click on it and select Remove.
15. Right-click on B and select Edit, or simply click on B, indicated in Figure 14–64, to have a look at the mapping settings via the texture palette.

    Here you will find that the bitmap is mapped repeatedly along the U and V directions of the surface for ten times. More details about mapping, including mapping channel, will be discussed later in this chapter.
16. Click on the OK button.

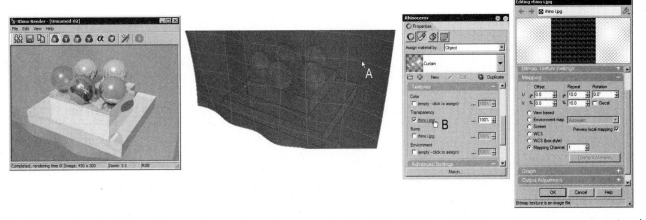

FIGURE 14–64 *From left to right: curtain turned on and scene rendered, curtain selected, curtain's transparency texture map assigned, and texture palette*
Source: Robert McNeel and Associates Rhinoceros® 5

17. Render the scene with transparency texture map assigned to the curtain.

    You will find in Figure 14–65 that the curtain is made partly transparent and the opaque portion of the curtain causes shadows behind.

18. While the curtain is being selected, check the bump box to assign a bump map.
19. Render the scene.

    Both bump and transparent effects are applied.
20. Uncheck the transparent map and render the scene.

    Only bump effect is applied.
21. Do not save your file.

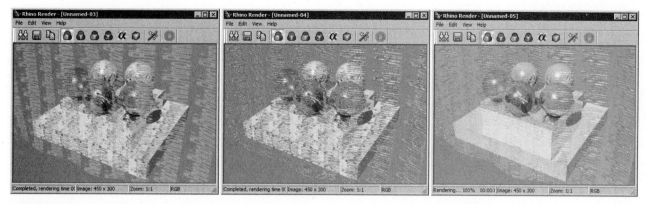

FIGURE 14–65 *From left to right: transparent map, transparent map + bump map, and bump map*
Source: Robert McNeel and Associates Rhinoceros® 5

Advanced Settings Area

This area assigns emissive effect, causing selected objects to look like they are emitting light and assigns ambient color to set the ambient environment color. Now perform the following steps.

1. Select File > Open and select the file Material02.3dm from the Chapter 14 folder on the companion CD.
2. Render the scene.
3. Select the ball and then click on the Object Properties button on the Standard toolbar.
4. In the Advanced Settings area of the Material tab of the Properties panel, assign red as the ambient color (Figure 14–66).
5. Assign red ambient color to the box.
6. Render the scene.

    The change in the rendered image is subtle, noticeable only when you zoom in the image
7. Assign yellow emission color to the ball.
8. Render the scene.
9. Do not save the file.

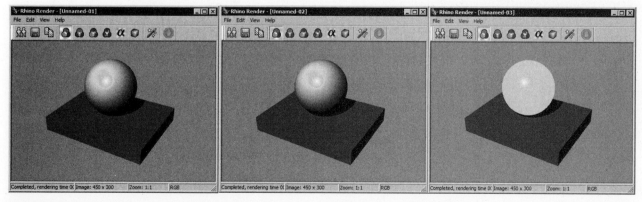

FIGURE 14–66 *From left to right: original, ambient color assigned, and emission color assigned*
Source: Robert McNeel and Associates Rhinoceros® 5

Material Editor

Apart from first selecting an object and then using the Properties panel to manage its material, you can first construct a material via the Material Editor and then assign the material to an object. Perform the following steps.

1. Select File > Open and select the file Material03.3dm from the Chapter 14 folder on the companion CD.
2. Select Render > Material Editor.

   In essence, this panel is similar to the Material tab of the Properties dialog box, as shown in Figure 14–67. Here in this file, three materials are already constructed.

3. Click on the + button X, shown in Figure 14–67.
4. Construct a new material called Gold Box and set its color to Gold, Gloss finish to 66 percent, and Reflectivity to 66 percent.
5. Click on Material Red Ball in the Material Editor and drag it to the viewport at A.

   Material Red ball is assigned to A.

6. Assign material Green Ball to object B, material Blue Ball to object C, and material Gold Box that you constructed to object D.
7. Render the scene.
8. Do not save your file.

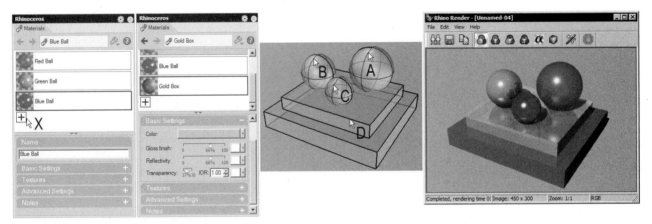

FIGURE 14–67 *From left to right: Material Editor, objects assigned with materials, and rendered scene*
Source: Robert McNeel and Associates Rhinoceros® 5

Texture Mapping and Texture Palette

Bitmaps are rectangular in shape, but surfaces on which bitmaps are mapped can be any shape: planar, spherical, or free-form. Therefore, if you simply put a bitmap on a surface that is not rectangular in shape, you may not get the desired outcome because the default way of placing a bitmap on a surface is to align the X and Y coordinates of the bitmap along the U and V directions of the surface.

To determine texture mapping methods, select an object, click on the Properties button on the Standard toolbar, and then click on the Texture Mapping button on the Properties panel, shown in Figure 14–68.

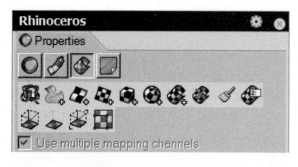
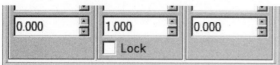

FIGURE 14–68 *Texture Mapping tab of the Properties panel*
Source: Robert McNeel and Associates Rhinoceros® 5

Using the Texture Mapping tab, you can apply the bitmap in one of the following ways.

- Unwrap a polysurface by specifying seams and flatten the surface on which bitmaps are mapped and then subsequently applied to the original polysurface
- Apply the bitmap in a custom way
- Apply the bitmap along the U and V directions of the surface
- Project the bitmap from a planar face onto the surface
- Place the bitmap on the faces of a box and project it toward the center of the box
- Place the bitmap on the face of a sphere and project it toward the center of the sphere
- Place the bitmap on the face of a cylinder and project it toward the center of the cylinder

Mapping Channel

Because you can assign more than one texture bitmap to an object and each texture bitmap can have different mapping methods, mapping channel, which is designated by a number, is used to serve the purpose of holding information about how a texture bitmap is to be applied. The default mapping channel of an object is 1. To reiterate, if there are no mapping channels assigned, texture mapping will be applied along the UV directions of the surface.

If you wish to assign more than one mapping method to an object, you have to check the Use multiple mapping channels button on the Texture tab of the Properties panel.

Surface and Planar Mapping

The simplest way to map a texture is along its U and V directions or to place the texture on a planar face and project the texture information onto the surface. To appreciate how surface and planar mappings are applied, perform the following steps.

1. Select File > Open and select the file Mapping001.3dm from the Chapter 14 folder on the companion CD.
   In the file, you will find two surfaces, in which the smaller one is a trimmed surface of the larger one.
2. Select surface A and then click on the Properties button on the Standard toolbar.
3. Click on the Material tab of the Properties panel.
4. In the Assign material by pull-down list box, select the object.
5. Click on the Color texture check box and select Checker04.tif from the Chapter 14 folder on the companion CD.
6. Repeat steps 2 through 5 to assign Checker03.tif from the Chapter 14 folder on the companion CD to surface B.
   As shown in the rendered viewport in Figure 14–69, the texture maps are, by default, assigned along the UV directions of the surfaces; surface A, being untrimmed, has the texture stretched along the UV directions evenly, and surface B, being trimmed, has part of the texture being invisible because the invisible portion is mapped onto the trimmed portion of the surface.

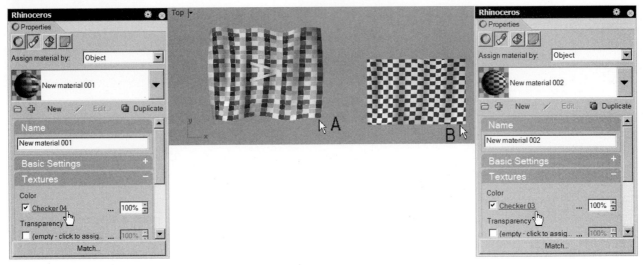

FIGURE 14–69 *From left to right: object A's texture assignment, object selection, and object B's texture assignment*
Source: Robert McNeel and Associates Rhinoceros® 5

7. Select surface A (Figure 14–70) and click on the Texture Mapping button of the Properties panel to display the Texture Mapping tab.
8. Check the Use multiple mapping channel button.
9. Click on the Apply Planar Mapping button and select endpoints C and D (Figure 14–70).
10. Click on UV on the command area.
11. Type 2 at the command area to assign the mapping to channel 2.
    There is no change to the mapping because the default mapping channel is 1 and is along the UV directions of the surface.
12. Click on the Material tab of the Properties panel.
13. Right-click on the Color bitmap and select Edit.
14. In the Editing Checker 04 dialog box, set Mapping Channel to 2.
    Mapping method, being assigned by channel 2, which is planar, is set.
15. Repeat steps 7 through 14 to assign surface B mapping channel 2 with planar settings to endpoints E and F.
    Mapping of surface B is changed.
16. Do not save your file.

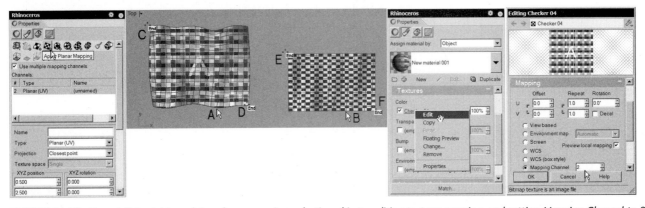

FIGURE 14–70 *From left to right: applying planar mapping, selecting objects, editing texture mapping, and setting Mapping Channel to 2*
Source: Robert McNeel and Associates Rhinoceros® 5

Box, Spherical, and Cylindrical Mapping

Depending on the shape of an object, you can place the bitmap on the walls of a box (six times, each time on each wall), the wall of a sphere, or the walls of a cylinder. Optionally, you can put only one bitmap on the cylindrical face of a cylinder or put two additional bitmaps on the caps of the cylinder. In each of these cases, texture is projected toward the object according to specified settings. To find out how to map via a box, sphere, and cylinder, perform the following steps.

1. Select File > Open and select the file Mapping002.3dm from the Chapter 14 folder on the companion CD.

   Here in this file, there are five spheres; you will apply box mapping on two spheres, cylindrical mapping on another two spheres, and spherical mapping on the fifth sphere.

2. Select sphere A (Figure 14–71) and click on the Properties button on the Standard toolbar.
3. Click on the Material tab and select object from the Assign material by pull-down list box.
4. Click on Color Texture check box and select Checker01.tif from the Chapter 14 folder on the companion CD.
5. Click on the Texture Mapping tab and select Apply Box Mapping.
6. Select points B, C, and D (Figure 14–71) to describe a box.
7. Click on No not to cap the box.

   Four texture bitmaps are placed on the walls of a bounding box without top and bottom faces.

8. Repeat steps 2 through 6 but select sphere E (Figure 14–71) and describe a bounding box by selecting points F, G, and H (Figure 14–71).
9. Click on Yes to cap the box.

   Six texture bitmaps are placed on the walls of a bounding box. Because the Use multiple mapping channels box is not checked, mapping will automatically be set to channel 1, which is the default mapping channel.

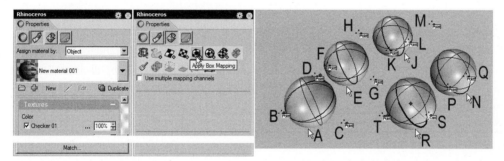

FIGURE 14–71 *From left to right: color texture selected, box mapping being applied, and spheres*
Source: Robert McNeel and Associates Rhinoceros® 5

10. While the Properties panel is still displaying, select sphere J (Figure 14–71), click on the Material tab, select object from the Assign material by pull-down list box, and assign Checker01.tif from the Chapter 14 folder on the companion CD as the color texture.
11. Click on the Texture Mapping tab and select Apply Cylindrical Mapping.
12. Click on points K, L, and M to describe a cylinder.
13. Click on No on the command area.

    A texture bitmap is placed on the cylindrical wall of the cylinder.

14. Repeat steps 10 through 12 to assign Checker01.tif color texture to sphere N (Figure 14–71) and select points P, Q, and M.
15. Click on Yes on the command area.

    Three texture bitmaps, one on the cylindrical wall and two on the top and bottom faces of the cylinder, are placed.

16. Repeat step 10 on sphere R.
17. Click on the Texture Mapping tab and select Apply Spherical Mapping (Figure 14–72).
18. Click on center S and point T to describe a sphere.

    A texture bitmap is placed on the wall of the sphere.

19. Render the image.

    Notice the differences of the mapping methods.

20. Do not save your file.

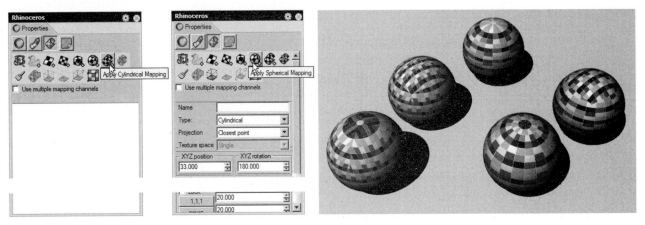

FIGURE 14–72 *From left to right: cylindrical mapping, spherical mapping, and rendered image*
Source: Robert McNeel and Associates Rhinoceros® 5

Custom Mapping

Sometimes, the shape of an object, in particular a polysurface, to be mapped may be so irregular that box, sphere, or cylinder is not a good geometry for placing a rectangular bitmap. In this case, you may first construct a simple single surface that roughly resembles the shape of the object to be mapped and use this simple surface as a custom mapping object. Now perform the following steps.

1. Select File > Open and select the file Mapping003.3dm from the Chapter 14 folder on the companion CD.
2. Select polysurface A, indicated in Figure 14–73, and click on the Properties button on the Standard toolbar.
3. Click on the Material tab of the Properties panel, select object from the Assign material by pull-down list box, and set color texture to Checker02.tif from the Chapter 14 folder on the companion CD.
4. Render the viewport

    Note that textures are mapped onto each of the individual surfaces of the polysurface and there is no correlation among them.

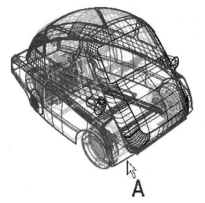

FIGURE 14–73 *From left to right: setting color texture, selecting surface A, and rendered viewport*
Source: Robert McNeel and Associates Rhinoceros® 5

5. Turn on the layer New Surface.

    Here a single surface that roughly resembles the polysurface to be mapped is constructed for your use.

6. Select surface B, indicated in Figure 14–74, and set its color texture to Checker03.tif from the Chapter 14 folder on the companion CD.
7. Render the viewport.

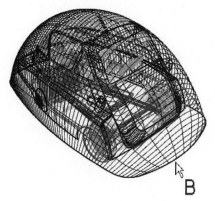

FIGURE 14–74 *From to left to right: setting selected surface's color texture, selecting surface B, and rendered image*
Source: Robert McNeel and Associates Rhinoceros® 5

Now surface B has a mapping applied, and you will use surface B as a custom mapping object to map surface A.

8. Select surface A, indicated in Figure 14–75, and select Custom mapping from the Texture Mapping tab of the Properties panel.
9. Select surface B, indicated in Figure 14–75.
10. Turn off the layer New Surface and render the viewport or delete surface B.

    Now surface A's mapping is customized, and the custom object B is not required. Even if you delete surface B, A's mapping will not be affected.

11. Do not save your file.

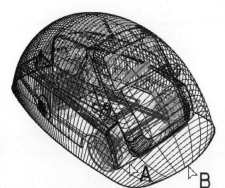

FIGURE 14–75 *From left to right: custom mapping being selected, surface B being used as custom surface for mapping surface A, and rendered viewport*
Source: Robert McNeel and Associates Rhinoceros® 5

Match Mapping

In essence, match mapping is equivalent to copying mapping coordinates from one object to another. To copy and match a series of surfaces/polysurfaces by using the same mapping method, perform the following steps.

1. Select File > Open and select the file Mapping004.3dm from the Chapter 14 folder on the companion CD.

    This is actually a continuation of the previous tutorial.

2. Select A, indicated in Figure 14–76, and set its color texture to Checker02.tif from the Chapter 14 folder on the companion CD.

3. Select Match Mapping from the Texture Mapping tab of the Properties panel.
4. Select B, indicated in Figure 14–76.

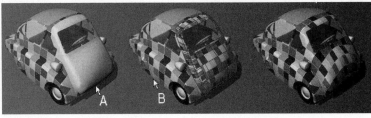

FIGURE 14–76 *From left to right: color texture setting, match mapping, object to be mapped, object to be used as the source of mapping, and the result of match mapping*
Source: Robert McNeel and Associates Rhinoceros® 5

5. Following steps 2 through 4, try by yourself to match C, indicated in Figure 14–77, as well.
6. Do not save your file.

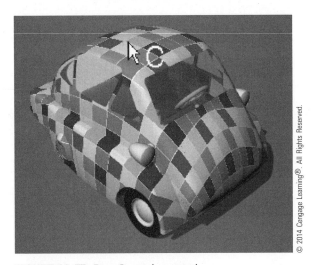

FIGURE 14–77 *Face C match mapped*

Unwrapping and UV Editing

A more complex way to place texture on a surface or a polysurface, you can unwrap the surface or polysurface onto a planar face and then assign a bitmap in accordance with the flattened face. To use this method of mapping, the polysurface must not be self-intersecting. Now perform the following steps.

1. Select File > Open and select the file Mapping005.3dm from the Chapter 14 folder on the companion CD.
2. Select A, indicated in Figure 14–78, and click on the Properties button on the Standard toolbar.
3. Click on the Material tab of the Properties panel and select object from the Assign material by pull-down list box.
   We will not select any color texture bitmap at this moment.
4. Click on Unwrap button on the Texture Mapping tab, indicated in Figure 14–78.
5. Select seams B and C, indicated in Figure 14–78, and press the ENTER key.
6. Click on the UV Editor, indicated in Figure 14–78.
7. Click on endpoints E and F.
   Points E and F here are for easy reference. As a matter of fact, exact location is unimportant. The mesh of the selected surface is constructed as shown in Figure 14–79.

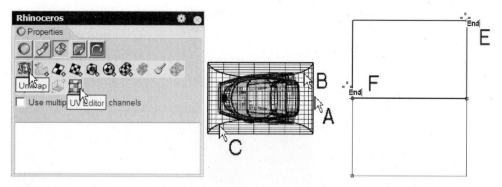

FIGURE 14–78 *From left to right: Unwrap and UV Editor buttons, polysurface and seams selected, and planar projection*
Source: Robert McNeel and Associates Rhinoceros® 5

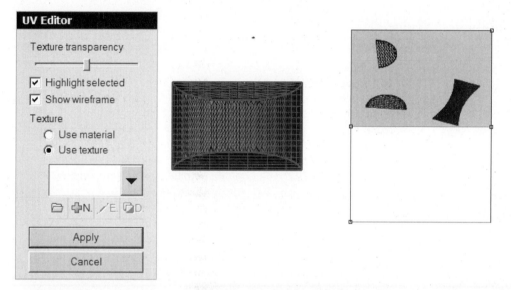

FIGURE 14–79 *From left to right: UV Editor, surface mesh, and planar mesh*
Source: Robert McNeel and Associates Rhinoceros® 5

To construct a bitmap from the flattened mesh, continue with the following steps.

8. Go to the Top viewport, zoom extended to the flattened mesh.
9. Select View > Capture > To File to capture the viewport and save the image.
10. Crop the screen capture to the rectangle area that the flatten mesh lies on in a 2D graphic program like Photoshop.
11. Put the cropped image to the background layer for reference to draw the texture.
12. Output the texture from the 2D graphic program and add it to a material in Rhino and apply the material to the object.
13. Click on the Apply button on the UV Editor dialog box.

To further appreciate flattening a surface without selecting any seam, continue with the following steps.

14. Turn off layer NewStuff.
15. Refer to steps 2 through 7 to flat surface A, indicated in Figure 14–80, but do not specify any seams and flatten the surface on rectangle BC.
16. Do not save your file.

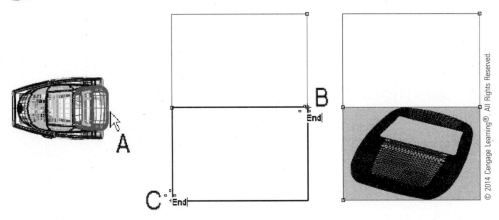

FIGURE 14–80 *From left to right: surface selected, rectangle being described, and planar mesh*

Managing Mapping Coordinates (Widgets)

Mapping coordinates can be managed in the following ways.

- You can select them from the channel list and delete them.
- You can edit a channel's number.
- You can show and subsequently hide a mapping widget.
- You can select a mapping widget and relocate it.

Now perform the following steps.

1. Select File > Open and select the file Mapping006.3dm from the Chapter 14 folder on the companion CD.
   This object has three mapping channels: box, spherical, and cylindrical.
2. Select the object and refer to Figure 14–81 to open the Texture Mapping tab.
3. Select Capped cylindrical from the Channels list and click on the Delete Mapping button.
   Channel 3 is deleted.

FIGURE 14–81 *From left to right: widget being deleted, widget's channel number being changed, and mapping texture's channel changed*
Source: Robert McNeel and Associates Rhinoceros® 5

4. Select Box from the Channels list and click on the Edit Channel button.
5. Change the channel number to 3.
   Because the color texture's channel is 1, you need to change its channel number unless you want to use the default UV mapping.
6. Select the Material tab and expand the Color mapping area.
7. Right-click on the Color texture bitmap and flowers, and select Edit.
8. Referring to Figure 14–81, change Mapping Channel to 3.

9. Select Box from the Channels list and click on the Show Mapping button.
10. Select A and then click on the Select Mapping Widgets button.
11. Click on B and drag it to a new location, indicated in Figure 14-82.

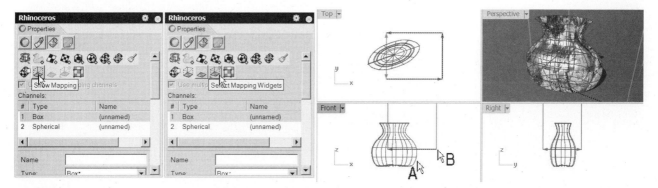

FIGURE 14–82 *From left to right: Show Mapping button, Select Mapping Widgets button, and mapping widget being selected and dragged*
Source: Robert McNeel and Associates Rhinoceros® 5

12. Click on the Hide Mapping button (Figure 14-83).
13. Do not save your file.

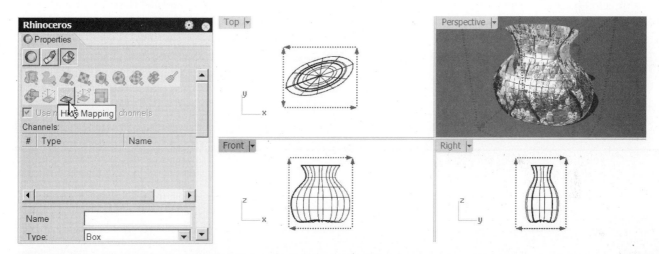

FIGURE 14–83 *Hide Mapping button (left) and mapping widget relocated (right)*
Source: Robert McNeel and Associates Rhinoceros® 5

Texture Palette

The texture palette serves the purpose of constructing a texture mapping for dragging and dropping onto the objects in the scene, including geometrical object and environment that will be explained later. Now perform the following steps.

1. Select File > Open and select the file Mapping007.3dm from the Chapter 14 folder on the companion CD.
2. Select Texture Palette from the Render or Panels pull-down menu.
3. With reference to Figure 14-84, click on the + sign A and select Checker03.tif from the Chapter 14 folder on the companion CD.
4. Click on the Checker 03 palette on the panel and click on the Enable filter button.
5. Click on the Checker 03 palette B and drag it on surface C, indicated in Figure 14-84.
   The texture is applied on the selected surface.
6. Do not save your file.

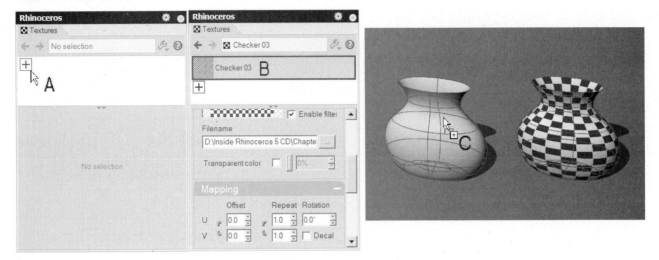

FIGURE 14–84 *From left to right: Textures panel, new texture added, texture being dragged to object, and object with color texture*
Source: Robert McNeel and Associates Rhinoceros® 5

Bitmap Map Filtering

Filtering is a technique in which the texture bitmap's individual pixel is reevaluated as the bitmap is stretched onto the surface where it is to be mapped, producing a better visual effect when the object being mapped is zoomed closely. Obviously, this takes a longer rendering time. Therefore, unless the camera is zoomed very close to the object, you may not need to check this box. To filter the bitmap, click on the Filter button.

Offset and Repeat

To better control how the bitmap (texture, transparency, bump, and environment) is wrapped onto objects in a layer or on selected objects, you manipulate two settings: offset and repeat.

- By default, the bitmap is stretched onto the entire surface. To have the bitmap repeated as tiles on the surface, you set U and V tiling. If you set the U value to 2, the bitmap is repeated twice in the U direction. Apart from repeating the bitmap, setting a negative value to U and V will mirror the bitmap about the U and V directions of the surface.
- By default, the bitmap starts the origin of the surface, with both U and V values equal to zero. To adjust the starting point where the map is to be applied on the surface, you adjust the U and V offsets.

Decals

To further add realism to an object, you may need to apply a decal on it. Perform the following steps.

1. Select File > Open and select the file Mapping008.3dm from the Chapter 14 folder on the companion CD.
2. Select Render > Texture palette.
3. Click on the + sign and select Rhinoceros R.jpg from the Chapter 14 folder on the companion CD.
4. Click on the Decal button.
5. Select A, indicated in Figure 14–85, and click on the Properties button on the Standard toolbar.
6. Open the Decal tab and click on the Add button.

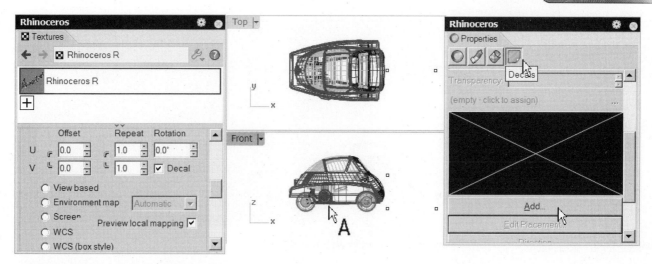

FIGURE 14–85 *From left to right: Texture palette, surface being selected, and Decals tab selected*
Source: Robert McNeel and Associates Rhinoceros® 5

7. Click on B from the Texture palette and click on the OK button.
8. Click on the OK button of the Decal Mapping Style dialog box.
9. Click on points C, D, and E (Figure 14–86).
   A decal is placed.

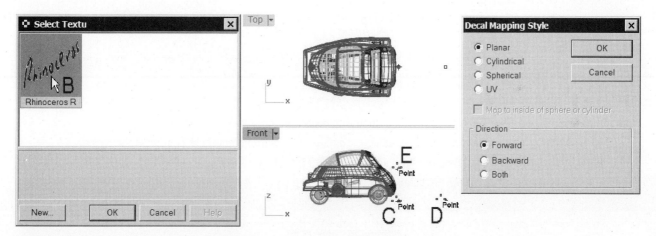

FIGURE 14–86 *From left to right: Decal selected, decal being placed, and placing method*
Source: Robert McNeel and Associates Rhinoceros® 5

Now you will reposition the decal.

10. Click on decal F, indicated in Figure 14–87, and drag it to position G and press the ENTER key.
11. Render the Perspective viewport.
12. Do not save your file.

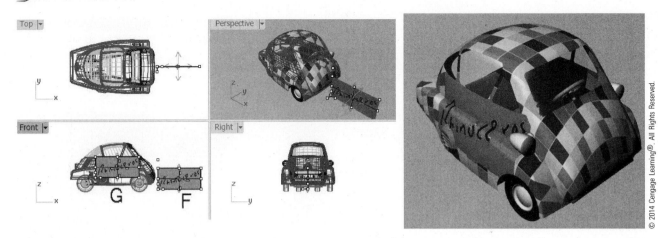

FIGURE 14–87 *Dragging the decal (left) and rendered Perspective viewport (right)*

Render Effects

Apart from using digital materials to enhance a rendered image, you can use four kinds of render effect tools: Piping, Displacement, Edge softening, and Shut lining.

Piping

If it is only for the purpose of simulating a piping in a rendered image, you can use the piping tool to construct a piping rendered mesh around a curve, as follows.

1. Select File > Open and select the file Effect008.3dm from the Chapter 14 folder on the companion CD.
2. Select Render > Effects > Curve Piping.
3. Select curve A, indicated in Figure 14–88, and press the ENTER key.
4. Select the Radius option on the command area and set it to 2.
5. Press the ENTER key.

   As the viewport is set to rendered, the curve piping render mesh is rendered. Note that the curve piping, being rendered mesh, can be seen only when the viewport is rendered.

To modify and to extract the mesh, continue with the following step.

6. Open the Properties panel and select B, indicated in Figure 14–88.

   You will find a Curve Piping tab in the panel; you can modify the parameters there to change the piping.

7. Type ExtractPipedCurve at the command area and select B, indicated in Figure 14–88.

   Now the render mesh is being extracted as real meshes, which will be displayed even when the viewport is not rendered.

FIGURE 14–88 *From left to right: curve being selected, Curve Piping tab of the Properties panel, and curve piping rendered mesh constructed*
Source: Robert McNeel and Associates Rhinoceros® 5

Displacement Mesh

A bump map simply produces the effect of displacement in rendering, and the surface mesh does not change at all. To construct a real displacement mesh, continue with the following steps.

8. Select Render > Effects > Displacement.
9. Select A, indicated in Figure 14–89, and press the ENTER key.
10. In the Displacement Settings dialog box, click on the empty space B and then the + sign to assign a texture.
11. Select Checker03.tif from the Chapter 14 folder on the companion CD and click on the OK button.
12. Click on the OK button of the Displacement Settings dialog box.

    Displacement is applied.

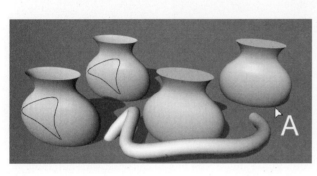

FIGURE 14–89 *Surface being selected (left) and Displacement Settings dialog box (right)*
Source: Robert McNeel and Associates Rhinoceros® 5

To modify the displacement settings and to extract the mesh from the displacement generated from the bitmap, continue with the following steps.

13. Open the Properties panel and select B, indicated in Figure 14–90.
14. Type ExtractRenderMesh at the command area.

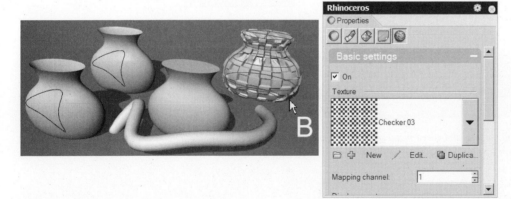

FIGURE 14–90 *Displacement applied (left) and Displacement tab of the Properties panel (right)*
Source: Robert McNeel and Associates Rhinoceros® 5

Edge Softening

Instead of actually filleting or chamfering the edges between contiguous surfaces, you can use the edge softening tool to construct filleted or chamfered render mesh. Continue with the following steps.

15. Select Render > Effects > Edge Softening.
16. Select A, indicated in Figure 14–91, and press the ENTER key.

17. Select Softening option on the command area and set it to 1.
18. Press the ENTER key.
   The sharp edge is softened.

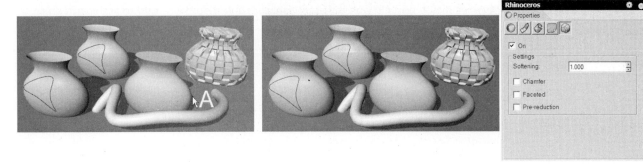

FIGURE 14–91 *From left to right: surface being selected, edge softened, and Edge Softening tab of the Properties panel*
Source: Robert McNeel and Associates Rhinoceros® 5

Shut Lining

To simulate the effect of having a shut lining in a rendered image, you can use shut lining tool to construct shut lining mesh. Continue with the following steps.

19. Select Render > Effects > Shut Lining.
20. Select A, indicated in Figure 14–92, and press the ENTER key.
21. Select the Radius option on the command area and set it to 1.
22. Select curve B, indicated in Figure 14–92, and press the ENTER key twice.
   Shut lining mesh is constructed.

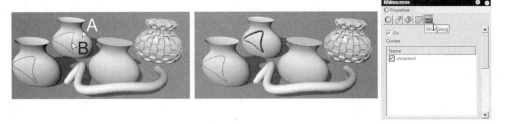

FIGURE 14–92 *From left to right: surface and curve being selected, shut lining constructed, and Shut Lining tab of the Properties panel*
Source: Robert McNeel and Associates Rhinoceros® 5

23. Select Render > Effects > Shut Lining.
24. Select A, indicated in Figure 14–93, and press the ENTER key.
25. Select the Raised option on the command area to change it to Yes.
26. Select curve B, indicated in Figure 14–93, and press the ENTER key twice.
   Raised shut lining mesh is constructed.
27. Do not save your file.

 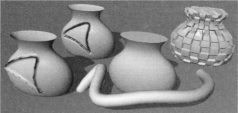

FIGURE 14–93 *Surface and curve being selected (left) and raised shut lining constructed (right)*
© 2014 Cengage Learning®. All Rights Reserved.

Flaming nXt Plug-in Material

Assigning material properties via a plug-in means that you take a material-properties source external to the Rhino program and bring it into a library (or construct a library). Material properties can then be assigned from that library. You need to use the Flamingo nXt renderer to be able to use the default material library or to construct a new material library of your own (Figure 14–94).

FIGURE 14–94 *From left to right: Plug-in option, Flamingo Materials dialog box, Flamingo Decals, and misc*
Source: Robert McNeel and Associates Rhinoceros® 5

ENVIRONMENT OBJECTS

Environment objects are objects you include in the scene to enhance its appearance. Using Rhino's basic renderer, you can adjust the ambient light and background and add a ground plane.

Ambient Light and Radiosity

In reality, the combined effect of all lighting in a scene causes objects not directly under any light source to be illuminated. This is done by reflection of lights in the entire scene. The effect is commonly known as radiosity. The effect of radiosity depends on the strength of the light sources and how the lights are reflected by the objects in the scene. For example, an object placed inside a white box will be better illuminated than an object placed inside a box of dark color. The reason is that more light is reflected by the white box.

Setting Ambient Light

Ambient light provides a means of approximating the radiosity effect. Using Rhino's basic renderer, you set the ambient light's color. To establish ambient lighting, perform the following steps.

1. Select File > Open and select the file Environment.3dm from the Chapter 14 folder on the companion CD.
2. Select Render > Render Properties.
3. In the Rhino Render tab of the Document Properties dialog box, select the Ambient Light color swatch.
4. In the Select Color dialog box, select orange color and then click on the OK button.
5. Close the Document Properties dialog box.
6. Render the scene, shown in Figure 14–95.

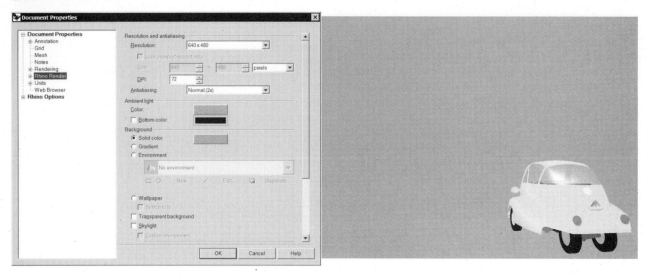

FIGURE 14-95 *Ambient light changed (left) and scene rendered (right)*
Source: Robert McNeel and Associates Rhinoceros® 5

Setting Background Color

A simple way to enhance the visual effect of a rendered image is to change the background color, as follows.

7. Select Render > Render Properties.
8. In the Rhino Render tab of the Document Properties dialog box, select the Ambient Light color swatch and then change its color to black (the default color).
9. Select the Background color swatch.
10. In the Select Color dialog box, select red and then click on the OK button.
11. Render the scene, shown in Figure 14-96.

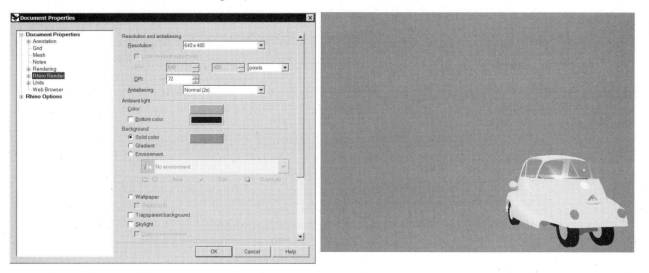

FIGURE 14-96 *Background color changed (left) and scene rendered (right)*
Source: Robert McNeel and Associates Rhinoceros® 5

Using Wallpaper

To simulate a real environment, an image placed in the background may save a lot of effort in object construction and material assignment. Continue with the following steps.

12. Right-click on the Perspective viewport's label and select Viewport Properties, or click on the Viewport Properties button on the Properties toolbar.

13. In the Viewport Properties dialog box, shown in Figure 14-97, click on the Browse button of the Wallpaper options area.
14. Select the file "nightscene.jpg" from the Chapter 14 folder on the companion CD.
15. Click on the Show wallpaper check box and clear the Show wallpaper as gray scale check box.
16. Click on the OK button to close the dialog box.

    A wallpaper is placed.

At this point, you can use the display tools to rotate, zoom, or pan the view to adjust the objects in the viewport to best suit the wallpaper.

17. Select Render > Render Properties.
18. Check the Wallpaper and Stretch to fit check boxes, and click on the OK button.
19. Render the scene. (See Figure 14-97.)
20. Do not save the file.

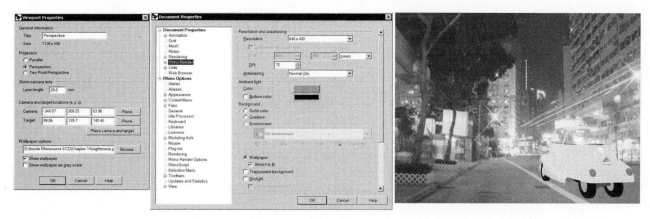

FIGURE 14-97 *From left to right: Viewport Properties dialog box, Document Properties, and rendered Perspective viewport*
Source: Robert McNeel and Associates Rhinoceros® 5

Perspective Matching Wallpaper

To add realism, you may place a wallpaper in a viewport, using it as a background image during rendering. In order to match the 3D objects with the 2D image, you have to adjust the camera, as follows.

1. Select File > Open and select the file PerspectiveMatch.3dm from the Chapter 14 folder on the companion CD.

    A bounding box is constructed, and lines are drawn on the wallpaper to help perspective match.

2. Select View > Set Camera > Match Wallpaper Image, or click on the Match Perspective Projection button on the Set View toolbar.
3. Click on location "1" and then select endpoint A (Figure 14-98).
4. Click on location "2" and then select endpoint B.
5. Click on location "3" and then select endpoint C.
6. Click on location "4" and then select endpoint D.
7. Click on location "5" and then select endpoint E.
8. Click on location "6" and then select endpoint F.

    A minimum of six points is required.

9. Press the ENTER key.

    The camera is adjusted.

FIGURE 14–98 *Perspective being matched*

10. Turn off the BoundingBox layer.

    The bounding box is not invisible. Alternatively, you may delete the bounding box.

11. Right-click on the Viewport's label and select Viewport Properties.
12. In the Viewport Properties dialog box, click on the Browse button of the Wallpaper options area.
13. Select the image "Toronto00.jpg" from the Chapter 14 folder on the companion CD, replacing the image with one without any reference lines.
14. Click on the OK button.

    Matching is complete. (See Figure 14–99.)

FIGURE 14–99 *Bounding box removed and wallpaper changed (left) and rendered image (right)*

Rendering with Wallpaper

To produce a rendered image including wallpaper using Rhino renderer, perform the following steps.

15. Select Render > Render Properties.
16. Click on the Wallpaper check button on the Rhino Render tab of the Document Properties dialog box. (See Figure 14–100.)
17. Select Render > Render.

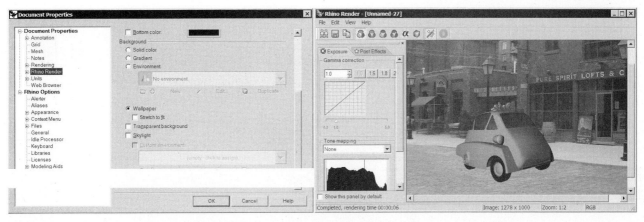

FIGURE 14–100 *Rhino Render tab of the Document Properties dialog box (left) and Rhino renderer (right)*
Source: Robert McNeel and Associates Rhinoceros® 5

If you are using Flamingo nXt, perform the following steps.

18. Select Flamingo nXt > Control Panel.
19. In the Environment tab of the Flamingo nXt control panel, select Image from the Background type pull-down list box.
20. Click on A, indicated in Figure 14–101, and then select the Toronto00.jpg image from the Chapter 14 folder on the companion CD.
21. Select Flamingo nXt > Render.
22. Do not save your file.

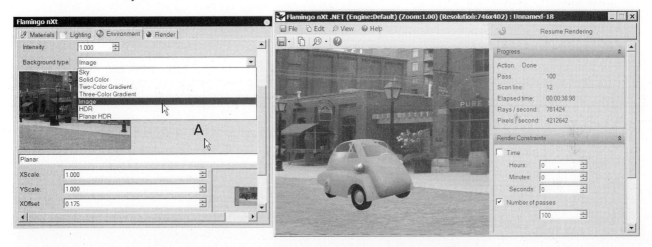

FIGURE 14–101 *Environment tab of Flamingo nXt control panel (left) and Flamingo renderer (right)*
Source: Robert McNeel and Associates Rhinoceros® 5

Ground Plane Panel

In previous versions, in order to simulate the effect of an object lying on the ground, it is a usual practice to construct a rectangular planar surface. Now you can use the Ground Plane panel instead. Perform the following steps.

1. Select File > Open and select the file Ground.3dm from the Chapter 14 folder on the companion CD.
2. Select Render > Panels > Ground Plane.
3. With reference to Figure 14–102, check the On box on the Ground Plane panel and click on the button A.
4. Click on B on the Front viewport.

   A ground plane is placed.

5. Render the scene.
6. Do not save your file.

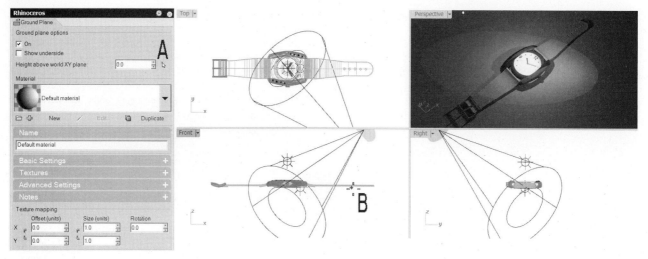

FIGURE 14–102 *Ground Plane panel (left) and viewports (right)*
Source: Robert McNeel and Associates Rhinoceros® 5

RENDER PROPERTIES

To produce a high-quality rendered image, you may need to go through various render options that affect the outcome of rendering.

Resolution

To manage output requirements, you can set the resolution of the rendered image, as follows.

1. Select File > Open and select the file Environment.3dm from the Chapter 14 folder on the companion CD.
2. Select Render > Render Properties.
3. In the Rhino Render page of the Document Properties dialog box, select Custom from the Resolution pull-down list box of the Resolution and Anti-aliasing area.
4. Set the size to 100 × 80 pixels.
5. Click on the OK button.
6. Render the viewport.
7. Repeat steps 2 through 7 to set the resolution to 400 × 320 pixels and render the viewport.
8. Compare the results, shown in Figure 14–103.
9. Do not save your file.

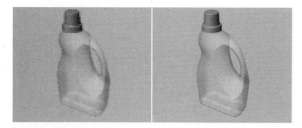

FIGURE 14–103 *100 × 80 resolution (left) and 300 × 240 resolution (right)*
© 2014 Cengage Learning®. All Rights Reserved.

Anti-Aliasing

Because each discrete pixel in an image has a unique color value, an inclined edge in a rendered image may look jagged. To minimize this jagged effect, adjacent pixels in the inclined edges are averaged. The process is called anti-aliasing. Anti-aliasing produces a blurred edge to mask the jagged effect. For example, in a red box on a white background image, pixels are either red or white. Anti-aliasing produces a set of pinkish pixels between the red and white pixels. Continue with the following steps.

10. Select Render > Render Properties.
11. In the Rhino Render page of the Document Properties dialog box, set the resolution to 100 × 80 pixels.
12. Select None from the Antialiasing pull-down list box and click on the OK button.
13. Render the viewport.
14. Repeat steps 11, 13, and 14; select High (10×) from the Antialiasing pull-down list box, and render the viewport.
15. Compare the results, shown in Figure 14–104.
16. Do not save your file.

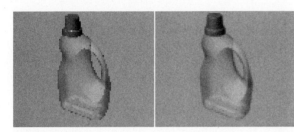

FIGURE 14–104 *No anti-aliasing (left) and high anti-aliasing (right)*
© 2014 Cengage Learning®. All Rights Reserved.

Miscellaneous

There are four check. boxes at the bottom of the Rhino Render page of the Document Properties dialog box. Their functions are explained below:

Use lights on layers that are off	Lights residing on layers that are off can be regarded as being turned off if this check box is cleared.
Render curve	Curves are not rendered unless this box is checked.
Render surface edges and isocurves	Edges and isocurves are not rendered unless this box is checked.
Render dimensions and text	Dimensions and text are not rendered unless this box is checked.

Render Details

Several factors affect the quality of a rendered image and the amount of time to produce the image. In essence, a higher-quality rendered image requires a longer time to render. To set render details, you click on the Render Details tab of the Document Properties dialog box.

Four major concerns are covered here: render acceleration grids, preventing self-shadowing, object and polygon BSP tree, and transparency.

Render Acceleration Grids

Two settings, screen grid cell size and spotlight shadow grid size, have significant effect on rendering time.

Screen Grid Cell Size. When Rhino constructs a rendered image, rays are shot from the camera point to the scene to intersect with the geometry in the scene in order to determine which polygon is closer to the camera point and which polygons are hidden behind other polygons. These rays are called eye rays because the camera point is located at the observer's eye position.

To accelerate rendering, the render viewport is divided into a number of rectangular regions, called grid cells, building a list of polygons in these regions and sorting the polygons in each cell in accordance with their distance from the camera point.

In the process of rendering, computation is accelerated in such a way that eye rays are needed only to intersect with the polygons in the cell which are closest to the eye, and that some polygons farther away from the camera point are skipped after an intersection is found.

In Rhinoceros, the rectangular grid cells are square in shape, and you can set the size of these cells by specifying the width and height in terms of pixel number.

If we set a small grid cell size, there will be many grid cells in the scene. Naturally, it will require a significant amount of memory and take time to build the grids. However, subsequent rendering time will be faster. The default size is 16 pixels, and it works for most scenes.

Spotlight Shadow Grid Size. When an intersection between the eye ray and the geometry in the scene is found, the ray is shot from the intersection to the light source. These rays are called shadow rays, for the construction of shadows in the scene. In practice, lengthy rendering time is usually the result of tracing these shadow rays.

To accelerate the computation of the spotlight's shadow rays, the light cone of the spotlight is divided into rectangular regions, sorting the polygons within each region.

Shadow grid size is defined in terms of the number of grid cells in the spotlight's cone. The default value is 256, dividing the cone into 256×256 cells.

Preventing Self-Shadowing

Due to numerical fuzz in computing the shadow ray, it is possible that the shadow ray may hit the very same polygon, placing the shadow in the wrong place. This phenomenon is called self-shadowing. To prevent self-shadowing, the shadow ray is intentionally offset a small value. The default value is 0.001.

Object and Polygon BSP Tree

BSP stands for binary space partitioning, a technique deployed for hidden face removal. It is a process in which the space is recursively subdivided until a stopping point is reached. In Rhino rendering, the stopping point is determined by the maximum tree depth value and the target node size. BSP tree can be used to accelerate rendering. However, building the BSP tree takes some time and memory. The bigger the depth and the smaller the node size, the more time and memory it will take to build.

Transparency

There may be a number of transparent objects in a scene. If the transparent objects are stacked, rendering time will increase significantly. In reality, if many transparent objects are stacked, it may be pointless to display objects at the very bottom. Therefore, you can, to save rendering time, set a limit on the number of stacked transparent objects for the rays to pass through. This value is called max bounces. For example, if you set max bounces = 5 and there are six transparent objects stacked together, the sixth transparent object will not be regarded as transparent. The default maximum bounce is 12, which is adequate for most situations.

Render Mesh Quality

As explained earlier, a set of polygonal meshes is produced for the sake of rendering, even though the models are constructed by using NURBS surfaces. To have a higher-quality rendered image, you may need to have a finer rendered mesh instead of a coarse rendered mesh. To set rendering mesh quality, you can click on the Mesh tab of the Document Properties dialog box.

Render Options

By clicking on the Render Options of the Document Properties dialog box, the Render Options page is displayed. Here, you can click on Two Stage render to render the image in two stages, thus producing a more accurate image. If you have a multiple CPU computer, the rendering task can be divided among the CPUs by specifying the number of threads, such as 2 or 4, and also determining the priority of CPU usage.

CONSOLIDATION

To produce a photorealistic image from objects you construct in Rhino, you use one of Rhino's basic renderer or the Flamingo nXt renderer. There are five major steps in producing a photorealistic rendered image from the objects you construct in Rhino. You apply material properties to the objects, construct lights in the scene, include environment objects, set up a camera, and use a renderer to produce an image. Using the basic renderer, you add basic materials (basic color, reflective finish, transparency, texture, and bumpiness) to objects.

In a viewport, you can establish five kinds of lighting effects (spotlight, point light, directional light, rectangular light, and linear light). To enhance light bouncing effect, you can set up bouncing directional light, bouncing spotlight, and bouncing point light. If these lights are already constructed in a viewport, you can relocate them to produce bouncing light effect. In addition to material and lighting, you can add reality to the viewport by including three kinds of environment objects: ambient light, background color, and wallpaper.

Using Rhino's default animation scripts, you can create turntable animation, path animation, fly-through animation, and sun study animation.

REVIEW QUESTIONS

1. State the methods by which material properties are applied to an object.
2. Outline the ways by which lights can be established in a viewport.
3. What environment objects can be included in a viewport?
4. In what ways can a camera be set?
5. Briefly describe the kind of animations you can construct using Rhino.

INDEX

A

ACIS (.sat, .sab) files, 568
Adjust Surface Blend dialog box, 62, 63
Adobe Illustrator (.ai) files, 567
Aliases, command, 96
Align Curves button, 198–199
Aligned dimension, 556
Aligning
    curves, 464–465
    curves in loft surface construction, 198–199
    detail viewports, 545–546
    objects, 465–466
    shapes and sweeping, 194
    vertices on mesh objects, 452–453
Along Line tool, 127
Along Parallel tool, 128
ALT key
    dragging objects, 138
    moving objects, 341
    and Osnap, 127
Ambient light, 625–626
    setting, 625–626
Analysis mesh, extracting, 508
Analysis tools, 86–87
Anchor points, 224
Angle dimension, 557
Angles
    constraints, 121
    draft, 454–455, 507
    measurement of, 504–505
    ortho, 124–125
    rendering, 507
    rotation, 520
    taper, 181
    tolerance, 459
    twisting, 480–481
Animation
    digital, 576
    tools, 88
    turntable, 576–578
Annotation(s), 554–564
    arrow head, 562
    dots, 561
    kinds of, 554

Anti-aliasing, 574, 630–631
Appearance tab, 113
Apply Box Mapping, 613
ApplyCrv command, 312
Apply Cylindrical Mapping, 613
ApplyMeshUVN, 440
Apply Spherical Mapping, 613, 614
Arc, 261–262
    adding segment to a curve, 273–275
    converting free-form curve to segments, 292
    curve, 292–293
    modifying radius, 262
Area centroid, 511
Area centroid inquiry, 511
Area inquiry, 511
Area moments, 511
Area moments inquiry, 511
AroundFaces, 422, 424, 427, 428, 431
Array, 142–146
    along a curve, 144–145, 146
    along a curve on a surface, 146
    along surface UV directions, 145–146
    constructing, 23
    linear, 142–143
Arrow head, 562
Artistic mode, 106
Aspect ratio, 109–110, 448
    extracting mesh faces, 454
Assembly, 529–530
    bottom-up approach, 530–531
    components of, 529–530
    defined, 529
    simulation, 529
Attached files, 536, 539
    activation, 537–538
    refreshing, 538
Auditing, of Rhino files, 514
AutoChain, 380
Auto-complete (command area), 93
Axis
    and cone construction, 367–368
    coordinate, 119

    of a cylinder, 364–365
    of ellipsoid, 374–375
    and pyramid construction, 373
    revolving around, 187
    tapering along an, 482
    and truncated cone construction, 370–371
    of tube, 377–378
    twisting around, 480–481

B

Backbone curve, 490–492
Backbone surface, 492, 496
Background, 99–100
    color, 626
    tracing images, 235–236
    viewport's gradient, 99–100
Background Bitmap command, 235
Bad object report, 514
Base point, 529
Base surface, 315, 320
Basis spline. *See* B-spline
Bending, 481–482, 482–485
    of control points, 482–485
Between tool, 128
Bevelled edge, 329
Bézier curves/surfaces, conversion to, 566
Binary space partitioning (BSP), 632
Bitmap, 235, 606–608, 610–611, 620
BlendCrv command, 280–281
Blend edge, variable, 396–398
Blended surface, 39, 331–335
Blending, 33, 279–281
Block definition, defining, 533–534
Block Instance command, 75, 77
Block manager, 523–525
Blocks, 518–535
    defined, 516
    defining from existing objects, 519–520

deleting definition, 527
edit, 528
exploding, 527
insertion, 520–521
instances, 519
nested, 526
purging unreferenced, 527
replacing, 528–529
Blocks edit, 528
Body, bubble car, 27–51
Boolean 2 objects command, 392
Boolean operation, 11, 14, 21, 207, 390, 441–443
Boolean splitting, 405, 444
Border, 300–302
Boss, 398
Bottle opener, 53–55
curves, 53–54
Bottom-up construction process, 3, 530–531
Bounce lights, 601
Boundary
detaching from trimmed surface, 320
extending curve to, 272–273
face moving to, 409
manipulation, 314–328
trimming, 44–45
Boundary command, 135
Boundary edges, 213, 322
Bounding box, 148–149, 376, 503–504, 584, 627–628
Box, 357–360
Box editing, 486–487
Box mapping, 613–614
Brightness, 154
Brushing command, 134
B-spline, 222
BSP tree, 632
Bubble car
body, 27–51
chassis, 5–26
chassis design, 12–15
drive and steering, 22–23
engine unit, 16–18
exhaust system, 18–19
floor panel and seat, 24–26
frontal door, 36–40
front suspension, 21–22
fuel tank, 20–21
intake system, 19–20
main body, 30–36
mechanical components construction, 7–22

modeling project, 1
rear axle, 10–11
rear suspension, 15–16
skirts, 42–44
tires and wheels, 7–10
windows, 40–42
Bulge, end, 287, 337–338
Bump map, 607–608

C

Cage, release from, 489
Cage editing, 487–489
Cage type, 489
Calculators, 113
Camera, 582–586
Camera pyramid, 585–586
Capped surface, 13, 18, 44–45, 380, 389
Cartesian coordinates, 119–121
Cascading toolbars, 105–106
Case sensitivity, 92
Case studies, 1–78
Center
for box construction, 360
of circle, 258–259
of ellipse, 263
of ellipsoid, 374–375
for sphere construction, 361
ChainContinuity, 380
ChainEdges option, 193
Chamfering, 277–278, 333–335
solids, 394–395
surface, 329–330
Chassis, 5–26
design, 12–15
CheckNewObjects command, 515
Circle, 258–260
around a curve, 259–260
circumscribed, 374
constructing, 118–119
curvature, 499–500
deformable, 260
inscribed, 374
tangent, 259
vertical, 259–260
Circular zone command, 135–136
Circumference, 258
Circumscribed circle, 374
Cleaning up, 404
Clickable options (command area), 93–94

Clipping planes, 110–111
Closed curve, 230–231
Closed-loop curve, 201, 224
Closed-loop polysurface, 389–390
Closed planar curve, 434–435
Closed polysurfaces, 17, 58, 69
Collaboration, design, 535–539
Collapsing, 446–449
Color, 606
background, 626
gradient, 99–100
of normal directions, 263
property, 18
setting in layer, 154–155
setting methods, 154–155
Color texture map, 607
Combining
mesh objects, 441–445
Rhino solids, 390–393
Command
aliases, 96
interaction, 106
scripting, 95–96
Command Area, 92–96
Command History dialog box, 94–95
Command List, 93, 95
Commands
complete list of, 95
most used, 94
recently used, 94
Comma-Separated Value (CSV), 567
Computer modeling
purpose of, 80
Rhinoceros and, 79–88
tools for 3D, 80–81
Cone, 91, 367–370
Cone, truncated, 370–373
Conic, 266
Conic rectangle, 255
Connecting, surfaces, 335
Construction plane, 46, 117–119
Cartesian system, 120
constructing a box with base vertical to, 359–360
coordinate system, 119–121
line vertical to, 251
manipulating, 119, 164–172
mobile, 170–172
named, 168–169
native, 164
polar system, 120
projecting to, 478–479

Construction plane *(continued)*
   remapping objects from one CPlane to another, 475–476
   setting orientation, 37
   universal, 169–170
Contiguous curves, 281
Contiguous surfaces, 28
   constructing solid from, 390
   continuity between, 179, 329–340
   joining, 36, 338
   merging, 37, 261–262
Continuity, 277
   between contiguous surfaces, 179, 329–340
   curves, 37, 177–178
   defined, 177
   between edge rails and sweep 2 rails surface, 197–198
   geometric, 502
   surface, 31, 177–178
   in sweeping, 192–193
   types of, 177–178, 335
Contour lines, 305–307
Control point curve, 223–224
   compared with interpolated curve, 224
   manipulation of, 226–228
Control points. *See also* Points
   adjusting weight, 227–228, 265
   bending of, 482–485
   constructing polygon mesh, 437
   and deformation operations, 482–485
   deleting, 226–227
   extracting, 309–310
   fitting a curve through, 238–239
   hiding, 268
   increasing, 266
   insertion and removal, 227, 267–268
   manipulating, 263, 266
   moving, 226
   shearing of, 482–485
   tapering of, 482–485
   twisting of, 482–485
Control polygon, 268
Converting
   to Bézier curves/surfaces, 566
   mesh faces, 456–458
ConvertToBeziers command, 566
Coordinate Display, 104

Coordinates, 119–121
Coplanar curves, 282–283
Co-planar faces, 404
Copying, 138
   holes, 416
   objects from one layer to another, 156
   page layout, 549
CopyLayout command, 549
Corner points surface, 205–206
Corners, 357–358
Corner-to-corner rectangular surface, 210
CPlanes toolbar, 169
CPU and Save (status bar panes), 104
Creases, 324
CreateUVCurves command, 312
Crossing command, 133
Cross-section profiles, 297–298
CSV. *See* Comma-Separated Value (CSV)
Culling, 461
Cult 3D (.cd) files, 567
Cursor movement, 124
Curvature circle, 499–500
Curvature continuity. *See* G2 continuity
Curvature graph, 498–499
Curvature radius, 500
Curvature rendering, 506
CurveHandle, 61
CurveHandleSpline, 61, 62
Curve network surface, 36–37, 39, 61, 201
Curves, 3. *See also* specific curve i.e. Periodic curve, 3D curves
   adjusting end bulge, 287
   aligning, 198–199, 464–465
   array along, 144–145
   basic construction, 222–223
   blending, 33, 53, 279–281
   Boolean, 207
   bottler opener, 53–54
   bounding box of, 503–504
   building, 8–9
   changing polynomial degree of, 225–226
   circle around, 259–260
   closed loop, 201, 224
   constructing, approaches for, 175–176
   constructing surfaces from, 175–177

   construction from point cloud, 243–244
   construction of, 53–54, 56–58
   continuity, 37, 177–178
   converting free-form to line/arc segments, 292
   as cutting object, 444
   deleting portion of, 276–277
   deviation between, 502
   direction matching of, 501–502
   direction of, 500
   electric cable, 56–58
   ellipse around, 263–264
   ellipsoid on, 376
   from existing objects, 293–313
   extending length of, 272–275
   extending on a surface, 275
   extruding, 13, 63
   extruding along, 182–184, 190–192
   extruding faces along, 411
   fairing, 290
   fitting through a set of points, 238
   fitting through control points of a surface, 239
   fitting through vertices or control points of a polyline, 238–239
   fixed length edit, 288
   flowing along, 65, 490–492
   free-form, 220–245
   helix around, 268
   improving smoothness of, 485–486
   interpolating, 295–296
   intersection of, 304
   length of, 504
   manipulation of, 271–313
   matching, 283–285
   modifying, 225–226, 277–285
   offsetting, 293–295
   open-loop, 224
   orient on, 473–474
   points near and along, 240–242
   polygon around, 256–258
   projecting, 33, 298–300
   pulling to surface, 298–299
   rebuilding, 286–287
   refinement methods, 286–293
   refitting, 291
   of a regular pattern, 246–270
   reparameterizing, 272

replacing portion of with a line segment, 290
revolved, 16
revolving, 186
shortest between two points on a surface, 239–240
simplifying, 290–291
soft editing, 287–288
sorting, 202
sphere around, 362–363
spiral around, 269
splitting, 276
for surface modeling, 220–222
tools, 81
torus around, 382–383
trimming, 38–39, 275–276
and tube construction, 378–379
CurveSpadeCurveOriginal, 60
CurveSpadeFrontProfile, 61
CurveSpadeTopProfile, 61
Custom mapping, 614–615
Cutting, 317
Cutting object, 318, 322, 444
Cutting plane rectangular surface, 211–212
Cut volume, 512–513
Cylinder, 364–367
 tangent, 425–426
Cylindrical mapping, 613–614

D

Data analysis, 497–515
 diagnostics tools, 513–515
 dimensional analysis tools, 503–506
 general tools, 497–502
 mass properties tools, 510–513
 surface analysis tools, 506–510
Database listing, 514
Data exchange, 87–88, 564–568
Decals (mapping), 620–622
Default light, 586
Deformable circle, 260
Deformable ellipse, 264
Deformation, 476–490
 applied to control points, 482–485
 cage editing, 487–489
 maelstrom editing, 490
 point-setting, 476–478
 shearing, 479–480
 smoothing, 485

stretching, 486
and translation, 490–495
Degree 1 curve, 254
Degree 2 curve, 254, 260, 261, 263, 265, 266
Degree 3 curve, 266, 268
Degree of polynomial. See Polynomial degree
Delete command, 138
Deleting
 block definition, 527
 holes, 416
 mesh faces, 450
 portion of a curve, 276–277
Density, 265, 419
Derived surfaces, 213–218
Design, 3–4, 535–539
Detaching, files, 320, 536–537
Detailing, 4, 393–403
Detail viewports, 544–551. See also Viewports
 display modification, 548–549
 manipulation of, 544–549
Developable surface, 199, 216–217
Deviation, 502, 509–510
Diagnostic tools, 513–515
Diagonal, 358
Diameter
 of circle, 258–259
 constructing a sphere, 361
 and cylinder construction, 364–365, 367
 of ellipse, 263
 of ellipsoid, 375
 of torus pitch circle, 381–382
 of tube, 379, 380
Diameter dimension, 558
Difference, 391, 442
Diffuse color, 606
Digital animation, 576–581
 fly-through animation, 580
 path animation, 578–580
 sun study, 580–581
 turntable animation, 576–578
Digital lighting, 586–596
 directional lighting, 592–593
 linear light, 595–596
 point light, 591–592
 rectangular lighting, 593–594
 spotlights, 586–591
Digital material, 604–625
Digital rendering, 569, 570–575

Digitizer, drawing line with, 252
Digitizing, 176, 243
Dimensional analysis tools, 503–506
Dimensioning, 554–564
Dimension(s)
 aligned, 556
 angle, 557
 diameter, 558
 exploding, 562
 kinds of, 554
 linear, 556
 ordinate, 559
 radial, 558
 rotated, 557
 styles, 554–555
Direct construction (surface modeling), 176
Direction, 500–501
 constraints when constructing a tube, 377–378
 constraints when constructing cone, 367–368
 constraints when constructing cylinder, 364–365
 constraints when constructing mesh cylinder, 424–425
 constraints when constructing pyramid, 373
 constraints when constructing truncated cone, 370–371
 flip normal, 262–263
 tangent, 232
 unifying normal, 458–459
Directional lighting, 592–593, 596–597
 highlighting, effect of, 596–597
Directory, working, 536
Direct X (.x) files, 566
Disjoint mesh, 461
Displacement mesh, as render effect tool, 623
Displacement Settings dialog box, 623
Display control, 107–108
Displaying toolbars, 104
Display methods, advanced, 111–112
Display mode, viewport, 106–107
Display order command, 112
Distance (status bar panes), 104, 504
Dividing, surface, 324
Docked toolbars, 104

Docked viewports, 101–102
Document properties, 113
Dollying, 108
Dots, 561
  conversion of, 562
Downstream operations, 564–565
Draft angle, 244–245, 454–455, 507
Draft angle rendering, 507
Drag and drop, 540
Dragging, 273
Drag mode, 137–138, 263
Drape points, 242–243
Drape surface, 208–209
Drawing aids, 121–132
Draw order command, 112–113
Drive unit, 22–23
Dropping, 494–495
Duplicating, 138
  border of individual surfaces of polysurface, 301
  objects from one layer to another, 156–157
  surface edge and border, 300–301

E

Edge curves
  flattening, 216–217
  and loft surface tangency, 199–200
  surface, 203–204
Edge length, 449, 455–456
Edge rails, 197–198
Edges, 101
  bevelled, 329
  chaining, 193–194
  continuity between edge rails and sweep 2 rails surface, 197–198
  filleting, 9, 18, 21, 49
  inclined, 630
  joining, 36, 275–276
  matching in curve network surface construction, 202–203
  matching on meshed object, 452
  merging, 274–275
  moving, 413–415
  and pyramid construction, 374
  rearranging mesh faces by swapping, 458
  rebuilding, 276
  rotating, 413
  rounding off, 42
  scaling, 414
  showing, 272–273
  splitting, 31, 32, 273–274
  trimmed, 325–326
  unjoining, 407–408
  untrimmed, 324–325, 335–337, 414–415
  unwelded, 456
  variable blend, 396–398
  variable chamfer, 394–395
  variable fillet, 395–396
  welding, 459–460
Edges, boundary. See Boundary edges
Edges, surface. See Surface edges
Edge softening, as render effect tool, 623–624
Edge splitting, 449
Edge treatment, 394–398
Editing
  box, 486–487
  cage, 487–489
  light, 596
  maelstrom, 490
  polygon meshes, 445–462
  soft, 269–270, 287–288
  solids, 404–417
  text, 560–561
Edit Material Properties button, 604
Edit point
  manipulation, 231
Effects panel, 571–573
Electric cable, 56–58
  curves, 56–58
Elevator mode, 123–124, 222
Ellipse, 263–264
  around a curve, 263–264
  by corners, 263
  deformable, 264
  vertical, 263–264
Ellipsoid, 374–376
E-mail, 115
Embed option, 522
End bulge, 287, 337–338
End cap, 380
Engineering drawing, 542–564
  functions of, 542
  tools, 87–88
Engine unit, 16–18

Engine vents, 45–51
Environment map, 507, 607
Environment objects, 625–630
  ambient light, 625–626
  Ground Plane panel, 629–630
  radiosity, 625–626
  wallpaper, 626–629
Erasing. See Delete command
Errors (objects), 514
Evaluate point, 503
Exhaust system, 18–19
Exploding
  block instance, 527
  to create solid, 390
  dimensions, 562
  joined curve, 282
  meshes, 443
  polysurfaces, 261
Exporting
  block definition, 534–535
  objects, 566–568
  Rhino files, 565
Export options, 114
Export with Origin command, 565
Exposure tab, 571–572
Extending
  in making a sweep surface, 327–328
  surface to make periodic, 326–327
  trimmed edge, 325–326
  untrimmed edge, 324–325
External files, 521–523
  attaching, 536
  detaching, 536–537
  exporting block definitions, 534–535
  inserting, 521
  reference to, 522–523
  type of, 521
Extracting
  control points, 309–310
  duplicated mesh face, 461
  isocurves, 308–309
  mesh elements, 453–456
  mesh face edge, 302–303
  mesh faces bound by un-welded edges, 456
  sub-curve of a polycurve, 310
  surface from a solid or polysurface, 416–417
  wireframe from isocurves, 309
Extruded polysurface, 15, 18
Extruded solid, 20, 21, 45

Extruded surfaces, 180–186
Extrude straight option, 180–181
Extruding, 13
    along a curve, 182–184, 190–192
    curve, 63
    faces, 410–411
    free-form surface, 387–388
    lightweight objects, 417
    normal to a surface, 186
    planar curve, 386–387
    to a point, 182
    a ribbon, 184–186
    surface, 63–64
    a surface to form solid objects, 386
Extrusion surface, 218
Eye rays, 631

F

Faces
    extrusion, 410–411
    folding, 412–413
    mesh, manipulating, 445–462
    moving, 408, 409
    rotating, 409–410
    shearing, 410
    of solid, 408–411
    splitting, 412
Fairing, 290
Falloff value, 463
Files
    attaching, 536
    detaching, 536–537
    exporting, 114, 565
    formats, 566–568
    formats Rhino can open and/or save, 566–568
    importing, 114, 565–566
    inserting, 539
    saving, 114–115
    sending via e-mail, 115
Files, external, 521–523
    attaching, 536
    detaching, 536–537
    exporting block definitions, 534–535
    inserting, 521
Files location tab, 113
File type, 521
Filleting, 9, 18, 21, 47, 49, 333–335
    curves, 278–279

    edges, 395–396
    surface, 330–331
Filter (status bar panes), 104
Filtering, bitmap map, 620
Fin command, 186
Findtext command, 561
1stDirFaces, 429
Fixed length curve edit, 288
Flaming nXt plug-in material, 625
Flamingo nXt rendering, 569, 575–576
Flamingo rendering farm, 582
Flat spiral, 268–269
Flattening, 216–217
Flip normal direction, 262–263
Floating toolbars, 104
Floating viewports, 101–102
    advantages, 101
Floor panel, 24–26, 44–45
Flowing, 65, 66
    along a curve, 490–492
    along a surface, 492–494
Fly-through animation, 580
Focal points, 242
Foci, 263, 375
Folding, 412–413
Four-point line, 249–250
Frame Buffer Controls toolbar, 571
Free-form curves
    constructing, 236–240
    converting to line/arc segments, 292
    on polygon mesh, 237–238
Free-form solid objects, 386–390
Free-form surfaces, 27, 179–209
From tool, 128–129
Frontal door, 36–40
Front suspension, 21–22
Front viewports. *See* Viewports
Fuel tank, 20–21
Full render, 574
Full screen display, 109

G

Gamma correction, 571–572
Gas cap, 45–51
G0 continuity, 177, 329, 335
G1 continuity, 178, 279, 329, 335
G2 continuity, 178, 329, 335, 396
G3 continuity, 178, 329
G4 continuity, 178, 329

General Hydrostatic System (.gf, .pm) files, 566
Geometric continuity, 502
Geometric objects
    copying, 138
    deleting, 138
    manipulating, 132–150
    moving, 137–138
    properties, 162
    rotating, 138–139
    selection methods, 132–136
    solids of, 357–386
German Association of Automobile Industries, 567
GetUserText command, 540
Ghosted display, 106
GHS files, 566
G-Infinity (blending technology), 177
Gloss color, 606
Graph, curvature, 498–499
Graphics area, 97–103
Graph paper, curves construction and, 175
Grid mesh, 44
Grid Snap (status bar panes), 104
Ground Plane panel, 629–630
Groups, 516–518
    adding and removing elements, 517
    assigning names, 517–518
    combining, 518
    constructing, 516–517
    defined, 516
Guide rail, 188
Gumball (status bar panes), 104, 149–150

H

Handlebar Editor, 232, 265–266
Handle curve, 232–233
Hatching, 552
    modifications in, 552–553
Height
    adding to sweep 2 rails surface, 196–197
    constructing a box, 358
Heightfield from image surface, 207–208, 437–438
Helixes, 53, 56–57, 266–268
    around curve, 268
    with straight axis, 266–267
    vertical, 267

Helper line, 596, 600–601
Help system, 113–114
Help topics, 114
History Manager, 94–95, 150–153, 179
    and extruded surfaces, 186
    and rail revolve surface, 189–190
    and relationship between original and symmetric objects, 152–153
    and revolved surfaces, 188
Holes, 49, 399–400
    arraying on planar face, 402–403
    constructing round, 75, 401
    copying, 416
    deleting, 416
    filling on meshed object, 451
    kinds of, 399
    mirror, 403
    moving and rotating, 415–416
    placing multiple, 400
    planar, 13, 74, 389
    revolved, 401–402
Hollow object, constructing, 393–394
HSB (hue, saturation, and brightness) color system, 154, 606
Hue, 154
Hyberbola, 265
Hybrid approach, 535
Hydrostatics, 513
Hyperlink, 113

I

IGES. *See* Initial Graphics Exchange Specifications (IGES)
IGES (.igs, .iges) files, 568
Importing, 565–568
Importing, page layout, 549
Import options, 114
Inclined edges, 630
Incremental save, 115
Index of refraction (IOR), 605, 606
Initial Detail Count, 544
Initial Graphics Exchange Specifications (IGES), 568
Inscribed circle, 374
Insert dialog box, 75, 77
Insert File Options dialog box, 75, 77

Insertion
    advantages of, 529
    blocks, 520–521
    files, 539
Insertion base point, 519
Intake system, 19–20
Intensity, of bitmap mapping, 607–609
    bump map and, 607–608
    color texture map and, 607
    environment map and, 607
    transparency map and, 608–609
Interpolate, points, 56
Interpolated curve, 223
    compared with control point curve, 224
    on a surface, 236–237
Interpolating, curves, 295–296
Interpolating surfaces, between two surfaces, 215
Intersecting surfaces, 42, 304
Intersection, 391
    of curves, 304
    of polygon meshes, 304–305, 442–443
    sets, 392
    of surfaces, 42, 304
INT object snap tool, 323
Invalid mesh, 461
Isetta, 6
Isocurves, 85–86, 265, 308–309
    density, 163–164
    display, 163
Isometric view, 109
Isoparametric curves, 130

J

Jewelry, 63–67
Joined curve, 282
Joining, 16. *See also* Welding
    contiguous surfaces, 338
    to create solid, 390
    edges, 36, 275–276
    meshes, 443
    surface, 40, 43
Joints, 329

K

Kinked rail, 327–328
Kink point, 229, 260, 263, 286, 324

Knots, 222, 228–229, 266–267, 269
Knot vectors, 485

L

Lasso command, 133–134
Layer dialog box, 154, 604
Layer group, 158–159
Layer Manipulator, 104, 153–161
Layers
    changing, 155–156
    copying and duplicating, 156–157
    managing objects in, 153–154
    material assignment, 155
    turning on/off, 157
Layer State Manager, 160–161
Layer states, 538–539
Leader, 559–560
Length
    of a curve, 504
    extending in a curve, 272–275
    manipulating curve's, 271–275
    surface edge, 504
Lens setting, 582
Lighting
    ambient, 625–626
    default, 586
    digital, 586–596
    directional, 592–593
    linear, 595–596
    managing, 602–604
    purpose of, 586
    rectangular, 593–594
    relocating, 601–602
    skylight, 603–604
    sunlight, 603–604
    turning off, 604
Light objects properties, 161–162
Lights, 45–51
Light Wave (.lwo) files, 567
Lightweight extrusion objects, 218–219
Lightweight objects, extruding, 417
Linear, array, 142–143
Linear dimension, 556
Linear fluorescent light, 595–596
Linear light, 595–596
Lines, 246–252
    adding a segment to a curve, 273
    at an angle, 249–250
    bisecting 2 lines, 249–250

connected segments, 254
constrained at an angle, 43
contour, 305–307
drawing with a digitizer, 252
four-point, 249–250
normal to a surface, 251
passing through set of points, 251–252
perpendicular/tangent to curves, 247–248
section, 305–307
segments, 246–247, 290, 292
single, 246–247
on surfaces, 130
tangent to a curve and perpendicular to another curve, 249
types, 553–554
vertical to a construction plane, 251
Link and Embed option, 522
Link check boxes, 75, 77
Link option, 522
Locking
objects in a layer, 159–160
Lofted surfaces, 198–204
and curve alignment, 198–199
and edge curves tangency, 199–200
to a point, 200–201
style in, 199
Loft Options dialog box, 71
LookAbout toolbar, 583–584

M

Maelstrom editing, 490. *See also* Deformation
Main body, 30–36
Main pull-down menu, 90–92
Maintain Height command, 196–197
Manipulating toolbars, 27
Mapping
box, 613–614
channels, 611
custom, 614–615
cylindrical, 613–614
match, 615–616
planar, 611–612
polygon meshes, 439–441
spherical, 613–614
surface, 611–612
texture, 610–611

Mapping coordinates (widgets), managing, 618–619
Mass properties tools, 510–513
Match Curve dialog box, 283–285
Matching
curves, 283–285
edges on meshed object, 452
untrimmed surface edge with another surface edge, 335–337
Match mapping, 615–616
MatchMeshEdge, 452
Match Surface dialog box, 336
Material
assigning properties, 604–605
digital, 604–625
Editor, 610
Rhino, 605–609
setting in layer, 155
Material Editor dialog box, 604–605
Material tab, of Properties panel, 605–609
advanced settings area, 609
basic settings area of, 605–606
Rhino materials and, 605–609
texture, 606–609
Max bounces, 632
Measurement
of angle, 504–505
of distance, 504–505
of length, 504–505
Measurement, units, 114
Mechanical components construction, 7–22
Memory, 320, 519
Menu, right-click, 131
Merging, 37
contiguous untrimmed surfaces to form single surface, 261–262
co-planar faces, 404
edges, 274–275
non-manifold, 261
Mesh area, 447–448, 453–454
Mesh box, 421–422
Mesh cone, 427–428
Mesh cylinder, 424–427
Mesh density
mesh box, 421–422
mesh cone, 427–428
mesh cylinder, 424
mesh ellipsoid, 429–430
mesh rectangular plane, 420
mesh sphere, 422–424

mesh torus, 431–433
mesh truncated cone, 428–429
reducing, 445–446
Mesh difference, 442
Mesh draft angle, 454–455
Mesh edge, 301–302, 447
Mesh element, 447–448
Mesh ellipsoid, 429–430
Meshes, 418–462. *See also* Polygon mesh(es)
density setting, 419
duplicating edge of, 301–302
filling holes, 451
matching edges, 452
rebuilding, 461
setting grid, 122
3D face, 419
Mesh face
collapsing, 447
converting, 456–458
deleting, 450
extracting, 453–456, 461
increasing, 449
manipulating, 445–462
non-planar quadrilateral, 456–457
patching, 450–451
rearranging by swapping edges, 458
triangular, 456–457
Mesh intersection, 442–443
Mesh patch, 435–436
Mesh rectangular plane, 420
Mesh repair, 461
Mesh sphere, 422–424
Mesh structure, 439
Mesh torus, 430–433
Mesh truncated cone, 428–429
Mesh union, 441
Mesh vertex, 446
Millimeters.3dm template, 75, 77
Mirror holes, 403
Mirroring, 14, 16, 35, 40, 72, 74, 146–147
Mobile construction plane, 170–172
Modeling methods, solid, 356–357
Moray UDO (.udo) files, 567
Mouse (command interaction), 106, 233–234
left-clicking effects, 106
right-clicking effects, 106

Move command, 137–138
MoveUVN command, 264
Moving
    edges of solid, 413–414
    holes, 415–416
    soft, 463–464
Moving target point, 584
Mud guards, 12
Multiple knots, 269

N

Naked edge, 273
Named construction plane, 168–169
Named position, 148, 535
Native construction plane, 164
    options of, 164–168
N direction, 500–501
Nested block, 526
Non-manifold merge, 261
Non-manifold object, 393
Nonuniform, scaling, 141
Nonuniform rational B-spline. *See* NURBS
Normal direction, 458–459
Notes window, 115
Nudge keys, 226, 263
NumSides option, 257
NURBS curves, 3
    defined, 222
    free-form, 220–245
NURBS surfaces, 3, 179–219. *See also* Surface(s); Surface model/modeling
    advantages of, 84
    constructing polygon mesh from, 436–437
    converting to solid volume, 86
    described, 83
    difference from polygon mesh, 436
    for engineering designs, 83
    isocurve display, 163

O

Object, solid text, 385–386
Object Properties (.csv) files, 161–164, 567
Objects
    aligning, 465–466
    environment, 625–630
    geometric. *See* Geometric objects
    imported or exported, 566–568
    light, properties. *See* Light objects properties
    orienting, 466–474
    uniformity of, 485–486
Object Snap. *See* Osnap
Object transformation, 463–496
Object type, selection by, 136
Offsetting, 213–215, 293–295, 386, 620
    mesh, 438–439
On Curve tool, 130
1D scaling, 140–141
On Surface tool, 130
OpenGL rendering library (SGI), 567
Open-loop curve, 224
Options, Rhinoceros, 113
Ordinate dimension, 559
Organic toolbar, 264
OrientCurveToEdge command, 474
Orienting, 466–474, 531–532
    curves, 473–474
    curve to edge, 474
    objects, 466–474
    polygon meshes, 473–474, 531–532
    surfaces, 473–474
OrientOnCurve command, 473–474
OrientOnSurface command, 470
OrientPerpendicularToCurve command, 471–473
Orient 2 Points command, 468–469
Orient 3 Points command, 76, 78
Ortho (drawing aid), 104
Ortho angle, 124–125
Orthogonal curves, 201
Ortho mode, 124–125, 559
Osnap (drawing aid), 104
    Along Line tool, 127
    Along Parallel tool, 128
    On Curve tool, 130
    dialog box, 126–127
    enabling/disabling, 127
    from, perpendicular from, and tangent From, 128–129
    Project Mode, 130–131
    On Surface tool, 130
    Between tool, 128
    tools, 125–131
Osnap dialog box, 76
Outline, 2D, 303–304

P

Page layout
    constructing, 544
    hiding and showing objects in detail views of, 550–551
    importing, 549
    manipulating layers in detail views of, 549–550
Pan command, 106, 107, 108
Parabola, 264–265
Paraboloid, 377
Parameterization, 272
Parasolid (.x_t) files, 568
Patch, 33
    mesh, 435–436
    single mesh face, 450–451
    surface, 62, 204–205
Patch command, 389
Patch Surface Options dialog box, 62
Path animation, 578–580
Pen mode, 106
Periodic curve, 230
Periodic surface, 326–327
Perpendicular From tool, 128–129
Persistence of Vision (POV), 567
Perspective match, 627–628
Perspective viewport, 582, 583. *See also* Viewports
Photorealistic images, 569
Picture frame rectangular surface, 212–213
Pipe, 380–381
Pipe-shaped voume command, 135
Piping, as render effect tool, 622
Planar (drawing aid), 104
Planar curves
    applying on surfaces, 311–312
    constructing 3D curve from, 296
    extruding, 386–387
    used to construct polygon mesh, 434–435
Planar face, 402–403
Planar hole, 13, 389
Planar mapping, 611–612
Planar mode, 123, 222
Planar surface, 44, 55, 73–74, 210–213
Plane, shearing along, 479–480
Plug-in material, 625

Point clouds, 240–245
  construction, 243
Point files, 566
Point grid surface, 205–206
Point light, 591–592, 598–599
  highlighting effect of, 598–599
Point objects, 220–245
Points, 240–245. *See also* Control points
  along a curve, 240–242
  and box construction, 359
  of circle, 258–259
  and cone construction, 369–370
  constructing, approaches for, 175–176
  constructing polygon meshes from, 434
  constructing polygon mesh through NURBS surface, 437
  constructing surfaces from, 174–175
  and cylinder construction, 366–367
  deviation from a surface or curve, 509–510
  draft angle, 244–245
  editing, 231–240, 270
  of ellipsoid, 375
  evaluating coordinates, 503, 509
  from existing objects, 293–313
  extruding to, 182
  focal, 242
  interpolate, 56
  line passing through a set of, 251–252, 253
  and lofted surfaces, 200–201
  and mesh cylinder, 426–427
  multiple, 240
  near a curve, 240–242
  orienting, 466–469
  projected, 38–39
  setting, 476–478
  for sphere construction, 361–362, 363–364
  on a surface, 509
  sweeping to, 194–195
  tools, 81
  and truncated cone construction, 372–373
  and tube construction, 379–380
  when building a torus, 384–385
Point-setting (deformation), 476–478
Polar array, 144
Polar coordinates, 119–121
Polycurve, 288–289, 310
Polygon, 256–258, 373–374
Polygon count, 462
Polygon mesh(es), 3, 418–462. *See also* Meshes
  area inquiry, 511
  bounding box of, 503–504
  combining and separating, 441–445
  constructed from closed planar curve, 434–435
  constructed from closed polyline, 435
  constructing by offsetting a mesh, 438–439
  constructing 2D outlines of, 303–304
  constructing from existing objects, 434–439
  constructing from image's heightfield, 437–438
  constructing from NURBS surface, 436–437
  constructing from set of points, 434
  constructing primitives, 418–433
  constructing through control points of NURBS surface, 437
  as cutting object, 444
  described, 82
  difference from NURBS surface, 436
  draft angle value of, 507
  drawback of, 82
  duplicating edges of, 301–302
  free-form sketch curve on, 237–238
  intersection of, 304–305
  joining and exploding, 443
  manipulating faces, 445–462
  mapping, 439–441
  overview, 418
  polyline on, 253–254
  projecting and pulling a curve, 299–300
  reducing mesh count, 446–449
  splitting, 443–445, 461
  tools, 83
  trimming, 445
  welding, 459–461
Polyline, 252–254
  closing, 277
  constructing polygon mesh from, 435
  filleting vertices of, 278–279
  fitting a curve through, 238–239
  moving segments of, 288–289
  offsetting, 386
  on polygon mesh, 253–254
  refitting a curve from, 291
  through a set of points, 253
Polynomial degree, 206, 225–226, 270
  changing, 272
  of a polygon, 256
  and polyline through a set of points, 253
  refitting, 272
  unifying, 290–291
Polynomial spline, 221
  tangent vector and, 221
Polysurfaces, 3, 40, 356–417
  capping, 44–45, 389
  changing color property, 18
  closed, 17, 58, 69
  closed-loop, 389–390
  constructing, 13
  defined, 316
  duplicating border of individual surfaces of, 301
  exploding, 261
  extracting a surface from, 416–417
  extruded, 15, 18
  forming, 338
  making two closed, 10–11
  mirrored, 16
  revolved, 15, 45
  splitting, 41, 404–407
  tools, 85–86
  trimming, 404–407
Positional continuity. *See* G0 continuity
Post Effects tab, 572–573
POV. *See* Persistence of Vision (POV)
POV-Ray Mesh (.pov) files, 567
Preview render window, 573–574
Primitives, 418–433
Printing, 563–564

Print Setup dialog box, 563
Project button, 126
Projecting
    to construction plane, 478–479
    curve to a polygon mesh, 299–300
    curve to a surface, 33, 298–299
Projection viewport, 582
Project Mode, 130–131
Prompt check box, 75, 77
Properties
    geometric object, 162
    light object, 161–162
    matching, 162–163
    viewport, 101–102, 161
Pull-down menu, 90–92
Pulling, 298–300
Purging, of unreferenced blocks, 527
Pyramid, 373–374
    camera, 585–586

R

Radial dimension, 558
Radiosity, 625–626
Radius
    of circle, 258–259
    curvature, 500
    of curvature, 274
    of a cylinder, 364–365
    and cylinder construction, 367
    modifying, arc, 262
    of torus pitch circle, 381–382
Rail revolve, 188–189
Raw Triangles (.raw) files, 568
Rays, eye, 631
Realism (object), 620
Rear axle, 10–11
Rear suspension, 15–16
Rebuilding
    edges, 276
    mesh, 461
    surface, 71, 72, 270–271
Rebuild Surface dialog box, 72
Recently used commands, 94
Record History, 104
Rectangle, 254–255
Rectangular array, 143
Rectangular box zone command, 135–136
Rectangular lighting, 593–594

Rectangular planar surfaces, 210–212
Redo command, 147
Reference check boxes, 75, 77
Reference geometry limit, 539
Reference UV curves, 311–312
Refitting, polynomial degree, 272
Reflection color, 606
Refreshing, attached files, 538
Regions, separating, 393
Regular pattern curves, 246–270
Relative construction plane Cartesian system, 120
Relative construction plane polar system, 120
Remapping, 475–476
    objects from one CPlane to another CPlane, 475–476
Rendered mode, 106
Rendered Shadows mode, 106
Render effect tools, 622–625
    displacement mesh, 623
    edge softening, 623–624
    piping, 622
    shut lining, 624–625
Rendering, 84, 569–633
    acceleration grids, 631–632
    BSP tree and, 632
    curvature, 506
    details, 631
    digital, 569, 570–575
    draft angle, 507
    environment map, 507
    Flamingo nXt, 575–576
    full, 574
    overview, 569
    preview, 570
    properties, 630–632
    reducing time for, 573
    resolution, 570
    tools, 88
    with wallpaper, 628–629
    zebra stripes, 508
Rendering meshes, 84
RenderMan (.rib) files, 567
Render meshes, quality of, 632
Render Options, 632
Render preview, 570
Render resolution, 570
Repair tools, 461–462
Re-parameterizing, 272, 607
Resolution, of rendered image, 630
Revcloud command, 561

Revert command, 115
Revolved curves, 16, 186
Revolved holes, 401–402
Revolved polysurface, 15, 45
Revolved solids, 20
Revolved surfaces, 186–190
Revolving, 187
RGB (red, green, blue) color system, 154–155, 606
Rhinoceros
    advantages of, 179
    analysis tools, 86–87
    basic operation methods, 117–172
    computer modeling and, 79–88
    data analysis, 497–515
    data exchange and engineering drawing tools, 87–88
    file exporting, 565
    file formats that can be opened/saved, 565
    functions and user interface, 79–116
    NURBS surfaces, 83, 179–219
    options and document properties, 113
    overview, 79
    panel, 89–90
    points and curves tools, 81
    polygon meshes tools, 83
    polysurface tools, 85–86
    rendering and animation tools, 88
    solid modeling method, 356–357
    solid tools, 85–86
    transformation tools, 86–87
    user interface, 88–89
    utilities and help system, 113–114
Rhino Render dialog box, 570
Rib, 399
Ribbon, 184–186
Ring, 63–64
Rotated dimension, 557
Rotation, 47, 138–139
    angle, 520
    edges of solid, 413
    of faces, 409–410
    holes, 415–416
    viewports, 107–108
Rounded rectangle, 255
Rounding off, 42

S

Safe frame, setting, 581–582
Saturation, 154
Save As command, 115
Save as Template command, 115, 565
Save command, 114
Save methods, 114–115
Save small command, 114–115
Save with preview image, 115
Saving, worksession file, 538
Scale By Plane, 142
Scale factor, 234, 520
Scaling, 109, 139–142, 414
    nonuniformly, 141
    in 1D, 140–141
    by Plane, 142
    in 2D, 140
    uniformly in 3D, 140
Screen grid cells, 631–632
Scripting, command, 95–96
Seam point, 230–231
Seat, 24–26
2ndDirFaces, 429
Section lines, 305–307
Selecting objects, 132–136
Selection filter, 136
Self-shadowing, preventing, 632
SelNakedMeshEdgePt command, 567
Sensitivity, case, 92
Separating, mesh, 441–445
Set intersection, 392
SetUserText command, 540
Set View Speed command, 109
Shaded mode, 106
Shading, of object, 569
Shadowing (rendering), 632
Shape alignment, 194
Shearing, 410, 479–480, 482–485. *See also* Deformation
    of control points, 482–485
Shortcut keys, 113
Show edges, 272–273
Shrinking, 320–322
Shut lining, as render effect tool, 624–625
Silhouette, 307–308
Simplifying, 290–291
Sketch curve, 233–236, 237
Sketches, 175–176
    curves construction and, 175–176
    ways of using, 175–176

Sketching, 233–236
Skirts, 42–44
Slab, 388–389
Slash, 196
Slice (.slc) files, 566
Small objects, curve construction, 52–67
SmartTrack (status bar panes), 104
Smart tracking, 131–132
Smashing, 216
Smooth factor, in smoothing, 485
Smoothing, 485. *See also* Deformation
Smooth surface, 270, 315
Smooth wires, 315
Snap (drawing aid), 58, 104
Snap option, 122
SnapToLocked command, 160
Soft editing, 269–270, 287–288
Soft moving, 463–464
Solid model
    design calculation, use of, 85
    forming, 86
    tools, 85–86
    vs. surface model, 80–81
    vs. wireframe model, 80–81
Solids, 3, 54–55, 69, 74–75, 356–417
    from closed-loop polysurface, 389–390
    combining, 390–393
    detailing, 393–403
    editing, 404–417
    extracting a surface from, 416–417
    extruded, 20, 21, 45
    hollowing, 393–394
    manipulating faces of, 408–411
    of regular geometric shapes, 357–386
    revolved, 20
    rotating, moving, scaling edges of, 413–415
    splitting and trimming, 404–407
    wire-cutting, 406–407
Solid text object, 385–386
Sorting, manual, 202
Space morphing, 492
Spatial Technology, 568
Sphere, 45, 361–364
Spherical mapping, 613–614
Spherical zone command, 135–136
Spiral, 268–269

around curve, 269
with straight axis, 268–269
Spline
    bending along a, 481–482
    B-spline, 222
    converting to lines and arcs, 292
    described, 221
    nonuniform rational B-spline, 222
    polynomial, 221
    segments, 228–229
Splitting, 38–39
    along U and V lines, 323
    an edge, 273–274
    by a curve, surface or polygon mesh, 444–445
    curves, 276
    disjoint mesh, 461
    edges, 31, 32, 449
    a face, 412
    meshes, 443–445
    polysurface or solid, 404–407
    polysurfaces, 41
    surface, 322–324
Spoon, 58–63
    handle, 61–63
    spade, 58–61
Spotlights, 586–591, 597–598, 632
    edit light by looking, 589–590
    highlighting effect of, 597–598
    setting to a specified view, 590–591
Spotlight shadow grid, 632
Squishing, 217–218
Squishing back, 217–218
Standard ACIS Binary (.sab) (file format), 568
Standard ACIS Text (.sat) (file format), 568
Standard for the Exchange of Product (STEP) model data, 566
Star polygon, 257, 373–374
Startup template dialog box, 88–89
Status bar, 104
Steering, 22–23
STEP (.stp, .step) files, 566
Step Size toolbar, 583
Stereolithography (STL), 567
StereoLithography Contour (.slc) files. *See* Slice (.slc) files
STL (.stl) files, 567

Straight extrusion, 181
Stretching, 486. *See also* Deformation
Subassembly, 75–77
Sub-curve, 182–184, 310
Sublayers, 158–159
Sun roof, 41
Sun study animation, 580–581
Surface(s), 3
    adjusting end bulge, 337–338
    applying planar curves on, 311–312
    area centroid of, 511
    area inquiry, 511
    area moments of, 511
    backbone, 492, 496
    blend, 39, 62
    blended, 331–335
    boundaries, 163
    bounding box of, 503–504
    capped, 18
    common free-form, 179–209
    concerns while constructing, 179
    connecting, 335
    constructed from edge curves, 204
    constructing a chamfer, 329–330
    constructing a fillet, 330–331
    constructing 2D outlines of, 303–304
    continuity, 31, 177–178, 192–193
    contour lines and section lines, 305–307
    curve network, 39
    cutting, 317
    as cutting object, 444
    detaching trimmed boundary from, 320
    directions of, 500–501
    dividing along its creases, 324
    draft angle value of, 507
    dropping onto, 494–495
    extending curve on, 275
    extending to make it periodic, 326–327
    extension, 327–328
    extracting from a solid or polysurface, 416–417
    extruding, 63–64, 186
    fitting a curve through the control points of, 239
    flowing along, 66, 492–494
    free-form, extruding, 387–388
    free-form sketch curve on, 237
    improving smoothness of, 485–486
    interpolated curve on, 236–237
    interpolating between two surfaces, 215
    intersecting, 42, 304
    joining, 40, 43
    line normal to a, 251
    lines on, 130
    making uniform, 271
    manipulation, 314–355
    merging, 37
    mirroring, 35, 40
    offsetting a curve, 294–295
    points on, 509
    profile manipulation, 262–272
    projecting and pulling curve to, 298–299
    rebuilding, 270–271
    reparameterizing, 272
    shortest curve between two points on, 239–240
    shrinking, 320–322
    silhouette of, 307–308
    smooth, 315
    soft editing of, 269–270
    splitting, 38–39, 322–324
    sweep, 18–19, 34, 54–55, 58
    swept. *See* Swept surfaces
    target, 492
    treating two or more, 328
    trimming, 38–39, 42, 43, 70–71, 72–73, 316–319
    untrimmed, 261–262
Surface, planar. *See* Planar surface
Surface analysis tools, 506–510
Surface boundary, 314–328
Surface edges
    duplicating, 300–301
    and edge matching, 202–203
    manipulating, 272–276
Surface From Curve Network dialog box, 61, 68, 202
Surface mapping, 611–612
Surface model/modeling, 4
    approaches, 174–177
    concepts, 173–174
    described, 82
    limitation, 84–85
    manipulating, 177
    overview, 173
    presentation methods, 82–84
    steps in, 174
    tools, 83–84
    used in photorealistic rendering/animation, 82
    vs. solid model, 80–81
    vs. wireframe model, 80–81
Surface profile, 263–272
Surface UVN, 440–441
Sweeping, 12–13, 192–193, 194–195
Sweeping rail, 193–194
Sweep 2 rails surface, 32, 71, 72, 195–198
Sweep 1 rail surface, 36, 54–55, 58, 190–192
Sweep surface, 34, 327–328
Swept surfaces, 13–14, 18–19, 54–55, 190–198
Symmetric objects, 152
Synchronization, 103

T

Tab, viewport, 102
Tangency
    adjusting, 265
    changing, 232
    and cone construction, 368–369
    and cylinder construction, 365–366
    line perpendicular/tangent to 1 or 2 curves, 247–248
    line tangent to a curve and perpendicular to another curve, 249
    between loft surface and edge curves, 199–200
    specifying for sphere construction, 362
    of torus pitch circle, 383–384
    and truncated cone construction, 371–372
    of tube, 378–379
Tangent circle, 259
Tangent continuity. *See* G1 continuity
Tangent cylinder, 425–426
Tangent From tool, 128–129
Tangent vector, polynomial spline and, 221
Taper angle, 181
Tapering, along an axis, 482–485
Technical mode, 106

Template file, 7, 29–30, 89, 97
Text, 540
    editing of, 560–561
Text box, 560
Text dialog box, 560
Text object, solid, 385–386
Textscale command, 561
Texture map, 607–609, 610–611
    bump, 607–608
    color, 607
    environment, 607
    texture palette and, 610–611
    transparency, 608–609
Texture palette, 610–611, 619–620
Thickening. *See* Extruding
Thickness, evaluating, 510
3D curves
    constructing on a surface, 298–299
    construction from planar curves, 296
    extruding, 180–181
3D mesh face, 419
3D objects, 543–544
3D point, 587
3D rotation, 138–139
3D studio files, 568
3D uniform scaling, 140
3 points rectangular surface, 210
Through points rectangular surface, 211
Tiling (bitmap), 620
Tires, 7–10
Tolerance value, 452, 459
Tone mapping, 571–572
Tone mapping algorithm, 571
Toolbar, 104–106, 131, 264
    cascading, 105–106
    displaying, 104
    manipulating, 105
Tools
    data analysis, 497–502
    diagnostic, 513–515
    dimensional analysis, 503–506
    mass properties, 510–513
    surface analysis, 506–510
Top-down approach, 3
    of component construction, 532–533
Top-down thinking, 174
Top viewport. *See* Viewports
Torus, 381–385
To surface orientation, 585
Toy car
    assembly, 68–78
    body, 69–75
    components installation, 77–78
    rendered images, 68
    subassembly, 75–77
Tracing, 235–236
Tracking lines, 131–132
Transformation, object, 463–496
Transformation tools, 86–87
Translation, 463–476, 490–495
    and deformation, 490–495
Transparency (objects), 606, 632
    color, 606
    level, 606
    map, 608–609
Transparent command, 108
TrimCurves, 62
Trim edge, 320
Trimming, 38–39
    boundary, 44–45
    bubble car components, 44–45
    curves, 275–276
    methods, 316–319
    polygon meshes, 445
    polysurface or solid, 404–407
    surface, 42, 43, 62, 70–71, 72–73, 314–319
    and wire-cutting a solid, 406–407
Truncated cone, 370–373
Tube, 377–380
Turntable animation, 576–578
Turntable display, 110
Twisting, 480–481, 482–485
Two-arc fillet, 282
2D curves, 216–217
2D drawings
    constructing from scratch, 551
    constructing page layout and detail viewports, 544–551
    generating from 3D objects, 543–544
    methods to construct, 542–543
    orienting, 475
    output and data exchange, 542–568
    overview, 542
2D surface, 215–216, 303–304
2D flat pattern, 215–218
2D rotation, 138
2D scaling, 140

U

U coordinates, of a point, 509
U curves, 311–312, 323
U direction, 145–146, 271, 500–501
UDO. *See* User-Defined Objects (UDO)
Undo command, 109, 147
Undo View Change command, 109
Ungrouping, 518
Uniform, surface, 271
Uniformity
    of objects, 485–486
    of surfaces, 485–486
Uniform offset surface, 213–214
Unifying, direction, 458–459
Union, 11, 14, 21, 390, 441
Units of measurement, 114
Universal construction plane, 169–170
Unjoining edges, 407–408
Unreferenced blocks, 527
Unrolling, 215–216
Untrimming, 319–320
Unwelding, 460–461
Unwrapping (mapping), 616–618
Upstream operations, 565–566
User-Defined Objects (UDO), 567
Utilities, 113–114
UV Editing (mapping), 616–618
UVN, surface, 440–441

V

Vacuum forming, 208–209
Variable blend edge, 396–398
Variable chamfer, 333–335, 394–395
Variable fillet, 333–335, 395–396
Variable offset surface, 214–215
V coordinates, of a point, 509
V curves, 311–312, 323
V direction, 145–146, 271, 500–501
Verband Der Automobileindustrie (.vda) files, 567
Vertex count, 439
Vertical circle, 259–260
Vertical ellipse, 263–264
VerticalFaces, 422, 424, 427, 428, 431
Vertical helix, 267

Vertical rectangular surface, 210–211
Vertical torus, 381–382
Vertices
   aligning on mesh object, 452–453
   fitting a curve through a polyline, 238–239
   welding, 460
View Capture, 112
Viewport Properties dialog box, 582
Viewports, 97. *See also* Detail viewports
   configuration, 97–99
   display mode, 106–107
   docked, 101–102
   floating, 101–102
   4 Default, 106
   front, 56, 58–59, 62
   gradient background, 99–100
   isometric, 109
   and lens setting, 582
   parallel projection, 582
   perspective, 47, 53, 56, 71, 75, 77, 97, 98–99, 108, 582, 583
   properties, 100–101, 161, 582
   rotating, 108–109
   silhouette of a surface in, 307–308
   synchronization, 25
   tab, 102
   top, 56, 58–59, 61
   view capture, 112
   zoom aspect ratio, 109–110
Virtual Reality Modeling Language (VRML), 567
Visibility (layers), 159–160

Volume, 86, 390, 391
   centroid, 512
   cut, 512–513
   inquiry, 511–512
   moments, 512
   and polygon meshes, 441, 442
Volume centroid inquiry, 512
Volume moments inquiry, 512
VRML (.wrl, .vrml) files, 567

W

Walkabout toolbar, 583
Wallpaper, 626–629
   perspective matching, 627–628
   rendering with, 628–629
   using, 626–627
Web browser, 113
Weight, control points, 227–228, 265
Welding, 459–461
What command, 497
Wheels, 7–10
Window Metafile (.wmf) files, 567
Windows, 40–42
   car body, 41
Windows title bar, 89
Window zone command, 133
Wire-cutting, 406–407
Wireframe, 309
Wireframe mode, 106
Wireframe model
   applications of, 80
   described, 80, 81
   tools, 81

   vs. solid model, 80–81
   vs. surface model, 80–81
Wires, smooth, 315
Working directory, 536
Worksession file, saving, 538
Worksession manager, 535–539
World Cartesian system, 120
World coordinate system, 119
World/Cplane toggle switch, 104
World polar system, 120
World relative Cartesian system, 120
World relative polar system, 120

X

X coordinates, 223, 559
XFaces, 420, 421
XGL files, 567
X-ray display, 106

Y

Y coordinates, 223, 559
YFaces, 420, 421

Z

Zebra stripes rendering, 508
ZFaces, 421
Zoom, 106, 107–108, 273
   aspect ratio, 109–110
   scale, 234
Zoom 1:1 Calibrate, 234
Zoom Dynamic, 107